AF179701

NOTES ON THE HERPETOFAUNA OF SURINAM IV

BIOGEOGRAPHICA

Editor-in-Chief

J. SCHMITHÜSEN

Editorial Board

L. BRUNDIN, Stockholm; H. ELLENBERG, Göttingen; J. ILLIES, Schlitz; H. J. JUSATZ, Heidelberg; C. KOSSWIG, Istanbul; A. W. KÜCHLER, Lawrence; H. LAMPRECHT, Göttingen; A. MIYAWAKI, Yokohama; W. F. REINIG, Hardt; S. RUFFO, Verona; H. SICK, Rio de Janeiro; H. SIOLI, Plön; V. SOTCHAVA, Irkutsk; V. VARESCHI, Caracas; E. M. YATES, London

Secretary

P. MÜLLER, Saarbrücken

VOLUME IV

DR. W. JUNK b.v., PUBLISHERS, THE HAGUE 1973

NOTES ON THE HERPETOFAUNA OF SURINAM IV

The lizards and amphisbaenians of Surinam

by

M. S. HOOGMOED

Rijksmuseum van Natuurlijke Historie
Leiden, The Netherlands

DR. W. JUNK b.v., PUBLISHERS, THE HAGUE 1973

ISBN 90 6193 205 X
© 1973 by Dr. W. Junk b.v., Publishers, The Hague
Cover design Max Velthuijs, The Hague
Printed in the Netherlands
by Koninklijke Drukkerij Van de Garde, Zaltbommel

CONTENTS

I. INTRODUCTION

Although biological exploration in Surinam started about 300 years ago, a thorough account of the Surinam herpetofauna has not hitherto been published. Several publications on the subject have been issued (see below), but none of them was comprehensive. However, there are several papers dealing with a part of the Surinam lizards. The main reason for the absence of a complete review of the Surinam lizards (and, for that matter, of the entire herpetofauna) has been the scarcity of the material available in museum collection although some of the more common species reached museums in considerable numbers. Until about 1900 the interior of Surinam was virtually *terra incognita* and most collecting took place in a restricted, cultivated area along the coast and the farthest inland point reached was Berg en Dal on the Suriname River. From 1900 exploration of the interior was actively encouraged by the Dutch Government, and under the auspices of the Koninklijk Nederlandsch Aardrijkskundig Genootschap several geographical expeditions explored the main river systems. Although the main purpose of these expeditions was to map the visited areas, usually the physician accompanying the expedition was entrusted also with the collecting of zoological and botanical specimens. Interesting material from the interior started to reach the Dutch museums in ever increasing numbers after the late nineteen thirties when Dr. D. C. Geijskes arrived in Surinam. The Surinam Expedition 1948–49 in which Dr. Geijskes and Mr. P. H. Creutzberg took part as zoologists was of special herpetological interest. Creutzberg concentrated on collecting Surinam lizards, including the smaller, secretive species. During 'Operation Grasshopper' Geijskes collected a large number of interesting specimens during the clearing of airstrips in the interior. As a result of the 'Brokopondo Project' Dr. M. Boeseman, Dr. G. F. Mees and Mr. (now Dr.) H. Nijssen went to Surinam to investigate the fish fauna of the Suriname River and other rivers but they also collected a considerable number of herpetological specimens.

Most of the material from Surinam collected after 1900 was deposited in the Rijksmuseum van Natuurlijke Historie, Leiden (RMNH) and has been available for the present study. Material from Surinam and adjacent regions in the collections of the museums listed below was also examined. The following abbreviations are used in the text.

AMNH: American Museum of Natural History, New York.
ANS: Academy of Natural Sciences, Philadelphia.
BM: British Museum (Natural History), London.
CM: Carnegie Museum, Pittsburg.

FMNH:	Field Museum of Natural History, Chicago.
KM:	Universitetets Zoologiske Museum, Copenhagen.
KU:	Museum of Natural History, University of Kansas, Lawrence.
LACM:	Los Angeles County Museum, Los Angeles.
MCZ:	Museum of Comparative Zoology, Cambridge, Mass.
MP:	Muséum National d'Histoire Naturelle, Paris.
NMW:	Naturhistorisches Museum, Wien.
SM:	Surinaams Museum, Paramaribo.
SMNS:	Staatliches Museum für Naturkunde in Stuttgart, Ludwigsburg.
UF:	University of Florida, The Florida State Museum, Gainesville.
UMMZ:	University of Michigan, Museum of Zoology, Ann Arbor.
USNM:	United States National Museum, Washington, D.C.
YPM:	Yale-Peabody Museum, New Haven.
ZMA:	Zoölogisch Museum, Amsterdam.
ZMB:	Institut für Spezielle Zoologie und Zoologisches Museum, Berlin.

II. ACKNOWLEDGEMENTS

I am most grateful to several of my colleagues who allowed me to study material in their care. They are: Dr. W. Auffenberg (UF), Dr. J. A. Böhlke (ANS), Dr. F. W. Braestrup (KM), Dr. W. E. Duellman (KU), Dr. J. Eiselt (NMW), Dr. J. Guibé (MP), Miss A. G. C. Grandison (BM), Dr. D. Hillenius (ZMA), Dr. R. F. Inger (FMNH), Dr. A. G. Kluge (UMMZ), Dr. E. V. Malnate (ANS), Dr. H. Marx (FMNH), Dr. C. J. McCoy (CM), Dr. C. W. Myers (AMNH), Dr. G. Peters (ZMB), the late Dr. J. A. Peters (USNM), Dr. T. M. Uzzell (YPM), Dr. H. Wermuth (SMNS), Dr. E. E. Williams (MCZ), Dr. J. W. Wright (LACM) and Dr. R. Zweifel (AMNH).

I wish also to express my thanks to several of my colleagues in the RMNH who have been very cooperative in many respects: Drs. M. Boeseman, D. C. Geijskes and G. F. Mees supplied me with additional notes on, and/or slides of specimens collected by them. Prof. Dr. L. B. Holthuis assisted by searching the archives and providing valuable advice concerning nomenclatorial problems, Dr. D. C. Geijskes also identified the stomach contents. Most persons dealt with in the list of collectors cooperated by providing personal data. Field work in Surinam in 1968 was made possible by a grant provided by the Netherlands Foundation for the Advancement of Tropical Research (WOTRO, W 956–2), in 1970 by the financial support of the Foundation Jan Joost ter Pelkwijk Fund and the legacy from Miss A. M. Buitendijk. Grants from WOTRO (WR 956–4), the Foundation Jan Joost ter Pelkwijk Fund and the legacy from Miss A. M. Buitendijk enabled me to visit American and European museums.

A grant from the Foundation for the Advancement of Herpetology covered part of the costs for publication of the photographs.

While in Surinam I was helped in all possible ways by several Government Services, among which the Surinam Forest Service (Ir. F. Bubberman and Dr. J. P. Schulz) should be especially mentioned. Further facilities, in transportation and lodging, were provided by the Police (S. Lapré), the Geological Mining Service (R. Cambridge) and the Surinam Aluminum Company (J. J. Jansen). During my intermittent stays in Paramaribo I was given hospitality in the Surinaams Museum (P. Bolwerk), and also working space.

A special word of thanks should be directed to Miss A. G. C. Grandison for the correction of my English text. Of course, as it is now, the text is wholly my responsibility.

In 1968 W. N. Polder and in 1970 J. J. P. Paats were my field-companions. The photographs were made by Chr. Hoorn Jr. of the Rijksmuseum van Natuurlijke Historie.

I thank my wife for her patience and understanding when carrying out fieldwork, for her help during part of the fieldwork and for her encouragement during the preparation of the manuscript.

III. GENERAL PART

1. A short history of the study of Surinam lizards.

As far as I am aware, the first author dealing with Surinam reptiles was KEYE (1659), who gives a description of some turtles; this account, without any changes also appears in his 1667 book. The next author writing about reptiles was WARREN (1667, 1669). He mentions turtles and tortoises, snakes, crocodiles, toads and frogs, but omits the lizards. The Dutch version of this book (WARREN, 1669) served as a source for several later publications. Another author (Anonymus, 1676) copied KEYE's description of the turtles with slight alterations. VAN BERKEL's (1695) book consisted of two parts, the first being a description of a journey to the Berbice River, the second being a description of a journey to Surinam. In the first part he mentions tortoises, crocodiles, snakes, lizards, iguanas and toads, in the second part only turtles and tortoises, crocodiles, snakes, toads and frogs. The second part of this book ('Reis na Suriname') has been copied almost literally from WARREN (Dutch version 1669), only some slight alterations having been made. The first author to provide not only descriptions but also good illustrations of Surinam lizards is MERIAN (1705a, b) who figured *Tupinambis nigropunctatus* Spix ('Sauvegard') and *Cnemidophorus l. lemniscatus* (Linnaeus) ('blauwe Hagedis'). Moreover, she figured and described two snakes and two frogs; all of them can be identified. HERLEIN (1718) added two other lizards, namely *I. i. iguana* (Linnaeus) ('Leguanen') and *A. a. ameiva* (Linnaeus) ('Hagedissen'). However, it should be kept in mind that the part on the iguana was copied nearly literally from the Berbice part of VAN BERKEL (1695) and that the part on the ameivas can be found there also, although in different words. Other passages seemingly have been taken from WARREN (1669). SEBA (1734) described and depicted three lizards from Surinam, namely *A. a. ameiva* (Linnaeus), *Tupinambis nigropunctatus* Spix and *I. i. iguana* (Linnaeus). Moreover, in his 1734 and in his 1735 volume, SEBA depicted a total of six species of lizards now known to occur e.g., in Surinam, but which, however, according to the accompanying text, are stated to have come from localities where they certainly do not occur, or just from the general area 'West-Indies'. SUNDIUS (1749) only mentions *Tupinambis nigropunctatus* Spix. A few years later GRONOVIUS (1756) published excellent descriptions of 29 species of Surinam reptiles and amphibians, among them eight lizards, all identifiable. LINNAEUS (1758) described 12 species of lizards known to occur in Surinam, but for none of these is Surinam mentioned as the place of origin. The type locality for all these species will be discussed below. Apart from HOUTTUYN (1764)

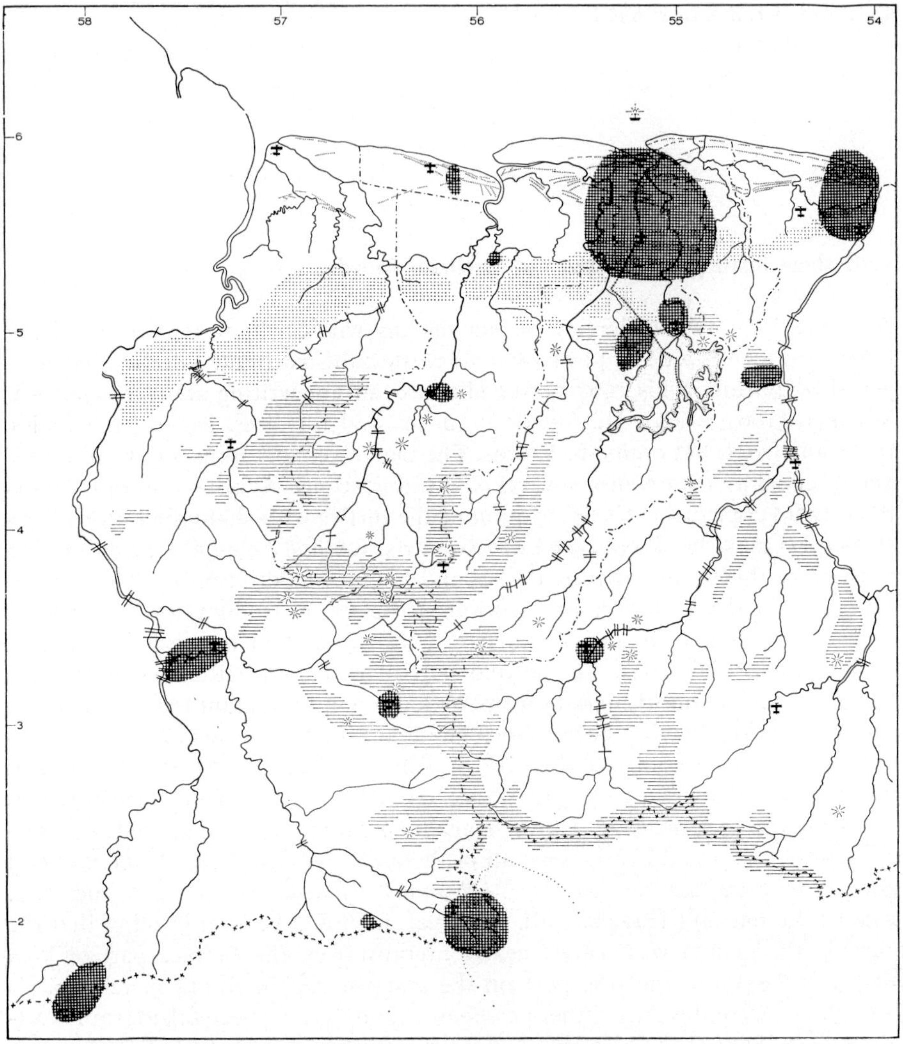

Fig. 1. Map of Surinam, showing areas where large or fairly large collections of lizards were assembled (cross-hatched). Dotted areas represent savannahs, dotted lines near the coast ridges, hatched areas mountainranges, dashes perpendicular to rivers cataracts, asterisks mountaintops. The boundaries between districts are indicated by lines of alternating dots and dashes. The border with Brazil is indicated by the line of alternating dashes and crosses.

only some popular works of little scientific value (PISTORIUS, 1763; FERMIN, 1765; 1769; 1770; BANCROFT, 1769a, b; HARTSINCK, 1770), general works mentioning species occurring in Surinam only as a sideline or narratives (STEDMAN, 1796a, b; 1799a, b; 1800a) appeared between 1758 and 1802. In this last year LATREILLE described four and DAUDIN 12 species of lizards from

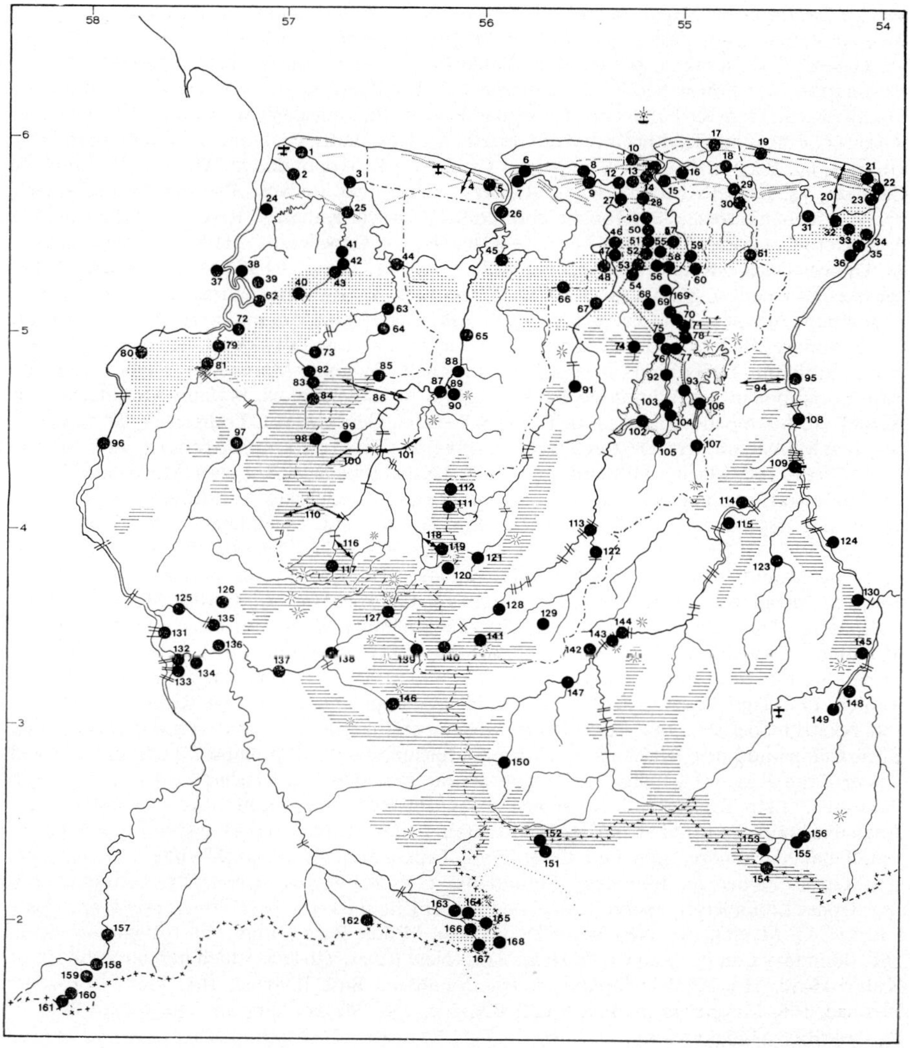

Fig. 2. Map of Surinam, showing localities where lizards were collected. For legend see fig. 1. 1. New Nickerie, 2. Nanni Creek, 3. Wageningen, 4. Coronie road, Surinam Expedition 1948–49, 5. 20 miles E. of Coronie, 6. Plantation Caledonia, 7. Boskamp, 8. Tambahredjo, 9. Groningen, 10. Weg naar Zee, 11. Plantation Clevia, 12. Uitkijk, 13. Santo Boma, 14. Paramaribo, 15. Plantation Peperpot, 16. Plantation Marienburg, 17. Matapica Canal, 18. Charlottenburg, 19. Bigisanti beach, 20. Moengo Tapoe and Wia Wia transects of Surinam Expedition 1948–49 (including Third Encampment, Djai Creek and Wane Creek N.), 21. Eilanti beach, 22. Galibi, 23. Langaman kondre and Christiaan kondre, 24. Forest Service Encampment MacClemen, 25. Cupido, 26. Batavia, 27. Santigron, 28. Lelydorp, 29. Potribo, 30. Kramakka, 31. Moengo, 32. Moengo Tapoe, 33. Mooi Wane Creek, 34. Albina, 35. Papatam, 36. Bigiston, 37. Oreala, 38. Forest Service Encampment Wakay, 39. Kaboeri Creek near Fajastik, 40. Kaboeri Creek, 2nd bivouac, 41. Awarra savannah, 42. Snake Creek, 43. Maratakka River, 2nd savannah, 44. Forest Service Encampment Arawarra, 45. Tibiti, 46. Matta, 47. Troeli Creek, 48. Forest

Service Encampment Moeroe Creek, 49. Onverwacht, 50. Republiek, 51. Zanderij, 52. Forest Bivouac Zanderij, 53. Berlijn, 54. Railway km 62, 55. Saramacca Creek, 56. Carolina Creek, 57. Powakka, 58. Krakka, 59. Between Blakkawatra and Jodensavanne, and 60 mi. SSE. of Paramaribo, 60. Forest Service Encampment 8, 61. Pininica, 62. Washabo, 63. Stondansi Cataracts, 64. Transect Fallawatra-Coppename River (first locality), 65. Camp 1, Coppename River, 66. Forest Service Encampment Goliath, 67. Loksihatti, 68. Donder Mountain, railway km 91.5, 69. Berg en Dal, 70. Compagnie Creek, 71. Brokopondo, 72. Matapi, 73. Lombok Cataract, 74. Brown's Mountain, 75. Railway km 121, 76. Between Brownsweg and Kabel, 77. Aroesoebanja soela, 78. Afobaka and Koffiekamp, 79. Kabalebo River, 80. Mataway, 81. Avanavero Cataracts, 82. Upper Nickerie River (H. van Capelle), 83. Blanche Marie Cataract, 84. Upper Nickerie River (Van Aerde, 15–IX–1964), 85. Transect Fallawatra-Coppename River (second locality), 86. Transect VI, D. C. Geijskes, 87. Raleigh Cataracts, 88. Coppename Expedition, 1901, 89. Transect VII, D. C. Geijskes, 90. Voltz Mountain, 91. Awarradam, 92. Gania kondre, Bedotti and between Gansee and Lombee, 93. Sikakamp, 94. Nassau Mountains, Surinam Expedition 1948–49, 95. French Guiana, Marowijne River, opposite Base Encampment, 96. Wonotobo Falls, 97. Airstrip Kabalebo, 98. Bakhuis Mountains (van Aerde), 99. Adampadda Creek, 100. Transect V, D. C. Geijskes, 101. Transect IV, D. C. Geijskes, 102. Mamadam, 103. Dessiong, 104. Marowijne Creek, 43 km S. Afobaka, 105. Middle Gran Creek, 106. Maripa Hill and Dam, 107. Langatabbetje, 108. Lower Marowijne River, 150 km inland, 109. Poeloegoedoe, 110. Transect II, D. C. Geijskes, 111. Base Bivouac Emma Mountains 1959, 112. Hendrik Mountain, 113. Goddo, 114. Drietabbetje, 115. Saniki, 116. Transect I, D. C. Geijskes, 117. Linker Coppename River, 118. Transect III, D. C. Geijskes, 119. Tafel Mountain, km 26, 120. Kappel savannah, 121. Base Bivouac near De Kock Mountain, 122. Mirahoedoe, 123. Gonini Expedition 1903, November, 124. Cottica Mountains, 125. Lucie River, December 1910, 126. Expedition Wilhelmina Mountains 1926, 127. Wilhelmina Mountains Expedition, Maguire 1963, 128. Suriname Expedition, 14–XI–1908, 129. Ananas Mountain, 130. Benzdorp, 131. King Frederik Willem IV Cataracts, 132. Barrington Brown Rapids, 133. Tigri Encampment, 134. Gonini Encampment, 135. Lucie River, 2–XII–1908, 136. Airstrip Coeroeni, 137. Lucie River, November 1910, 138. Lucie River, 9–XI–1910, 139. Lucie Camp and Lucie River, 10–X–1910, 140. Encampment 5, Suriname Expedition 1908 and Upper Gran Rio, 7–IX–1910, 141. Bivouac near Franssen Herderschee Peak, 142. Airstrip Paloemeu, 143. Tapanahoni River near Magneet Rock, 144. Kodobakoe Cataracts, 145. Anapaike kondre, 146. Airstrip Kayser Mountains, 147. Between Tepoe and airstrip Paloemeu, 148. Janamale kondre, 149. Feti Creek, 150. Tapanahoni River, 22–X–1904, 151. Apisike, 152. Upper Paloemeu, Boundary Mountains, 153. Waremapan Creek, 154. Temomairem, 155. Upper Litani River, 10–XII–1903, 156. Upper Litani River, 29–XI–1903, 157. Upper New River (J. G. Myers), 158. New River Depot, 159. Middle Base Camp, 160. Navigation Head, 161. Boundary Camp, Itabu Creek Head, 162. 'New River, 750 feet' which in reality should be Kutari Head, 163. Airstrip Sipaliwini, 164. Sipaliwini Base Bivouac, 165. Vier Gebroeders Bivouac, 166. Meyers' farm, 167. South Bivouac, 168. Meyers' airstrip, 169. 65 km SSE. of Paramaribo, Afobaka Road.

Surinam. The next work of importance for Surinam saurology was the 'Erpétologie Générale' of DUMÉRIL and BIBRON, appearing between 1836 and 1839. These authors based their descriptions on material (from Surinam) from the Leiden museum, and recorded ten species of lizards from that country. KAPPLER (1854a, b; 1881) dealt with the Surinam fauna and in 1881 included a list of the lizards sent to the Stuttgart museum. In this list 17 names of lizards appear, of which 13 can be readily allocated as the material is extant in Stuttgart. The remaining four names (*Polychrus marmoratus* Cuv., *Anolis reticulatus* Gray, *A. lucius* Gr., *A. bullaris* L.) remain doubtful and cannot be interpreted with certainty. A few years later, the same author (KAPPLER 1885, 1887) dealt more extensively with the Surinam fauna and described eight species of lizards. Also in 1885 Boulenger mentioned eight species of lizards occurring in Surinam, including *Anolis ortonii* which had not previously been reported from Surinam. VAN LIDTH DE JEUDE (1904) studying material collected by several topographical expeditions to the interior mentioned 17 species of lizards, of which three were new to science. In 1905/1906 a series of newspaper articles appeared, written by the PENARD brothers and mainly concerning Surinam mammals and snakes. Just after they started to treat the lizards, publication of the series inexplicably ceased. Nevertheless they recorded six species of lizards. Their biological data are very unreliable because both brothers, being leprosy sufferers, were confined to their home and they had to rely on Indians and Negroes for data on the specimens. In 1917 VAN LIDTH DE JEUDE published a compilation of what was known about the reptiles and amphibians of Surinam and mentioned a total of 24 species of lizards, three of which are now known to have been based on one species, and in another instance two names are synonyms, leaving a total of 21 different species. Since 1917 several authors (L. MÜLLER, 1923; BRONGERSMA, 1927; 1928; 1932a, b; 1935; 1946a, b; UZZELL, 1966; ETHERIDGE, 1968; 1970) have either described new species or dealt with a single genus of which representatives occur in Surinam, but no serious compilation of the Surinam herpetofauna has appeared. Only D'AUDRETSCH (1965–1968) provided a compilation but it is superficial and does not aim at completeness.

2. Collectors.

HOLTHUIS (1959) dealt extensively with collectors of Surinam Crustacea, HUSSON (1962: 10/11) gave additional data on HELLER. Most of the collectors they mentioned also collected herpetological specimens. Collectors of herpetological specimens who have been fully discussed by HOLTHUIS (1959) are (in parentheses the page number of HOLTHUIS' paper): BLEYS (32), BOLTEN (27–28), BRUIJNING (33), BUITENDIJK (28), COSSEE (30), DAHLBERG (17–19), DIEPERINK (21–23), GEIJSKES (31–32), HAVERSCHMIDT (32), HELLER (29), C. HERING (23–24), C. J. HERING (25–26), HERMANS (31), VAN HEURN (29), HEYDE (33), HUMMELINCK (33–34), VAN DER KAMP (34), TEN KATE (26), KRAEPELIN & HOLM (25), MÖSCHLER (25), ROLANDER (19–21), SANDERSON (30), SCHMIDT (30–31), SPITZLY (26–27), WYMAN (24–25). Other persons have collected reptiles and amphibians in Surinam and they are treated below in the same way as HOLTHUIS did; however, they are arranged alphabetically instead of chronologically.

EDWARD BARTLETT, son of ABRAHAM DEE BARTLETT, sold specimens of birds from the Upper Amazon and Mexico to the British Museum (Natural History) in 1866–74. During this same period some reptiles and amphibians from Surinam bearing his name arrived in the same institution. As he himself probably never visited Surinam it is very likely that most or all of this Surinam material was collected by KAPPLER, who in this period sold his natural history specimens. BARTLETT participated in natural history expeditions to Syria, Palestine and the upper Amazon. For some years he was curator at the Maidstone Museum in Kent and afterwards curator at the Sarawak Museum at Kuching.

HARRY ANDREW BEATTY (born St. Croix, Virgin Islands, January 10, 1902) is a professional collector. He arrived in Surinam August 20, 1960 and left again February 10, 1962. During this period he visited several places in the interior to collect zoological specimens for the Field Museum of Natural History, Chicago. September 24, 1960 he left Paramaribo and went to the airstrip Kayser Mountains. Kayser Camp was set up on the Zuid River and abandoned on February 14, 1961. On March 15, 1961 he left for the airstrip Paloemeu, where he stayed till August 24. October 1, 1961 he left again for the airstrip Kayser Mountains, but this time he attempted to reach the Wilhelmina Mountains by travelling down the Zuid River and up the Lucie River. An unknown tributary was navigable as far as the outlying foothills of the mountains, and the Wilhelmina Camp was set up, at 800 meters altitude, on the 19th. He struck camp on January 27, 1962 and went down the Lucie River to the airstrip Coeroeni, from where he flew back to Paramaribo. During this trip representatives of many groups were collected, among which a collection of 1100 birds is important. Also a considerable number of reptiles and amphibians was collected, mainly during his stay near the airstrip Kayser Mountains and while he was in Paramaribo (between trips). All of this material is in the Field Museum of Natural History, Chicago.

EMMET R. BLAKE (born Abbeville, South Carolina, U.S.A., November 29, 1908) was assistant curator, division of birds, of the Field Museum of Natural History in Chicago from 1935 to 1947, associate curator from 1947 to 1955 and curator since 1955. In his capacity of assistant curator he made a collecting trip to Guyana (formerly British Guiana) where he arrived August 10, 1938 and left early January, 1939. He acted as the leader of the 1938 Sewell Avery Expedition to British Guiana. Other members of this expedition were Mr. RICHARD BALDWIN, in charge of transportation, HABIB RASOOL, a preparator, and thirteen Negroes and Indians who manned the boats. The expedition followed the Corantijn and the New River to near the source of the latter (see map in BLAKE, 1950). On the way up- and downstream collecting took place in several localities along the rivers. Localities visited were from north to south: King Frederik Willem IV Falls (Surinam: August 16–20), Ashiru Falls (Surinam: October 27), Oronoque Base Camp (Surinam: August 27, October 26), Calf Bird Camp (Guyana: August 30), Pairima Camp (Guyana: October 24), Phantom Falls (Surinam: October 22–23), New River Depot (Guyana: September 3), Middle Base Camp (Surinam: September 4–14; October 21), Navigation Head (Surinam: October 20–21), Boundary Camp (Surinam, 1° 40′ N, 58° W: September 20, October 15). The actual collecting of zoological specimens took place between August 16 and October 15. The total collections included 987 bird skins, 135 bird skeletons, 150 mammals, 87 reptiles, about 800 fishes and about 500 insects. During the descent of the 'King William Falls' (probably King Frederik Willem IV Falls) a boat was lost and with it about half of the total collection including 487 bird skins, 135 skeletons, 99 mammals, but no reptiles. Narrative: BLAKE 1950. All material collected by the 1938 Sewell Avery Expedition to British Guiana is in the Field Museum of Natural History, Chicago.

MARINUS BOESEMAN (born Enkhuizen, June 22, 1916) is curator of fishes at the Rijksmuseum van Natuurlijke Historie, Leiden. In that capacity he made collecting trips to Surinam (November 13, 1963 to early October 1964 and in 1971). He arrived on November 28, 1963 at the village of Brokopondo on the Surinam River, which he used as a base and where he spent the time between trips. From Brokopondo trips were made up the Suriname River to the Gran and Pikien Rio, and downstream to Jodensavanne. Also a trip on the Saramacca River was made from Kwakoegron to Posoegroenoe and Boslanti. A certain amount of collecting took place along the East-West Road between Albina and Paramaribo. Itinerary: BOESEMAN 1964a, b, c. In 1971 he took part in the expedition to northwestern Surinam, staying in the countiy from January 19 till April 26.

LEO DANIËL BRONGERSMA (born Bloemendaal, May 17, 1907) was director of the Rijksmuseum van Natuurlijke Historie. From September 8 to September 22, 1963 he visited Surinam. During his stay he visited Paramaribo, Afobaka, Albina, Eilanti, Galibi and Bigiston. Only in these two last-named localities was material collected which is now present in the Rijksmuseum van Natuurlijke Historie, Leiden.

CORNELIS CORNET (born Leiden, March 18, 1942) is a technician in the division of invertebrates at the Rijksmuseum van Natuurlijke Historie. He accompanied W. VERVOORT to Surinam from April 13, 1966 to July 10, 1966.

PETER HANS CREUTZBERG (born Altona, June 1921) took part in the 1948 to 1949 Surinam Expedition. In 1956 he again went to Surinam, now as a film-producer and in that capacity took part in the 1959 Coeroeni Expedition. In 1963 he left Surinam. The material he collected between 1948 and 1949 and in 1959 is in the Rijksmuseum van Natuurlijke Historie, Leiden.

JOHN WILLIAM DALY (born Portland, Oregon, U.S.A., June 8, 1933) is chief of Pharmacodynamics, Laboratory of Chemistry, National Institute of Arthritis, Metabolism and Digestive Diseases. He made several trips to Central and northern South America in connection with his work on the skintoxins of Dendrobatid frogs. All trips were made in the company of C. W. MYERS. For a detailed itinerary of DALY's stay in Surinam see under MYERS.

PIETER HENDRIK VAN DOESBURG (born Semarang, April 10, 1921) worked as an entomologist at the Landbouwproefstation in Paramaribo from April 1957 to April 1963. During this period he made numerous collecting trips in the coastal area, covering the region between Nickerie and Albina, south to Kabel. Most of his collecting took place in the environs of Paramaribo. After he left the Landbouwproefstation he made a four-month collecting trip together with VAN DER VECHT in the coastal area and to the Raleigh Cataracts-Voltz Mountain area. Moreover, he made a trip to the airstrip Sipaliwini. In August 1963 he left Surinam for the Netherlands, where he is now curator of Hemiptera, Orthoptera and some smaller insect groups at the Rijksmuseum van Natuurlijke Historie, Leiden. In November 1968 he attended the IV Congreso Latinoamericano de Zoología in Caracas, Venezuela and paid visits of a week to Rancho Grande and of two weeks (November/December) to Surinam, where he collected in the environments of Paramaribo. During his spells in Surinam he collected mainly Heteroptera and Syrphidae, but other insect groups, mammals, reptiles and amphibians were also collected. Most of his material is in the Rijksmuseum. van Natuurlijke Historie, Leiden, whereas smaller parts of his collections are in the Surinaams Museum, Paramaribo and in the Field Museum of Natural History, Chicago (mainly mammals).

RUDOLPH FREUND (born Philadelphia, April 8, 1915; died Norristown October 9, 1969) was a preparator and a free lance scientific illustrator. Accompanied by his wife he visited Surinam from November 18 to December 17, 1961. They visited several places in the interior: airstrip Kayser Mountains (November 23 to December 6, 1961), airstrip Sipaliwini and Sipaliwini savannah (December 7–11, 1961). Herpetological specimens were also collected in Paramaribo. The specimens they collected are preserved in the Yale-Peabody Museum, New Haven, Conn. More extensive information on their Surinam trip is given by HUMPHREY & FREUND (1962); however, it should be kept in mind that the positions of the airstrips Paloemeu and Kayser Mountains are incorrectly sited on their map.

MURRAY HENRY DE LA FUENTE (born Paramaribo, March 5, 1945) as a boy developed an interest in taxidermy and learned to prepare skins in the Surinaams Museum and from the Arowak Indians of Powakka. With a training grant from the Carnegie Museum he stayed in Pittsburg from November 1965 to April 1966. From April 1966 to October 29, 1966 he made a collecting trip to Surinam on behalf of the Carnegie Museum. During this trip he visited Kramakka near the confluence of Commewijne and Cassewinica Rivers, Potribo north of Kramakka, Powakka, Uitkijk, Paramaribo and environs, Nieuw Nickerie, Tonentei opposite Oreala and Mataway on the Corantijn River, Avanavero Cataracts in the Kabalebo River, Republiek and Zanderij. Another short collecting trip was made from December 29, 1968 to January 15, 1969, during which the environs of Paramaribo, Powakka, Bigi Poika and Moengo were visited. During the 1966 trip he collected 24 mammals, 344 birds, 358 reptiles and amphibians and 110 fishes. The second trip yielded 40 birds and 80 reptiles and amphibians. All the material he collected during the trips mentioned above is in the Carnegie Museum, Pittsburgh, whereas part of the material collected by him before his departure to U.S.A. is in the Rijksmuseum van Natuurlijke Historie, Leiden.

DIRK CORNELIS GEIJSKES (HOLTHUIS, 1959: 31) returned to the Netherlands in 1965. In 1971 he took part in the expedition to northwestern Surinam, staying in the country from January 19 till April 23.

PHILIP HERSHKOVITZ (born Pittsburgh, October 12, 1909) is research-curator of mammalogy at the Field Museum of Natural History, Chicago. He went to Surinam in November 1961, together with JACK FOODEN, to collect mammals. Fieldwork was initated in Carolina Creek, about 50 km south of Paramaribo on November 22. On December 5 they installed themselves at Loksihatti on the left bank of the Saramacca River. There they stayed for one month after which they returned to the coastal region were collecting took place from mid--January to early February. During this period they made succesful collections at three stations: Lelydorp, La Poule and Dirkshoop. By mid-February HERSKOVITZ and FOODEN left Surinam for Chicago. The material collected by them, mainly consisting of mammals, is in the Field Museum of Natural History, Chicago.

WILLEM CORNELIS VAN HEURN (born The Hague, February 20, 1887, died Deventer, June 6, 1972). Information on his collecting activities in Surinam has been given by HOLTHUIS (1959). Additional information will be given by HOLTHUIS & HUSSON (1973, in press).

MARINUS STEVEN HOOGMOED (born Rotterdam, March 19, 1942) is curator of reptiles and amphibians at the Rijksmuseum van Natuurlijke Historie, Leiden. From April 24 to November 20, 1968 (HOOGMOED, 1969a) and from January 10 to February 18, 1970 he stayed in Surinam; during the first trip he collected all over the country while during the second trip collecting took place mainly on the Sipaliwini savannah (January 15 to February 14, 1970) and in the environs of Paramaribo (January 10–14, February 15–17). During part of the first trip (October 22 to November 20, 1968) he was assisted by his wife (RIA HOOGMOED née VERSCHOOR) who took part in a few trips (Albina, Raleigh Cataracts-Voltz Mountain area, Brown's Mountain) and during the second trip he was accompanied by J. J. P. PAATS. In 1968 approximately 2500 specimens of reptiles and amphibians, together with a number of mammals, birds, fishes and insects, were collected. In 1970 approximately 500 specimens of reptiles and amphibians, together with 80 birds and a few mammals and fishes were collected. All of his material is preserved in the Rijksmuseum van Natuurlijke Historie, Leiden. Itinerary: HOOGMOED 1969a.

CUTHBERT ARTHUR HUDSON (born London, January 24, 1889) served with the Royal Engineers as a civil engineer and surveyor. He was a member of the Boundary Expedition that between 1934 and 1938 established the border between British Guiana and Brazil. He arrived in British Guiana early in 1934 and and left late in 1938. During his work for the Boundary Commission he collected zoological specimens in his spare time. Most of his time in Guiana was spent in the triangle between the New River, the Coeroeni River and the Surinam-Brazil boundary, an area which by the Guyanese government is claimed to be Guyanese, but which the Surinam government claims to belong to Surinam. He made several trips in the area: Corantijn River to New River Depot and back (May 1934 to February 1935); Oronoque Base Camp to Kutari headwaters (November 1935 to May 1936); Oronoque Base Camp to New River and back (April to October 1937); King Frederik Willem IV Falls to New River headwaters (November 1937 to May 1938). All of his material, mostly insects but also including a sizeable (and very interesting) collection of reptiles and amphibians, is preserved in the British Museum (Natural History), London. According to the labels his material came from 'New River, 750 feet, British Guiana', but in correspondence Mr. HUDSON told me that this was only the place from where it was shipped and that the actual collecting site most probably is Kutari Headwaters. In the distribution maps this locality has been used.

ANTONIUS MARIE HUSSON (born 's Gravenhage, June 8, 1913) is curator of mammals at the Rijksmuseum van Natuurlijke Historie, Leiden. He made a collecting trip to Surinam from December 8, 1962 to July 9, 1963, accompanied by P. STAFFELEU. HUSSON's itinerary reads as follows. Surroundings of Paramaribo with which are meant the following localities: Botanical Garden, plantation Clevia, road Paramaribo—Zanderij and road Paramaribo—Uitkijk: December 9, 1962 to January 31, 1963; February 2–27; March 15–31; April 12–28; May 12 to June 4, June 19 to July 8. Collecting also took place in Nieuw Amsterdam (February 1), Albina and Langamankondre (February 28 to March 5), Jodensavanne (March 14), Raleigh Cataracts-Voltz Mountain area (April 1–11), surroundings of Pininica Creek (April 29 to May 11), Wageningen (June 5–18). The material collected during this trip, comprising about 1200 mammals and approximately 80 other vertebrates (most of them reptiles and amphibians), is preserved in the Rijksmuseum van Natuurlijke Historie, Leiden.

JACOB VAN DER LAND (born Kollum, April 4, 1935) is curator of Vermes at the Rijksmuseum van Natuurlijke Historie, Leiden. From February 11 to March 19, 1967 he visited Surinam to collect small freshwater organisms. He collected along a north-south line from the beach (Weg naar Zee) to Sabakoe Creek near Zanderij. Among these freshwater organisms are a number of tadpoles. Apart from these, some other herpetological specimens were collected. His material is preserved in the Rijksmuseum van Natuurlijke Historie, Leiden.

FRANCOIS LEVAILLANT (born Paramaribo, 1753—died La Nomé (near Sezanne) November 22, 1824) was the son of parents who were keenly interested in nature. With them he frequently went out hunting in the jungle. He soon started his own collection. In 1763 the family returned to France (STRESEMAN, 1951). Material from Surinam collected by 'LEVAILLANT' is in the Muséum National d'Histoire Naturelle, Paris. It is not known whether this material is from FRANCOIS or from his parents.

THOMAS HOWARD LEWIS (born Red Lodge, Montana, July 28, 1919) is in the Medical Corps of the U.S.Navy as a psychiatrist. He is interested in vertebrate zoology and anthropology. He paid visits to Surinam in August 1961 (Paramaribo and environs), August 1963 and 1964 (Paramaribo and airstrip Paloemeu). During these trips small collections of reptiles, amphibians and fishes were made. These collections are preserved in the United States National Museum, Washington, D.C.

PAULUS JOHANNES MARIA MAAS (born Arnhem, February 27, 1939) is a scientific worker at the Instituut voor Systematische Plantkunde (Institute for Systematic Botany) at Utrecht. He made a botanical collecting trip to Surinam between November 28, 1964 and July 1965. During this trip most of the time was spent in the western part of the country. A simplified itinerary is as follows: Kabalebo airstrip (December 12, 1964 to January 12, 1965), Bakhuis Mountains (January 28 to March 17), Winana Creek and Corantijn River (April 5–25), Snake Creek and Maratakka River (May 6 to June 2), Nickerie River, i.a. Blanche Marie Cataracts, Kamisa Cataracts and Paris Jacob Creek (June 14 to July 6). During the intervening periods he paid several visits to the Brokopondo-artificial lake. Although most of his time was devoted to botanical collecting, a small collection of reptiles and amphibians was assembled, which is in the collections of the Rijksmuseum van Natuurlijke Historie, Leiden.

GERLOF FOKKO MEES (born Santpoort (municipality of Velsen), June 16, 1926) is curator of birds at the Rijksmuseum van Natuurlijke Historie, Leiden. In this capacity he made collecting trips to Surinam from March 1965 to April 1966, and from November 15, 1971 to November 16, 1972. During his first stay in Surinam he was stationed in Brokopondo and was assisted in collecting by his wife. Several trips up the Surinam River were made, the southernmost point reached being Awarradam in the Gran Rio. Also several trips to other parts of the country were made: Avanavero on the Kabalebo River (September 7–25, 1965); Paloemeu (November 16 to December 3, 1965); Matapica (January 7–13, 1966); Sipaliwini (January 23 to February 22, 1966) and Albina (April 6–10, 1966). His main interests when collecting in Surinam were birds and fresh-water fishes, but apart from these groups, he assembled a considerable number of herpetological specimens. In 1971 he again went to Surinam. He arrived in Paramaribo on November 15 where he stayed till December 4. On that day he started to make trips to the coast and to the interior, in between trips he spent some time in Paramaribo (February 29—March 3, 1972; March 30—April 14; May 9–13; July 12–18; August 10–16; September 1–9; September 20–24; October 21–26; November 5–15). During his stay in Paramaribo he made short trips to Powakka on April 2 and 12. In the interior the following places were visited: Brown's Mountain (December 4, 1971—February 29, 1972); Raleigh Cataracts (March 3–30) with an interruption during the period March 22–24 when he visited Voltz Mountain; Matapica (April 14–25); Krofajapassie (April 25–27); Bigisanti Beach (April 27 – May 8), Great Constantia (May 8 and 9), Sipaliwini (May 13–July 12), Nassau Mountains (July 18–August 10); Brown's Mountain (August 16–September 1); Singrilanti (September 9–20); Airstrip Tafel Mountain (September 24–October 21); Powakka (October 26–November 5). On November 16, 1972 he returned to Holland. All of the material collected during his Surinam trips is in the Rijksmuseum van Natuurlijke Historie, Leiden. Itinerary: MEES 1967a, b.

HARRY MAX VAN MEEUWEN (born Braunschweig, Germany, July 8, 1943). He arrived in Surinam on February 25, 1969. Until April 21 he stayed in Paramaribo, making trips in the surrounding area, then left for the airstrip Paloemeu, where he stayed until May 10, interrupting this stay for short excursions, e.g. to the airstrip Kayser Mountains (April 23–27, 28, May 2), to the airstrip Oelemari (May 3–4) and to the airstrip Sipaliwini (May 5–6). From Paloemeu he made day-trips to several points by helicopter. He returned to Paramaribo May 10, staying there till May 28 when he left for Moengo (May 28 to July 11). From Moengo a trip up the Patamacca Creek was made. An excursion to Stoelmanseiland, Drietabbetje and Cottica (July 11–17) followed and a period (July 17–30) in Paramaribo completed his stay in Surinam. He left Surinam July 30 and returned to the Netherlands. Herpetological specimens were collected in most of the places visited and sent alive to the Netherlands. Part of this material is now in the Rijksmuseum van Natuurlijke Historie, Leiden.

CHARLES W. MYERS (born St. Louis, Missouri, U.S.A., March 4, 1936) is Assistant Curator in the Department of Herpetology of the American Museum of Natural History in New York. He was in Surinam between February 8 and 26, 1972, where he collected in the vicinity of Paramaribo (February 8–14), on Brown's Mountain (February 15–17) and near the Raleigh Cataracts and Voltz Mountain (February 18–25). The trip to Surinam was made in the company of J. W. DALY. Most of the material they collected during this trip is in the American Museum of Natural History in New York, a smaller part was left with the Surinam Forest Service and is in the custody of the Surinaams Museum, Paramaribo. Some specimens are in the collections of the Rijksmuseum van Natuurlijke Historie, Leiden.

HAN NIJSSEN (born Amsterdam, October 28, 1935) is curator of fishes at the Zoölogisch Museum in Amsterdam. He made a collecting trip to Surinam between April 18, 1966 and June 30, 1967, using Brokopondo as a base and from there he made collecting trips to other parts of the country, visiting five different river-systems. Chronologically these trips are as follows: Lower Marowijne River (May 30 to June 3, 1966); mouth of Coppename and Saramacca Rivers (June 21–24, 1966); Gran Rio, south to Awarradam (January 26 to February 2, 1967); Kleine Saramacca River (February 26 to March 2, 1967); Nickerie River, south to Stondansie (April 5–6, 1967); upper Marowijne, lower Lawa and lower Tapanahoni Rivers (April 17–26, 1967); Linker Coppename River to the northern reaches of the Wilhelmina Mountains (May 3–30, 1967), the actual collecting took place between May 9 and 25. The time between trips was spent in Brokopondo from where trips were made on the articicial lake, to creeks in the surroundings and on the Suriname River. His main interest was fishes, but a number of reptiles and amphibians was also collected. All of his material is in the Zoölogisch Museum, Amsterdam. Itinerary: Nijssen, 1967a, b, c; 1969a, b, c.

JOHANNES JACOBUS PHILIPPUS PAATS (born Leiden, August 25, 1943), a technician in the division of reptiles and amphibians at the Rijksmuseum van Natuurlijke Historie, Leiden, accompanied M. S. HOOGMOED on his second trip to Surinam (January 9 to February 18, 1970).

WOUTER NICOLAAS POLDER (born Amsterdam, November 21, 1929) is a biology teacher. He left for Surinam on September 16, 1965 and stayed until August 31, 1968. On that date he left for the Netherlands. While in Surinam he made many excursions in the surroundings of Paramaribo, but also extended trips to the Voltz Mountain, the Saramacca River and Albina. In 1968 many trips were made together with M. S. HOOGMOED, e.g. to Brown's Mountain. The specimens he collected are in the Rijksmuseum van Natuurlijke Historie, Leiden.

DUNCAN GERARD REEDER (born Paramaribo, May 19, 1948) served with the

Surinam Forest Service (Dienst L.B.B.) from January 26, 1967 until May 31, 1971. During this period he was stationed at the following places: Bigisanti Beach, Krofajapassi Creek, Eilanti Beach, Coppename Mouth Nature Reserve, Brinckheuvel Nature Reserve, Voltz Mountain and Raleigh Cataracts, Brown's Mountain, Blakkawatra, Carolina, Jodensavanne, Cassipoera, Encampment 8, Moeroe Creek, Bigi Poika, Zanderij, Auca, Encampment Goliath and Kwakoegron.

He was a member of the Expedition to northwestern Surinam from January 26 to April 19, 1971. Since September 15, 1971 he is a technician in the division of mammals of the Rijksmuseum van Natuurlijke Historie, Leiden.

During the periods he spent on the several stations he collected botanical and zoological specimens. Among the zoological specimens reptiles and amphibians formed the main part. Zoological material collected by REEDER is in the Florida State Museum, Gainesville, in the Carnegie Museum, Pittsburgh (only insects), in the Surinaams Museum, Paramaribo, in the Rijksmuseum van Natuurlijke Historie, Leiden and in his private collection.

THOMAS ANTHONY RENSSEN (born Arnhem, August 12, 1920) arrived in Surinam in 1958 and worked on several plantations, namely: Marienburg (1958 to 1959, 1965 to 1970), Peperpot (1959 to 1961, 1963 to 1964) and Slootwijk (1961 to 1962). He collected in the coastal area between the Coppename and Marowijne Rivers, south to Afobaka and on the Sipaliwini savannah between January 15 and February 25, 1970. He left Surinam in 1970. Most of his material (birds and bird eggs) reached the Rijksmuseum van Natuurlijke Historie via F. HAVERSCHMIDT, part of the birds and mammals was sent to the Zoologisch Laboratorium of the Free University in Amsterdam. Several mammals and part of his herpetological collection are in the Rijksmuseum van Natuurlijke Historie, Leiden, some reptiles and amphibians are in the Zoölogisch Museum, Amsterdam.

LOUIS FRANCOIS EMMANUEL ROUSSEAU (born Belleville, 1788; died Paris 1868). Material from Surinam bearing his name is in the Paris Museum. Nothing is known of his travels in Surinam.

PIETER STAFFELEU (born Leiden, October 6, 1940, exhibition technician and taxidermist of the Rijksmuseum van Natuurlijke Historie, Leiden, accompanied A. M. HUSSON (curator of mammals at the same museum) on a collecting trip to Surinam between 1962 and 1963. The itineraties of HUSSON and STAFFELEU do not completely coincide, because on several occasions they split parties and worked separately. STAFFELEU arrived in Paramaribo on December 7, 1962. The periods between extended trips were spent in Paramaribo, which served as a base for collecting-trips in the environs. The following periods were spent in Paramaribo: December 7, 1962 to January 6, 1963; January 16–21; February 2–19; March 13–31; April 12–30; May 12 to June 5; June 18 to early July. Extended trips were made to Matta (January 7–15), the environs of Gansee (January 22–30), Brownsweg (February 1–2), Sipaliwini (February 20 to March 12), Raleigh Cataracts-Voltzberg area (April 1–11), Commewijne River, Pininica Creek, Java (May 1–11) and Wageningen (June 5–18). He returned to Amsterdam on

July 25, 1963. The material collected, comprising about 1200 mammals and approximately 80 other vertebrates (most of which reptiles and amphibians), is preserved in the Rijksmuseum van Natuurlijke Historie, Leiden. From January 19 until early June 1971 he took part in an expedition to northwestern Surinam. Between April 19 (the end of the expedition) and early June he collected in the surroundings of Paramaribo. He returned to the Netherlands on June 19, 1971. He paid a third visit to Surinam from November 1, till December 10, 1972. He returned in the Netherlands on December 11, 1972. During his stay in Surinam he was stationed in Paramaribo and from there made regular trips between Nieuw Nickerie and Albina. Moreover, he made some short trips to places further removed from the coast, e.g., Moeroe Creek, Bigi Poika, Carolina Creek and Matta. The main objective of this trip was to collect mammals, but a few reptiles were collected as well.

JACOBUS VAN DER VECHT (born Den Haag, July 5, 1906) was curator of Hymenoptera at the Rijksmuseum van Natuurlijke Historie, Leiden, from January 1, 1956 until April 1, 1964. He visited Surinam from April 20, 1963 to late July 1963, making collecting-trips in the coastal area and to the Raleigh Cataracts-Voltz Mountain area with P. H. VAN DOESBURG. During these trips mainly Hymenoptera and other insects were collected, but also a few reptiles and amphibians. The material collected during his stay in Surinam is preserved in the Rijksmuseum van Natuurlijke Historie, Leiden.

WILLEM VERVOORT (born Schiedam, June 12, 1917) is director of the Rijksmuseum van Natuurlijke Historie at Leiden. His first trip to Surinam started on March 10, 1966 and ended July 11, 1966. Most of this time was spent collecting on the continental shelf north of Surinam aboard H.Nl.M.S. Snellius (March 25 to May 13). The rest of the time he collected in the coastal region between Boskamp and Albina, in the lower Marowijne River, in the Gran Creek and around Brokopondo. During part of this trip he was accompanied by C. CORNET. His second visit to Surinam lasted from March 13 to May 2, 1969. Most of this time was again spent collecting on the continental shelf aboard H.Nl.M.S. Luymes. Only short visits were paid to the coastal region. During both trips marine and freshwater organisms were collected in considerable numbers but reptiles and amphibians were collected only during the first trip. All of his material is preserved in the Rijksmuseum van Natuurlijke Historie, Leiden.

JEFFRIES WYMAN was thought to have visited Surinam in 1856 (HOLTHUIS 1959: 24). However data in the Proceedings of the Boston Society for Natural History for the years 1856 to 1859 indicate that this is not correct. WYMAN was president of this society and thus presided at the meetings. It appears from the Proceedings that the last meeting presided by him was that of February 4, 1857, the next meeting on February 18 was presided by the vice-president. The meeting of September 2, 1857 was again presided by WYMAN, who, during that meeting gave 'a brief account of some facts noticed during a recent visit to Surinam'. In the meeting of September 16, WYMAN gave a more elaborate account of some peculiar modes of gestation in fishes and amphibians observed during his stay in Surinam. In a subsequent paper WYMAN (1859) states that in

May 1857 he was in 'the country inhabited by the Bush Negroes, above Sara Creek on the upper Surinam River'. The name of one of his companions was JOHN GREEN. From the above it may be concluded that WYMAN and his pupils visited Surinam from about February to August 1857.

3. Expeditions.

HOLTHUIS (1959) deals extensively with a number of expeditions to the interior of Surinam during which decapods were collected. All the expeditions he mentions (except the 1957 'Coquette' Investigations) also yielded herpetological material. However, the total number of such expeditions is obviously greater than listed by HOLTHUIS and those that collected herpetological specimens are added here. The reptiles and amphibians collected during the 1900 Nickerie Expedition, the 1901 Coppename Expedition, the 1902 to 1903 Saramacca Expedition, the 1903 to 1904 Gonini Expedition, the 1910 to 1911 Corantijn Expedition, the 1948 to 1949 Surinam Expedition, and the 1952 Medical Expedition are in the Rijksmuseum van Natuurlijke Historie, Leiden. The reptiles and amphibians collected during the 1922 Expedition to the Hendrik Mountain and by the 1926 Expedition to the Wilhelmina Mountains are in the Zoölogisch Museum, Amsterdam.

1904 Tapanahoni Expedition. Personnel: Lieut. A. FRANSSEN HERDERSCHEE, officer of the Topographic Service of the Netherlands East Indian Army (leader, geography), Lieut. C. H. DE GOEJE (geography), G. M. VERSTEEG (physician, botany, zoology). The expedition left Paramaribo by boat on July 5, 1904 and arrived in Albina, near the mouth of the Marowijne River the next day. The confluence of Lawa and Tapanahoni was reached on July 17. From there the Tapanahoni was followed upstream until the mountain Tepoe was reached on August 7. For surveying purposes this mountain was climbed. From there the journey was continued some 14 km to the west, where Magneet Rock was climbed on August 19. The expedition members separated into two parties on August 29. FRANSSEN HERDERSCHEE and DE GOEJE reached the Indian village Intelewa near the confluence of the Tapanahoni and the Paloemeu the next day. From there they continued up the Paloemeu to near the Kassikassima mountain which was climbed on September 13. VERSTEEG joined them on September 19. The party was again split on October 1, DE GOEJE continued up the Paloemeu to near its source and eventually crossed (on October 13) the divide and the next day reached the Paru River in Brazil. He was back in Albina on November 30 and in Paramaribo on December 7. FRANSSEN HERDERSCHEE and VERSTEEG returned to Intelewa and from there followed the Tapanahoni (October 9–21) to about 2°50′ N. They reached Albina on November 17 and Paramaribo on November 26. The zoological material resulting from this expedition is in the Rijksmuseum van Natuurlijke Historie, Leiden. Narrative: FRANSSEN HERDERSCHEE, 1905b.

1908 Suriname Expedition. Personnel: Lieut. J. G. W. J. EILERTS DE HAAN (leader, geography), Lieut. R. H. WIJMANS (geography), J. H. A. T. TRESLING (physician, botany, zoology). The expedition left Paramaribo by boat on June 30, 1908, following the Suriname River upstream, reaching Gansee on July 7. Goddo at the confluence of the Gran and Pikien Rio was reached on July 17. From Goddo the expedition first followed the Pikien Rio to about 3°30'N (July 22 to August 6). From about 3°38'N a transect was constructed in a southerly direction to the Ananas Mountain which was climbed for surveying purposes. The party returned to Goddo August 15. On August 19 the expedition started to follow the Gran Rio. On September 1 they reached a point from where EILERTS DE HAAN and WIJMANS cleared a transect in a southeasterly direction towards the Franssen Herderschee Peak. The top of this peak was reached on September 7 and several observations were made. On September 19 the journey up the river was continued and on the 22nd the Final Camp was reached at a latitude of 2°55'33" N. WIJMANS and TRESLING were able to follow the Gran Rio to near its source (September 23–24) which is about 8 km south of the Final Camp. From the Final Camp several overland trips were made to mountains in the neighbourhood. Between October 6 and 30 a trip was made in a westerly direction to a mountain reaching 820 m. During this trip the Lucie River was discovered. The Final Camp was left on November 5, Goddo was reached on November 11 and Paramaribo on November 20. The zoological material of this expedition is in the Zoölogisch Museum, Amsterdam. Narrative: EILERTS DE HAAN, 1910.

1938 Sewell Avery British Guiana Expedition. Personnel: E. R. BLAKE (leader, zoology), R. BALDWIN (transportation), HABIB RASOOL (preparator) and 13 labourers. For more details of this expedition see under E. R. BLAKE in the list of collectors. The zoological material acquired by this expedition is in the Field Museum of Natural History, Chicago. Narrative: BLAKE, 1950.

1939 Waremapan Expedition. Personnel: Dr. D. C. GEIJSKES. GEIJSKES left Paramaribo on June 24 by steamer for Albina. From there the Marowijne River was followed to Stoelmanseiland, thence by the Lawa and Litani Rivers to the Oayana Indian village of Janemalé, which was reached on July 10. A few days were spent there, before the trip farther sourth was continued on July 15. The boats were left in the Waremapan Creek on July 23; from here a trip on foot in a southerly direction into the Tumuc Humac Mountains, lasting until July 31, was made. Janemalé was reached again on August 5 and another period was spent there until August 24, during which short trips were made to the Oelemari River and the Loë Creek. Albina was again reached on September 4 and Paramaribo on September 9, 1939. The zoological material is in the Rijksmuseum van Natuurlijke Historie, Leiden. Narrative: GEIJSKES, 1942; 1957.

1943 to 1944 Coppename Hevea Expedition. Personnel: Dr. D. C. GEIJSKES. The expedition left Paramaribo on July 5, 1943 by corial and went up the Coppename River. On August 12, 1943 the most southerly point in the Linker Coppename River attainable by boat was reached. Here clearing of the first transect (Transect I) was started in a southeasterly direction for 18 km. This transect reached the northern Wilhemina Mountains where a hill of 600 m was climbed. From this hill several branch lines with a total length of 6 km were cleared. One month later Transect II was started at a point one day downstream from where Transect I started. It was cut in a northwesterly direction for 24 km where it ended at the top of a mountain reaching 780 m in the southern Bakhuis Mountains. In October the party sailed down the Linker Coppename River and went up the Rechter Coppename River to a point just south of the Emma Mountains and there Transect III was cleared in an east-southeasterly direction, towards the Tafel Mountain, which was reached after 24 km. The mountain then was climbed and investigated for two days. This project finished, the expedition went down the Rechter Coppename River and pitched a bivouac at the confluence of the Linker and Rechter Coppename River. Simultaneously Transects IV and V were cleared from here, the first one to the Kroetoe mountain in an easterly direction for 19 km, the second one in a westerly direction via the Hebiweri mountain towards the Bakhuis Mountains for 28 km. On December 20, 1943 this bivouac was left and another bivouac was pitched downstream from Langadam soela from where Transect VI was cut in a northwesterly direction for 21 km into the basin of the Nickerie River, on the other side of the Bakhuis Mountains. In January 1944 this traject was ready and the expedition left for the Raleigh Cataracts from where a final traject (Transect VII) was cut towards the Voltz Mountain in a southeasterly direction for 7 km. This work being completed the expedition returned to Paramaribo where it arrived on January 20, 1944. The zoological material collected during this survey is in the Rijksmuseum van Natuurlijke Historie, Leiden. Narrative and maps: GEIJSKES, 1946.

1944 New York Botanical Garden Expedition to Tafel Mountain. Personnel: B. MAGUIRE (leader, botany), L. SCHMIDT. The expedition left Paramaribo on June 12 and travelled up the Saramacca to where the Toekoemoetoe Creek joins it. This creek was followed upstream to its right bank tributary, the Tafelberg Creek, which was followed for 8 km; there a base camp was set up on July 9. From this camp a trail was cut in a southwesterly direction to the foot of Tafel Mountain which was reached on July 27 after a journey of 23 km. GEIJSKES' 1943 Transect III was crossed between km 19 and 20. The mountain was first climbed on July 29, via a chimney one km east of North Ridge. The porters, however, reached the top of the mountain via the North Ridge, where the escarpment that surrounds the rest of the mountain is absent. On the mountain plateau two camps were set up, the first two km from the upper end of North Ridge, the second south of Arrowhead Basin. From these camps trails were cut in all directions, thus enabling an exploration of the eastern, southern, western and central parts of the plateau. The northeastern part remained unvisited. The expedition stayed on the plateau for 54 days. The main purpose of the expedition was to collect botanical specimens, but SCHMIDT also did some zoological collecting. The herpetological material collected is in the Rijksmuseum van Natuurlijke Historie, Leiden. Narrative and map: MAGUIRE, 1945.

1948 to 1949 Surinam Expedition. Apart from the narrative by BAKKER & LANJOUW, 1949 referred to by HOLTHUIS (1959) there is another publication concerning this expedition (LINDEMAN, 1953) in which there are detailed maps of the first three transects of the expedition.

In 1957 a project, named 'Operation Grasshopper' was started with the intention of opening up the interior by establishing small airstrips in places that seemed to be of geological interest. Three of the exploration parties that set out from Paramaribo by boat to achieve this object were led by Dr. D. C. GEIJSKES, who collected zoological specimens during the clearing of the future airstrips Tafelberg, Coeroeni and Sipaliwini. As a result of this project the following airstrips were established between 1958 and 1961:

1958 Tafel Mountain

1959 Coeroeni, Paloemeu, Kayser Mountains, Kabalebo, ?Oelemari

1961 Sipaliwini

Only the three expeditions led by Dr. D. C. GEIJSKES are of interest here because he collected zoological specimens; these are dealt with in some detail.

1958 Tafel Mountain Expedition. Personnel: Dr. D. C. GEIJSKES (leader, zoology), Ir. H. BECKERING VINCKERS (geology), H. R. MASSINK (aero-mapping), R. KAPPEL (aviator) and 27 labourers among which the cooks RADIN and VAN GEENE. The expedition left Paramaribo on March 3, 1958 by car for Zanderij, thence by train to Kwakoegron. In Kwakoegron, on the Saramacca River, the corials in which the rest of the trip was to be made were waiting. The Saramacca River was followed upstream to near its source where the Base Bivouac was reached on March 11. Southwest of this bivouac a small savannah, enclosed by rain-forest, occurred, on which an airstrip was to be constructed. Construction started on March 16 and eight days later the first Cessna landed. GEIJSKES started a trip to the Tafel Mountain on March 26, which was climbed on the 29th. GEIJSKES and his companions stayed there until April 3. Then they returned to the Base Bivouac. On April 5 the last members of the expedition left this bivouac. The expedition returned to Paramaribo on April 10, 1958. The herpetological specimens collected are in the Rijksmuseum van Natuurlijke Historie, Leiden. Narrative: GEIJSKES, 1959; 1961.

1959 Coeroeni island Expedition. Personnel: Dr. D. C. GEIJSKES (leader, zoology). S. FRÄSER (topography), H. P. PIJPERS (radio-telegraphy), P. H. CREUTZBERG (film-producer), W. R. VAN AERDE (mechanic), 28 labourers. The expedition left Paramaribo for Nickerie in separate parties. The first group consisting of VAN AERDE, PIJPERS and the labourers left Paramaribo by sea on July 13. GEIJSKES left Paramaribo by plane on July 15 and CREUTZBERG also came by sea to Nickerie on July 16. On July 17 the party was complete and left Nickerie. The Corantijn was followed upstream to the Wonotobo falls where on July 21 a permanent camp was set up at the foot of the falls. Eight days were spent transporting the boats and the freight over the falls and on July 31 the expedition left the Wonotobo falls and reached the Frederik Willem IV falls in the afternoon of August 2. Six days were spent in conquering these falls. On August 10 Coeroeni island was at last reached. Between August 10 and September 2 a start was made to clear the forest in preparation for the construction of an airstrip. On September 3 GEIJSKES, PIJPERS, CREUTZBERG and VAN AERDE left Coeroeni-island together with six Bush-Negroes to travel along the Lucie and Zuid Rivers to the airstrip under construction near the Kayser Mountains. They arrived there on September 6. GEIJSKES made a short trip by air to Paramaribo on September 8 and 9 and the party left again for Coeroeni island on September 10. Coeroeni island was reached on September 13. In the meantime work on the airstrip had continued and now was carried on with renewed energy. As a result the first plane landed on the airstrip Coeroeni on October 10 and GEIJSKES returned with it to Paramaribo. FRÄSER and PIJPERS left the airstrip in the same way on October 16. CREUTZBERG and VAN AERDE left the airstrip on October 19 by descending the Corantijn. They arrived in Nickerie on October 30. On November 4 all the expedition members were back in Paramaribo. The zoological material collected during this expedition is divided between the Rijksmuseum van Natuurlijke Historie, Leiden and the Surinaams Museum, Paramaribo.

1961 Sipaliwini Expedition. Personnel: D. C. GEIJSKES (leader, zoology), RIJHINER (topography), H. P. PIJPERS (radio-telegraphy), 13 labourers. The expedition left Paramaribo by air on January 25 and in two flights was flown to the airstrip Coeroeni. From there the journey to the Sipaliwini savannah was made by boat. The airstrip was left January 27 and on February 5 the edge of the large savannah was reached where a camp was set up. On February 7 a small savannah, surrounded by forest, 4 km west of the large savannah was chosen as the place where the airstrip could be constructed and the expedition set up a new camp there. During the period between February 8 and 23 part of the small savannah was cleared and levelled and on February 23 the first plane landed. But it was not until March 12 that the strip was improved and finished. On March 12 the members of the expedition were flown back to Paramaribo. The herpetological specimens obtained are in the Rijksmuseum van Natuurlijke Historie, Leiden. Narrative: GEIJSKES, 1961; HOEDEMAN, 1969.

1959 Emma Mountains Expedition. Personnel: F. P. JONKER (leader, botany), J. J. WENSINK (geomorphology), A. G. H. DANIËLS (botany), H. T. L. VAN EYK (geology), 17 labourers. The expedition left Paramaribo on July 10 by train for Kwakoegron, where corials were boarded, and the journey was continued up the Saramacca River and the Toekoemoetoe Creek. A Base Camp had been set up previously in the upper reaches of the Toekoemoetoe Creek by a military expedition under Captain F. BÖSCHEN and Lieutenant P. BERDEN. The expedition reached this camp on July 14, and stayed there until July 20. VAN EYK broke an arm and returned to Paramaribo on July 17. The other members of the expedition left Base Camp on July 20 and reached Main Camp at the foot of the eastern slope of the Emma Mountains the next day. From July 21 until August 15 they stayed there and explored in westerly and easterly directions, and to the north reaching the Gonggrijptop. On August 14, they moved on to North Camp between the Hendriktop and the Staheltop and stayed there until August 26. From this camp the Hendriktop was climbed. Between August 26 and September 12 the expedition again moved to Main Camp. The southern part of the Emma Mountains was explored between September 12 and 29 with

South Camp as a base. When this exploration was completed the end of the expedition drew near and the party again moved into Main Camp, staying there from September 29 to October 5. Between October 5 and 10, the staff stayed in Camp 15, between Main Camp and Base Camp. The expedition remained at Base Camp from October 10–19. On this last day they left and went down the Toekoemoetoe Creek and the Saramacca River, arriving in Kwakoegron on October 29. The expedition staff reached Paramaribo that same day by train, and the labourers the next day. The main object of this expedition was a botanical and geological survey of the Emma Mountains. However, some herpetological specimens also were collected which are now in the Rijksmuseum van Natuurlijke Historie, Leiden. Itinerary: JONKER (1961a, b, c, d), JONKER & WENSINK (1960).

1963 Wilhelmina Mountains Expedition. Personnel: Dr. B. MAGUIRE (first half, botany), IRWIN (second half, botany), N. HOLMGREN (botany), T. SOEDERSTROM (botany), G. WESSELS BOER (first half, botany), Dr. J. P. SCHULZ (first half, botany), S. LIGORI (zoology), H. P. PIJPERS (radio-telegraphy). The expedition arrived on the airstrip Kayser Mountains on June 29, 1963. From there the Zuid River was followed downstream to where it empties into the Lucie River, which then was followed upstream to about 15 km before its confluence with the Oost River. Here Base Camp was set up. From this base a trail was cut towards the Juliana-top in a northwesterly direction, which turned northeast at km 14 and north at km 19. From km 19 a side-trail was cut in an easterly direction for 3.5 km. Camps were set up at km 9, 14, 19, 21 and halfway along the eastern trail. The members of the expedition returned to the airstrip Kayser Mountains during the third week of September. The last to leave was PIJPERS, who left on September 24. Although the main object of this expedition was a botanical survey, zoological specimens were collected by LIGORI and PIJPERS. The zoological material is in the Rijksmuseum van Natuurlijke Historie, Leiden.

1968–1969 Sipaliwini Expedition. Personnel: R. NORDE (botany), F. H. F. OLDENBURGER (botany) (with some interruptions), H. T. RIEZEBOS (geomorphology) for the entire period. The expedition was joined for varying periods by several people: M. S. HOOGMOED (zoology: August 22 to October 7), Dr. J. P. SCHULZ (botany: October 7 to December 4), Ir. F. BUBBERMANN (botany, archeology: November 13 to December 4). A varying number of labourers was present throughout the duration of the expedition, but mostly they numbered about 7. The expedition left Paramaribo by plane on August 22, 1968 and arrived at Sipaliwini airstrip the same day. Between August 22 and September 10 and between November 24 and January 10, 1969 the expedition stayed in the Base Bivouac, 5 km east of the airstrip, near the border between savannah and forest. Between September 10 and November 2 the Vier Gebroeders Bivouac on the western slope of the Vier Gebroeders Mountain was occupied. On November 2 the expedition moved to a new camp near the Morro Grande Mountain. From there NORDE, SCHULZ and BUBBERMAN made a trip by foot to Pouso Tirio, an Indian village on the Brazilian part of the savannah (between November 18 and 23). This Morro Grande Bivouac was left on November 24. A new camp was set up on January 10, 1969 in a deserted 'farm' (Meyers' farm) in the southern part of the savannah by RIEZEBOS and, for a short time, OLDENBURGER. This camp was left on January 30 and the members of the expedition returned to the airstrip. NORDE, OLDENBURGER and RIEZEBOS arrived in Paramaribo on January 31, 1969. The zoological material collected during this expedition is in the Rijksmuseum van Natuurlijke Historie, Leiden. Itinerary: HOOGMOED, 1969 (only first part), RIEZEBOS, 1970, and an unpublished account by OLDENBURGER, RIEZEBOS and NORDE.

1970 Sipaliwini Expedition. Personnel: Dr. J. P. SCHULZ (leader, botany), M. S. HOOGMOED (zoology), F. H. F. OLDENBURGER (botany), A. T. RENSSEN (ornithology), J. J. P. PAATS (assistant, zoology), J. TAWJOERAN (assistant, botany), and four labourers. The expedition left Paramaribo by air on January 15 and arrived at Sipaliwini airstrip at about noon. The same afternoon SCHULZ, HOOGMOED, RENSSEN and PAATS left for the Base Bivouac. SCHULZ left the next day for Meyers' farm, situated approximately 9 km southeast of the airstrip on the savannah near the Sipaliwini Creek. There he met OLDENBURGER and TAWJOERAN coming directly from the airstrip. HOOGMOED, PAATS and RENSSEN stayed in the Base Bivouac until January 19 and then went to Meyers' farm, which by then had been vacated (January 17) by SCHULZ and OLDENBURGER. The two botanists had set up a new camp (South Bivouac I) in a forest-island along the Middle Sipaliwini Creek, approximately 15 km southeast of the airstrip. OLDENBURGER stayed there until January 24 and then left for South Bivouac II. PAATS and RENSSEN stayed on Meyers' farm from January 19 to February 6, HOOGMOED interrupted his stay there and joined SCHULZ in South Bivouac I on January 24. The next day SCHULZ, HOOGMOED and TAWJOERAN left for Meyers' airstrip on the border with Brazil, reaching it the same day. They left again on January 28 and arrived in South Bivouac I together with OLDENBURGER. HOOGMOED left South Bivouac I and returned to Meyers' farm the next day. SCHULZ and OLDENBURGER arrived at Meyers' farm in the afternoon of February 1. HOOGMOED, PAATS and RENSSEN left Meyers' farm on February 6 and stayed in Vier Gebroeders Bivouac until February 8. That day they left and returned to the airstrip, where they were joined by SCHULZ and OLDENBURGER. The entire party stayed on the airstrip until February 14, when SCHULZ, HOOGMOED, OLDENBURGER and PAATS left by plane for Paramaribo. RENSSEN stayed on the airstrip for a few more days. All zoological material of this expedition is in the Rijksmuseum van Natuurlijke Historie, Leiden. Itinerary: Anonymus 1970.

1971 Expedition to northwestern Surinam. Personnel: Dr. D. C. GEIJSKES and Dr. M. BOESEMAN (leaders, zoology), P. STAFFELEU (assistant, zoology), D. G. REEDER (assistant, botany and zoology), and a varying number of labourers. The expedition left Paramaribo on January 26 by car for Boskamp and from there continued the journey to Arawarra by boat. From there the Nickerie River was followed upstream to the Blanche Marie Cataracts, via camps at the Stondansi and Lombok Cataracts. The expedition returned at Arawarra on February 20, and from there went to Wageningen where several days were spent.

Between February 25 and March 8, a trip was made to the upper reaches of the Maratakka River. Again a period (March 9–14) was spent in Wageningen and another (March 15–21) in Nieuw Nickerie. Nieuw Nickerie was left on March 22 for a trip up the Corantijn and Kabalebo Rivers. From March 25 to April 4 a tributary of the Corantijn, the Kaboeri Creek was investigated. A camp was set up at the Avanavero Cataracts in the Kabalebo River where the period between April 6 and 12 was spent. The expedition returned in Nieuw Nickerie on April 16 and in Paramaribo on April 19. All zoological material collected by this expedition is in the Rijksmuseum van Natuurlijke Historie, Leiden.

4. Faunal relationships.

PARKER (1935b) arranged the species of frogs, lizards and snakes at the time known from the Guianas (in the wide sense of the term) in six groups. Group A contained the species endemic in the region from Trinidad to Pará, as well as a few species occurring slightly beyond this region's limits. Group B comprised wide-ranging neotropical species, occurring from Colombia or Central America across the whole cis-Andean-region to Brazil or still farther south; some even crossing the Andes. Group C consisted of species ranging from Central America or Colombia through Venezuela to the Guianas, rarely extending farther southwards into eastern Brazil. Group D consisted of species of the Amazon basin, ranging from southeastern Colombia through eastern Ecuador, eastern Peru and eastern Bolivia to Matto Grosso and eastwards to the Guianas in the north and the state of Rio in the south; with a few extending farther southwards to Paraguay and northern Argentina, but rarely reaching the Uruguay-Sta. Catharina region. Group E contained only a few species that were supposed to represent relicts with Antillean affinities. And to conclude, forming what is here called group F, the circumtropical gecko *Hemidactylus mabouia* (Moreau de Jonnès) was listed. From these data PARKER concluded that the Guianas are situated on the border-line between two zoogeographical regions, viz., Venezuela and Amazonia, with a high percentage of endemics (Group A). However, PARKER realised that a number of these eventually might have to be transferred to his group D, comprising the Amazonian species. According to PARKER 30% of the herpetofauna is endemic, 40% Amazonian, 25% belongs to the wide-ranging neotropical species, 4% to species reaching their eastern limit in the Guianas, 1% is relict and 0.5% cosmopolitan. Since then considerable changes have taken place in our knowledge, which are demonstrated, though only for the lizards and amphisbaenians, in table 1.

A list of the species now known for certain to occur in the Guianas (in the wide sense of the term), and arranged in the groups distinguished by PARKER, is given in table 2. A comparison of these two tables with PARKER's lists shows that changes took place in groups A and D. A total of 13 species was added to PARKER's lists, of which six were described between 1952 and 1970. In two instances (*Gonatodes varius* (Duméril) and *Neusticurus tatei* (Burt & Burt))

Table 1. Numbers of species belonging to zoogeographical groups.

PARKER, 1935 b.	A	B	C	D	E	F	total
Frogs, lizards and snakes.	62(30%)	52(25%)	8(4%)	84(40%)	3(1%)	1(0.5%)	210
Lizards and amphisbaenians only.	25(40%)	8(13%)	5(8%)	22(35%)	2(3%)	1(2 %)	63
HOOGMOED, 1973.							
Surinam lizards and amphisbaenians.	8(19%)	7(16%)	1(2%)	26(60%)	0	1(2 %)	43
Guiana lizards and amphisbaenians.	17(29%)	7(12%)	5(8%)	29(49%)	0	1(2 %)	59

Table 2. Surinam species of lizards arranged in the groups of PARKER (1935b), with an indication of their distribution. V. = Venezuela; G. = Guyana; S. = Surinam; F. = French Guiana; B. = Brazil.

	Present classification	Distribution V.	G.	S.	F.	B.
PARKER's group A						
Coleodactylus meridionalis (Boulenger)	*Coleodactylus meridionalis* (Boulenger)	×	×			
Gonatodes annularis Boulenger						
Gonatodes beebei Noble	*Gonatodes annularis* Boulenger	×	×	×	×	×
Gonatodes booni Van Lidth de Jeude						
not recorded	*Gonatodes varius* (Duméril)				×	
Pseudogonatodes guianensis Parker	*Pseudogonatodes guianensis* Parker		×	×		
Sphaerodactylus molei Boettger	in group C					
Gymnodactylus geckoides Spix	not present in Guiana					
not recorded	*Hemidactylus palaichthus* Kluge	×	×	×		
Anolis aeneus Gray	*Anolis aeneus* Gray	×	×			
Anolis chrysolepis Duméril & Bibron	in group D as					
Anolis lentiginosus O'Shaughnessy	*Anolis chrysolepis* Duméril & Bibron					
Anolis nitens (Wagler)						
Enyalius bibroni Boulenger	not present in Guiana					
not recorded	*Amapasaurus tetradactylus* Da Cunha					×
Arthrosaura kockii (Van Lidth de Jeude)	in group D					
Arthrosaura versteegii Van Lidth de Jeude	in group D as *Arthrosaura reticulata* O'Shaughnessy					
Bachia boettgeri (Boulenger)	*Bachia schlegeli* (Duméril & Bibron)		×			
Bachia cophias (Schneider)	in group D					
Bachia parkeri Ruthven	in group C					
Euspondylus leucostictus (Boulenger)	*Euspondylus leucostictus* (Boulenger)	×	×			
recorded in group B	*Gymnophthalmus underwoodi* Grant	×	×	×	(×)	
Iphisa elegans Gray	in group D					
Kentropyx intermedius (Gray)	*Kentropyx borckianus* W. Peters	×	×	×		
Kentropyx striatus (Daudin)	in group D					
not recorded	*Leposoma guianense* Ruibal		×	×	×	×
not recorded	*Neusticurus racenisi* Roze	×				×
Neusticurus dejongi Brongersma						
Neusticurus rudis Boulenger	*Neusticurus rudis* Boulenger	×	×	×	×	
Neusticurus surinamensis Müller						
not recorded	*Neusticurus tatei* (Burt & Burt)	×				
Amphisbaena stejnegeri Ruthven	*Amphisbaena stejnegeri* Ruthven		×			
not recorded	*Amphisbaena vanzolinii* Gans		×	×		

	Present classification	Distribution V.	G.	S.	F.	B.
PARKER's group B						
Thecadactylus rapicauda (Houttuyn)	*Thecadactylus rapicauda* (Houttuyn)	×	×	×	×	×
Iguana iguana (Linnaeus)	*Iguana iguana* (Linnaeus)	×	×	×	×	×
Polychrus marmoratus (Linnaeus)	*Polychrus marmoratus* (Linnaeus)	×	×	×	×	×
Mabuya mabouya (Lacépède)	*Mabuya mabouya* (Lacépède)	×	×	×	×	×
Ameiva ameiva (Linnaeus)	*Ameiva ameiva* (Linnaeus)	×	×	×	×	×
Cnemidophorus lemniscatus (Linnaeus)	*Cnemidophorus lemniscatus* (Linnaeus)	×	×	×	×	×
Gymnophthalmus laevicaudus Cope	in group A as *Gymnophthalmus underwoodi* Grant					
Amphisbaena fuliginosa Linnaeus	*Amphisbaena fuliginosa* Linnaeus	×	×	×	×	×
PARKER's group C						
recorded in group A	*Sphaerodactylus molei* Boettger	×	×			
Norops auratus (Daudin)	*Anolis auratus* Daudin	×	×	×	×	×
Anolis gibbiceps Cope	*Anolis gibbiceps* Cope	×	×			
Basiliscus basiliscus (Linnaeus)	not present in Guiana					
Tropidodactylus onca (O'Shaughnessy)	*Tropidodactylus onca* (O'Shaughnessy)	×				
recorded in group A	*Bachia parkeri* Ruthven		×			
Leposoma dispar Peters	not present in Guiana					
PARKER's group D						
not recorded	*Coleodactylus amazonicus* (Andersson)	×	×	×	×	×
Gonatodes humeralis (Guichenot)	*Gonatodes humeralis* (Guichenot)	×	×	×	×	×
not recorded	*Lepidoblepharus festae* Peracca					×
Anolis scypheus Cope and recorded in group A	*Anolis chrysolepis* Duméril & Bibron	×	×	×	×	×
Anolis fusco-auratus d'Orbigny	*Anolis fuscoauratus* d'Orbigny	×	×	×	×	×
Anolis ortonii Cope	*Anolis ortonii* Cope	×	×	×	×	×
Anolis punctatus Daudin	*Anolis punctatus* Daudin	×	×	×	×	×
Plica plica (Linnaeus)	*Plica plica* (Linnaeus)	×	×	×	×	×
Plica umbra (Linnaeus)	*Plica umbra* (Linnaeus)	×	×	×	×	×
Tropidurus torquatus (Wied)	*Tropidurus torquatus* (Wied)	×	×	×	(×)	×
Urocentron azureum (Linnaeus)	*Uracentron azureum* (Linnaeus)		×	×	×	×
Uranoscodon superciliosa (Linnaeus)	*Uranoscodon superciliosa* (Linnaeus)	×	×	×	×	×
not recorded	*Alopoglossus angulatus* (Linnaeus)		×	×	×	×
recorded in group A	*Arthrosaura kockii* (Van Lidth de Jeude)		×	×	×	×
recorded in group A	*Arthrosaura reticulata* O'Shaughnessy		×	×	×	×
recorded in group A	*Bachia cophias* (Schneider)	×	×	×	×	×
Cercosaura ocellata Wagler	*Cercosaura ocellata* Wagler		×	×	(×)	×
Crocodilurus lacertinus (Daudin)	*Crocodilurus lacertinus* (Daudin)				×	×
Dracaena guianensis Daudin	*Dracaena guianensis* Daudin				×	×
recorded in group A	*Iphisa elegans* Gray		×	×	×	×
Kentropyx calcaratus Spix	*Kentropyx calcaratus* Spix	×	×	×	×	×
recorded in group A	*Kentropyx striatus* (Daudin)	×	×	×	(×)	×
Leposoma percarinatum (L. Müller)	*Leposoma percarinatum* (L. Müller)	×	×	×	(×)	×
Leposoma scincoides Spix	not present in Guiana					
Neusticurus bicarinatus (Linnaeus)	*Neusticurus bicarinatus* (Linnaeus)	×	×	×	×	×
Neusticurus ecpleopus Cope	not present in Guiana					
not recorded	*Prionodactylus argulus* (W. Peters)	(×)	×	×	×	×
not recorded	*Ptychoglossus brevifrontalis* Boulenger			×		

	Present classification	Distribution V.	G.	S.	F.	B.
Calliscincopus agilis Ruthven	*Tretioscincus agilis* (Ruthven)		×	×	×	×
Tupinambis nigropunctatus Spix	*Tupinambis nigropunctatus* Spix	×	×	×	×	×
Tupinambis teguixin (Linnaeus)	not present in Guiana					
Amphisbaena alba Linnaeus	*Amphisbaena alba* Linnaeus	×	×	×	×	×
Parker's group E						
Anolis sagrei Cocteau	not present in Guiana					
Ameiva punctata Duméril & Bibron	not present in Guiana					
Parker's group F						
Hemidactylus mabouia (Moreau de Jonnès)	*Hemidactylus mabouia* (Moreau de Jonnès)	×	×	×	×	×

Parker overlooked the species and the remaining five cases concern extensions of species ranges that became known only recently.

Nine of the species reported by Parker as occurring in the Guianas actually do not exist there. *Gymnodactylus geckoides* Spix, *Enyalius bibroni* Boulenger, *Tupinambis teguixin* (Linnaeus) and *Leposoma scincoides* Spix occur far more to the east and to the south; *Neusticurus ecpleopus* Cope is restricted to the western and southern periphery of the Amazon drainage; *L. rugiceps* (Cope) occurs in Columbia and Panama only; *Basiliscus basiliscus* (Linnaeus) does not occur farther east in Venezuela than the state of Mérida, which is about 1100 km west of Venezuelan Guiana. According to Baskin & Williams (1966) the records of *Anolis sagrei* Cocteau from Guyana are based on misidentifications. One of the four specimens on which this record was based proved to belong to *A. aeneus* Gray, since long known from Georgetown, the other three represent *A. oculatus* (Cope) from Dominica. Baskin & Williams (1966) also clarified the *Ameiva major* Duméril & Bibron—*A. punctata* Gray problem. There has been much confusion concerning these names and the first to shed some light on this problem was Parker (1935a), who tentatively suggested that *A. major* Duméril & Bibron came from Martinique and not from Cayenne and Trinidad as stated in the original description. Baskin & Williams (1966) made it clear that this is quite probable for three of the four original syntypes. The remaining specimen was described by Bocourt (1874) as *A. major* var. *flaviceps*, a name shown by Baskin & Williams (1966) to be a synonym of *A. erythrocephala* (Daudin), a species occurring on Nevis, St. Kitts and St. Eustatius. In the same way *A. punctata* Gray, which was described from Demerara, is a synonym of *A. erythrocephala* (Daudin).

From the data in table 1 it is clear that the Surinam lizard fauna (Sauria + Amphisbaenia) is composed mainly of three elements, the most important being the Amazonian (60%), less important the widespread species (16%) and the endemics (19%). For the future I expect that the number of endemics will decrease still further when the Amazon basin fauna is investigated more intensively. Here,

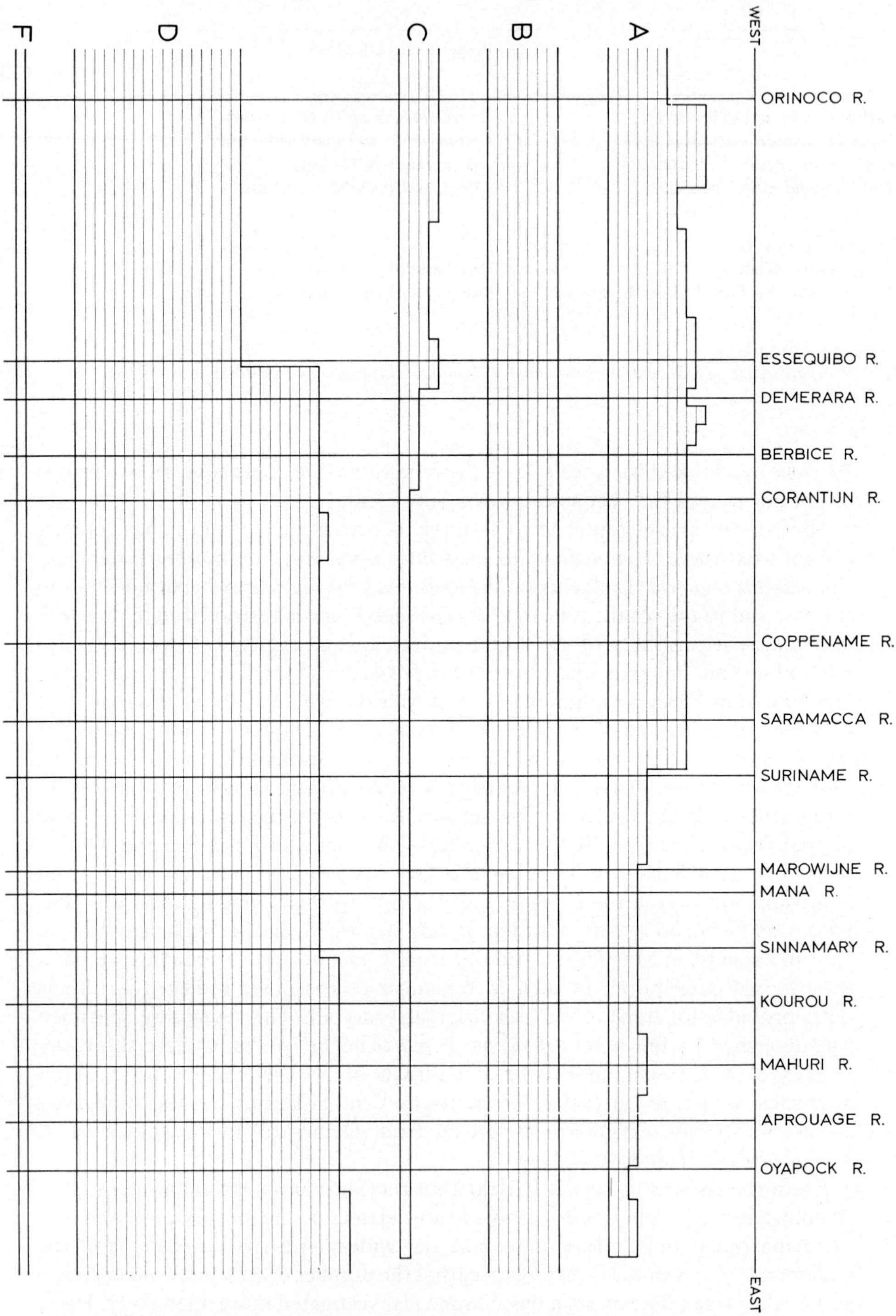

Fig. 3. Diagram giving an impression of the importance of the zoogeographical barriers in a W.-E. direction for the Guianan lizards and amphisbaenians. The west-east bars represent (from top to bottom) the groups of PARKER (1935b). Note that west-east discontinuities do not represent distributional gaps of a given species.

it may be observed that the group of Amazonian species can be subdivided into a few minor groups, viz., species with a peripheral distribution, species present only in the eastern part of the Amazon basin, species present in the eastern and western part of the Amazon basin, and species occurring in the entire area. I do not pursue this matter here, because possibly these groups are only artificial and induced by our scanty knowledge of the herpetofauna of the Amazon basin. In the group of the endemic species, most seem to be western forms, reaching their eastern limit in Guyana, Surinam or French Guiana. Only three of these are present in the Brasilian part of Guiana, Amapá, which is relatively well known. *Amapasaurus tetradactylus* Da Cunha is the only species restricted to this area. *Leposoma guianense* Ruibal seems to be restricted to the eastern part of Guiana and to reach its western border somewhere in central Guyana.

For the entire Guiana area 59 species are now known. Of these, 43 are present in Surinam (see above). The remaining 16 may be allotted to four groups. Ten species, viz., *Sphaerodactylus molei* Boettger, *Anolis aeneus* Gray, *A. gibbiceps* Cope, *Tropidodactylus onca* (O'Shaughnessy), *Bachia schlegeli* (Duméril & Bibron), *B. parkeri* Ruthven, *Euspondylus leucostictus* (Boulenger), *Neusticurus racenisi* Roze, *N. tatei* (Burt & Burt) and *Amphisbaena stejnegeri* Ruthven, only occur in the western part of Guiana. Four species, namely *Gonatodes varius* (Duméril), *Amapasaurus tetradactylus* Da Cunha, *Crocodilurus lacertinus* (Daudin) and *Dracaena guianensis* Daudin are restricted to the eastern part of Guiana. One species, *Coleodactylus meridionalis* (Boulenger), occurs in eastern Venezuela, Guyana and in eastern Brazil, the two areas being separated by a wide zone from which the species appears to be absent and where it apparently is replaced by *C. amazonicus* (Andersson) (VANZOLINI, 1968b). And finally, *Lepidoblepharus festae* (Peracca) occurs both near the mouth of the Amazon River and at its headwaters; it is not known from the intermediate area.

Of the 43 Surinam species, 33 are also known to occur in Guyana and French Guiana, five occur in Guyana and Surinam and very probably also in French Guiana, one (*Ptychoglossus brevifrontalis* Boulenger) until now has been found only in Surinam and Ecuador, and four (*Hemidactylus palaichthus* KLUGE, *Pseudogonatodes guianensis* Parker, *Kentropyx borckianus* W. Peters, *Amphisbaena vanzolinii* Gans), seem to reach their eastern limit in Surinam (figs. 3 and 4). Thus, Guyana has 42 species in common with Surinam and in addition nine not occurring there, making a total of 51 species; Surinam has 43 species of which one is not known from Guyana, but very probably also occurs there; French Guiana has at least 33 and very probably 38 species in common with both Surinam and Guyana, and in addition has three species not occurring in these two countries. From these facts it may be concluded that eastern Surinam and western French Guiana form a wide transitional zone for a small part of the lizard fauna of Guiana, separating eastern and western components. In Surinam a north to south line approximately coinciding with the Coppename River, forming an important phytogeographical border, has been observed (JONKER, 1966). For the lizards a similar line may be distinguished along the Suriname River, as three of the four species having their eastern distributional

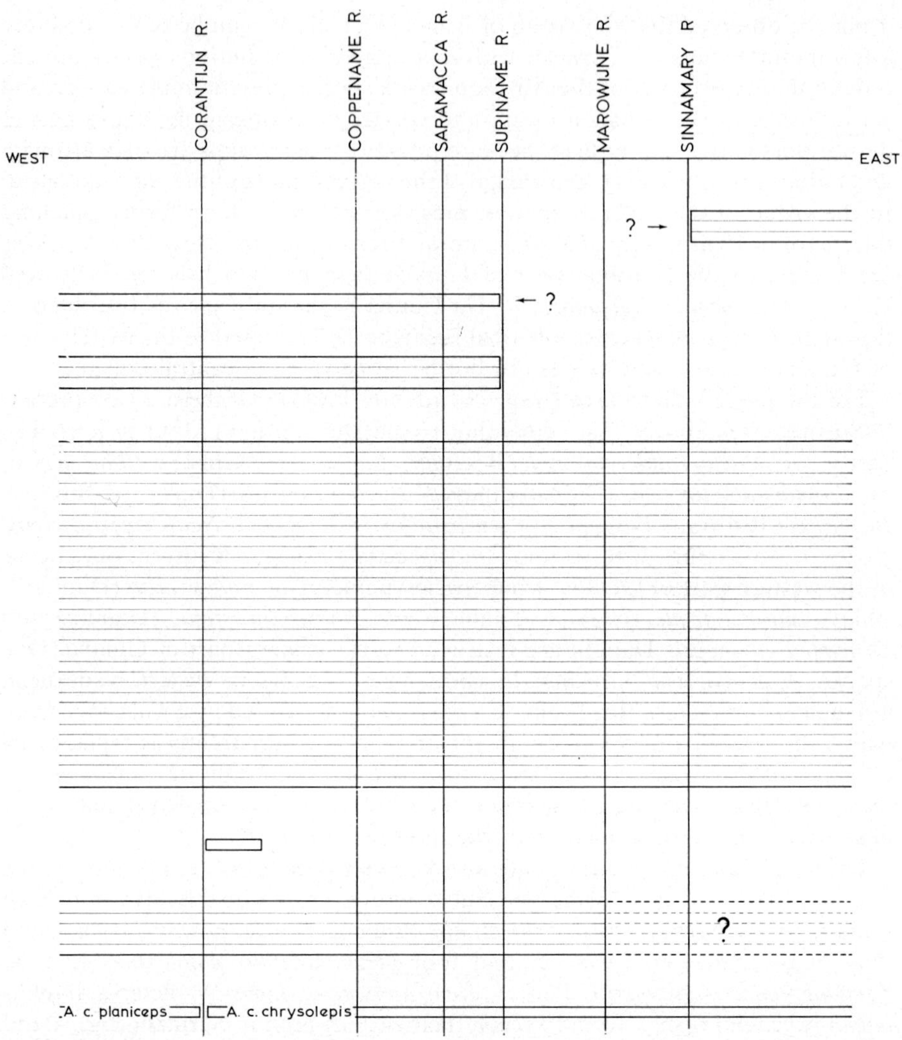

Fig. 4. Diagram giving an impression of the importance of zoogeographical barriers within Surinam, demonstrated on the species of lizards and amphisbaenians and, independantly on one subspecies of lizard.

limit in Surinam do not occur east of this river. For the fourth species (*Hemidactylus palaichthus* Kluge) no exact data on the distribution in Surinam are available. Also from French Guiana no definite locality records for *Crocodilurus*, *Dracaena* or *Gonatodes varius* (Duméril) are available, and it is impossible to indicate their western limits; however, considering the information obtained on frogs, the boundary may be along the river Sinnamary. In that case the area between the Suriname River and the river Sinnamary would form an area in which

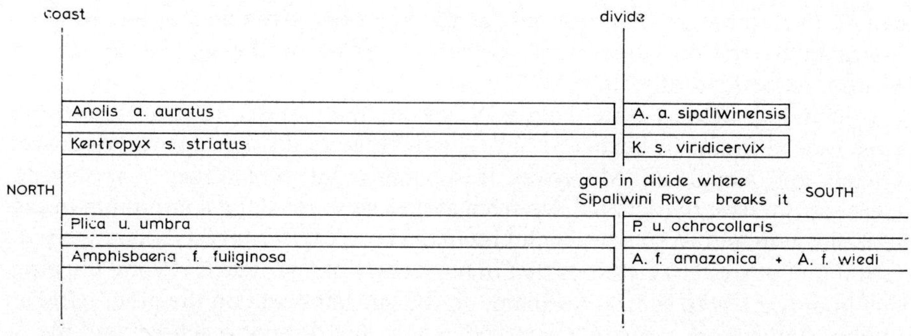

Fig. 5. Diagram giving an impression of the importance of the divide between Amazonia and Surinam as a zoogeographical barrier. Only subspecies for which this divide is a barrier have been indicated.

occur only the 38 species that are widespread in the Guianas. However, it should be kept in mind that this is merely a hypothesis, based on a few facts.

The picture slightly changes when we take into account the subspecies occurring in the Guianas. The following subspecies occur throughout the area: *I. i. iguana* (Linnaeus), *A. a. ameiva* (Linnaeus), *Cnemidophorus l. lemniscatus* (Linnaeus), *Mabuya m. mabouya* (Lacépède), *Amphisbaena f. fuliginosa* Linnaeus, *Anolis a. auratus* (Daudin), *A. fuscoauratus kugleri* Roux, *A. p. punctatus* Daudin, *Plica. u. umbra* (Linnaeus), *Tropidurus torquatus hispidus* (Spix), *Arthrosaura reticulata versteegii* Van Lidth de Jeude and *Cercosaura o. ocellata* Wagler. Only five species have different subspecies in the different countries here considered, viz., *Anolis c. chrysolepis* Duméril & Bibron occurs in Surinam, French Guiana and northeastern Brazil, *A. c. planiceps* Troschel in Guyana and Venezuela; *A. auratus sipaliwinensis* nov. subsp. and *Kentropyx striatus viridicervix* nov. subsp. are only known from a large inland savannah on the Surinam/Brazil border. These last named two subspecies, together with the subspecies *Amphisbaena f. fuliginosa* Linnaeus and *Plica u. umbra* (Linnaeus), present some evidence for a separating line between Amazonia and Guiana, which approximately coincides with the divide. To the south of this divide there exist other subspecies. North of the divide *Anolis a. auratus* Daudin, *Plica u. umbra* (Linnaeus), *Kentropyx s. striatus* Daudin and *Amphisbaena f. fuliginosa* Linnaeus are present. On the divide or south of it the following subspecies occur: *A. a. sipaliwinensis* nov. subsp., *P. u. ochrocollaris* (Spix), *K. s. viridicervix* nov. subsp., *A. f. amazonica* Vanzolini and *A. f. wiedi* Vanzolini (see fig. 5). This can be easily explained by a forest refuge on the Tumuc Humac and Boundary Mountains, which spread much faster to the north than to the south. Apparently the Amazonian forest refuges (see below) started to spread earlier than the forest refuge on the Tumuc Humac Boundary Mountains, thus blocking the way to the south for some subspecies that had originated in the Tumuc Humac/Boundary Mountains forest refuge, because in the Amazonian forests the available niches were already occupied by related subspecies. Consequently the only way left to extend their range of distribution was to the north. Another border, which is indicated by the distribution of the subspecies of *A. chrysolepis* Duméril & Bibron, is the Corantijn River.

Concluding we may state that on the species level an east to west transitional zone can be found in Surinam, but that there is no distinct partition between Guiana and Amazonia. Moreover, it is evident that the number of species decreases from west to east and also from east to west, reaching a minimum in eastern Surinam and western French Guiana. This pattern suggests a relatively recent influx of western elements that in most cases did not reach beyond Guyana, and in only a few to central Surinam, on the one hand and on the other hand an unimportant recent influx of eastern elements that did not reach beyond the river Sinnamary. On the subspecies level, however, in only one case can a distinct border running from north to south be found, but in four other cases a distinct border running from east to west, coinciding with the divide between Guiana and Amazonia, is found.

5. Notes on the savannah-inhabiting lizards.

In these notes I consider as savannahs, without any further discrimination, all types of open vegetation in which grasses and sedges play a dominant role while trees are of minor importance. In this category fall the beaches, some ridges, the coastal savannah belt, the isolated savannahs in the interior (airstrips Kayser Mountains and Tafel Mountain, Saniki), the Sipaliwini savannah and open spaces in towns and villages. There are only three lizards that occur in (nearly) all these habitats and, moreover, are confined to them, namely: *Anolis auratus* Daudin, *Cnemidophorus lemniscatus* (Linnaeus) (not on the Sipaliwini savannah) and *Kentropyx striatus* (Daudin) (not on the beaches). In some of the savannahs a number of other lizards can be found, which, however, are not typically restricted to this habitat. These include five species, namely *Iguana iguana* (Linnaeus), *Ameiva ameiva* (Linnaeus), *Gymnophthalmus underwoodi* (Grant), *Tupinambis nigropunctatus* Spix and *Mabuya mabouya* (Lacépède). These species also can be found in open places in the rain-forest: in clearings, on banks of creeks and rivers, along trails, etc. Apart from the lizards mentioned above, two amphisbaenians, viz., *Amphisbaena alba* Linnaeus and *A. fuliginosa* Linneaus, occasionally are found in the savannah, but these also occur in the rainforest. In the coastal savannah belt another lizard may be encountered, namely *Tropidurus torquatus* (Wied). The species is well known outside Surinam for its very diverse habitats. In Guyana it occurs in savannahs, in Brazil it occurs on open places in towns (Belém do Pará), in the caatingas of the northeast, the cerrados of central Brazil, the pantanais of the Mato Grosso, the stony mountains in Minas Gerais, while in forested places it occurs in exposed habitats (VANZOLINI, 1963). To some extent the same diversity in habitats can be found in Surinam, where the species occurs in several places in the coastal savannah-belt, but also isolated in the rain-forest of the interior, on inselbergs and boulders along rivers. However, too few inselbergs and small savannahs in the interior have been explored to gain a good insight into its actual distribution in Surinam. For instance, I am quite certain that it does not occur on the savannahs surrounding the airstrip Kayser Mountains or on the Sipaliwini savannah. On the one hand specimens are known from the Tafel Mountain, but on the other hand nothing is known about their occurrence on the airstrip Tafel Mountain. Furthermore, it has not yet been seriously looked for on the mountains with large exposed rockfaces in the major mountain ranges of the interior. Therefore, in the following paragraphs I will leave this species out of consideration.

Consequently, there remain only three lizards that may be termed typical savannah dwellers and these may serve as indicators of the history of savannahs in Surinam. Recently the problem of the extension of the savannahs has received much attention, as well from botanists, as from zoologists, pedologists, geologists, and archeologists. About the Surinam savannahs there is an extensive literature, mostly dating from after 1936. LANJOUW (1936a, b; 1954) made clear that in Surinam the savannahs are not primarily the result of climatological influences, as, for instance, there is no difference in rainfall on the savannah and on the adjacent rain-forest. According to him the main causes for their origin are: edaphic factors (flat divides), repeated burning and, as an additional factor, the climate, which helps in leaching the soil and in providing possibilities for the xeromorph vegetation to develop. It is possible to discern a number of types in the Surinam savannahs (COHEN & VAN DER EIJK, 1953), but these do not seem to be of importance in herpetology. Until recently, the opinion that the South-American savannahs were recent and man-made was widely disseminated (GONGGRIJP & BURGER, 1948). Some authors (DARLINGTON, 1957; SCHWABE, 1968) believed that the glacial periods had only a slight influence on the flora and fauna of tropical lowlands, and they considered the flora and fauna to be old and the high number of species to be the consequence of 'a long ecological evolution which must have taken place in the absence of any destructive natural catastrophes' (SCHWABE, 1968). However, SCHWABE supposed that before the Tertiary large areas of the Hylaea may have been covered with steppes and savannahs, from which the former woodland vegetation had withdrawn into refugia, and from where it subsequently invaded the Amazon basin. Recently it has become clear that the processes SCHWABE described for ancient times continued until very recently. VAN DER HAMMEN (1966) showed that in the Holocene considerable changes in the extent of the savannahs and forest in Guiana took place. Other data have been assembled by VANZOLINI & WILLIAMS (1970), who discerned between botanical, zoological, geomorphological and palynological evidence for recent climatic changes. All of their examples are quite convincing. P. Müller (1968; 1969; 1970) and P. MÜLLER & SCHMITHÜSEN (1970) provided additional evidence for a former connection between the atlantic forest in southeastern Brazil and the Hylaea and between the savannahs among themselves. From all these data it becomes evident that between 7000 and 1000 B.C. (WILLIAMS & VANZOLINI, 1970; P. MÜLLER, 1971b) a large part of South America experienced a dry period, which had important consequences for the dispersal of both animals and plants. During this dry period large parts of Amazonia must have been covered with savannahs, with local forest refugia. VANZOLINI & WILLIAMS (1970) and VANZOLINI (1970a), working on *Anolis chrysolepis* Duméril & Bibron, came to the conclusion that this species probably survived in six of these refugia and, after a change of climate, spread from there over Amazonia. HAFFER (1969, 1970) working with Amazonian birds also came to the conclusion that there must have been six forest refuges in cis-Andean South America. Three of these coincide with those suggested for *Anolis chrysolepis* Duméril & Bibron namely the Guianan (Tumuc Humac and Boundary Mountains), the Napo and the East Peruvian refuges. Moreover, there is some geomorphological evidence for the former existence of forest refuges, indicating that certain areas have been continuously forested, consequently also during the dry periods.

Taking into account the information given, we may assume that about 3000 years ago the greater part of Surinam was covered with savannahs and that only the Tumuc Humac and Boundary Mountains were covered with rain-forest. During this dry period, savannah-dwelling species had a wide distribution and occurred all over Surinam. After this dry period the forest again spread over Surinam starting from the Tumuc Humac and Boundary Mountains. Certain places apparently were unfavourable for reforestation and remained savannahs. These are the presently existing savannahs (sensu strictu) in Surinam. Evidence that the developments outlined above took place in the assumed sequence may be deduced from certain facts concerning *Anolis auratus* Daudin and *Kentropyx striatus* (Daudin). Both lizards only occur in a savannah type habitat, the first can be found everywhere in this kind of habitat, the second is absent on the beaches and in gardens. Both are represented on the Sipaliwini savannah by populations distinctly diverging from those occurring in the remaining parts of Surinam.

This well agrees with the assumption that formerly, on the Tumuc Humac and Boundary Mountains, there existed a forest refuge that subsequently spread from there. This having been the case, and assuming a normal pattern of dispersal, a part (or parts) of the Sipaliwini savannah must have been the first of all the savannahs in inland Surinam to become isolated by being enclosed by rain-forest, and the populations there of *A. auratus* Daudin and *K. striatus* (Daudin) at that time must have become well isolated. These populations now can be recognised as different subspecies. The next area to become enclosed by the spreading rain-forest must have been the area around the airstrip Kayser Mountains. Nowadays three species of savannah-dwelling lizards are present in this area: *A. auratus* Daudin, *K. striatus* (Daudin) and *Cnemidophorus l. lemniscatus* (Linnaeus), but the fact that none of these differs from the populations in the coastal area, suggests that the separation between the two occurred relatively recently, at least after the isolation of the Sipaliwini savannah took place. Another point in favour of this point of view is the fact that *C. l. lemniscatus* (Linnaeus) is absent from the Sipaliwini savannah. Recent information from Brazil (VANZOLINI, 1970b) suggests that this species may be in the process of expanding its range along the main course of the Amazon towards the west.

The species probably originated in Venezuela (DUELLMAN, 1966), from where it spread to the north and the east and recently reached the mouth of the Amazon by way of the coastal savannahs of the Guianas and Amapá. As an offshoot of this dispersal route it may have reached the savannahs around the airstrip Kayser Mountains when these were still connected with the coastal savannah-belt. The Sipaliwini savannah was not reached because a connection with it no longer existed. Consequently I think that the occurrence of subspecies, clearly differentiated from the coastal populations, on the Sipaliwini savannah serves to strengthen the views held by HAFFER and by VANZOLINI & WILLIAMS, and points to the former existence of a forest refuge on the Tumuc Humac and Boundary Mountains.

More arguments to support this hypothesis of a forest refuge on the Tumuc Humac and Boundary Mountains that started to spread recently can be taken from the occurrence of forest islands in the Sipaliwini savannah. There is geomorphological evidence (ZONNEVELD, 1967) indicating that recently a larger part of the Sipaliwini savannah was covered by forest, with only small scattered patches of savannah in some places, and that it reached its present extent as a result of deforestation by human interference. Here and there in the present-day Sipaliwini/Paru savannah area, smaller or larger patches of savannah surrounded by forest, which served as nuclei from where savannah-dwelling species populated the entire area must have been present. These places were qualified by edaphic factors (ZONNEVELD, 1967). After the extension of the savannah, forest-islands remained on several places, mostly (the large ones) in valleys, sometimes (the small ones) on hilltops. As pointed out before (HOOGMOED, 1969b) the forest-islands are residues of the rain-forest formerly covering and now surrounding the savannah. This is confirmed both by the floral composition, which shows only slight differences, and by the herpetofauna. The species of lizards inhabiting the forest-islands do not show any differences from the populations inhabiting the surrounding rain-forest, indicating that until recently there has been contact. The only difference from the surrounding rain-forest is the lower number of species present. This is quite understandable considering the small size of the forest-islands and the low population density of certain species in the rain-forest. Another fact indicating a recent connection between the forest-islands is the presence of a blue frog, *Dendrobates azureus* Hoogmoed (1969b), in a number of them. This may easily be explained by considering it as a relict, as the species is unable to cross the open savannah and thus cannot have migrated from one forest-island to another. Moreover, it seems to be absent from the rain-forest surrounding the savannah, where only relatives appear to be present. However, no data concerning the forest east and north of the savannah are as yet available.

Also botanically, there is evidence that the savannahs in the interior of Guiana are relatively old (HUECK, 1966) and not entirely man-made. According to VAN DONSELAAR (1968) 4% of the plants of the Sipaliwini/Paru savannah is endemic, whereas no endemics occur in the savannahs of northern Surinam. This also indicates that the Sipaliwini/Paru savannah, or at least part of it, was separated from other savannahs relatively early.

It is intriguing to note that a number of birds on the Sipaliwini savannah have their nearest relatives in the campos of Brazil (MEES, 1968), several hundreds of kilometres to the south, indicating that during the dry period a number of campos species migrated northward. On the other hand, all savannah-dwelling lizards do not extend farther south than the Amazon, and they seem to have migrated from the west to the east, possibly from a savannah refuge in southern Venezuela, where now the Llanos are situated.

IV. SYSTEMATIC PART

1. Methods.

As confusion may arise concerning the names of the scales, it seems useful to point out that I adopted the nomenclature as used by SMITH (1946) and PETERS (1964). For nomenclature of head scales I may refer to several drawings in SMITH (1946). I subsequently present an alphabetical list of the families of lizards and amphisbaenians, adding for each the methods used to obtain the various scale counts.

Gekkonidae: KLUGE's (1967, 1969) method of counting scales is followed here.

Iguanidae: Apart from the vertebral count ETHERIDGE's (1968, 1970) method is adopted here. The vertebral count is taken from the occipital to the level of the posterior margin of the thighs (=base of tail).

Scincidae: Scales were counted the same way as in the Teiidae.

Teiidae: For this family the terminology and method of counting of RUIBAL (1952) and UZZELL (1966) was followed when feasible. The number of scales around midbody was determined by counting all scales (dorsal, lateral and ventral) in an annulus around the body at the point halfway between the forelimbs and the hind limbs. The number of transverse rows of dorsals (only when enlarged) was counted from the first transverse row following the parietals or nuchals to and including the row of which the posterior scale-margins are at the level of the posterior surface of the thighs (though, in some species a slightly different method was used, as will be explained in the appropriate place). The number of transverse rows of ventrals was assessed by starting to count immediately posterior to the collar (= gular fold of SMITH (1946)), or, when a collar was absent, at the level of the anterior margin of the forelimbs, counting along the median line to (and including) the last clearly defined transverse row of ventrals. This method excludes the scale row separating the series of femoral pores (which in some species is perforated by preanal pores). In *Bachia* a slightly different method was used. The number of longitudinal dorsal scale rows was counted only in those species in which the dorsals are distinctly enlarged and easily distinguishable from the laterals. To assess the number of longitudinal rows of ventrals all enlarged scales in an annulus at midbody were counted when they were well marked off from the laterals. When the laterals were of nearly the same size as the ventrals, only the number of longitudinal rows between the bases of the forelimbs was counted. The collar scales are the distinctly enlarged scales on the border of throat and chest. In some species all transverse gular scale rows were counted between the posterior margin of the last pair of chinshields and the collar. In others only the transversely enlarged gular rows were counted. The supralabials were counted from the rostral to the angle of the mouth, the infralabials were counted from the mental to the angle of the mouth. However, in some species this method could not be used because of the small size of the posterior scales, of which it is not clear whether or not they still belong to the lips. In these cases, only the enlarged labials were counted.

Amphisbaenidae: The terminology of GANS & ALEXANDER (1962) was followed. Thus, the number of body annuli is the number of annuli from the posterior margin of the posterior infralabial to and including the annulus containing the preanal pores. The number of tail annuli is assessed by starting to count at the first uninterrupted annulus behind the postcloacals and continuing to the last clearly defined annulus. The blunt tip of the tail is indicated by '+1' added behind the number of tail annuli. The lateral rows in the cloacal region are the rows between the body annulus containing the preanal pores and the first uninterrupted tail annulus. The dorsal and ventral scales are counted at midbody (measured) between the lateral sulci. Adding both values provides the number of scales around midbody.

For all lizards the following may be stated. Subdigital scales were counted under the fourth finger and toe. Wherever the plantar and the subdigital scales merge, the counting was started at the point where the third finger or toe joined the fourth. In the event of the subdigital scales being strikingly different from the plantar scales, either in size or in structure, the first scale showing this character was taken to be the first in the count, even if it is among plantar scales.

All measurements are in mm. The length of the head is the distance between the tip of the snout and the posterior border of the ear-opening. The width of the head is the distance between the most projecting parts of the temples. The height of the head is taken at the same level as the head width. This measurement, however, is not very reliable, because it depends on whether the animal has its mouth opened or not, its throat bulging or not.

The ranges of variation given in the text are based only on the Surinam specimens of the species under consideration. Tables with counts and measurements have been compiled only for species of which just a few specimens are known or of which until recently few specimens were known.

In the following descriptions some expressions are used in connection with size. It seems useful to explain them here.

very small	=	20–30 mm
small	=	30–70 mm
medium	=	70–130 mm
large	=	130–300 mm
very large	=	300–360 mm

In figs. 1 and 2 all localities where lizards and/or amphisbaenians have been collected are indicated. In the lists of material the localities have been arranged from north to south and from west to east. In compiling the range of the species

I used as sources DA CUNHA (1961), DONOSO-BARROS (1968), J. A. PETERS (1967), J. A. PETERS & DONOSO-BARROS (1970) and UNDERWOOD (1962).

It seems appropriate to provide a guide to the pronunciation of the vernacular names.

a	sounds like the a in arm, or like the last a in Guyana.
oe or u	sounds like the oe in shoe.
o	sounds like the o in two.
ie	sounds like the ee in sheet.
i	sounds like the i in this.
oi	sounds like the oy in joy.
ai	sounds like the i in high.
e	may sound like the a in snake, like the e in the or like the ea in head.

It is customary in the Division of Reptiles and Amphibians at the Rijksmuseum van Natuurlijke Historie, Leiden, to number lots. When it is necessary to indicate individual specimens these are labelled with letters, written on small labels and attached to the specimen.

2. Notes on Linnaean names.

Among the lizards Linnaeus described in 1758 there are 14 species from South America, 12 of which occur in Surinam, the two exceptions being *Lacerta Basiliscus* and *L. Teguixin*. The latter name however covered a mixture of *Tupinambis teguixin* (Linnaeus) and *T. nigropunctatus* Spix (occurring in Surinam) as was shown by LÖNNBERG (1896) and ANDERSSON (1900). It is possible to arrange these species in three groups, namely:

1. Without any references added.
2. Of the earlier Linnaean papers only the Museum Adolphi Friderici (LINNAEUS, 1754) is mentioned.
3. Of the earlier Linnaean papers the Museum Adolphi Friderici and the Amoenitates Academicae (LINNAEUS, 1749) are mentioned.

The first group comprises four species, namely *Lacerta bicarinata* (no. 7), *L. angulata* (no. 18), *L. umbra* (no. 31) and *L. plica* (no. 32). Of these *L. angulata* undoubtedly comes from Surinam (see below). According to THUNBERG (1787) and HOLM (1957) there is one specimen of *L. bicarinata* and one of *L. umbra* in the Uppsala collection, both originating from the Donatio Adolphi Friderici in 1745. However, in BALK's thesis (1746) these two species are not recorded and the first description of these species is in the Systema Naturae. What exactly was the reason for their absence from the 'Museum Adolpho-Fridericianum' is not known. But, relying upon THUNBERG (1787) we may assume that both specimens were present from 1745 on. Thus, of this first group only two species remain, namely *L. angulata* and *L. plica*, which at the time were recent acquisitions and on which original descriptions were based. They probably came into LINNAEUS's possession after the return of ROLANDER from Surinam and the type locality of both can be restricted to coastal Surinam. Details of the type-locality for *L. angulata* are given below; the most likely type-locality for *L. plica* would be the confluence of the Cottica River and the Perica Creek, Surinam, because DAHLBERG, who was ROLANDER's host in Surinam, had two plantations there which were visited by ROLANDER (HOLTHUIS, 1959). Types of three of the species are extant: *L. bicarinata* (Uzzell, 1966: Zoological Institute, University of Uppsala no. 70); *L. plica* (Etheridge, 1970: Naturhistoriska Riksmuseet, Stockholm KaF 1900: 112) and *L. umbra* (Andersson, 1900; ETHERIDGE, 1970: Naturhistoriska Riksmuseet, Stockholm 1900: 111; HOLM, 1957: Zoological Institute University of Uppsala no. 73).

The second group also contains four species, namely *Lacerta superciliosa* (no. 3), *L. azurea* (no. 12), *L. lemniscata* (no. 39) and *Amphisbaena alba* (no. 239). These species probably all originate from DAHLBERG (HOLTHUIS, 1959) who visited Sweden in 1754 when he presented King Adolf Fredrik with botanical and zoological collections from Surinam. This material was studied by LINNAEUS round about the end of September 1754 (LOVÉN, 1887) and has been incorporated into the Museum Adolphi Friderici (LINNAEUS, 1754). It is now in the Naturhistoriska Riksmuseet in Stockholm (ANDERSSON, 1899; 1900) and may be considered the type material on which LINNAEUS based his descriptions. The type locality for *L. azurea* has been restricted by ETHERIDGE (1968) to the vicinity of Paramaribo, Surinam, and the same can be done with the type localities for the other species of this section, but here again I prefer narrowing it down to: the confluence of the Cottica River and the Perica Creek, Surinam.

The third and last group again contains four species, namely *Lacerta Ameiva* (no. 14), *L. Iguana* (no. 28), *L. marmorata* (no. 33) and *Amphisbaena fuliginosa* (no. 230). They are all described in the Museum Adolphi Friderici and in the Amoenitates Academicae. The first three species have two Amoenitates references, namely one to Amphibia Gyllenborgiana (respectively nos. 13, 11 and 15) (HAST, 1745) and one to the Museum Adolpho-Fridericianum (respectively nos. 19, 12 and 13) (BALK, 1746). *A. fuliginosa* only has one reference, namely to the Museum Adolpho-Fridericianum (no. 22). There are no references to SUNDIUS' (1749) paper, which also appeared in the Amoenitates, although part of his species are identical with those recorded in the aforementioned papers in the Amoenitates, to which Sundius refers.

Part of the material belonging to the Donatio Caroli Gyllenborg, 1744, the Donatio Adolphi Friderici, 1745 and to the Donatio Claudii Grill, 1746 is still preserved in the Zoological Institute of the University of Uppsala (HOLM, 1957); the remainder seems to have been lost. HOLM stated that one specimen of *Lacerta Iguana* and one of *L. Ameiva* of the Gyllenborg donation are still present and are the types; moreover, in Uppsala there are two specimens of *L. marmorata*, one of the Gyllenborg donation and one of the Adolf Fredrik donation, which are syntypes. Furthermore there is one specimen of *L. bicarinata* and one of *L. umbra*, both of the Adolf Fredrik donation (see above), which are the types of these species. Thus, ETHERIDGE's (1970) remark concerning the syntypes of *Plica umbra* should be amplified with the information that another syntype of this species is present in Uppsala.

Concerning the origin of the material donated by GYLLENBORG, ADOLF FREDRIK and GRILL, the following can be said. According to HOLM (1957) GYLLENBORG presented the University of Uppsala with 'a collection consisting mainly of snakes and lizards from North America'. From the material described it is evident that this is not true, for none of the described species occurs there. Most of his species originate from northern South America, some from Europe and some from southern Africa. In 1745, both South-Africa and Surinam were in Dutch hands, so it seems probable that GYLLENBORG received his specimens from a Dutch source and that the American material actually came from Surinam. The type-localities of the American material in his collection can therefore be restricted to the vicinity of Paramaribo, Surinam. No source for ADOLF FREDRIK's material is available, but it does not seem unreasonable to suppose that the American material in his collection also came from Surinam, in which case the type locality also can be restricted to the vicinity of Paramaribo, Surinam. GRILL's collection came from Surinam and the locality of his material can also be restricted to the vicinity of Paramaribo.

Concluding we have the following results for the Surinam lizards described by LINNAEUS in 1758:

Lacerta angulata (*Alopoglossus angulatus*), type lost, neotype designated below.

Lacerta plica (*Plica plica*), type Naturhistoriska Riksmuseet Stockholm KaF 1900: 112, type-locality restricted by ETHERIDGE (1970) to vicinity of Paramaribo, Surinam, but her restricted to the confluence of the Cottica River and the Perica Creek, Surinam.

Lacerta umbra (*Plica u. umbra*) syntypes Naturhistoriska Riksmuseet Stockholm 1900: 111 and one specimen in Uppsala, type-locality restricted by ETHERIDGE (1970) to the vicinity of Paramaribo, Surinam, here narrowed down to the confluence of the Cottica River and the Perica Creek, Surinam.

Lacerta bicarinata (*Neusticurus bicarinatus*), type Zoological Institute University of Uppsala, no. 70, type locality here restricted to the vicinity of Paramaribo, Surinam.

Lacerta superciliosa (*Uranoscodon superciliosa*), syntypes, 2 ex., Naturhistoriska Riksmuseet Stockholm, type locality here restricted to the confluence of the Cottica River and the Perica Creek, Surinam.

Lacerta azurea (*Uracentron azureum*), syntypes Naturhistoriska Riksmuseet Stockholm KaF 1900: 113, type locality restricted by ETHERIDGE (1968) to the vicinity of Paramaribo, Surinam, but here restricted to the confluence of the Cottica River and the Perica Creek, Surinam.

Lacerta lemniscata (*Cnemidophorus l. lemniscatus*), syntypes, 1♀, 2♂♂, Naturhistoriska Riksmuseet Stockholm, type locality here restricted to the confluence of the Cottica River and the Perica Creek, Surinam.

Amhisbaena alba, syntypes, 2 ex. in Naturhistoriska Riksmuseet Stockholm, type locality here restricted to the confluence of the Cottica River and the Perica Creek, Surinam.

Lacerta Ameiva, (*A. a. ameiva*), syntypes, 2 ex. in Naturhistoriska Riksmuseet Stockholm. These specimens were described in 'Museum Adolphi Friderici' and probably originate from Dahlberg. Another specimen from Gyllenborg's collection is present in Uppsala and also can be considered a syntype. The type locality here is restricted to the confluence of the Cottica River and the Perica Creek, Surinam.

Lacerta Iguana (*I. i. iguana*), syntypes, 1 ex. in Naturhistoriska Riksmuseet Stockholm, 1 ex. from the Gyllenborg collection in Uppsala. The type-locality here is restricted to the confluence of the Cottica River and the Perica Creek, Surinam, from whence the Stockholm specimen probably came.

Lacerta marmorata (*Polychrus marmoratus*), syntypes, 2 ex., in Uppsala, one from the Gyllenborg and one from the Adolf Fredrik collection. Type-locality here restricted to the vicinity of Paramaribo, Surinam.

Amphisbaena fuliginosa, type Naturhistoriska Riksmuseet Stockholm, a specimen in Uppsala from Adolf Fredriks collection is considered not to be a type by Holm (1957). Type locality here restricted to the confluence of the Cottica River and the Perica Creek, from whence the Stockholm specimen probably came.

3. Notes on the species.

The order Squamata is represented in Surinam by three suborders, namely the Sauria or lizards, the Amphisbaenia or amphisbaenians and the Serpentes or snakes (which are not dealt with here). They can be distinguished as follows.

Key to the suborders of Surinam Squamata

1. Legs present; eyes well developed, not covered with an opaque scale . . . . Sauria, p. 44
   Legs absent 2
2. Body covered with rectangular scales which are arranged in annuli, eyes poorly developed, covered by opaque scales . . . . Amphisbaenia, p. 371
   Scales on body never rectangular, not arranged in annuli, eyes poorly to well developed and covered with opaque or transparent scales . . . . Serpentes (not dealt with in this paper).

Suborder SAURIA

The lizards are represented in Surinam by four families (Gekkonidae, Iguanidae, Scincidae and Teiidae), which can be separated with the following key.

Key to the families of Surinam lizards

1. Head covered with small scales or granules; eyes covered with an immovable, transparent lower eyelid . . . . Gekkonidae, p. 45
   Head covered with at least some large scales or plates; eyes usually with free

movable lower eyelids; when the lower eyelid is transparent and has grown over the eye, the head is covered with large scales only 2

2. More than four supraocular scales, either a nuchal crest, a dewlap, a gular pouch, or a combination of these characters . . . . Iguanidae, p. 97
At most four supraocular scales, no nuchal crest, no dewlap or gular pouch 3

3. Scales on the head juxtaposed; large and small scales on the chin; scales on the body without underlying osteoderms . . . . Teiidae, p. 214
Scales on the head imbricate; scales on the chin small, all of the same size; scales on the body with underlying osteoderms . . . . . Scincidae, p. 207

Family GEKKONIDAE

The main external character by which this family is readily recognisable is the apparent absence of the lower eyelid, which has grown over the eye and forms a transparent scale over it. Moreover all Surinam representatives have at least the basal part of the digits more or less dilated and bearing enlarged lamellae. The majority of the geckos is noctural, but in Surinam only three species are nocturnal and four are diurnal. These two groups represent the two subfamilies of geckos in Surinam. The family is almost cosmopolitan.

Key to the subfamilies of GEKKONIDAE

1. Pupil vertical, with lobed margins, digits strongly dilated at the base or over their entire length . . . . Gekkoninae, p. 45
Pupil round, digits only slightly dilated at the base
Sphaerodactylinae, p. 65

Subfamily GEKKONINAE

This subfamily has a world wide distribution and is represented in Surinam by two genera and three species, of which one is almost cosmopolitan, the other two are restricted to South and Central America. They are easily recognised by the shape of their pupils, being the only Surinam lizards with vertical pupils. The digits are strongly dilated. The dilated part bears two rows of large subdigital lamellae. Usually two eggs are laid, either at the same time or with a short interval. Postanal slits and bones are present or absent.

Key to the genera of Surinam Gekkoninae

1. Digits dilated at the base, the distal part long, rising from within the extremity of the dilated part; subdigital lamellae rounded, in two series; digits without a basal web; scales on the back of unequal size, small granules intermixed with larger tubercles . . . . *Hemidactylus*, p. 46
Digits entirely dilated; subdigital lamellae transversely enlarged, in two series; digits connected by a basal web; scales on the back of equal size, small, granular . . . . *Thecadactylus*, p. 55

Genus **Hemidactylus** Gray

DIAGNOSIS.—Small to medium-sized lizards with distinctly depressed head and body. Body and tail covered with granules and longitudinal rows of larger tubercles. Digits free, dilated at the base. Distal part of digits cylindrical, long, rising from within the extremity of the dilated part. Subdigital lamellae enlarged, rounded. Claws not retractile. About 70 species in Africa, Europe, S. Asia, Oceania and America. Six species in South America. In Surinam two representatives of this genus occur.

Key to the species of Surinam *Hemidactylus*

1. Fourth toe with eight enlarged subdigital lamellae, starting some distance from the base of the toe, body tubercles small *Hemidactylus mabouia*, p. 46
   Fourth toe with ten enlarged subdigital lamellae, starting from the base of the toe, body tubercles large . . . . . . *Hemidactylus palaichthus*, p. 53

Hemidactylus mabouia (Moreau de Jonnès), (fig. 6, pls. 1, 3; table 3)

? Sterrepootje-Haagdis' Hartsinck, 1770: 100 (partly).
Gecko Mabouia Moreau de Jonnès, 1818:138.
Hemidactylus mabouia: Penard & Penard, 1906: no. 91 (partly).

TYPE-LOCALITY.—Antilles, restricted to St. Vincent Island, Lesser Antilles, by Stejneger (1904).

MATERIAL. – Surinam.
Paramaribo: 1♀, 1 juv., 2 eggs, RMNH 16188, 21–VIII–1968, leg. M. S. Hoogmoed; 1♀, RMNH 16207, 3–V–1968, leg. F. Westerling; 1♂, ZMA 12658, 11–VIII–1966, leg. H. Nijssen.
Nieuw Nickerie: 2♀♀, RMNH 16458, 15–III–1971, leg. D. C. Geijskes & D. G. Reeder.
Maratakka River: 1 juv., RMNH 16447, 6 hours upstream from Cupido, 1–III–1971, leg. P. Staffeleu.
Forest Service Encampment Wakay, Corantijn River: 1 juv., RMNH 16483, 25–III–1971, leg. D. G. Reeder.
Surinam/French Guiana.
Lower Marowijne River: 1♀, KM. R. 34138, 150 km inland, February-June 1955, leg. H. Larsen.

DIAGNOSIS.—A small gecko, with distinctly depressed body. Body covered with small granules and 13–16 longitudinal rows of enlarged tubercles. Basal part of fingers and toes strongly dilated, eight enlarged lamellae under the fourth toe, starting some distance from the base. Colour greyish white to grey-brown with some more or less distinct transverse bands.

DESCRIPTION.—Head large, representing 28–30% of the snout-vent length in adults of both sexes and 30–35% in juveniles; longer than wide, distinctly de-

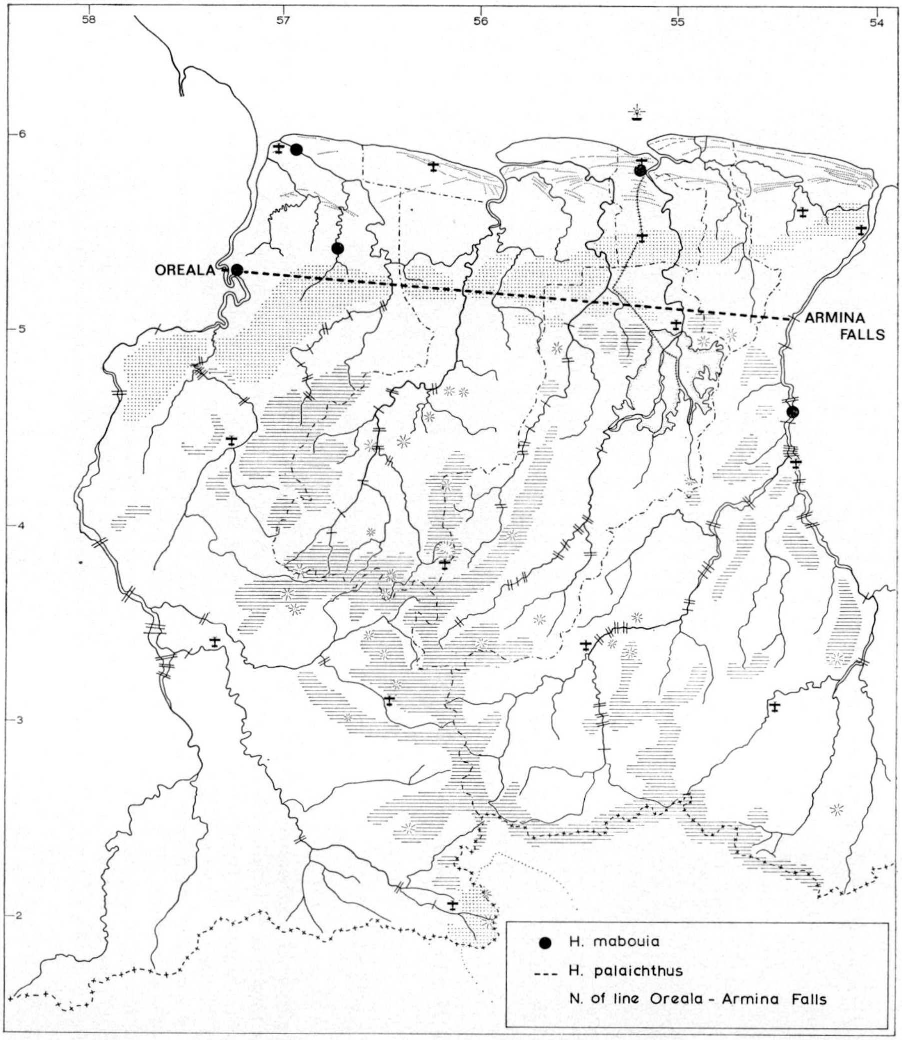

Fig. 6. Distribution of *Hemidactylus mabouia* (Moreau de Jonnès) and *H. palaichthus* Kluge in Surinam.

pressed, 1.6–1.8 times as wide as deep in females and in the male and 1.5–1.6 times in juveniles; distinctly wider than the neck, which is only slightly narrower than the body. Snout pointed with a rounded tip, not projecting over the mouth, concave in profile; frontal region strongly concave; loreal region straight, gradually sloping towards the mouth.

Tongue lanceolate, with slightly nicked tip, villose. Teeth conical, subequal.

Rostral large, roughly rectangular, twice, or slightly less than twice as wide as

Table 3. Counts and measurements of *Hemidactylus mabouia* (Moreau de Jonnès).

Reg. no.	A	B	C	D	E	F	G	H	I	J	M	O	P	J'	V	W	X	C'	D'	I'
RMNH																				
16188a	♀	61	40(4)	16.8	11.6	6.6			41	121	37	9-9	8-8	19-20	2	8-8	8-8	3	13	3-2
b	juv.	20	18	6.6	4.5	3.0			35	?	?	9-10	8-9	20-16	2	8-8	8-8	3	16	2-2
16207	♀	41	45	12.0	7.9	4.6	13	17	40	127	42	11-11	9-9	15-17	2	8-8	8-8	3	14	2-2
16458a	♀	63	54(4)	17.4	12.3	7.8			36	134	41	12-12	9-9		2	9-9	8-8	3	14	2-2
b	♀	50	51(18)	14.9	10.3	6.2			33	126	40	11-12	9-8		2	8-8	8-8	3	13	2-2
16477	juv.	22	22	7.6	4,9	3.2			36	?	39	11-12	10-9		2	8-8	8-8	3	13	2-2
16483	juv.	24	21(21)	7.7	5.2	3.2			37	?	40	13-11	9-9		2	8-8	8-8	3	14	2-2
ZMA																				
12658	♂	55	41(41)	15.4	10.2	6.3	18	23	33	119	40	11-11	9-8		2	8-8	8-8	4	14	2-2
KM.R.																				
34138	♀	39	18(18)	11.2	7.5	3.8			33	128	42	11-11	10-9		2	8-8	8-8	2	14	2-2

Table 4. Counts and measurements of *Hemidactylus palaichthus* Kluge. For explanation see table 3.

Reg. no.	A	B	C	D	E	F	G	H	I	J	M	O	P	J'	V	W	X	C'	D'	I'
RMNH																				
7584	♀	52	45(29)	14.8	11.0	7.0			38	113	35	11-11	10-10	14-16	2	8-8	10-10	3	16	2-2

Explanation of tables 3–30.

A, sex; B, snout-vent length; C, taillength, when regenerated or broken, the length of the original tail left is between brackets; D, head length; E, head width; F, head depth; G, length of forelimb; H, length of hind limb; I, femoral/preanal pores; J, scales around midbody; K, transverse rows of dorsals; L, transverse rows of ventrals; M, longitudinal rows of ventrals; N, collarscales; O, supralabials; P, infralabials; Q, supraciliaries; R, supraoculars; S, suboculars; T, scales in preanal plate; U, transverse rows of gulars; V, pairs of chinshields; W, lamellae under fourth finger; X, lamellae under fourth toe; Y, longitudinal rows of dorsals; Z, lateral caudals per two ventrals; A', supraciliaries outer margin; B', supraciliaries inner margin; C', postrostrals; D', longitudinal rows of body tubercles; E', distance between the anterior corner of the eye and the tip of the snout; F', distance between the posterior corner of the eye and the ear-opening; G', diameter eye; H', postmentals; I', postnasals; J', loreals; K', body annuli; L', tail annuli; M', lateral rows; N', postcloacal scales; O', postgenials; P', autotomy level of tail; Q', length of tibia; R', canthal scales; S', scales in supraorbital row; T', median azygous scale present (+) or absent (—); U', internasals. All measurements are in mm. For ranges of variation given in the text, only Surinam material was used.

Plate 1. Hemidactylus mabouia (Moreau de Jonnès). Figs. 1–3. Dorsal, lateral and ventral aspect of ♀, RMNH 16188, Paramaribo.

deep; a median cleft extending forward from the posterior border, which is indented at the beginning of the cleft; the rostral visible from above; bordered posteriorly by two to four, mostly three, small scales of which the median one(s) is (are) much smaller than the laterals. Nostril bordered by the rostral, a lateral scale from the postrostral row (supranasal), two postnasals and the first supralabial. Supralabials 9–13, mostly 11, decreasing in size posteriorly, the eighth or ninth under the centre of the eye. Loreal region with small, oval, juxtaposed, conical scales, but larger than on the rest of the head. Canthus rostralis absent. Temporal region with small, round, juxtaposed, convex scales, mixed with some inconspicuously enlarged tubercles. Ear-opening vertically oval, small, with smooth margins; situated behind and at the same level as the commissure of the mouth; external auditory meatus long, tympanum not visible. Granules on the snout directly behind the rostral slightly enlarged, as large as the loreals; on the rest of the snout and on the supraorbital region small, round, subequal, juxtaposed, flat to convex granules. Granules on the posterior part of the head like those on the anterior part, but mixed with larger conical scales, more or less arranged in longitudinal rows. Eye large, with a lobed vertical pupil, which, when contracted forms four rhomboid openings in a vertical series.

Mental large, triangular, as wide as long or slightly wider, bordered posteriorly by a pair of postmentals. Two pairs of enlarged scales behind the mental, the anterior pair (postmentals) forming a short median suture behind the mental, the second pair widely separated, smaller than or as large as the anterior pair. Eight to nine, rarely ten, infralabials, decreasing in size posteriorly. Two or three rows of small sublabials. Scales on chin and throat small, round, flat, juxtaposed anteriorly, imbricate posteriorly. No collar or subgular folds.

The commissure of the mouth is horizontal to below the centre of the eye, where it turns dorsally and posteriorly, parallels the orbit for a short distance and then again resumes a horizontal or even slightly declining position.

Scales on the nape and on the side of the neck identical to the dorsals. A short, feeble fold halfway between the ear-opening and the insertion of the forelimbs.

Dorsals and laterals small, round, juxtaposed, subequal, increasing in size on the flanks towards the ventrals; mixed with them are larger, striated trihedral tubercles forming 13–16, mostly 14, longitudinal rows, part of which are continued on the base of the tail. Ventrals larger than the dorsal tubercles, round to hexagonal, flat, imbricate, in oblique and in 37–42 longitudinal rows. A narrow zone of transition between the laterals and the ventrals. Total number of scales around midbody 119–134. A continuous series of femoral and preanal pores comprising a total of 33 pores in the male; a continuous series of enlarged femoral and preanal scales without pores in the females in the same place, comprising a total of 33–41 scales. Preanal plate covered with scales similar to the ventrals; only some rows immediately anterior to the vent consisting of minute granules.

Scales on the tail small, rhomboid, striated, imbricate, with six longitudinal rows of enlarged pointed tubercles among them; scales under the tail imbricate, large, with one median row of transversely enlarged and two accompanying rows of rhomboid scales on each side. Tail divided into very distinct verticils with two ventral and about ten dorsal scales; the enlarged pointed tubercles also form transverse rows along the posterior margin of the verticils, three tubercles on each side of the median. Breaking points between the verticils. Regenerated tails lack the enlarged tubercles and are covered on their dorsal and lateral surfaces with rhomboid, imbricate scales which are larger than those on the original tail, although smaller than the tubercles. The ventral part of a regenerated tail shows the same enlarged scales as the original tail. Base of the tail in the male distinctly swollen, showing two bulbous areas. Tail round in cross-section.

Dorsal and anterior surface of the forelimbs, anterior surface of the hind limbs and posterior surface of the lower legs with moderately large, rhomboid, imbricate scales; dorsal surface of the hind limbs with small granules mixed with larger tubercles; posterior and ventral surfaces of the forelimbs and posterior surface of the thighs with small, round, juxtaposed granules; ventral surface of the hindlimbs with large, rounded, imbricate scales. Fingers and toes consisting of a proximal transversely dilated part, from which extends the terminal compressed phalanx. The proximal part paddle-shaped with a row of enlarged, rounded subdigital lamellae, partly single, partly paired. Under the fourth finger the row starts at the base of the finger with a single plate, which is followed by six pairs of plates and is terminated by another single plate. Under the fourth toe, the composition of the row of lamellae is identical; only, in this case it starts some distance from the base of the toe. Palms and soles with small, round to rhomboid, slightly imbricate scales. Fourth finger and toe longest.

Colour in life not known, but probably not very different from the colour of preserved specimens.

Colour in preservative greyish white to grey-brown (light-brown in a juvenile) with some more or less dinstinct dark-brown transverse bands. One band on the back of the head, three on the body. Lips with or without brown spots. There may or may not be a nearly continuous dark-brown lateral band from the neck to the groin. Toes with transverse brown bands. Dorsal surface of the tail with transverse grey-brown bands; the anterior part of each band lighter than the posterior margin, which is dark-grey. All ventral surfaces white, immaculate. Under the tip of the tongue two oval, bluish grey patches.

A small lizard with a distinctly depressed body and moderately long tail. Maximum snout-vent length in females 63 mm, 55 mm in the single male. The undamaged tail in a female is 1.1 times as long as the snout-vent length, in two juveniles it is 0.9 and 1.0 times as long as that distance. Limbs moderately long and stout, hind limb 1.2–1.3 times as long as forelimb. When laid along the body forelimb and hind limb do overlap, the hind limb not quite reaching the axilla.

HABITAT.—Found on the walls of brick-houses in Paramaribo and Nieuw Nickerie and in some human dwelling places in the forest.

Plate 2. Hemidactylus palaichthus Kluge. Figs. 1–3. Dorsal, ventral and lateral aspect of ♀, RMNH 7584, Surinam.

NATURAL HISTORY.—Probably mostly noctural. Small, round eggs were found in August. Recently hatched juvenlies were found in March and August (snout-vent length + tail length respectively 22 + 22, 24 + 21 (broken) and 20 + 18 mm). No stomach contents were examined.

RANGE.—A widely spread species, occurring in Africa, Madagascar, South America and the Caribbean. In Africa it occurs south of 10°N, in the Caribbean it is present in the Lesser Antilles, in South America it occurs in a narrow belt from Uruguay, along the east and north coast of the continent to the Orinoco delta, with a considerable inland extension along the Amazon. Detailed maps of the distribution of this species are given by KLUGE (1969). In Surinam the species is known from two places in Paramaribo, near the harbour, from Nieuw Nickerie, and from inland localities on the Maratakka, the Corantijn and the Marowijne Rivers (fig. 6).

Hemidactylus palaichthus Kluge (fig. 6; pls 2, 3; table 4)

? 'Sterrepootje-Haagdis' Hartsinck, 1770; 100 (partly).
Hemidactylus mabouia: Penard & Penard, 1906: no. 91 (partly); Van Lidth de Jeude, 1917g: 437.
Hemidactylus palaichthus Kluge, 1969: 39.

Type-locality.—Kurupukari, Guyana (4°N, 59°25′W).

MATERIAL. – Surinam.
Surinam: 1♀, RMNH 7584, leg. H. F. C. ten Kate.

DIAGNOSIS.—A small gecko with distinctly depressed body. Body covered with small granules and 16 longitudinal rows of strongly enlarged tubercles. Basal part of fingers and toes strongly dilated, ten enlarged lamellae under the fourth toe, starting from the base. Colour greyish brown with some more or less distinct transverse bands.

DESCRIPTION.—Head large, representing 28% of the snout-vent length, longer than wide, distinctly depressed, 1.6 times as wide as deep; distinctly wider than the neck, which is only slightly narrower than the body. Snout pointed with a rounded tip, not projecting over the mouth, concave in profile; frontal region concave; loreal region straight, slightly steeper than in *H. mabouia* (Moreau de Jonnès).

Tongue and teeth as in *H. mabouia* (Moreau de Jonnès).

In most scale characters this species resembles *H. mabouia* (Moreau de Jonnès). Only the points in which it differs and some scale counts are mentioned below.

There are three postrostrals; two postnasals; 11 supralabials, of which the ninth is under the centre of the eye; the number of loreal scales is slightly smaller than in *H. mabouia* (Moreau de Jonnès); temporal region with small scales, mixed with several, distinctly larger tubercles. Two pairs of enlarged scales be-

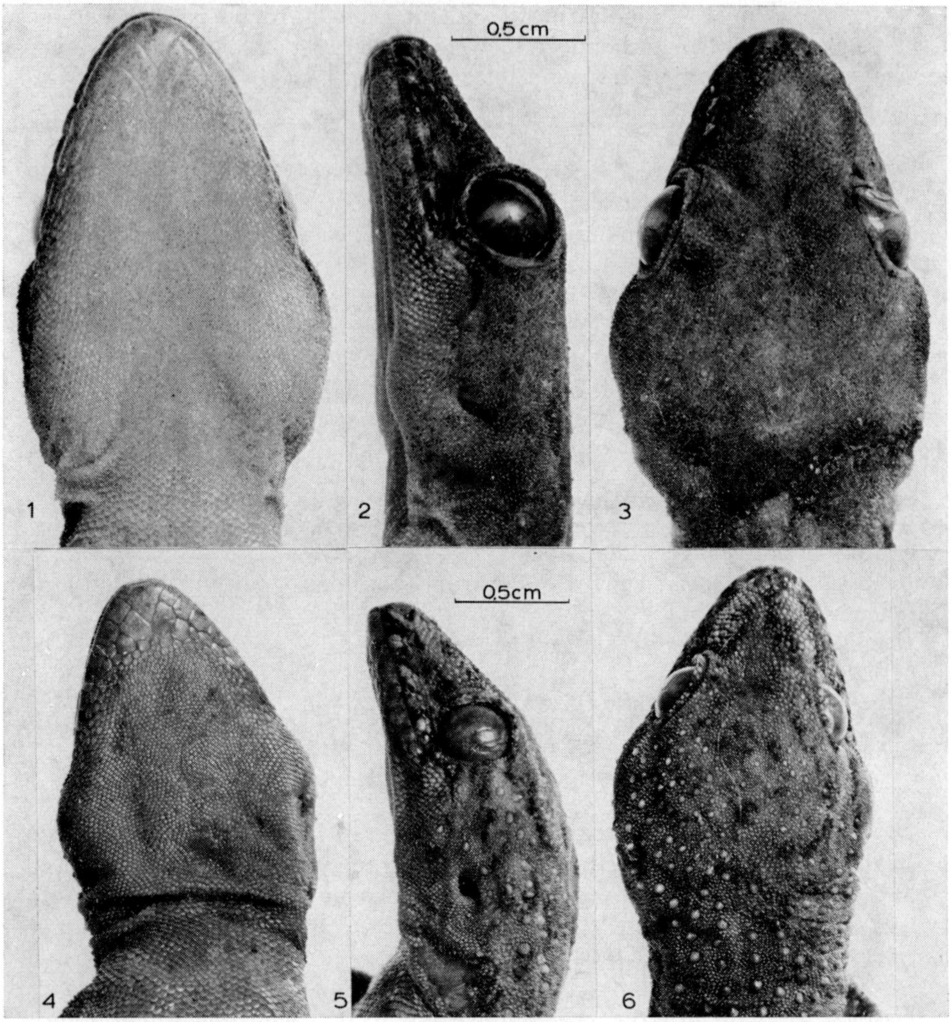

Plate 3. Figs. 1–3. *Hemidactylus mabouia* (Moreau de Jonnès); ventral, lateral and dorsal view of the head of ♀, RMNH 16188, Paramaribo. Figs. 4–6. *H. palaichthus* Kluge; ventral, lateral and dorsal view of the head of ♀, RMNH 7584, Surinam.

hind the mental; the anterior pair forming a suture, the second pair widely separated; ten infralabials.

Dorsals and laterals small, round to oval, juxtaposed, subequal; among them there are 16 longitudinal rows of large, striated, trihedral longitudinal rows. Total number of scales around midbody 113. No femoral or preanal pores, the scales in the row in which the pores would occur in a male scarcely enlarged, totalling 38. The distal part of the tail is regenerated, but from the original part that is left, it is clear that there are eight longitudinal rows of caudal tubercles, four on each side of the median line. The regenerated part of the tail with small, triangular, imbricate, irregularly arranged scales, larger than those on the original tail; enlarged tubercles are lacking.

Fingers and toes as in *H. mabouia* (Moreau de Jonnès) except for one very important difference in the number and arrangement of the lamellae under the fourth toe. In *H. palaichthus* the row of transversely enlarged, rounded lamellae starts at the base of the toe with three single lamellae, followed by six paired lamellae and is terminated by a single lamella.

Colour in life unknown.

Colour in preservative greyish brown above, with some dark-brown transverse bars, one anterior and one posterior to the insertion of the forelimbs very distinct, four others on the body indistinct. A brown stripe on the loreal region to the eye and two narrow stripes from the posterior margin of the ear to the temporal region. Some brown spots on the supralabials. Ventral parts immaculate, pale brown.

A small lizard with distinctly depressed body. The snout-vent length of the single available female is 52 mm. Limbs moderately long and stout, hind limb longer than forelimb. When laid along the body forelimb and hind limb overlap, the hind limb not quite reaching the axilla.

Habitat and natural history.—No information available.

Range.—Surinam, Guyana, northern Brazil, central and northeastern Venezuela, Trinidad and St. Lucia. It is not known where the species occurs in Surinam, because the label of the single available specimen merely states 'Surinam'. However, it was collected by Ten Kate, who stayed in Surinam for nine months and who during this period made extensive trips in the coastal region (Anonymus, 1886; Bonaparte, 1886; Holthuis, 1959). Thus we may restrict the locality to coastal Surinam, north of the line Oreala-Armina Falls (fig. 6).

Genus **Thecadactylus** Oken

Diagnosis.—See diagnosis of the species. Only one species is known to belong to this genus.

Thecadactylus rapicauda (Houttuyn) (Fig. 7; pls. 4, 5)

? 'Gekko' Hartsinck, 1770: 102.

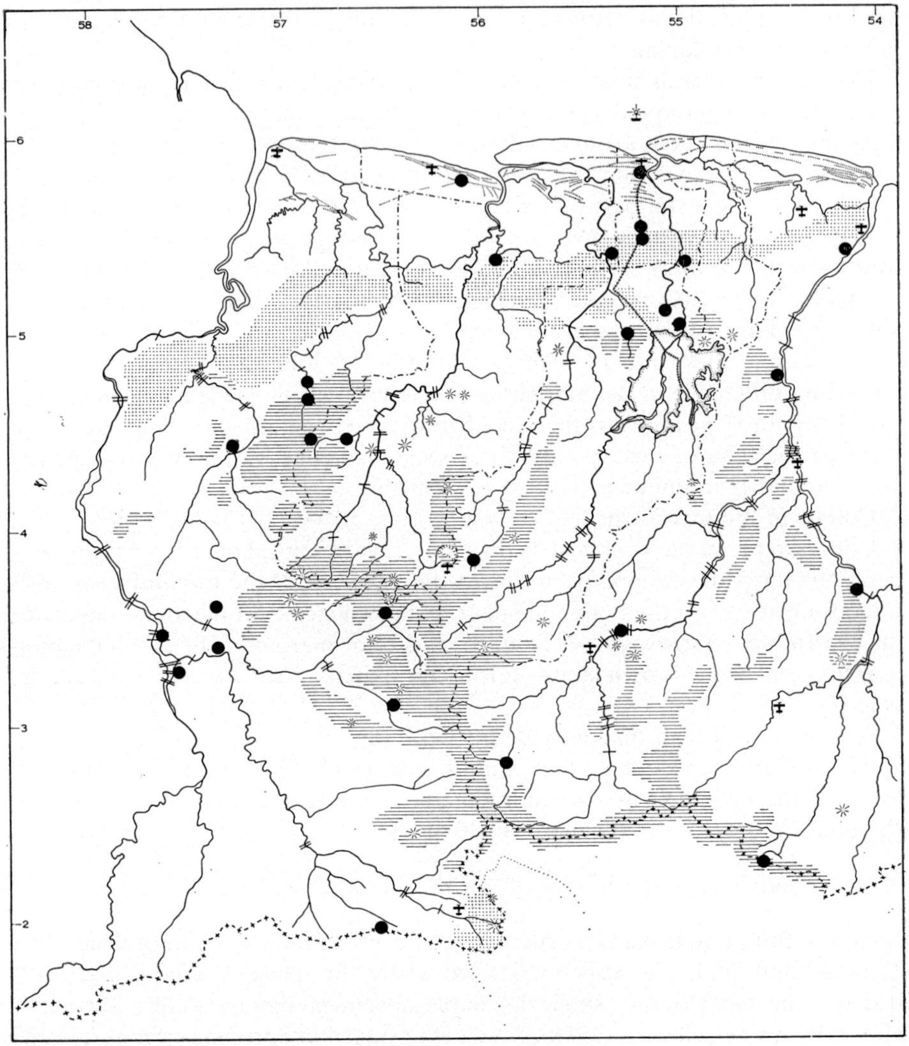

Fig. 7. Distribution of *Thecadactylus rapicauda* (Houttuyn) in Surinam.

Gekko Rapicauda Houttuyn, 1782: 323.
Gecko surinamensis Daudin, 1802b: 126; Cloquet, 1820: 271.
Gecko rapicauda: Daudin, 1802b: 141.
'Der surinam'sche Gecko' Schneider, 1812: 41.
Gecko laevis Cuvier, 1816: 48; Cloquet, 1820: 271; Merrem, 1820: 42; Schinz, 1822: 76; Schinz, 1833: 74; Anslijn, 1827: 95; Cuvier, 1829: 55; Cuvier, 1834: 36; Cuvier, 1836: 393.
Lacerta rapicauda: Bory de St. Vincent, 1828: 127.
Gecko levis Griffith & Pidgeon, 1830: 229.

'Kwakwa' Teenstra, 1835: 442.
Platydactylus Theconyx Duméril & Bibron, 1836: 306.
Thecodactylus laevis: Fitzinger, 1843: 98; P. L. Martin, 1882: 105.
Platydactylus theconyx: Van der Hoeven, 1855: 537.
Hemidactylus rapicauda: Schlegel, 1858: 543, pl. 1 fig. 23.
Thecadactylus rapicaudatus Kappler, 1881: 166; Kappler, 1885: 817; Kappler, 1887: 127.
Thecadactylus rapicaudus: Boulenger, 1885a: 111; Boettger, 1893: 27; Van Lidth de Jeude, 1904: 88; Van Lidth de Jeude, 1917g: 437; D'Audretsch, 1965: 53; J. A. Peters & Donoso-Barros, 1970: 261.
Thecadactylus rapicauda: Strauch, 1887: 17; Vanzolini, 1968a: 68.
Platydactylus rapicauda: Penard & Penard, 1906: no. 81, 91.
'Kwakaslang (Gecko sp.)' Fernandes, 1927: 249.
'Kwakwa snekie' Walsh & Cannon, 1967: 66.

TYPE-LOCALITY.—West-Indies, restricted to Chichén Itzá, Yucatan, Mexico, by SMITH & TAYLOR (1950), here restricted to Paramaribo, Surinam.

MATERIAL. - Surinam.

Surinam: 2 ex., RMNH 2646; 2 ex., RMNH 2656, leg. Henkel; 1 ex., RMNH 7580, leg. P. Buitendijk; 1 ex., RMNH 7582, 1883, leg. C. J. Hering; 2♂♂, RMNH 16261, 1864, leg. F. F. Steenberghe; 1♂, RMNH 16274, February 1888, leg. J. Spitzly; 1♂, RMNH 16275; 1♀ RMNH 16276, leg. J. Th. Noordijk; 1 ex., ZMA 12324; 6 ex., ZMA 12326–31; 1♂, ZMA no number, 1922, leg. J. W. Gonggrijp & G. Stahel, 1♂, BM 70.3.10.64, leg. A. Kappler; 1 ex., ANS 7520, leg. Wilson.

Paramaribo: 1♂, RMNH 5651, November 1911, leg. Samuels; 1♀, RMNH 16259, 5–VI–1941, leg. D. C. Geijskes; 1 ex., ZMB 26377, 1♂, NMW 18753, 1910, both leg. A. O. C. Heller.

Coronieroad: 1♀, RMNH 16269, picket 216, 23–XII–1948, leg. Surinam Expedition 1948–49.

Tibiti: 1♀, RMNH 16270, 19–I–1949, leg. Surinam Expedition 1948–49.

Troeli Creek: 1♀, RMNH 15035, 9–V–1968, leg. M. S. Hoogmoed.

Republiek: 1♂, RMNH 16267, 4–XII–1955, leg. D. C. Geijskes; 1♂, RMNH 16279, 7–V–1963, 1♂, RMNH 16280, 6–V–1963, both leg. P. H. van Doesburg; 1 ex., CM 44333, 23–IX–1966, leg. M. H. de la Fuente.

Zanderij: 1♂, BM 1946.4.2.72, 10–IX–1938, leg. I. T. Sanderson.

60 mi. SSE of Paramaribo: 1 ex., AMNH 73844, 12–XII–1953, leg. D. Bergomini.

Bigiston: RMNH 16281, September 1963, leg. L. D. Brongersma.

Berg en Dal: 1 ex., ZMB 8035, leg. H. B. Möschler.

Brokopondo: 1♀, ZMA 12664, 10–IX–1966, 1 ex., ZMA 13200, 4–III–1967, 1♂, ZMA 13214, 27–IV–1967, all leg. H. Nijssen.

Brown's Mountain: 1♀, RMNH 16257, near Irene Fall, 13–VII–1968, leg. M. S. Hoogmoed; 1♀, SM 19, 4–XII–1970, leg. D. G. Reeder.

Blanche Marie Cataract, Nickerie River: 1♀, RMNH 16466, 12–II–1971, leg. D. G. Reeder.

Upper Nickerie River: 1♂, RMNH 16277, 15–IX–1964, leg. R. van Aerde.

Bakhuis Mountains: 1♀, RMNH 16265, January 1963, leg. R. van Aerde.

Adampadda Creek: 2♂♂, RMNH 16268, September 1963, leg. R. van Aerde.

Nassau Mountains, Base Encampment: 1♀, RMNH 16271, 12–II–1949, 1♀, 1 juv., RMNH 16272, 14–II–1949, 1♂, RMNH 16273, 25–II–1949, all leg. Surinam Expedition 1948–49.

Upper Saramacca River: 1♀, RMNH 16278, Base Bivouac near De Kock Mountain, 12–III–1958, leg. D. C. Geijskes.

Wilhelmina Mountains: 1 ex., ZMA 12325, 1926, leg. Wilhelmina Mountains Expedition; 1♂, RMNH 16266, June-September 1963, leg. H. P. Pijpers (Expedition Maguire).

King Frederik Willem IV Cataracts, Corantijn River: 2 ex., FMNH 30915–16, August-October 1938, leg. E. R. Blake.

Tigri Encampment, New River: 1♂, RMNH 16258, 30–VI–1968, leg. M. S. Hoogmoed.

Airstrip Coeroeni: 2♀♀, 1♂, RMNH 16262, August 1959, leg. D. C. Geijskes.

Kodobakoe Cataracts, Tapanahoni River: 1♂, RMNH 16263, 3–VI–1952, leg. D. C. Geijskes.

Benzdorp: 1♂, RMNH 16264, May 1963, leg. A. Hermans.

Airstrip Kayser Mountains: 2 ex., FMNH 128948, 128997, 1–II–1961, leg. H. A. Beatty.

Tapanahoni River: 1♂, RMNH 5695, 22–X–1904, leg. G. M. Versteeg (Tapanahoni Expedition).

Temomairem: 1♂, RMNH 16260, 28–VII–1939, leg. D. C. Geijskes.

Surinam/Guyana.

New River, 750 feet: 1♀, BM 1939.1.1.66, leg. C. A. Hudson.

Surinam/Brazil: 2♀♀, 2♂♂, NMW 18738.

DIAGNOSIS.—A medium-sized lizard with depressed head and body. Entire digits strongly dilated, connected by a basal web. Subdigital lamellae in two rows, separated by a median groove which widens and deepens towards the tip of the digit. Claws retractile in a sheeth opening into the median groove. Dorsal scales uniform, granular. Limbs short and stout. The anterior part of a regenerated tail is considerably swollen. Dark- to light-grey above with black and white markings, lips white with black edges to the labials, a white stripe from eye to ear.

DESCRIPTION.—Head large, its length 26–31% of the snout-vent length in females, 25–29% in males, 32% in a juvenile; 1.3–1.6 times as long as wide; distinctly depressed, 1.5–1.8 times as wide as deep in females, 1.6–1.8 times in males, 1.5 times in a juvenile. Distinctly wider than the neck, which is as wide as, or narrower than the body. Snout pointed with rounded tip, concave in profile, not projecting over the mouth. Frontal and intraorbital region concave.

Tongue lanceolate, villose; tip broadly rounded, slightly nicked. Teeth subequal, conical with blunt tip.

Rostral large, rectangular, twice as wide as deep; with a short, median cleft extending from the posterior border; rostral distinctly visible from above; bordered posteriorly by two large, quadrangular scales. Nostril bordered by the rostral, the first supralabial, three to five small postnasals and a postrostral. Nine to sixteen, mostly nine to twelve rectangular supralabials; decreasing in size posteriorly, the posterior scales minute; seven to ten supralabials to below the centre of the eye. Ear-opening moderately large, smaller than the eye, obliquely to horizontally oval, with smooth margins; behind, and slightly below the corner of the mouth; external auditory meatus long, tympanum hardly visible. Scales on dorsal and lateral surfaces of the head subequal, small,

Plate 4. Thecadactylus rapicauda (Houttuyn). Figs. 1–3. Dorsal, lateral and ventral aspect of ♂, RMNH 16258, Tigri Encampment.

granular. Canthus rostralis distinct, rounded. Loreal region sloping steeply towards the mouth. Margin of upper eyelid with a row of slightly enlarged conical scales, forming a denticulate margin; in the posterior part up to 13 (mostly five to seven) of these scales are elongated and form spikes. In the single available juvenile, these spikes are absent. No other correlation with age could be found. Eye large, pupil vertically oval with lobed margin. When contracted the pupil forms a vertical row of four small apertures.

Mental small, triangular, mostly wider than long; bordered posteriorly by two, rarely one, postmentals. Nine to fourteen, mostly nine to eleven, quadrangular infralabials, decreasing in size posteriorly, the posterior scales minute. A row of sublabials. Scales on chin and throat small, granular.

Scales on the nape and the side of the neck identical with the dorsals.

Dorsals and laterals small, granular; in indistinct oblique rows. Ventrals larger than the dorsals, round to oval, smooth, imbricate, in oblique rows, 95–129 between the insertion of the forelimbs and the vent. Total number of scales around midbody 140–177. A gradual transition between laterals and ventrals. Scales on the preanal plate like the ventrals, except a few rows immediately anterior of the vent, which are smaller and more or less granular. No femoral or preanal pores. An elaborate description of the hemipenis is given by Dowling, Majupuria & Gibson (1971b).

Scales on the proximal part of the tail like the dorsals, on the distal part slightly larger, flat, slightly imbricate, in transverse rows. Scales under the tail larger than the ventrals, quadrangular, flat, imbricate, in transverse rows. Tail divided into indistinct verticils, each containing five scalerows. Regenerated part of the tail with small, round to oval, convex, juxtaposed, irregularly arranged scales. Of 36 specimens examined for this character only two have retained their original tail, one has half the tail regenerated, 32 have it regenerated, or miss it, from the base. The regenerated tail has a peculiar turnip-shaped form, its anterior part being fleshy and much wider than the base of the tail, while distally it tapers off into a blunt tip. The base of the tail in males shows two bulbous areas. Postanal slits present or absent. One to three enlarged postanal scales.

Scales on the upper, anterior and lower surfaces of the forelimbs and on the anterior and lower surfaces of the thighs small, round to rhomboid, flat, slightly imbricate; on the lower surface of the lower legs large, lanceolate, flat, strongly imbricate; on the posterior surface of the forelimbs, on the posterior and upper surfaces of the thighs and on the posterior, upper and anterior surfaces of the lower legs small, granular. Fingers and toes with two rows of transversely enlarged subdigital lamellae, in contact proximally, separated by a deep longitudinal sulcus distally; 16–24, mostly 19–22, under the fourth finger; 18–25, mostly 20–23, under the fourth toe. Claws retractile in a hood-shaped sheath covered with small scales only. Palms and soles with small, round to oval, pointed, slightly imbricate scales. Fingers and toes depressed, paddle-shaped, with dorsally a narrow median elevation which is highest near the distal end (for figure see Brongersma, 1934: 178). Basal parts of fingers and toes connected by a web.

Colour in life: dark- to light-grey above with black and white markings, a white stripe from the eye to the ear being very conspicuous. Lips white, frequently with black margins to the labials. Tongue blue, oral cavity lined with orange. Iris golden. BEEBE (1944a) gives an excellent description of the colour of living specimens and describes the differences between day and night colour.

In preservative the colour of the back ranges from pale brown to grey. Head and back with or without black markings. When present, they are V-shaped, the apex pointing towards the tail, arranged more or less in transverse series. The middorsal region often lighter than the rest of the back. A distinct, white to pale orange stripe from the eye, over the ear and the insertion of the forelimb onto the flank, where it becomes indistinct and fades in most specimens, but in a few it can be traced, with some difficulty, to the groins. Another, but vague stripe from the corner of the mouth to the insertion of the forelimb. The zone between these two light stripes frequently dark-brown to black. Lips dirty white, with or without black seams to the labials. Undamaged tail with dark--brown to black, wide, transverse bands, the posterior one widest. Regenerated tail sometimes with this same pattern, but mostly with numerous, short, irregularly arranged, longitudinal, dark-brown stripes. Ventral parts immaculate or with small spots, white to dark-brown. Tongue grey anteriorly, white posteriorly.

A medium-sized lizard with depressed body and short conical tail. Maximum snout-vent length both in females and males 121 mm. The undamaged tail, only present in a juvenile (snout-vent length 35 mm) and a small female (60 mm), is respectively 0.8 and 0.7 times as long as the snout-vent length. Limbs short and stout, hind limb 1.2–1.3 times as long as forelimb. When laid along the body fingers and toes slightly overlap.

HABITAT.—An arboral lizard, living on the trunks of trees in the forest, frequently on the lower part between the buttresses, but found to at least 6m above the ground. It has also settled in human habitations, both of wood and of stone, in the forest and in villages, abandoned or still in use. In Venezuela it also seems to be present in savannahs (DONOSO-BARROS, 1968; MARCUZZI, 1954). VANZOLINI (1968b) regards it as a forest dweller.

NATURAL HISTORY.—A nocturnal species (BEEBE, 1944a; RAND & HUMPHREY, 1968; VANZOLINI, 1968b), which may, however, be seen at large in daytime, basking on treetrunks, in front of slits that open into the interior of the tree. The three specimens collected by me were all taken in daytime. One, collected at 7.30 a.m. was disturbed in its hiding-spot and was not active at the time. The other two, collected at 11.45 a.m. and 3.30 p.m. respectively, were actively basking on the lower part of treetrunks. BEEBE (1944a) apparently observed no daytime activity, he saw specimens come out of their hiding-spots only during the evening to start forageing on the insects attracted by the lamp. The same author reports as haunts of this species decayed logs, hollow trees, cavities behind leaf--stalks of palms and houses. KAPPLER (1887) reports this species only from hou-

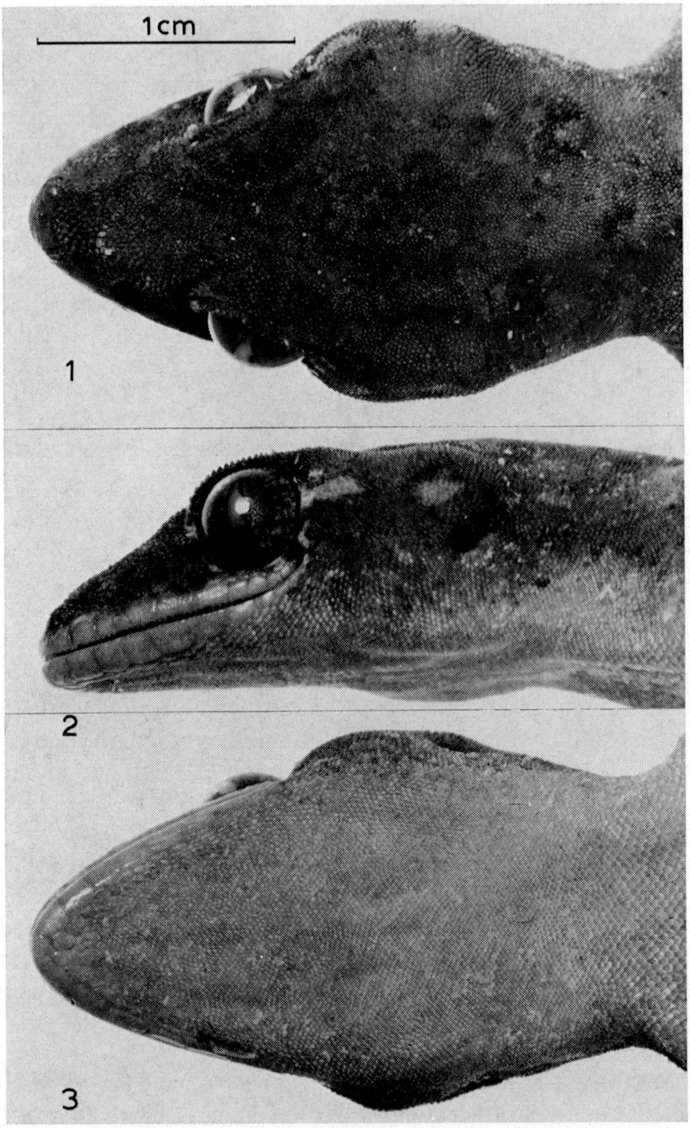

Plate 5. Thecadactylus rapicauda (Houttuyn). Figs. 1–3. Dorsal, lateral and ventral view of the head of ♂, RMNH 16258, Tigri Encampment.

ses. One female was found to contain a single, mature egg, another contained one mature and one immature egg. The first specimen was collected in August, the second, unfortunately is without collecting date. A female collected in September contained a small, immature egg. Beebe kept a female in captivity which laid one egg in August. From these scanty data it might be concluded that the normal egg complement is two and that a considerable interval elapses between the laying of each egg. Nothing is known on the breeding season; it may be restricted to the long dry period, but it is also possible that it extends throughout the year.

Normally this is a slowly moving lizard, but when pursued it can develop considerable speed. BEEBE (1944a) gives an excellent description of the way this gecko moves. At night this lizard vocalises (BEEBE, 1944a; GEIJSKES, 1970: personnal communication; KAPPLER, 1887). The call is described by BEEBE as 'a high, rapidly reiterated chick-chick-chick or chack-chack-chack, fifteen or twenty times repeated, much like the note of some insect', by KAPPLER as 'Toké-Toké'. It seems to aid in establishing territories (KLUGE, 1967). Of seven specimens the stomach and/or intestine contents were examined. Five specimens contained grasshoppers (fairly large). Other arthropods found included two cockroaches, a beetle, a termite, a homopteron, a caterpillar and a spider. The presence of a small flowerbud in the stomach of one individual is puzzling, but it may have emanated from the stomach of a grasshopper found in the lizard's digestive tract. From the above data a preference for grasshoppers might be concluded, but this is not very probable as other sources (BEEBE, 1944a) suggest that they eat everything they can catch, which also is substantiated by the other items found.

RANGE.—Surinam, Guyana, French Guiana, Brazil (Amazonas, Amapá, Pará, Maranhao), Venezuela (Zulia, Falcon, Carabobo, DF, Monagas, Sucre, Aragua), Colombia, Equador, Bolivia, Peru, Central America, north to Mexico, Lesser Antilles. In Equador and Colombia on both sides of the Andes, in Peru and Bolivia only east of the Andes. From VANZOLINI's (1968a) data it is clear that the distribution of this species extends farther to the east than was previously believed. Most checklists (J. A. PETERS & DONOSO-BARROS, 1970; WERMUTH, 1965) give the Guianas as the eastern border of the distribution of this species, but this can be extended to central Maranhao, and probably the eastern border coincides with the border between Amazonian rain-forest and cerrado (fig. 7).

REMARKS.—BRONGERSMA (1934) examined a large number of geckos to establish the presence or absence of postanal slits and ossicles. He divided them into three groups, of which the first and the second can be united in the light of recent studies (KLUGE, 1967). His first group contains species in which males exhibit postanal slits and ossicles and includes *Thecadactylus rapicauda* (Houttuyn). I examined 39 specimens for this character. In one, a juvenile, I could not decide whether slits were or were not present. In the other 38 (18♀♀, 20♂♂) slits were present in 21 specimens (9♀♀, 12♂♂), absent in 17 specimens

(9♀♀, 8♂♂). From this it is clear that the situation is not as simple as BRONGERSMA stated. In this species about half the females and 40% of the males lack the slits. In males, both with and without slits, ossicles are present, in females they are absent. What exactly is the function of these postanal slits and ossicles has not yet been solved, though there are indications that they may play a role during mating in certain species. GREENBERG (1943) reports that in *Coleonyx variegatus* (Baird) they are used by the male to enlarge the cloacal aperture of the female, thus facilitating the introduction of the hemipenis. In *T. rapicauda* (Houttuyn), however, this function could be accomplished in only 60% of the males and consequently it cannot be very important in this species.

KLUGE (1967) made a thorough study of the family Gekkonidae and concluded that it could be divided into four subfamilies, the Eublepharinae, the Diplodactylinae, the Gekkoninae and the Sphaerodactylinae. In the Eublepharinae and Diplodactylinae postanal slits are present in all genera. Most Gekkoninae have postanal slits, but in seven genera they are absent, as in all Sphaerodactylinae. Among the gekkonine genera listed by KLUGE as lacking the postanal slits *Thecadactylus* is absent. As far as I know *Thecadactylus* is the first genus in which a variable situation has been proved beyond doubt. KLUGE only mentions *Lygodactylus* as '(variable?)'. KLUGE observed that all gekkonine genera lacking postanal slits are African, with the exception of *Aristelliger* (West Indies and mainland of Central America). Recently, the existence of a species of *Lygodactylus* (Vanzolini 1968a) in Brazil has been proved. Thus, in America now three genera, namely *Aristelliger*, *Lygodactylus* and *Thecadactylus* are known in which postanal slits are absent, in two of them (*Aristelliger* and *Lygodactylus*) ossicles are absent too, in *Thecadactylus* ossicles are present. According to KLUGE loss of postanal sacs and bones occurred only once, but the variable situation found in *Thecadactylus* might be interpreted as a second, independant loss of the sacs. The loss of postanal sacs and bones in KLUGE's opinion probably occurred in Africa, and the American genera without postanal sacs and bones reached the New World by rafting across the Atlantic Ocean.

As, in my opinion, there exists no close relationship between *Aristelliger* and *Thecadactylus* on the one hand and African geckos on the other hand, KLUGE's opinion seems a little far fetched, and I would suggest that the loss of postanal sacs and bones occurred independantly, several times, both in Africa and America.

The type-locality of *T. rapicauda* (Houttuyn) was restricted to Chichen Itzá, Yucatán, Mexico by SMITH & TAYLOR (1950), but they did not give their reasons. VANZOLINI (1968a) doubted the validity of this restriction. In the original description HOUTTUYN (1782) only says 'Deeze *Westindische Gekko* verschilt zeer van de voorgaande...' (This Westindian Gecko differs much of the preceding ...), thus indicating as type-locality the West Indies. Of course, this area can be broadly interpreted as SMITH & TAYLOR probably did. However, in the Netherlands from the 17th to the early 20th century the term West Indies had a fairly restricted meaning and comprised only the Dutch West Indian islands (Aruba, Bonaire, Curaçao, Saba, St. Eustatius, St. Maarten) and Surinam, and during part of that period comprised also Demerara, Essequibo and Berbice.

The specimen on which the description was based was in HOUTTUYN's possession as is evident from the sentence: 'Ik heb, wegens de zeldzaamheid van dit Voorwerp, waar van ik nog geene Afbeelding had gezien, hetzelve naar het myne, volkomen in de Natuurlyke grootte en gestalte getekend...' (Due to the rarity of this specimen, of which I had not yet seen a picture, I have drawn the same after mine exactly in natural size and shape). This specimen may have come either from one of the above mentioned islands or from Surinam. However, there is circumstancial evidence that points to Surinam as place of origin of the type specimen. First, there is the fact that the American material in Dutch cabinets in the 17th and 18th century largely originated from Surinam. Secondly, it is evident that HOUTTUYN had in his cabinet material from Surinam, although he does not definitely say so. This last piece of evidence can be found in HOUTTUYN (1764) where, on p. 432 when discussing *Amphisbaena fuliginosa*, he says: 'Het zelfde kan ik van den mynen verzekeren...' (The same I can ensure of mine) and 'Nog heb ik 'er een van 11 Duim lang,...' (Also I have one of 11 inches long). Again on p. 434 when discussing *A. alba* he says: 'In de myne zyn ze...' (In mine they are...). Also from several remarks concerning the snakes and fishes (personal communication Dr. M. BOESEMAN) it is evident that HOUTTUYN possessed specimens from Surinam. The examples cited above concerning *Amphisbaena* are conclusive as both species do not occur on the islands, but are fairly common in coastal Surinam. Therefore, I think, it is not too far fetched to assume that the type of *Gekko Rapicauda* Houttuyn came from Surinam. As in the 18th Century only coastal Surinam was inhabited by Europeans who did not travel far inland, it seems reasonable to suppose that the type came from Paramaribo, a place where it still occurs. Thus, I here change and restrict the type-locality of *Gekko Rapicauda* Houttuyn to Paramaribo, Surinam. I have not been able to find out what happened to the type of this species after HOUTTUYN's cabinet was sold. Possibly it is still in existence.

VERNACULAR NAME.—This well known and feared lizard in the Surinam language is called 'Kwakwasnekie' and is considered highly poisonous by the inhabitants (Indians, Bush-Negroes and White people).

Subfamily SPHAERODACTYLINAE

This subfamily is known only from the New World tropics and in Surinam is represented by three genera and four species. These geckos are of a small size. The pupil is round with a smooth margin. The digits are only slightly dilated at the base and this dilated part has only one row of enlarged subdigital lamellae. Only one egg is laid. Postanal slits and bones are absent.

Key to the genera of Surinam Sphaerodactylinae

1. Scales on the back imbricate, of equal size; claw retractile in an asymmetrical sheath consisting of four scales . . . . *Coleodactylus*, p. 67

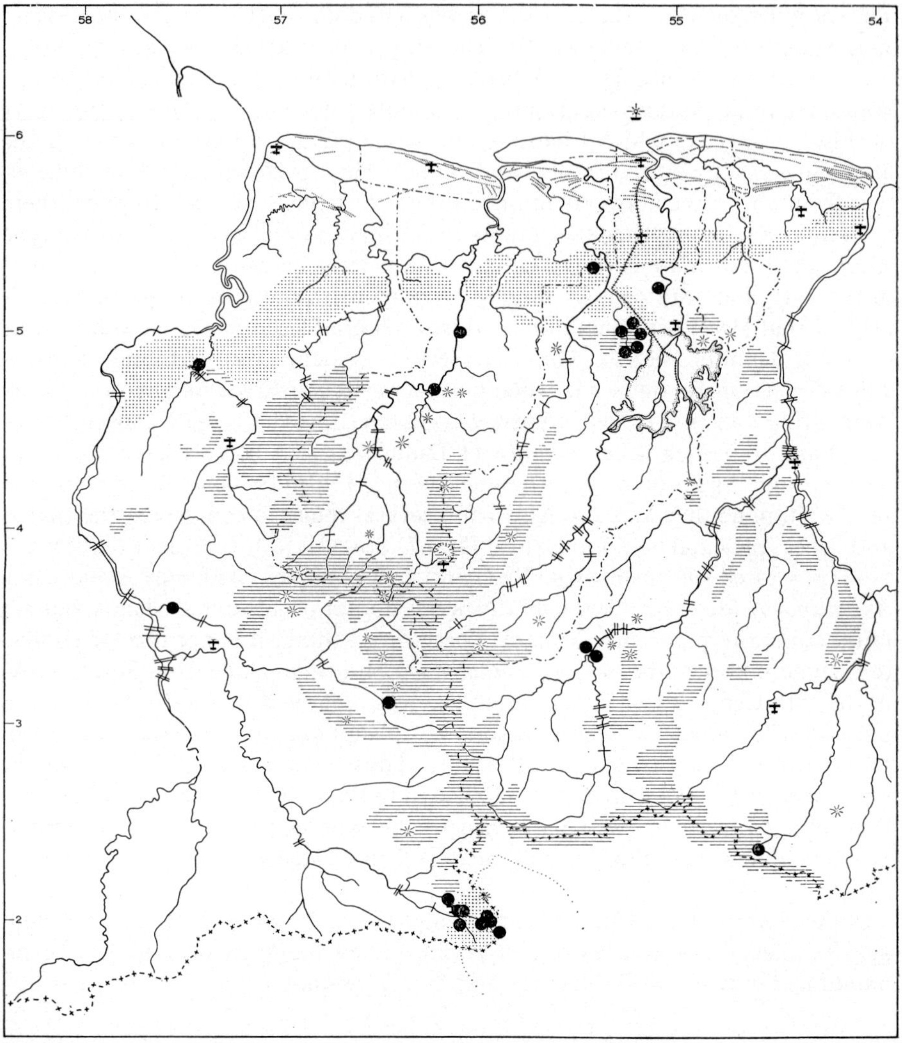

Fig. 8. Distribution of *Coleodactylus amazonicus* (Andersson) in Surinam.

Scales on the back juxtaposed, of equal or inequal size; claw retractile in a symmetrical sheath or not retractile 2

2\. Claw not retractile, between two scales, a small superior and a larger inferior . . . . *Gonatodes*, p. 72

Claw retractile in a symmetrical sheath composed of five scales, a pair of very large infero-laterals, a pair of smaller superiors and a small terminal scale . . . . *Pseudogonatodes*, p. 90

Genus **Coleodactylus** Parker

DIAGNOSIS.—Very small lizards with a cylindrical body, short legs and a short tail. Dorsal scales imbricate, smooth or keeled. Ventral scales imbricate, smooth. Claws obliquely retractile in a more or less asymmetrical sheath composed of four or five scales, viz., two infero-laterals and two or three dorsal elements. Upper eyelid always without a spine, forming a projecting flap that, in its anterior part, bears enlarged scales. Teeth minute, conical, subequal. Pupil round. Four species in northeastern South America, one of which is present in Surinam.

Coleodactylus amazonicus (Andersson) (fig. 8; pls. 9, 10; table 5)

Sphaerodactylus amazonicus Andersson, 1918: 1.

TYPE LOCALITY.—Lago Poraquêcuare, Manaus, Amazonas, Brazil.

MATERIAL. – Surinam.

Encampment Moeroe Creek: 1♂, RMNH 16217, 3 km S., 24–VII–1968, leg. M. S. Hoogmoed.

65 km SSE. of Paramaribo, Afobaka Road: 3♀♀, 1♂, AMNH 108755–58, 19/20–II–1972, leg. C. W. Myers & J. Daly.

Camp 1, Coppename River: 1 juv., BM 1946.4.2.80, 27–VI–1938, leg. I. T. Sanderson.

Brown's Mountain: 1♀, 2 ex., RMNH 16218, near Irene Fall, 13–VII–1968, leg. M. S. Hoogmoed; 1♂, RMNH 16226, 25–VIII–1968, leg. W. N. Polder; 1♂, RMNH 16233, foot, 19–VIII–1968, 4♀♀, RMNH 16237, foot, 20–VIII–1968, 6♀♀, 3♂♂, RMNH 16240, near Irene Fall, 11–XI–1968, 2♀♀, 3♂♂, 1 juv., RMNH 16241, E. flank, 12–XI–1968, 3♀♀, RMNH 16243, 11–XI–1968, all leg. M. S. Hoogmoed; 2♀♀, RMNH 16244, S. part, 11–I–1970, 4♀♀, RMNH 16245, Mazaroni top, 11–I–1970, all leg. M. S. Hoogmoed & J. J. P. Paats; 1 ex., RMNH 16255, 13–VI–1969, 1 ex., RMNH 16256, 12–VI–1969, both leg. H. M. van Meeuwen; 1♀, AMNH 108759, 15/16–II–1972, leg. C. W. Myers & J. Daly.

Avanavero Cataracts, Kabalebo River: 1♀, RMNH 16480, 9–IV–1971, 1♀, RMNH 16481, 10–IV–1971, both leg. D. G. Reeder.

Raleigh Cataracts, Coppename River: 4♀♀, 1♂, RMNH 16242, 4–XI–1968, leg. M. S. Hoogmoed; 1♀, 1♂, AMNH 108760–61, 19/20–II–1972, leg. C. W. Myers & J. Daly.

Lucie River: 1 ex., RMNH 16253, December 1910, leg. J. F. Hulk (Corantijn Expedition).

Airstrip Kayser Mountains: 1♂, RMNH 16220, E., 8–VIII–1968, 1♂, RMNH 16221, NE., 2–VIII–1968, 2♀♀, RMNH 16222, NE., 9–VIII–1968, 2♂♂, RMNH 16223, NE., 3–VIII–1968, 1♂, RMNH 16228, SW., 5–VIII–1968, all leg. M. S. Hoogmoed; 1 ex., FMNH 128993, 1–II–1961, leg. H. A. Beatty.

Airstrip Paloemeu: 1 ex., RMNH 16215, S., 28–V–1968, 1♂, RMNH 16216, N., 30–V–1968, both leg. M. S. Hoogmoed.

Airstrip Sipaliwini: 1♀, 1 juv., RMNH 16232, 1–1.5 km E., 5–IX–1968, leg. M. S. Hoogmoed; 3♀♀, 1♂, RMNH 16247, between airstrip and Meyers' farm, 30–I–1970, 2♀♀, RMNH 16249, 4 km NW., 9–II–1970, both leg. M. S. Hoogmoed & J. J. P. Paats.

Sipaliwini, Base Bivouac: 1♀, 1♂, RMNH 16229, 1–IX–1968, 2♀♀, 2♂♂, RMNH 16235, 0–5 km N., 2–IX–1968, 4♀♀, 2♂♂, RMNH 16236, 0–5 km N., 4–IX–1968, all leg. M. S. Hoogmoed; 2♀♀, 1♂, RMNH 16252, 2–4 km N., 16–I–1970, leg. M. S. Hoogmoed & J. J. P. Paats.

Sipaliwini, Vier Gebroeders Bivouac: 2 ex., RMNH 15248, 12–IX–1968, 1♂, RMNH 16224, 15–IX–1968, 1♀, RMNH 16227, 11–IX–1968, 9♀♀, 3♂♂, RMNH 16239,

16–IX–1968, all leg. M. S. Hoogmoed; 1♂, RMNH 16248, 6–II–1970, leg. M. S. Hoogmoed & J. J. P. Paats.

Sipaliwini, Vier Gebroeders Mountain: 1♂, 1 ex., forest-island 1 km N. of top, 12–IX–1968, 2♂♂, RMNH 16230, forest-island 3 km N., 20–IX–1968, 1♀, RMNH 16231, forest-island on N. flank, 21–IX–1968, 2♀♀, 2♂♂, RMNH 16234, forest-island 2 km N., 19–IX–1968, 6♀♀, 4♂♂, RMNH 16238, forest-island 3 km N., 23–IX–1968, all leg. M. S. Hoogmoed; 6♀♀, RMNH 16251, 7–II–1970, leg. M. S. Hoogmoed & J. J. P. Paats.

Sipaliwini, N. part of Lange Dijk: 2♂♂, RMNH 16250, forest-island, 27–I–1970, leg. M. S. Hoogmoed & J. J. P. Paats.

Sipaliwini: 1♀, RMNH 16225, September 1968, leg. M. S. Hoogmoed.

Waremapan Creek: 1 ex., RMNH 16254, 1–VIII–1939, leg. D. C. Geijskes.

Guyana.

Marudi: 15 ex., AMNH 77099, April 1938, leg. R. Snedigar.

French Guiana.

Sophie: 3 ex., MCZ 77506–08, 8–VIII–1963, leg. T. Monath.

Trail from Sophie to La Greve: 1 ex., MCZ 77505, 10–VIII–1963, leg. T. Monath.

Brazil.

Sipaliwini, S. part of Lange Dijk: 5 ex., RMNH 16246, forest-islands, 27–I–1970, leg. M. S. Hoogmoed & J. J. P. Paats.

Diagnosis.—A very small ground-dwelling, diurnal gecko with a cylindrical body, short legs and a short tail. Dorsal scales rhomboidal, keeled, imbricate. Anterior margin of the upper eyelid forming a projecting flap which bears several enlarged scales. Claws obliquely retractile in a slightly asymmetric sheath composed of four scales. Light- to dark-brown above with a yellowish brown dorsolateral stripe in the sacral region. A white stripe on the back of the thighs, continued on the proximal part of the tail.

Description.—Head small, representing 20–25% of the snout-vent length in both sexes, 26% in a single juvenile; 1.3–1.5 times as long as wide in both sexes; distinctly depressed. Head slightly wider than the neck, which is as wide as the body. Snout short, pointed with round tip, straight to concave in profile; not projecting over the mouth.

Tongue arrowhead-shaped, covered with flat, imbricate, irregularly arranged, cycloid papillae; tip round, slightly nicked. Teeth minute, conical, subequal.

Rostral very large, distinctly visible from above, with a median cleft extending from the posterior border; posterior part with a circular depression, forming a distinct canthus with the vertical part of the rostral; bordered posteriorly by four to five, rarely six, small scales, of which the two laterals are largest and also border on the nostril. Nostril bordered by the rostral, the first supralabial, one postnasal and a supranasal (= the lateral of the postrostral scales). Three to four, mostly four, supralabials, the anterior one largest; the others decreasing in size posteriorly; the fourth minute, or absent; posterior border of first supralabial just before the anterior margin of the orbit; third supralabial under the centre of the eye. Canthus rostralis rounded, indistinct. Ear-opening small, vertically oval with smooth margins; much smaller than the eye; behind and at the same level as the commissure of the mouth; external auditory meatus long, tympanum invisible. Scales on the snout and loreal region polygonal, flat, juxtaposed, distinctly larger than on the posterior part of the head; decreasing in size posteriorly. Behind the supranasal there is a large scale, as large as or slightly smaller than the supranasal, larger than the other scales on the snout. Scales on the rest of the head granular, except on the supraocular region. Scales on the supraocular region as large as those on the snout; rhomboid to round, flat, imbricate; anterior margin of the upper eyelid forming a projecting flap, best developed over the centre of the eye, ending just posteriorly to the former point; upper margin of flap with four to seven enlarged scales, lower margin with six to ten enlarged scales. Eye large, with round pupil.

Mental very large, horseshoe-shaped, posterior margin concave; bordered posteriorly by four to seven, mostly five, small scales, which are as large as the other scales on the chin. Three to four, mostly four, infralabials; the anterior one very large, its posterior border below the anterior margin of the orbit; the second smaller; the third and fourth very small. Sublabials absent. Scales on the chin round, flat, imbricate. Scales on the throat larger, similar to the ventrals. Scales on the nape larger than those on the head, rhomboidal, keeled, imbricate. Scales on the side of the neck granular.

Dorsals and laterals larger than the scales on the nape, rhomboidal to quadrangular, keeled, imbricate, in oblique rows. Ventrals larger than the dorsals, rhomboidal with rounded tips, smooth, imbricate, in oblique rows; 29–40 between the insertion of the forelimbs and the vent. No distinct border between ventrals and laterals. Total number of scales around midbody 40–48. Scales on the preanal plate identical to the ventrals, except in a narrow zone just anterior of the vent where they are smaller. No preanal or femoral pores.

Scales on the tail similar to the dorsals, but smooth; under the tail like the ventrals, in oblique and longitudinal rows. Scales on the regenerated part of the tail scarcely different from those on the original tail, maybe slightly smaller. Base of the tail in males slightly swollen. Tail round in cross-section.

Forelimbs and hind limbs, except the posterior surface of the thighs, with rhomboidal, smooth, imbricate scales in oblique and longitudinal rows; posterior surface of the thighs with granular scales. Six, rarely seven, enlarged, smooth, rounded lamellae under the fourth finger; seven, rarely six or eight, similar lamellae under the fourth toe. The subdigital lamellae are followed distally by a large, round to oval scale (outer infero-lateral), forming part of the sheath in which the claws can be retracted. Claws retractile in a sheath consisting of four scales: a large asymmetrical, outer infero-lateral, a smaller inner infero-lateral, a large inner supero-lateral and a small median dorsal (for figure see VANZOLINI, 1957). Palms and soles with small, rhomboidal to triangular, smooth, imbricate scales, larger near the base of the fingers and toes. Fingers and toes cylindrical. Third finger and toe longest.

Colour in life light- to dark-brown above. With or without whitish spots on the flanks. Back mottled with black. A yellowish brown dorsolateral stripe in the sacral region. Knee and lower leg with orange bands, a white stripe on the back of the thighs, continued on the proximal part of the tail. Chin, throat and belly white, immaculate or with small brown spots. A black line from the corner of the mouth to the base of the forelimb, separating the brown dorsal and the white ventral colour. Iris reddish brown. Tip of tongue blue, rest pink.

Colour of the back in preservative grey- to dark-brown with or without numerous, small, black spots, which may form a zigzag on the tail. A short creamish dorsolateral stripe, bordered below by a black line, may or may not be present over the hind limbs. A light-brown, transverse spot on the knee in all specimens, sometimes two similar spots on the lower legs. In all specimens an oblique white line on the back of the thighs, starting in the hollow of the knee, continued on the sides of the tail, in juveniles to approximately halfway the tip, in adults less far. Above and below, this line is bordered by black lines. The regenerated part of the tail is devoid of pattern and is brown above, slightly lighter brown below. Often some scales of the original part of the tail, just anterior of the regenerated part are whitish, thus emphasizing the border between original and regenerated part of the tail. A dark-brown line from the nostril to the eye, another one from the corner of the mouth obliquely down towards the insertion of the forelimbs. A dark-brown to black line along the lower margin of the orbit. Chin and throat immaculate, white; belly and ventral surface of limbs creamish, immaculate or scales with brown posterior edges. Ventral surface of the tail dark-brown with a wide, creamish zone down the middle on the proximal part. Tongue anteriorly with two grey, oval areas, separated by a midventral white line, rest of the tongue white.

A very small lizard with cylindrical body and short, conical tail. Maximum snout-vent length in females 24 mm, in males 23 mm. The tail is undamaged in only 24 out of the 126 specimens examined. The undamaged tail is 0.8–0.9 times as long as the snout-vent length in both sexes. The limbs are short and stout, hind limb hardly longer than forelimb. When laid along the body there is considerable overlap between fingers and toes; the hind limbs reaching to approximately halfway the axilla.

HABITAT.—An inhabitant of the rain-forest floor, where it lives on dry ground among leaf-litter. I never found it on the edges of creeks, where *Pseudogonatodes guianensis* Parker is present. It never ventures far from the ground, occasionally it may be seen on the base of a tree or on a fallen branch, but this is exceptional. Mostly it is present in thick layers of fallen leaves, none was seen on bare ground.

NATURAL HISTORY.—A diurnal species, which was taken between 8.00 a.m. and 6.00 p.m. No basking was ever observed, all specimens were in shady spots between fallen leaves. This agrees with the scanty information provided by VANZOLINI (1968a, b). When fallen leaves were shuffled *Coleodactylus* would try to hide under leaves, but never attempt to escape into cracks or hollows. They easily lose large patches of skin when touched. Pregnant females, each containing one large, mature, subovoid egg, have been found in August, September, November, January and February. Probably, because of space problems, only a single egg develops at a time. No recently hatched juveniles are available. The above data may indicate that breeding takes place throughout the year, except in the long wet period (May–July). However, it should be observed that although no females collected in May–July were pregnant, only a small amount of material was obtained at that time, most specimens having been collected in August–November and January–February. Whether the fact that in May–July 1968 only a few specimens could be collected while in August –November 1968 numerous were obtained reflects a real difference in abundance, a rest period during the wet season, or my being more agile in collecting this species, remains unsolved. The stomachs and intestines of seven specimens were examined. Five stomachs contained Collembola, two exclusively so, in two stomachs they were absent. Apart from the Collembola the stomachs contained: sow-bugs, a juvenile myriapod, a pseudoscorpion, a mite, termites, a beetle, a bug, and the remains of a caterpillar or beetle larva. Thus, there seems to be a preference for Collembola, but other small arthropods are not despised.

RANGE.—Surinam, Guyana, French Guiana, Brazil (Amazonas, Pará, Amapá), Venezuela (Amazonas). When more material becomes available it will probably become clear that this species is distributed throughout the entire Amazon basin. In Surinam the species is widely distributed south of the coastal savannah belt in rain-forest. In the Sipaliwini area it is present in the rain-forest bordering on the savannah and in all forest-islands visited. It occurs from about 20 to at least 500 m above sea-level (fig. 8).

REMARKS.—VANZOLINI (1957, 1963, 1968b, 1970a) discusses the distribution of the species of this genus and in accompanying maps plots all known localities. Since his last paper additional data on Surinam, Guyana (see above) and Venezuela (DONOSO-BARROS, 1968) have become available and the range of *C. amazonicus* (Andersson) thus can be extended to include most of Surinam, southern Guyana and southern Venezuela. The absence of the species from northern Guyana probably is real and might be caused by the presence of *C. meridionalis* (Boulenger), which is present in at least part of that area. The specimens present in the forest-islands on the Sipaliwini-savannah do not differ from those in the vast area of rain-forest in the rest of Surinam.

The genus *Coleodactylus* contains four species, all of them restricted to northeastern South America. They can be divided into two groups, one, containing *C. meridionalis* (Boulenger) and *C. brachystoma* (Amaral), has smooth dorsal scales and nearly symmetrical digits; the other, containing *C. amazonicus* (Andersson) and *C. guimaraesi* Vanzolini (which may eventually prove to be identical), has keeled dorsal scales and strongly asymmetrical digits. PARKER (1926) and VANZOLINI (1957) figure a digit of *C. meridionalis* (Boulenger). In this species the ungual sheath consists of five scales: an internal and an external infero-lateral of about equal size, a terminal, a median dorsal and an internal supero-lateral. The sheath of *C. amazonicus* (Andersson) is probably derived from this type, in that the terminal and the outer infero-lateral have grown together to form the large, asymmetrical outer infero-lateral, typical of this species. Parallel to the outer margin, on the dorsal surface of this large scale, a feeble ridge is visible which most probably represents the line of fusion between the two original components.

Genus **Gonatodes** Fitzinger

DIAGNOSIS.—Small lizards with cylindrical body, well developed limbs and short to moderately long tail. Teeth conical, subequal. Tongue arrowhead--shaped, tip slightly nicked. Dorsal scales small, granular. Ventral scales large, imbricate. Base of digits slightly dilated with a single row of lamellae underneath. Claws not retractile; between two scales, a small superior and a large latero-inferior. Sexual dimorphism in colour pattern. Pupil round.

Key to the Surinam species of *Gonatodes*

1. Lamellae under the fourth toe narrower than the digit, 24–30; scales around midbody 90–99; 40–48 ventrals along the midventral line; scales on the snout as large as those on the posterior part of the head; light antehumeral bar present or absent . . . . *Gonatodes annularis*, p. 72
   Lamellae under the proximal part of the fourth toe as wide as the digit, 17–21; scales around midbody 108–135; 53–61 ventrals along the mid-ventral line; scales on the snout larger than those on the posterior part of the head; a light antehumeral bar, preceded or not by a black spot, is always present . . . . *Gonatodes humeralis*, p. 80

Gonatodes annularis Boulenger (fig. 9; pls. 6, 8; table 6)

Gonatodes annularis Boulenger, 1887b: 154; Rivero-Blanco, 1967: 107; Vanzolini, 1968a: 18; J. A. Peters & Donoso-Barros, 1970: 131; Da Cunha, 1971: 113.
Gonatodes Boonii Van Lidth de Jeude, 1904: 87.
'55, 58. Grondhagedissen' Fernandes, 1927: 249.
Gonatodes booni: Burt & Burt 1933: 2; Vanzolini, 1955: 122.
Gonatodes boonii: Wermuth, 1965: 43; Hoogmoed, 1968: 612; Vanzolini, 1968a: 18.

TYPE-LOCALITY.—Maccasseema, Pomeroon River, Guyana.

MATERIAL. – Surinam.
Adampadda Creek: 1♂, RMNH 16214, September 1963, leg. R. van Aerde.
Raleigh Cataracts, Coppename River: 1♀, RMNH 16211, 6–XI–1968, leg. R. Hoogmoed-Verschoor.
N.W. base of Voltz Mountain: 1♀, AMNH 108751, 22/23–II–1972, leg. C. W. Myers & J. Daly.
Coppename River basin: 1♀, RMNH 4462, 1901, leg. Coppename Expedition (holotype of *Gonatodes boonii* Van Lidth de Jeude).
Brown's Mountain: 1♂, RMNH 16208, September 1968, leg. Suralco; 1♂, RMNH 16209, 16–X–1968, 1♀, RMNH 16210, 12–XI–1968, 1 juv., RMNH 16212, 10–XI–

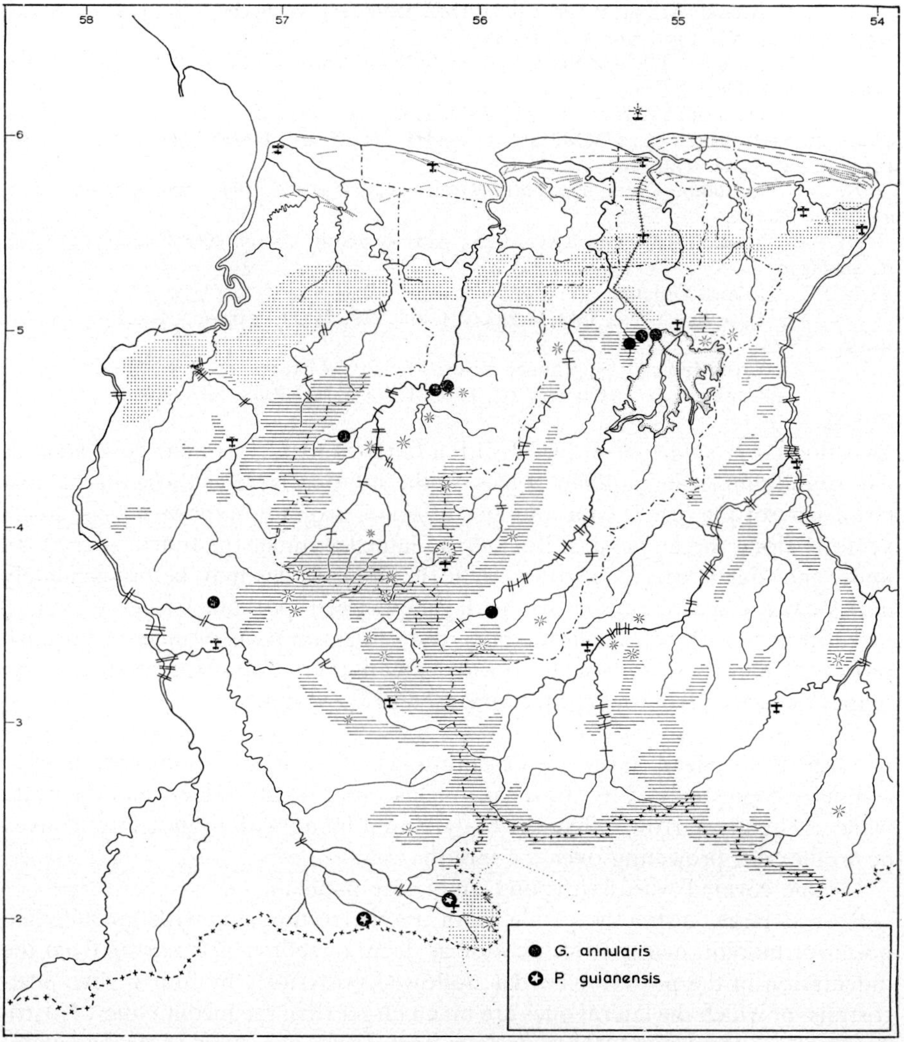

Fig. 9. Distribution of *Gonatodes annularis* Boulenger and *Pseudogonatodes guianensis* Parker in Surinam.

1968, all leg. M. S. Hoogmoed; 1♂, RMNH 17281, 7–XII–1971, leg. G. F. Mees; 2♀♀, SM 20, 27–X–1970, leg. D. G. Reeder.

Railway km 121: 1♀, RMNH 16213, July 1949, leg. C. Bleys.

Wilhelmina Mountains: 2♀♀, ZMA no number (collectors numbers 55, 58), 1926, leg. Wilhelmina Mountains Expedition.

Suriname River: 1♂, ZMA 12316, 14–XI–1908, leg. J. H. A. T. Tresling (Suriname Expedition).

Guyana.

Macasseema (Makasima), Pomeroon River: 2♀♀, BM 1946.8.22.97–98, leg. W. L. Sclater (types of *Gonatodes annularis* Boulenger).

Aremu (Arimu) River: 1♂, UMMZ 70066, leg. W. Beebe; 1 ex., UMMZ 83741, Haul Over, 24–XII–1937, leg. A. S. Pinkus.

Kartabo: 1♂, AMNH 21251, 26–VIII–1920, leg. W. Beebe (holotype of *Gonatodes beebei* Noble).

Dunoon: 1♀, 1♂, 1 juv., UMMZ 47644–45, 47647, 25–VIII–1914, 1♂, UMMZ 47648, 27–VIII–1914, 1 ex., UMMZ 47649, 4–VIII–1914, 1 juv., UMMZ 79970, S., 25–VIII–1914, all leg. A. G. Ruthven.

Hubudibu Creek, Demerara River: 2 ex., UMMZ 53894, 70056, 22–XI–1914, leg. E. N. Clarke.

Marudi, Kuyuwini River: 1 ex., AMNH 60975, 2°5′ N, 59°E, March 1938, leg. R. Snedigar.

French Guiana.

Cayenne: 1 juv., LACM 44444, 6 km SSE, Mont Cabassou, leg. P. H. Silverstone.

Creek Ipoucin, Approuage River: 1♀, LACM 44471, leg. P. H. Silverstone.

Lower Matarony River: 1♂, LACM 44454, leg. P. H. Silverstone.

Diagnosis.—A small, stout lizard with a fairly short tail. Tongue covered with flat, round, imbricate papillae. Scales on the snout as large as those on the posterior part of the head. Total number of scales around midbody 90–99, 40–48 ventrals along the midventral line; 24–30 lamellae under the fourth toe, 22–25 under the fourth finger, narrower than the digit. Males may be immaculately dark-brown above or dark-brown with yellow-green spots and stripes. Chin and throat orange with ochre-yellow spots, rest of ventral parts orange- to purplish brown. Females light-brown above with a creamy vertebral stripe and transverse stripes departing from it. A blue rim around the pupil.

Description.—Head large, representing 23–26% of the snout-vent length; distinctly depressed; 1.3–1.5 times as long as wide; slightly wider than the neck, which is slightly narrower than the body. Snout blunt, with rounded tip; convex in profile; not projecting over the mouth.

Tongue covered with flat, round, imbricate papillae.

Rostral large, distinctly visible from above, rectangular with an indented posterior margin; nearly twice as wide as deep; a median cleft arising from the indentation in the posterior border; followed posteriorly by four or five postrostrals, of which the lateral ones are much larger than the median ones. Nostril bordered by the rostral, three or four postnasals, one supranasal (= lateral postrostral) and sometimes by the first labial (or narrowly separated from the last-named). Six to seven supralabials of which the fifth or the sixth is under the centre of the eye; first largest, the rest decreasing in size posteriorly; the posterior one minute. Loreal region straight, sloping steeply to the mouth; with small, round, convex, juxtaposed scales, as large as those on the snout, 10–13, mostly 12–13, in a line between the postnasals and the anterior margin of the orbit. Canthus rostralis distinct, rounded. Ear-opening small, round to oval, with smooth margins; much smaller than the eye; behind and at the same level as the commissure of the mouth; external auditory meatus long, tympanum invisible. Scales on the snout small, round, convex, juxtaposed, not or only slightly larger than those on the posterior part of the head and on the back. Anterior margin of the upper eyelid with a series of seven to twelve enlarged scales, of which the one but last is largest, forming a distinctly projecting spine, just anterior of the centre of the eye; the other scales of the series are more or less distinctly conical. Eye large, its diameter less than the distance between the orbit and the nostril, equalling half the distance between the orbit and the ear; pupil round.

Mental large, subtriangular; wider than deep; posterior margin convex; bordered posteriorly by two (rarely three) enlarged postmentals. Six infralabials; the anterior one largest, its posterior margin reaching the level of the suture between the second and the third supralabial; rest of the infralabials decreasing in size posteriorly; the sixth minute. A row of enlarged sublabials bordering the infralabials. Scales on chin and throat equal, small, round, juxtaposed, conical.

Scales on the nape and on the side of the neck like the dorsals.

Dorsals and laterals subequal, small, round, juxtaposed, slightly conical on the back, very distinctly so on the flanks. Ventrals large, hexagonal to rhomboid, imbricate, flat; in oblique and in 18–21, mostly 19–20, longitudinal rows; 40–48 along the midventral line, between the insertion of the forelimbs and the anterior margin of the vent. Transition between laterals and ventrals fairly abrupt. Total number of scales around midbody 90–99. Scales on the preanal plate identical to the ventrals, except along the anterior margin of the vent where they are minute, round, juxtaposed. No preanal or femoral pores.

Scales on the tail small, slightly larger than the dorsals, rhomboid to hexagonal, slightly imbricate; on the anterior part of the tail in oblique and in transverse rows; on the posterior part in oblique rows only. Scales under the tail much larger, cycloid to hexagonal, imbricate, flat, of different sizes, largest scales in the midventral line. Scales on the regenerated part of the tail smaller than on the original part, rounded, subequal, flat, slightly imbricate, irregularly arranged; under the regenerated part a series of transversely arranged, imbricate scales is present, except near the tip, where the scales are like those on the dorsal part. Tips, both of original and of regenerated tails blunt. Base of tail in males slightly swollen. Tail round in cross-section.

Scales on all surfaces of the limbs, except on the ventral surface of the hind limbs, small, round, juxtaposed, conical, subequal; on the ventral surface of the hind limbs like the ventrals, smaller on the lower legs than on the thighs. Twenty two to 25 single, smooth lamellae under the fourth finger; 24–30 single, smooth lamellae under the fourth toe; lamellae narrower than the digits, slightly wider and larger under the proximal than under the distal part. Palms and soles with small, round, juxtaposed, flat to tubercular scales of different sizes. Fingers and toes slender, compressed, fourth finger and toe longest.

Colour in life in adult males deep chocolate brown on the dorsal surface of the head. Back and tail slightly lighter brown. Upper parts of the head and the back with yellow-green spots and stripes, most densely on the head, decreasing in number posteriorly. Belly plus the underside of the tail and limbs orange to purplish brown, immaculate. Chin and throat orange with ochre-yellow spots.

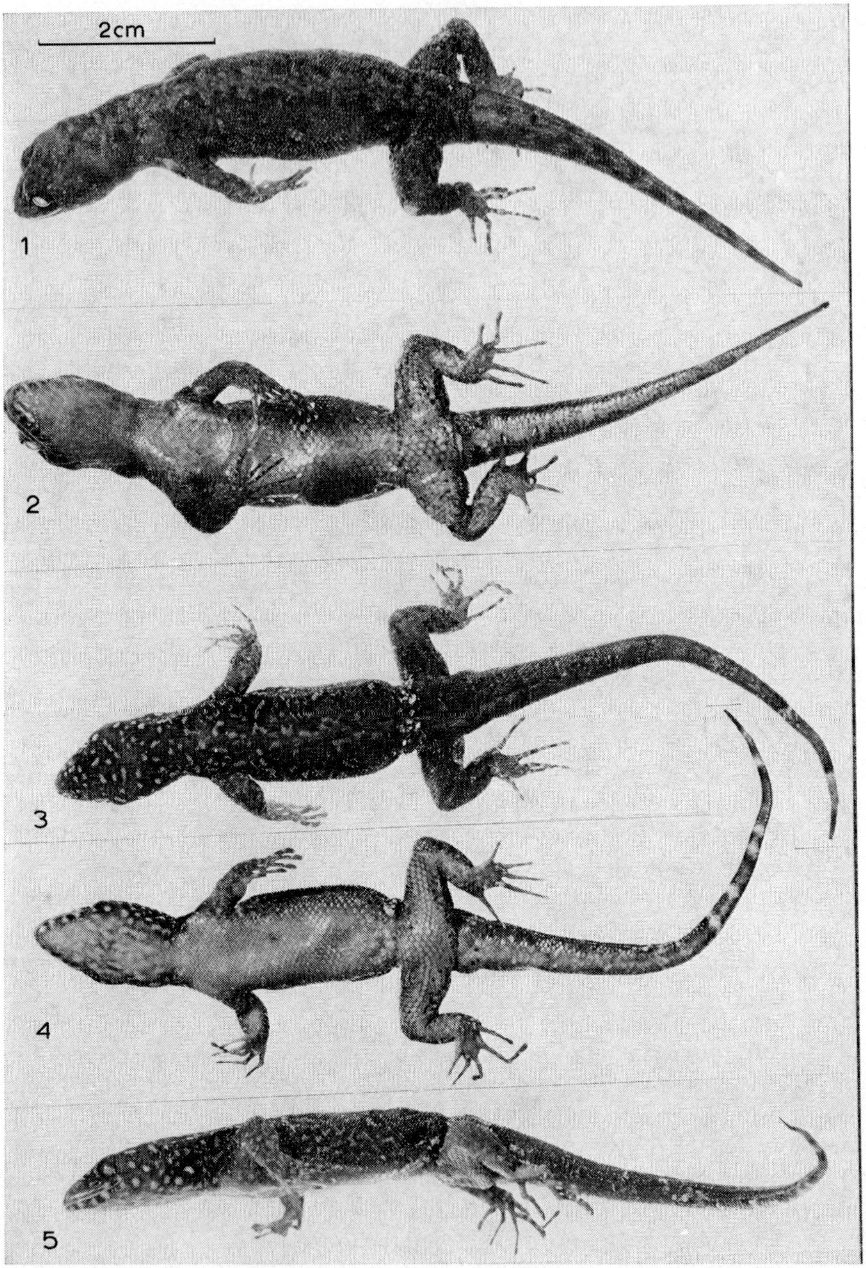

Plate 6. Gonatodes annularis Boulenger. Figs. 1, 2. Dorsal and ventral aspect of ♀, RMNH 16210, Brown's Mountain. Figs. 3–5. Dorsal, ventral and lateral aspect of ♂, RMNH 16209, Brown's Mountain.

Table 6. Counts and mesurements of *Gonatodes annularis* Boulenger. For explanation see table 3.

Reg. no.	A	B	C	D	E	F	G	H	I	J	L	M	O	P	Q	J'	V	W	X	C'	H'	I'
RMNH																						
* 4462	♂	50	54(21)	12.4	9.3	6.7			-	99	41	20	6-6	6-6	8-9	13-13	2	25-24	?-25	5		4-4
16208	♂	?	?	?	?	?			-	?	?	?	6-6	6-6	?	?	2	22-22	25-24	5		
16209	♂	48	57	12.1	8.2	5.4			-	94	41	21	7-7	6-6	8-8	13-12	2	24-24	26-26	5		4-4
16210	♀	55	46(4)	12.9	9.1	5.9			-	95	44	20	7-7	6-6	8-9	13-13	2	25-25	30-29	4		4-4
16211	♀	51	52(23)	13.0	8.9	6.7			-	90	42	19	7-7	6-6	8-10	12-12	2	23-25	26-28	4		3-4
16212	juv.	31	29	8.2	5.3	3.6			-	96	45	19	6-6	6-6	9-10	12-12	2	24-25	30-29	5		3-3
16213	♀	49	35(3)	12.5	8.5	6.3			-	93	41	20	7-7	6-6	9-10	12-13	2	24-25	28-26	4		3-3
16214	♂	50	52(16)	12.9	9.1	6.1			-	92	42	18	6-6	6-6	12-11	12-12	2	25-24	27-28	5		4-4
ZMA																						
12316	♂	46	55	11.4	7.7	5.6			-	94	41	20	6-6	6-6	10-10	13-12	2	23-25	30-29	5		4-4
-	♀	42	47	10.4	7.4	5.4			-	?	?	19	6-6	6-6	11-?	11-11	2	23-23	26-27	5		4-3
-	♀	48	41(12)	11.5	8.1	5.6			-	99	41	19	6-6	6-6	?-9	?-12	2	24-25	26-26	5		4-3
BM																						
* 1946.8.22.97	♀	40							-				7-7	6-6			2		27-25			
AMNH																						
* 21251	♀	47	44	11.7	7.3	5.1			-				5-5	6-5			2	24-20	25-26	5		
UMMZ																						
47644	♂	40	31	10.7	7.4	5.1			-				6-6	5-5			2	22-24	24-24			3-3
47645	♀	39	48	10.0	7.1	5.4			-				6-6	5-5			2	24-24	28-27			3-3
47647	juv.	26	27	7.4	4.9	3.9			-				6-6	5-5			2	23-22	26-25			3-3
47648	♂	?	?	11.0	7.4	5.5			-				6-6	5-5				23-23	26-26			3-3
47649	?	45	49	11.5	8.0	5.6			-				6-6	5-5				24-23	26-25			3-3
53894	?	36	12.5(12.5)						-				6-6	6-6				24-25	28-27			3-3
70056	juv.	26	19						-				6-6	6-6				24-23	24-26			3-3
70066	juv.	32	?	8.4	6.1	4.8			-				6-7					? -23	30-28	4		4-4
79970	juv.	20	25	5.8	4.5	3.3			-				6-6	5-5				21-?	23-25			3-3
83741	?	38							-				6-5	4-4				25-25	? -26			3-3
LACM																						
44444	♀	38	39	9.6	6.2	4.3			-	90	48	20	6-6	6-6	11-11	11-10		24-23	25-27	5		3-3
44454	♂	48	52(24)	12.1	7.9	5.4			-	?	40	?	6-6	6-6	10-10	13-13		23-22	29-27	4		4-4
44471	♂	37	26(26)	10.4	6.6	4.3			-	?	?	?	6-6	6-6	10-9	12-12		22-24	27-27	5		3-4

Distal part of tail with several white spots below and some grey-green spots above. Adult females light-brown above with a creamish vertebral stripe, two transverse, yellowish stripes on the neck and two spots of the same colour on the flanks. Rest of the back variegated with several shades of brown. Ventral parts pale orange. Iris both in males and females golden with a blue zone around the pupil. A juvenile (RMNH 16212) was light-brown with two rows of dark-brown paravertebral spots, which on the tail change into transverse bars. Tip of tail yellow. Throat brownish white with posteriorly converging brown lines, belly purplish brown, underside of tail with transverse dark-brown bands, separated by white spots. Iris in the juvenile brown.

Colour in preservative light- to dark-brown above. There are three distinct types of pattern, two of which are restricted to the males, one to the females and juveniles, and a modification of this last type.

The first type of male pattern, which I should like to call the '*beebei*-type' because the description of *Gonatodes beebei* Noble was based on a specimen of this type, occurs rarely in Surinam and I saw only one specimen (RMNH 16214). The specimen is immaculate dark-brown above, whitish on the chin, uniformly brownish white on the ventral parts.

The second type of male pattern is more numerous and all (four) other Surinam males exhibit it. The description of *G. boonii* Van Lidth de Jeude was based on a male with this pattern. The colour of the back is light- to dark-brown with a number of whitish spots or short stripes. Just anterior to the shoulder a short vertical stripe may be present, and a similar stripe may be present on the flanks. Number of spots decreasing posteriorly. Vertebral zone less densely spotted than the flanks. Fingers and toes with transverse whitish bands. Belly, lower surface of the limbs and proximal part of the ventral surface of the tail greyish white, immaculate. Distal part of the ventral surface of the tail with some round to oval white spots, separated by dark-brown bands. Throat very light- to dark-brown with well defined whitish spots in three, ill defined spots in one (ZMA 12316) specimen. The spots may be arranged haphazardly or in posteriorly converging lines (ZMA 12316). This type has been depicted by VAN LIDTH DE JEUDE (1904) and HOOGMOED (1968).

Most adult females have a pattern that is intermediate between that of *boonii*-type males and juveniles. Because of this I prefer to describe the juvenile pattern first. BOULENGER (1887) based his description of *G. annularis* on two subadult females and this pattern consequently is indicated as the *annularis*-type. This pattern is exhibited by RMNH 16212. It is brown above with 2 paravertebral series of dark-brown spots starting on the neck, just behind the head, and continued onto the tail. Farther down the tail the spots of both sides unite and form three transverse bands over the tail. The dark-brown spots are also arranged in transverse series, two on the neck, four on the body, one on the sacrum and one on the tail. In each transverse row on the body there are four spots. A transverse, U-shaped, dark-brown band extending from the posterior corner of the eyes; a dark-brown band from the eye to the shoulder. A tiny, indistinct, vertical, white line in front of the shoulder. Upper lips with dark-brown and dirty white bars. Limbs marbled with brown of different shades, fingers and toes with lighter transverse bands. Belly and ventral surface of limbs immaculate, dirty white. Throat dirty white with greyish brown, posteriorly converging lines, originating on the lower lips. Lower lips with distinct dark-brown and dirty white bars, corresponding with those on the upper lips. Ventral surface of the tail with seven dark-brown transverse bands, space between them in the proximal part dirty white, in the distal part bright white, tip of tail also bright white. Some adult females (e.g., ZMA no number, collectors number 58) retain this pattern practically without change. In others there are some changes, the paravertebral spots becoming less distinct and forming two indistinct, wavering dark-brown paravertebral lines, bordering a grey vertebral zone. Flanks brown. Head brown, variegated with grey; on the neck there are two more or less distinct vertical stripes. I saw only two females with undamaged tails and here the ventral surface of the distal part of the tail showed four bright white spots, separated by black.

A small lizard with cylindrical body and short, conical tail. Maximum snout-vent length in females 55 mm, in males 50 mm. The undamaged tail in females is 1.1 times as long as the snout-vent length, in two males it is 1.2 times as long and in a single juvenile also 1.1 times as long. Limbs moderately long, stout, hind limb slightly longer than forelimb. When laid along the body forelimb and hind limb probably show a considerable overlap.

HABITAT.—Probably a semi-arboreal lizard. All specimens collected by me were taken in rain-forest on the lower part of tree trunks with well developed buttresses or stilt roots. One specimen (RMNH 16211) was taken from a crack in a rock in rain-forest near the edge of a river. BEEBE (1944a) reports that the type of *G. beebei* Noble was captured on a fallen tree trunk in high jungle. Under '*G. booni*' (BEEBE, 1944a) additional notes are given: he saw at least two specimens of this colourphase 'beneath an overhanging bank of earth in the jungle', one of the specimens actually was dug from a 12 inch deep hole in which it had taken refuge.

NATURAL HISTORY.—A diurnal species which was taken between 2.30 and 6.30 p.m. I doubt whether these data reflect the actual activity pattern. The specimen taken at 6.30 p.m. was taken from a crack in a rock and probably had retired to this place to spend the night. An active specimen was taken just before sunset at 6.00 p.m. No basking behaviour was observed. No pregnant females, nor recently hatched juveniles are available. The number of available specimens was too small to examine stomach contents.

RANGE.—Surinam, French Guiana, Guyana, Eastern Venezuela, and Brazil (Amapá). In Surinam the species is present south of the coastal savannah-belt and reaches at least an elevation of 500 m above sea level (fig. 9).

REMARKS.—RIVERO-BLANCO (1967) concluded from his material, in which some transitional specimens were present, that the lizards described as *G. annularis* Boulenger, *G. beebei* Noble and *G. boonii* Van Lidth de Jeude belong to a single species for which the name *G. annularis* Boulenger had to be used. I have had the good fortune to examine the types of all three forms and am convinced that RIVERO-BLANCO is correct in synonymising these three. RIVERO-BLANCO reports two colour phases (*beebei* and *boonii*) for males and one (*annularis*) for females. In the Surinam and French Guianese material I had available there is also a second female phase, which has been described above. In the Guianese material I examined, no transitional cases as reported by RIVERO-BLANCO, were seen.

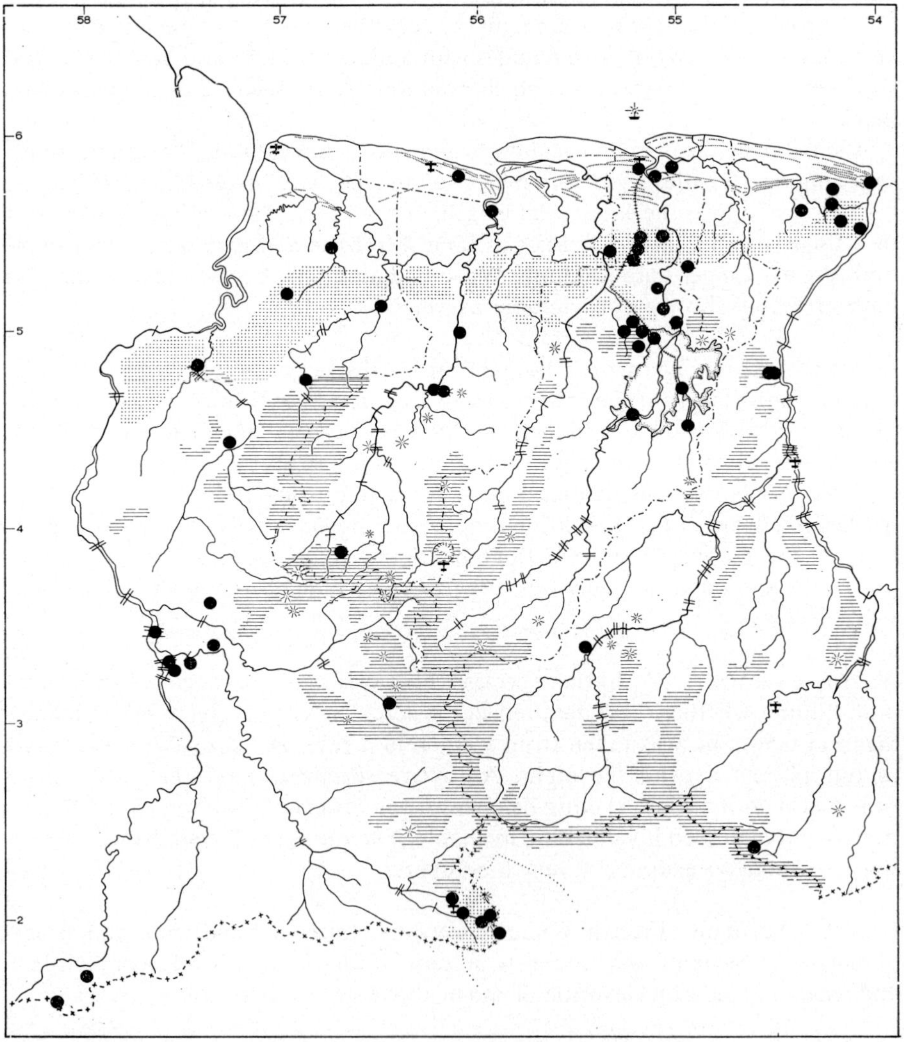

Fig. 10. Distribution of *Gonatodes humeralis* (Guichenot) in Surinam.

Plate 7. Gonatodes humeralis (Guichenot). Figs. 1–3. Dorsal, lateral and ventral aspect of ♂, RMNH 16292, Paramaribo. Figs. 4–6. Dorsal, lateral and ventral aspect of ♀, RMNH 16282, Berlijn.

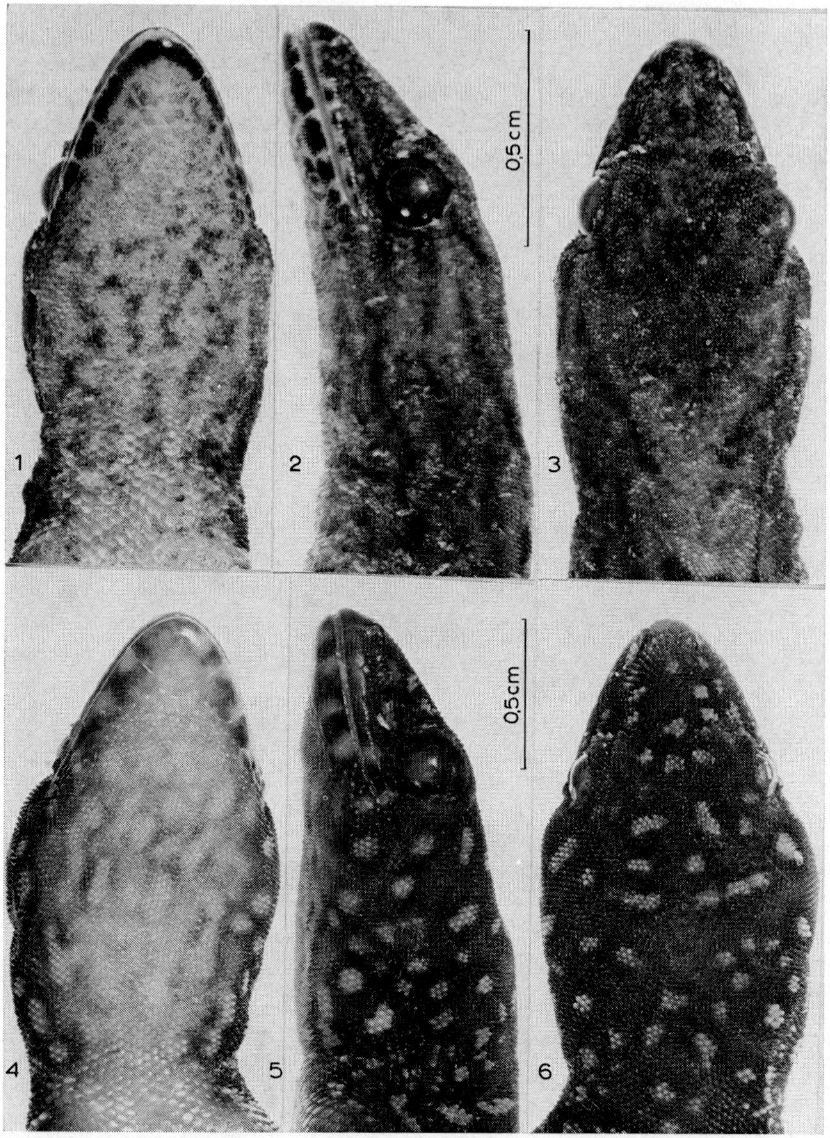

Plate 8. Figs. 1–3. *Gonatodes humeralis* (Guichenot); ventral, lateral and dorsal view of the head of ♀, RMNH 16282, Berlijn. Figs. 4–6. *G. annularis* Boulenger; ventral, lateral and dorsal view of the head of ♂, RMNH 16209, Brown's Mountain.

Under the name *G. annularis*, BEEBE (1944a) described a species quite different from the one described above. The species on which BEEBE's account was based clearly was *G. humeralis* (Guichenot).

Gonatodes humeralis (Guichenot) (fig. 10; pls 7, 8)

Gymnodactylus humeralis Guichenot, 1855: 13.
Gonatodes albogularis: Van Lidth de Jeude; 1904: 86, Van Lidth de Jeude, 1917g: 437 (partly).
'59, 60. Grondhagedissen' Fernandes, 1927: 249.
Gonatodes annularis: Burt & Burt, 1931: 243.
'Surinaamse daggecko' Nedermeijer, 1963: 34.
Gonatodes humeralis: D'Audretsch, 1967b: 6 (originally cited with question mark).
Gonatodes spec. D'Audretsch, 1967b: photo p. 6.
Gonatodes humeralis: Vanzolini, 1968a: 23.

TYPE-LOCALITY.—Río Ucayali, Mission de Sarayacu, Peru.

MATERIAL. – Surinam.

Surinam: 2♀♀, 1♂, RMNH 10274, leg. Coppename Expedition; 1♀, RMNH 16327, leg. H. F. C. ten Kate; 1♀, RMNH 16345, leg. Surinam Expedition 1948–49; 1♀, RMNH 16361, 1969, 1♂, RMNH 16368, 1969, both leg. H. M. van Meeuwen; 1 ex., ZMA 10156; 1 ex., ZMA 12315, 1887, leg. Van Brussel; 1 ex., ZMB 27017, leg. A. O. C. Heller.

Paramaribo: 3♀♀, 5♂♂, RMNH 15034, 3–V–1968, 1♀, 1 juv., RMNH 16285–86, 7–V–1968, 1♀, RMNH 16287, 6–VI–1968, 2♂♂, RMNH 16292, 17–VII–1968, 2♀♀, RMNH 16293, 4–V–1968, all leg. M. S. Hoogmoed; 2♀♀, 2♂♂, RMNH 16309, leg. W. N. Polder; 1 juv., RMNH 16317, October 1968, leg. K. W. R. Zwart; 1 juv., RMNH 16320, May 1965, leg. R. & F. Deuss; 4♀♀, RMNH 16324, 19–III–1939, 1♂, RMNH 16328, 17–III–1939, both leg. H. W. C. Cossee; 2♂♂, RMNH 16336, 1911, 1♂, RMNH 16337, August 1911, both leg. Jhr. W. C. van Heurn; 1♂, RMNH 16342, leg. P. H. van Doesburg; 1♂, RMNH 16344, July/August 1911, leg. Jhr. W. C. van Heurn; 1♂, 1 juv., RMNH 16362–63, 1969, 10♀♀, 6♂♂, 1 ex., 1 embryo, RMNH 16364–65, March 1969, 1♀, RMNH 16366, April 1969, all leg. H. M. van Meeuwen; 3 ex., ZMB 26083, 27031, 27103, leg. A. O. C. Heller; 1♀, 2♂♂, 1 juv., 2 ex., FMNH 121086–91, 20–IX–1960, leg. H. A. Beatty; 1♀, CM 44329, 29–VI–1966, 1♀, CM 44331, 16–IX–1966, 1♀, CM 44332, 5–X–1966, all leg. M. H. de la Fuente.

Coronie road: 1♀, RMNH 16329, picket 216, 17–XII–1948, 1♀, RMNH 16352, picket 216, km 0.1, 16–XII–1948, leg. Surinam Expedition 1948–49.

Plantation Peperpot: 1♂, ZMA 11626, 17–XI–1964, 1♀, 1♂, ZMA 11720, 12430, 18–XI–1964, both leg. T. A. Renssen.

Plantation Marienburg: 1♀, RMNH 16312, 9–X–1968, leg. T. A. Renssen.

Batavia: 1♀, BM 1946.4.2.75, 20 miles inland, 12–VI–1938, I. T. Sanderson.

Wia Wia transect: 1♂, RMNH 16332, 3rd swamp, km 16.6, 15–X–1948, leg. Surinam Expedition 1948–49.

Third Encampment: 1♂, RMNH 16347, km 14.9, 20–X–1948, 1♂, RMNH 16359, km 14.9, 26–X–1948, both leg. Surinam Expedition 1948–49.

Wane Creek N.: 1♀, RMNH 16330, km 5.2, 29–IX–1948, leg. P. H. Creutzberg (Surinam Expedition 1948–49); 1♂, RMNH 16348, km 5.3, 25–X–1948, leg. Surinam Expedition 1948–49.

Galibi: 1♀, 1♂, RMNH 16335, October 1963, leg. B. Malkin; 1♂, RMNH 16350, 6–XI–1948, 1♂, RMNH 16351, 5–XI–1948, both. leg. Surinam Expedition 1948–49.

Moengo: 1♀, RMNH 16360, 1969, leg. H. M. van Meeuwen.

Moengo Tapoe: 2♂♂, RMNH 16295, 10 km E., 8–VI–1968, leg. M. S. Hoogmoed.

Albina: 1♂, RMNH 16349, 2–XI–1948, leg. Surinam Expedition 1948–49.

Awarra savannah, Maratakka River: 1♂, RMNH 16449, 7–III–1971, leg. D. G. Reeder.

Maratakka River: 1♂, RMNH 16450, 6 hours upstream from Cupido, 1–III–1971, leg. D. G. Reeder.

Troeli Creek: 1♂, RMNH 16283, 8–V–1968, leg. M. S. Hoogmoed.

Zanderij: 1♂, RMNH 16339, 16–VI–1964, leg. D. C. Geijskes; 1♀, 1 juv., ZMA 12317–18, 12–IV–1956, leg. J. van der Kamp; 1♂, BM 1946.4.2.73, 5–X–1938, 1♀, BM 1946.4.2.74, 2–VIII–1938, both leg. I. T. Sanderson.

Powakka: 1♂, CM 44328, 11–V–1966, leg. M. H. de la Fuente.

Berlijn: 1♀, RMNH 16282, 10–VII–1968, 1♀, RMNH 16313, 15–X–1968, both leg. M. S. Hoogmoed.

Forest Service Encampment 8, Mapane area: 1♀, RMNH 16296, 13–V–1968, leg. M. S. Hoogmoed.

Kaboeri Creek, Corantijn River: 1♂, RMNH 16495, 2nd bivouac, 29–III–1971, leg. D. C. Geijskes.

Stondansi Cataracts, Nickerie River: 1♂, RMNH 16459, 30–I–1971, leg. D. G. Reeder.

65 km SSE. of Paramaribo, Afobaka Road: 2♀♀, AMNH 108753–54, 12/13–II–1972, leg. C. W. Myers & J. Daly.

Berg en Dal: 2 ex., ZMB 7664, leg. H. B. Möschler.

Camp 1, Coppename River: 1♂, BM 1946.4.2.76, 27–VI–1938, 1♂, BM 1946.4.2.77, 8–VII–1938, 1♀, BM 1946.4.2.78, 24–VI–1938, 1♂, BM 1946.4.2.79, 21–VI–1938, all leg. I. T. Sanderson.

Brown's Mountain: 5♀♀, 2♂♂, RMNH 16284, 13–VII–1968, 1♀, RMNH 16307, foot, 19–VIII–1968, 2♀♀, 1♂, RMNH 16308, north slope, 11–XI–1968, 1♂, RMNH 16310, east slope, 12–XI–1968, all leg. M. S. Hoogmoed; 1♀, RMNH 16369, Mazaronitop, 11–I–1970, 2♂♂, RMNH 16370, north slope, 11–I–1970, both leg. M. S. Hoogmoed & J. J. P. Paats; 1♂, AMNH 108748, Mazaronitop, 15/16–II–1972, leg. C. W. Myers & J. Daly.

Railway km 121: 1♂, RMNH 16318, 21–IV–1949, leg. D. C. Geijskes; 2♂♂, RMNH 16326, May 1949, 1♀, RMNH 16333, Mahami Creek, 30–X–1949, 1♀, 3♂♂, RMNH 16346, July 1949, all leg. C. Bleys.

Brokopondo: 1♀, RMNH 16322, 2–VII–1965, leg. G. F. Mees; 1♀, ZMA 12649, 12–V–1966, 1♂, ZMA 12660, 18–VIII–1966, 1♂, ZMA 12703, 28–IX–1966, 1♀, 1♂, ZMA 12707–08, 16–X–1966, 1♂, ZMA 12766, 12–XI–1966, 3♂♂, ZMA 12768–70, 18–XI–1966, 1♂, ZMA 13197, 15–II 1967, 1♀, ZMA 13204, 14–III–1967, 2♂♂, ZMA 13209, 28–III–1967, all leg. H. Nijssen.

Avanavero Cataracts, Kabalebo River: 1♀, RMNH 16321, 21–IX–1965, leg. G. F. Mees; 1♀, RMNH 16496, 8–IV–1971, leg. D. G. Reeder.

Blanche Marie Cataracts, Nickerie River: 2 juvs., RMNH 16445, 14–II–1971, leg. P. Staffeleu.

Raleigh Cataracts, Coppename River: 2♂♂, RMNH 16315, 4–XI–1968, 3♀♀, 1♂, RMNH 16316, airstrip under construction, 6–XI–1968, both leg. M. S. Hoogmoed; 1♀, RMNH 16338, 8/16–VII–1962, leg. P. H. van Doesburg; 1♂, RMNH 17293, March 1972, leg. G. F. Mees; 1♂, AMNH 108749, 18–II–1972, 1♀, AMNH 108750, 19/20–II–1972, both leg. C. W. Myers & J. Daly.

Voltz Mountain: 1♀, RMNH 16341, 7–IV–1963, leg. A. M. Husson & P. Staffeleu; 1♀, AMNH 108752, NW. base, 22/23–II–1972, leg. C. W. Myers & J. Daly.

Dam, Sara Creek: 1♂, RMNH 16319, 26–XII–1963, leg. M. Boeseman.

Nassau Mountains, Base Encampment: 2♀♀, 2♂♂, RMNH 16331, 12–II–1949, 1♂, RMNH 16353, 17–II–1949, 1♂, RMNH 16356, 25–II–1949, 2♀♀, RMNH 16358, 27–III–1949, all leg. Surinam Expedition 1948–49.

Nassau Mountains: 1♀, 3♂♂, RMNH 16325, km 0–0.6, 15–II–1949, 1♀, RMNH 16354, km 0.7–0.8, 17–II–1949, 1♀, RMNH 16355, km 1, 25–II–1949, 1♀, RMNH 16357, km 3.7, 11–III–1949, all leg. Surinam Expedition 1948–49.

Airstrip Kabalebo: 1♀, CM 44330, 2–III–1965, leg. M. H. de la Fuente.

Mamadam, Suriname River: 1♀, RMNH 16323, 22–VII–1965, leg. G. F. Mees.

Sara Creek: 1♀, ZMA 12709, 21 km S. of Dam, 10–X–1966, leg. H. Nijssen.

Wilhelmina Mountains: 1♀, RMNH 16340, transect I, 5–IX–1944, leg. D. C. Geijskes; 2♀♀, ZMA no number, 1926, leg. Wilhelmina Mountains Expedition (collector's numbers 59, 60).

King Frederik Willem IV Cataracts, Corantijn River: 1♀, BM 1936.4.2.40, 30–XII–1935, leg. J. G. Myers; 1♂, 1♀, FMNH 30917, 30921, 16/20–VIII–1938, leg. E. R. Blake.

Tigri Encampment, New River: 1♀, RMNH 16288, 2–VII–1968, 2♀♀, RMNH 16291, 29–VI–1968, 2♂♂, RMNH 16314, 2–VII–1968, all leg. M. S. Hoogmoed.

Gonini Encampment, Coeroeni River: 1♀, 1 juv., RMNH 16289, 26–VI–1968, 1♀, RMNH 16290, 24–VI–1968, both leg. M. S. Hoogmoed.

Airstrip Coeroeni: 1♀, 3♂♂, 1 juv., RMNH 16343, August 1959, leg. D. C. Geijskes.

Airstrip Kayser Mountains: 1♀, RMNH 16294, NE., 2–VIII–1968, 1♀, RMNH 16298, SW., 5–VIII–1968, 1♀, RMNH 16306, SE., 7–VIII–1968, all leg. M. S. Hoogmoed; 1♀, RMNH 16367, late April 1969, leg. H. M. van Meeuwen; 3♀♀, 3♂♂, 1 ex., FMNH 128990–92, 128994–96, 128998, 1–II–1961, leg. H. A. Beatty.

Airstrip Paloemeu: 1♂, RMNH 16297, 20–V–1968, leg. M. S. Hoogmoed.

Boundary Camp, Itabu Creek Head; 1♀, FMNH 30918, September/October 1938, leg. E. R. Blake.

Middle Base Camp, Itabu Creek: 1♀, FMNH 30920, 4/14–IX–1938, leg. E. R. Blake.

Airstrip Sipaliwini: 1♂, RMNH 16303, 5 km E., 1–IX–1968, leg. M. S. Hoogmoed; 1♂, RMNH 16372, 4 km NW., 9–II–1970, leg. M. S. Hoogmoed & J. J. P. Paats.

Sipaliwini, Base Bivouac: 1♀, RMNH 16299, 24–VIII–1968, 1♀, RMNH 16301, 6–IX–1968, 2♂♂, RMNH 16302, 0–5 km N., 2–IX–1968, all leg. M. S. Hoogmoed.

Sipaliwini, Vier Gebroeders Bivouac: 1♀, RMNH 15177, 14–IX–1968, 1 juv., RMNH 16300, 16–IX–1968, 1♂, RMNH 16310, 13–IX–1968, all leg. M. S. Hoogmoed; 1♀, RMNH 16436, 7–II–1970, leg. M. S. Hoogmoed & J. J. P. Paats.

Sipaliwini, forest-island 3 km N. of Vier Gebroeders Mountain: 1♀, RMNH 16304, 20–IX–1968, 1♂, RMNH 16305, 23–IX–1968, both leg. M. S. Hoogmoed.

Waremapan Creek, Tumuc Humac Mountains: 1♂, RMNH 16334, 1–VIII–1939, leg. D. C. Geijskes.

Surinam/Guyana.

Barrington Brown rapids, New River: 1♀, FMNH 30919, August–October 1938, leg. E. R. Blake.

Brazil.

Sipaliwini, Lange Dijk: 1♂, 1 juv., RMNH 16371, forest-island on S. part, 27–I–1970, leg. M. S. Hoogmoed & J. J. P. Paats.

Diagnosis.—A small, slender lizard with moderately long tail. Tongue villose. Scales on the snout larger than those on the posterior part of the head. Total number of scales around midbody 108–135, 53–61 ventrals along the midventral line; 17–21 lamellae under the fourth toe, 15–19 under the fourth finger, the proximal ones as wide as the digit. Males above brownish grey, two rows of red spots on the flanks, a yellow or white band from eye to eye via the occiput. Females brown with a more or less distinct light-brown vertebral band. In both sexes a whitish, vertical antehumeral bar is present, wider in males than in females; underside of tail orange and in most specimens a black stripe on the lower lips. Pupil brown.

DESCRIPTION.—Head small, long and narrow, representing 24–28% of the snout--vent length in both sexes; distinctly depressed; 1.5–1.8 times as long as wide in both sexes; wider than the neck, which is as wide as the body. Snout pointed, with rounded tip; straight in profile; not projecting over the mouth.

Tongue villose.

Rostral large, rectangular, clearly visible from above; about twice as wide as deep; with a median cleft extending anteriorly from the posterior border; followed posteriorly by three, rarely two or four postrostrals of which the lateral ones are much larger than the median one. Nostril bordered by the rostral, three postnasals, and one large supranasal (= lateral postrostral), narrowly separated from the first supralabial. Six to nine, mostly seven or eight, supralabials, of which the fifth or the sixth, rarely the fourth or the seventh is under the centre of the eye; first largest, the rest decreasing in size posteriorly; the posterior scales scarcely larger than the adjoining granules. Loreal region slightly convex, sloping steeply to the mouth, with polygonal, slightly convex, juxtaposed scales, larger than those on the snout, seven to eleven, mostly nine to ten, in a line between postnasals and the anterior margin of the orbit. Canthus rostralis indistinct, rounded. Ear-opening small, round to vertically oval or triangular, with smooth margins; much smaller than the eye; behind and at the same level as or slightly below, the corner of the mouth; external auditory meatus moderately long, tympanum visible. Scales on the snout small, polygonal anteriorly, round posteriorly; slightly convex, juxtaposed, larger than those on the posterior part of the head and on the back. Anterior margin of the upper eyelid with a series of one to six slightly enlarged, conical scales, of which one may form a small spine over the eye. Eye large, its diameter slightly less than the distance between the orbit and the nostril or the ear; pupil round.

Mental large, more or less rhomboidal; wider than long; posterior margin convex; bordered posteriorly by two (rarely three) enlarged postmentals. Five to eight, mostly six or seven infralabials; the anterior one largest, its posterior margin reaching the level of the suture between the second and the third supralabials or slightly beyond; rest of the infralabials decreasing in size posteriorly; the last one minute. No enlarged sublabials. Scales on the chin increasing in size towards the median, decreasing in size towards the throat, round, juxtaposed, flat. Scales on the anterior part of the throat small, round, juxtaposed, flat, on the posterior part larger, round, imbricate, flat.

Scales on the nape and on the side of the neck like the dorsals.

Dorsals and laterals subequal, granular, scales on the flanks only slightly more conical than those on the back. Ventrals large, cycloid, imbricate, flat, in oblique and in 16–23, mostly 19–21, longitudinal rows; 53–61 along the mid-ventral line between the insertion of the forelimbs and the anterior margin of the vent. Transition between laterals and ventrals fairly abrupt. Total number of scales around midbody 108–135, mostly 121–125. Scales on the preanal plate identical to the ventrals, except a few rows of minute, round, juxtaposed scales bordering the anterior margin of the vent. No preanal or femoral pores.

Hemipenis not divided, with a short, wrinkled stem and a large round head. Sulcus spermaticus deep, a straight line on the posterior surface. The entire head covered with a reticulum of low, serrated ridges, except two oblong zones on top of the head, one on each side of the sulcus spermaticus, which are covered with small pointed papillae.

Scales on the tail larger than the dorsals, round to polygonal, flat, slightly imbricate, in oblique rows only. Scales under the tail much larger, imbricate, irregularly shaped, with a median series of transversely enlarged scales. Scales on the regenerated part of the tail smaller than on the original part, quadrangular, subequal, flat, slightly imbricate, in indistinct oblique rows; scales under the regenerated part of the tail like those under the original tail, with a median row of transversely enlarged scales. Undamaged tail divided into indistinct verticils with three ventral and eight dorsal scales. Breaking points between the verticils. Tip, both of original and regenerated tail, blunt. Base of tail in males slightly swollen. Tail round in cross-section.

Scales on the ventral and the posterior surfaces of the upper arms, on the posterior, the dorsal and the anterior surfaces of the forearms, on the posterior and dorsal surfaces of the thighs and on the posterior, the dorsal and the anterior surfaces of the lower legs small, round, juxtaposed, convex; on the remaining surfaces large, round to subtriangular, imbricate, flat. Fifteen to 19, mostly 16–18, single, smooth lamellae under the fourth finger, the proximal four to six, mostly five, as wide as the digit, distinctly larger than the rest; 18–20, rarely 17 or 21, single, smooth lamellae under the fourth toe, the proximal six or seven, mostly six, as wide as the digit, distinctly larger than the rest. Palms and soles with small, round, slightly imbricate, tubercular scales. Fingers and toes slender, compressed, fourth finger and toe longest.

Colour in life: sexual dichromatism present, males much more brilliant. The back in adult males is mottled with brown, black, grey and yellow, giving a brownish grey impression. On the flanks there are two longitudinal rows of red spots; some scattered red spots may be present on the back. From the tip of the snout via the eye to the occiput a yellow or greyish white band, which, on the occiput is connected with its fellow of the other side. From the nostril to the eye this band is bordered ventrally by a red band which extends a short distance beyond the eye in the direction of the ear. Upper lips yellow or greyish white. Lower lips with or without an indistinct black stripe. This very distinct head-pattern is not present in all males, but at least a yellow or greyish white spot is present on the tip of the snout. All intermediate stages occur. A white to pale yellow vertical antehumeral bar. Dorsal surface of the tail with dark-brown transverse bands. Chin and throat yellow or grey, immaculate. Belly and ventral surface of the limbs grey. Ventral surface of the tail orange.

The back in adult females is brown, with a more or less distinct light-brown vertebral band of variable width. Five transverse series of black and beige spots may or may not be present. Antehumeral bar white to yellowish, in most females narrower than in males. Head brown above with oblique dark-brown stripes. Lower lips with a distinct black stripe. Chin and throat white with brown spots, irregularly arranged or in oblique, posteriorly converging series. Belly white, immaculate; ventral surface of the limbs white with brown spots. Ventral surface of the tail orange. Juveniles have the female pattern. Both males and females have a brown iris, tongue white with two bluish spots near the tip.

In preservative, males are brown to grey above, mottled with dark-brown. Antehumeral stripe white, bordered with dark-brown; a black spot in front of it or not. The red spots have completely disappeared. Pattern on the head whitish. From the nostril to the eye a dark-brown band. Ventral parts dirty white, immaculate. An indistinct black line may or may not be present on the lower lips.

Females are brown above with a light-brown vertebral band of variable width, with indented margins, bordered by dark-brown lines. Antehumeral stripe narrow, white. Just in front of the level of the antehumeral stripes there is a pair of black spots in the vertebral band. Flanks with two indistinct longitudinal dark-brown bands or series of spots; sometimes with some white, narrow vertical stripes. Back of the head with oblique, posteriorly converging dark--brown lines. A black line on the lower lips. Chin and throat with irregularly arranged brown spots, or spots in oblique, posteriorly converging series. Belly dirty white, immaculate, although sometimes it may have brown spots near the flanks. Ventral surface of the limbs dirty white with brown spots. Ventral surface of the tail dirty white, immaculate. Both in males and females the tail has dark-brown, transverse bands above, which are most distinct in the distal part. The regenerated part of the tail has the same colour as the body and shows irregular, waving, longitudinal, dark-brown lines of variable length. Tongue white with two oval, grey spots near the tip. Juveniles have the same colour and pattern as females.

A small lizard with cylindrical body and moderately long, tapering tail. Maximum snout-vent length in females 37 mm, in males 38 mm. The undamaged tail in females is 1.1–1.3 times as long as the snout-vent length, in males it is distinctly longer, 1.3–1.4 times and in a juvenile 1.2 times. Limbs moderately long and slender, hind limb slightly longer than forelimb. When laid along the body forelimb and hind limb show a considerable overlap, the hind limb not quite reaching the axilla.

HABITAT.—An arboreal lizard inhabiting forests. In some localities this species was collected on the outer walls of houses. It is present in various types of forest: rain-forest, ridge-forest, forest-islands surrounded by savannah, and planted cocoagroves. Mostly it is found on the lower part of thick treetrunks not higher than one or two metres (although I also found it once in a treetop), sometimes on big rocks in the forest and rarely on small twigs near the ground. RAND & HUMPHREY (1968) found this species in the same habitat in Belém. VANZOLINI (1968b) reports them as occurring on the lower part of treetrunks, to 1.50 m above the ground. BEEBE (1944a) mentions as favourite haunts 'old stumps, hollow and fallen trees and the débris behind half-rotten leaf spathes of palms'. Once he found two specimens at a height of thirty feet in a bromeliad.

NATURAL HISTORY.—A very common diurnal lizard which was never seen basking. All specimens collected by me were in the shade. RAND & HUMPHREY (1968) also found this species to be a nonheliotherm. Active specimens were collected between 8.00 a.m. and 5.30 p.m. A number of sleeping specimens was collected at night on the upper surface of leaves about 50–100 cm above the ground. All sleeping specimens were in bushes overhanging water, but probably this is not an unbiased observation, because they were collected during frog--hunting. According to VANZOLINI (1968a) RAND found this species in Belém, sleeping on leaves of Araceae. While asleep males have a very light, nearly white colour but unfortunately no notes on the colour of a sleeping female were made. Several times a male and a female were found on the same treetrunk, indicating pair formation. Recently hatched juveniles with a snout-vent length of 16–18 mm, are known from May, June, September and October. Pregnant females are known from every month, thus breeding takes place throughout the year. Most females contain one large, mature egg, some, however, also contain a small, immature one. Two mature eggs were never found in one female, and only one egg seems to be laid at a time. In March 1969 I received a number of live specimens from Paramaribo among which were several pregnant females. Two eggs were deposited on March 16. One of these was opened on June 17 and contained a living, nearly full-grown embryo. The other hatched on July 3. The eggs were kept at a constant temperature of 25°C in daytime and at about 20°C at night. This incubation period of 110 days may very well be longer than under natural circumstances, but no other data on the complete incubation time are available. BEEBE (1944a) reported on four eggs found close together on July 23, 1922. The first of these eggs hatched on August 17, two others were opened that same day and turned out to be one third and fourth-fifths developed, respectively. The fourth egg hatched on September 13, indicating a minimum period of development for this egg, under natural conditions, of 52 days. In Surinam eggs were found at the base of trees among dead leaves and behind loose flaps of bark. I did not find more than one egg at the same place. BEEBE (1944a) found six batches of eggs in Kartabo on four occasions. Thrice there was only one egg, thrice there were more than one. According to BEEBE, 'eggs are always hidden beneath soft, fine débris well below the surface'. He found them in decayed parts of palm and bamboo stems, in a hole among roots, and in a termite nest. Under *G. annularis* BEEBE (1944a) gives extensive data on the natural history of *G. humeralis* (Guichenot). In most respects they agree with my Surinam data. Two remarks of BEEBE certainly are not true for Surinam. He says 'they are less nocturnal than *Thecadactylus*', indicating a certain degree of nocturnal activity for *G. humeralis*. In Surinam this species is decidedly not nocturnal, specimens being found asleep already at 8.00 p.m. Between sunset and 8.00 p.m. probably no activity takes place. According to BEEBE this species is decidedly solitary, but as mentioned above, I observed pair-formation, also RAND & HUMPHREY (1968) in Belém occasionally found pairs together. The stomach contents of seven specimens were examined. Items present included small specimens of a sow-bug, spiders, a pseudoscorpion, a scorpion, a millipede, beetles, caterpillars, a butterfly, a wasp, a fly, a cicada, a mantis and a landsnail.

RANGE.—Surinam, Guyana, French Guiana, Brazil (Amapá, Roraima, Pará, Amazonas, Acre, Mato Grosso, Maranhao), Venezuela (Estado Sucre, Monagas, Territorio del Delta Amacuro), Peru, Equador and Bolivia. In Surinam the species is present throughout the country, and occurs from sea level to at least 700 m (fig. 10).

Genus **Pseudogonatodes** Ruthven

DIAGNOSIS.—Very small lizards with cylindrical body, short legs and a short tail. Dorsal scales juxtaposed or imbricate. Ventral scales imbricate, smooth. Claws vertically retractile in a symmetrical sheath composed of five scales, viz., two infero-laterals, two superior and one terminal scale. The superiors are in contact throughout their length. Toes and fingers short, third finger and toe longest. Upper eyelid always without a spine, forming a projecting flap that, in its anterior part, bears enlarged scales. Teeth minute, conical, subequal. Pupil round. Five species in northwestern South America, one of which is present in Surinam.

Pseudogonatodes guianensis Parker (fig. 9; pls. 9, 10; table 7)

Pseudogonatodes guianensis Parker, 1935b: 514.

TYPE-LOCALITY.—Upper Cuyuni River, Guyana.

MATERIAL. – Surinam.
Sipaliwini: 4♀♀, 1♂, RMNH 16206, 4 km NW. of airstrip, 10–II–1970, leg. M. S. Hoogmoed & J. J. P. Paats.
Guyana.
Upper Cuyuni River: 2♀♀, BM 1946.8.27.8–9, 28–VII–1933, leg. G. S. Carter (holo- and paratype).
Shudikar-Wau: 6 ex., AMNH 61433, leg. R. Snedigar.
Surinam/Guyana.
New River, 750 feet: 1♀, 1♂, BM 1939.1.1.67–68, leg. C. A. Hudson.

DIAGNOSIS.—A very small, ground dwelling diurnal gecko with cylindrical body and short conical tail. Dorsals and laterals granular; 93–103 scales around midbody. Ventrals large, imbricate, 37–44 between the forelimbs and the vent. Fourth toe with six or seven subdigital lamellae. Claws of fingers and toes can be retracted in a sheath consisting of two large infero-laterals, two large superior and one small terminal scale. Toes and fingers short. Granules bordering on the mental three to five. Dark-brown to grey, with a light-brown horseshoe-shaped mark on the occiput and an H-shaped, whitish figure on the sacrum.

Plate 9. Figs. 1–3. *Pseudogonatodes guianensis* Parker; dorsal, lateral and ventral aspect of ♀, RMNH 16206 b, Sipaliwini. Figs. 4–6. *Coleodactylus amazonicus* (Andersson); dorsal, lateral and ventral aspect of ♀, RMNH 16237 a, Brown's Mountain.

DESCRIPTION.—Head small, representing 20–22% of the snout-vent length in both sexes; 1.4–1.6 times as long as wide in both sexes; distinctly depressed; slightly wider than the neck, which is as wide as the body. Snout pointed, with a rounded tip; straight in profile; not projecting over the mouth.

Tongue lanceolate, covered with flat, imbricate, cycloid papillae; tip slightly nicked. Teeth conical, of equal size.

Rostral very large, nearly twice as wide as deep, clearly visible from above; with a median cleft projecting from the posterior border; bordered posteriorly by three to five, mostly four granules of which the two laterals are larger and also border on the nostril. Nostril bordered by the rostral, the first supralabial, two postnasals and one supranasal (= the lateral granule of the row of post-rostral granules). Four, rarely five, supralabials; the first very large, the posterior border reaching from halfway the nostril and the anterior margin of the orbit to just anteriorly of this margin; the rest decreasing in size; the third under the centre of the eye. Ear-opening small, obliquely oval, with finely lobed margins; much smaller than the eye; behind and at the same level as the commissure of the mouth; external auditory meatus long, tympanum invisible. Scales on the snout larger than on the rest of the head, oval, conical, arranged in transverse rows; on the rest of the head smaller, round. Upper margin of upper eyelid with three much enlarged, flattened scales, lower margin with five or six enlarged, flattened scales, decreasing in size posteriorly. Eye large, with round pupil.

Mental very large, horseshoe-shaped, posterior margin widely V-shaped, opening posteriorly; bordered posteriorly by three to five granules which are larger than the other granules on the chin. Four, rarely five, infralabials; the anterior one very large, longer than the anterior supralabial; its posterior border reaching the level of the anterior margin of the orbit. Granules on the chin small, round, juxtaposed, smooth, convex to flat. Scales on the throat larger, rhomboid with rounded tips, imbricate, smooth, flat.

Granules on the nape and on the side of the neck small, round to hexagonal, juxtaposed, conical.

Dorsals and laterals identical to the granules on the nape. A sharp border between the laterals and the ventrals. Ventrals large, rhomboid with rounded tips, imbricate, smooth, flat, in oblique and in 21–24, mostly 22 longitudinal rows; 37–44, mostly 41 ventrals between the insertion of the forelimbs and the vent. Total number of scales around midbody 93–103. Scales on the preanal plate identical to the ventrals, except in a few rows immediately bordering the vent, where they are much smaller. No preanal or femoral pores.

Scales on the tail larger than the dorsals, rhomboid with rounded tips, imbricate, smooth, flat, in longitudinal and oblique rows; under the tail slightly larger. Scales on the regenerated part of the tail identical to those on the original tail, only smaller. Base of the tail in males distinctly swollen. Tail round in cross-section.

Scales on the anterior and upper surfaces of the upper arms, on the anterior and lower surfaces of the forearms, on the anterior and lower surfaces of the thighs and on the lower and posterior surfaces of the lower legs large, rhomboid with rounded tips, imbricate, smooth; scales on the other surfaces of the limbs with small granules, round to oval; juxtaposed, conical. Five smooth, rounded lamellae under the fourth finger; six, rarely seven, identical lamellae under the fourth toe. The subdigital lamellae are followed distally by two long, narrow scales, forming part of the sheath in which the claws can be retracted. The claws can be retracted into a sheath formed of two very large infero-laterals, two large superior and one small terminal scale. Palms and soles with round, imbricate, smooth scales, smaller than the ventrals, larger than the dorsals. Fingers and toes very short, slightly compressed. The third finger and toe longest.

Colour in life dark-brown to grey above, with a light-brown horseshoe-shaped spot on the occiput. Narrow light-brown dorsolateral stripes on the sacral region and anterior part of the tail, connected by a narrow stripe of the same colour on the anterior part of the tail, thus forming an H-shaped figure. Throat white with some longitudinal brown markings, belly dirty white, lower surface of the tail black with an interrupted white median stripe. Iris orange-brown. Anterior part of the tongue blue-grey, posterior part whitish.

Colour in preservative dark-brown above with irregularly scattered dirty white spots. A whitish to greyish brown horseshoe- to W-shaped mark on the occiput, starting from the posterior corner of the eye. Upper surface of the snout to between the eyes grey. Labials black with white seams. A white, oblique stripe from beneath the eye to the throat, accompanied anteriorly and posteriorly by dark-brown stripes, of which the posterior one is continued onto the anterior surface of the upper arm, thus separating the dorsal and ventral colour. A second white, narrow, indistinct, slanting line from the posterior corner of the eye to the ear-opening. An H-shaped whitish figure on the sacrum. Chin white, with small, brown spots, throat white with approximately three longitudinal light-brown (dark-brown in the male RMNH 16206d) stripes. Belly light-brown, mingled with dirty white. Ventral surface of the tail dark-brown, with an irregular, interrupted median white stripe. Posterior surface of the thighs dark-brown with large white spots, which in the male RMNH 16206d form two horizontal stripes. Tongue anteriorly bluish grey, posteriorly whitish.

A very small lizard with cylindrical body and short conical tail. Maximum snout-vent length in females 29 mm (30 mm in the holotype), 27 mm in males. The tail is undamaged in only one specimen (♀ RMNH 16206b) and is 0.8 times as long as the snout-vent length. Limbs short and stout, hind limb longer than forelimb, when laid along the body the fingers and toes slightly overlap.

HABITAT.—This species is a ground dweller in rain-forest. All specimens collected by me, and some others that could not be collected, were found among fallen leaves in the bed of dry creeks in which a number of isolated pools were still present. The water in the pools was brown but clear. The beds consisted of coarse, white sand, the banks of the creeks were fairly steep, without vegetation, with numerous projecting roots, that offered the lizards plenty of cover. The deeper parts of the beds showed bare sand, the higher parts near the banks were

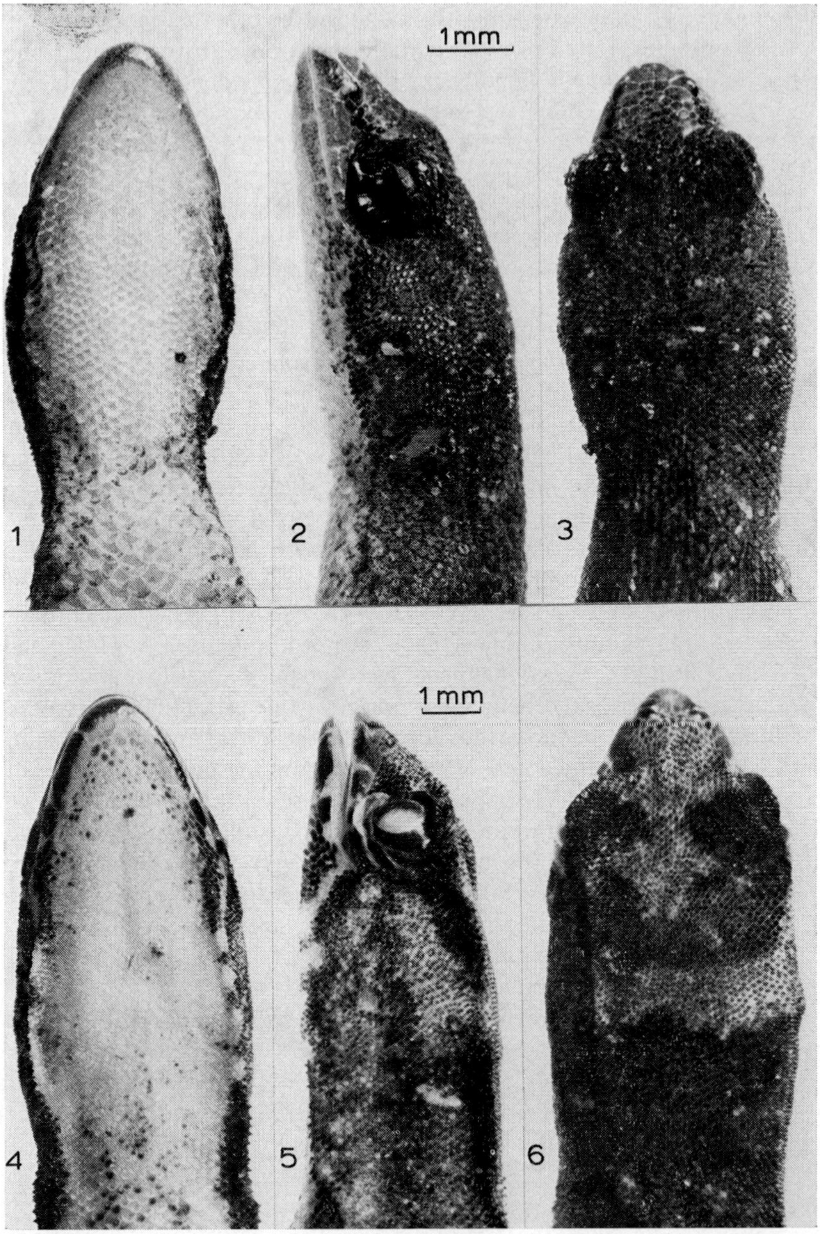

Plate 10. Figs. 1–3. *Coleodactylus amazonicus* (Andersson); ventral, lateral and dorsal view of the head of ♂, RMNH 16221, Airstrip Kayser Mountains. Figs. 4–6. *Pseudogonatodes guianensis* Parker; ventral, lateral and dorsal view of the head of ♀, RMNH 16206 b, Sipaliwini.

Table 7. Counts and measurements of *Pseudogonatodes guianensis* Parker. For explanation see table 3.

Reg. no.	A	B	C	D	E	F	G	H	I	J	L	M	O	P	E'	F'	G'	J'	V	W	X	A'	B'	C'	H'	I'
RMNH																										
16206a	♀	29	21(11)	6.0	3.9	2.9			-	98	41	24	4-4	4-4	2.4	1.8	1.7	9-9	-	6-6	8-7	3-3	6-5	4	4	2-2
b	♀	26	20	5.5	3.9	3.2			-	94	44	22	4-4	4-4	2.0	1.9	2.0	8-7	-	?-6	7-7	3-3	6-6	4	4	2-2
c	♀	27	9(9)	5.6	3.9	2.9			-	103	37	22	4-5	4-4	2.3	2.1	1.5	8-8	-	6-6	7-7	3-3	6-6	3	5	2-2
d	♂	26	7(7)	5.7	4.1	2.8			-	?	41	21	4-4	4-4	2.1	1.8	1.6	8-8	-	6-6	7-7	3-3	7-6	5	5	2-2
e	♀	22	2(2)	?	?	?			-	?	?	?	4-4	?	?	?	?	8-7	-	6-6	7-7	3-3	6-8	4	4	2-2
BM																										
1939.1.1.67	♂	27	12(2)	5.8	3.6	2.5			-	93	?	22	5-4	4-4	2.1	1.8	1.9	7-6	-	6-6	7-7	3-3	6-7	3	4	2-2
68	♀	26	3(3)	5.6	3.5	2.6			-	?	41	22	4-4	4-4	2.2	1.7	1.7	8-9	-	6-6	7-8	3-3	?-5	4	?	2-2
* 1946.8.27.8	♀	30	?	6.0	4.2		8	10	-			22	4-4	?	2.3	2.2	1.8		-	?	6-6	3-3		4	3	2-2
* 9	♀	27	?	5.5	3.5		7	9	-			22	4-4	?	2.1	2.2	1.6		-	?	6-6	3-3		4	3	2-2

covered with fallen leaves and in this zone *P. guianensis* Parker was present. In other creeks without sandy bed this species was absent. In this borderzone between water and forest-floor it probably replaces *Coleodactylus amazonicus* (Andersson), which is present on the forest-floor away from creeks.

NATURAL HISTORY.—*P. guianensis* Parker is diurnal, the specimens collected by me were taken between 12.00 a.m. and 1.15 p.m. No basking was observed. When the fallen leaves among which they were present were disturbed the lizards fled into the bare banks where they tried to escape by hiding under projecting roots or in fissures. When pursued they developed considerable speed, but when at ease they walk very slowly, lifting the belly and tail clear of the ground, moving the left forelimb and the right hind limb at the same time, etc. During this slow gait, the body bends laterally and the tail sways from one side to the other. Among the RMNH specimens there are no pregnant females. The holotype from Guyana, which was captured in July, contains one large, oval egg. The paratype seems to have deposited it recently. Probably only a single egg develops at a time, there being no more room available. Nothing is known about food. When touched they easily lost patches of skin.

RANGE.—Until now it is known only from extreme southern Surinam (Airstrip Sipaliwini, Serra Acarai) and southern and northern Guyana (Shudikar-wau and Upper Cuyuni River, respectively). If I am correct in my assumption (see below) that *P. amazonicus* Vanzolini is only a synonym of *P. guianensis* Parker, the species range is extended to Igarapé Belém, Rio Solimoes, Amazonas, Brazil. In this case, the species is probably distributed throughout the Amazon valley and adjacent regions. The scarcety of data is due to the lizard's small size and secretive habits and probably also to its restricted habitat (fig. 9).

REMARKS.—VANZOLINI (1967) described *P. amazonicus* on the basis of three specimens from Igarapé Belém, Rio Solimoes, Amazonas in western Brazil. In most respects this species agrees with *P. guianensis* Parker, the characters in which it is said to differ being the number of granules bordering the posterior border of the rostral and the form of the mental. It is clear from the specimens I have examined that in *P. guianensis* PARKER the number of granules bordering the rostral posteriorly varies from three to five, two specimens having three, one specimen having five and six specimens having four granules. In all specimens the two lateral scales, which also border on the nostril and thus can be named supranasals are larger than the median scale(s). In the specimens with three scales the situation is exactly as figured by VANZOLINI (1967, pl. 2 fig. 3). In his discussion (p. 6) and in the table (p. 11) he says that in *P. guianensis* PARKER there are two granules on the posterior border of the rostral. This, however, is not true as there are always more than two granules. If VANZOLINI meant only the median granules, this is true indeed for the majority of the specimens, but as pointed out above, the character is variable. Thus, this character falls within the range of variability in *P. guianensis* Parker. The second character, the form of the mental does not show any difference. VANZOLINI says that in *P. guianensis* Parker the posterior margin of the mental is transverse and that in *P. amazonicus* Vanzolini it forms a V, opening posteriorly. In all specimens of *P. guianensis* Parker I have examined, the mental is horseshoe-shaped, that is with a posterior margin forming a V, opening posteriorly. Unfortunately I was not able to examine a type of *P. amazonicus* Vanzolini myself but on the basis of the above, I am convinced that *P. amazonicus* Vanzolini is a synonym of *P. guianensis* Parker. The population of Igarapé Belém may represent a different subspecies slightly differing in number of postrostral granules from *P. guianensis* Parker.

Recently TEST et al. (1966) showed that there were too many differences between *P. furvus* Ruthven and *P. lunulatus* (Roux) to consider them conspecific, but that *P. lunulatus* (Roux) and *P. amazonicus* VANZOLINI showed many similarities and might prove to be conspecific. VANZOLINI (1968b) suggested that *P. amazonicus* VANZOLINI, *P. furvus* RUTHVEN, *P. guianensis* PARKER and *P. lunulatus* (Roux) are all races of a single species. In a recent paper HUEY & DIXON (1970) described a new species, *P. peruvianus*, and showed that the species in the genus could be divided into three groups, one with a high subdigital count and granular dorsals, containing *P. furvus* Ruthven and *P. peruvianus* HUEY & DIXON, another with a low subdigital count and granular dorsals and containing *P. lunulatus* (ROUX), *P. guianensis* PARKER and *P. amazonicus* VANZOLINI, and a third with a low subdigital count and large imbricate dorsals, containing *P. barbouri* (NOBLE).

I agree with the division proposed by HUEY & DIXON (1970), but in my opinion the second group only contains two species, namely *P. lunulatus* ROUX and *P. guianensis* PARKER, *P. amazonicus* VANZOLINI being a synonym of the latter. I am not certain whether *P. lunulatus* ROUX and *P. guianensis* PARKER will prove to be conspecific, as HUEY & DIXON (1970) and VANZOLINI (1968b) suppose. I think the differences between the two forms are large enough to warrant specific recognition.

Family IGUANIDAE

This family shows a relatively uniform appearance. All representatives of this family have well developed legs. They have crests on their necks and backs, gular pouches, dewlaps (gular fans) or a combination of these characters. The scales on the head are mostly small, of irregular size, and the enlarged scales that are present are not very uniform in appearance. The dorsal scales are always small, except on the vertebral region where they may be slightly or, in the case of a vertebral crest, strongly enlarged. The ventral scales are larger than the dorsals but they never form series of regularly arranged large scales.

Representatives are mostly arboreal in their habits, although one species has become terrestrial, one rupicolous and two are more or less adapted to an aquatic environment. All species are diurnal.

The iguanids occur only in North and South America, in Madagascar and on the Fiji and Tonga Islands.

Key to the genera of Surinam Iguanidae

1. Femoral pores present 2
   Femoral pores absent 3
2. Third and fourth toes of equal length, body strongly compressed, no dorsal crest, eyeballs covered for their greater part by the eyelids . *Polychrus* p. 175
   Fourth toe longer than the third, body cylindrical, dorsal crest present . . . . *Iguana* p. 148
3. Digits more or less dilated or depressed, with smooth transverse lamellae inferiorly, the distal joint narrower, cylindrical or compressed *Anolis* p. 98
   Digits not dilated, of equal width throughout 4
4. Caudal scales large and spinose, arranged in whorls . . . *Uracentron* p. 191
   Caudal scales small, not spinose, not arranged in whorls 5
5. Upper head scales small, subequal, pyramidal . . . . . *Uranoscodon* p. 199
   Upper head scales larger, of different sizes, not pyramidal 6
6. Interparietal very large, larger than the ear-opening, much wider than the distance between the orbits; suboculars five to eight, subequal . *Plica* p. 159
   Interparietal smaller, smaller than the ear-opening, only slightly wider than the distance between the orbits; suboculars two or three, the posterior one very long . . . . *Tropidurus* p. 184

Genus **Anolis** Daudin

DIAGNOSIS.—A genus of small to large lizards, with well developed legs. The large representatives confined to some of the larger Antillean islands, the small to medium-sized representatives to the South American mainland. Part of the subdigital lamellae of both fingers and toes have changed into transversely enlarged imbricate plates which form a very efficient climbing apparatus, the plates bearing minute recurved projections along their outer border which assure a firm hold even on nearly smooth surfaces. These plates may be well developed, causing part of the digits to be much wider than the rest, or only moderately, in which case the digits are of nearly equal width throughout. The tongue adheres to the floor of the mouth and cannot be protruded. It is thick, arrowhead-shaped, villose with a rounded tip. Anterior teeth conical; lateral teeth tricuspid, laterally compressed. Males of all species possess a gular fan (dewlap) which in excitement is extended. In females it is also present, but less well developed. Preanal or femoral pores are absent.

Key to the Surinam species of *Anolis*

1. Digital expansions of fingers and toes poorly developed, i.e. the third phalanx is not wider than the basal one; number of scales around midbody less than 88; 10–14 rows of distinctly enlarged, keeled dorsals . . . . *Anolis auratus* p. 99
   Digital expansions of fingers and toes moderately to well developed, i.e. the third phalanx is distinctly wider than the basal one; number of scales around

midbody at least 104; less than ten rows of enlarged dorsals, or dorsals small 2
2. At least two rows of dorsals slightly enlarged, keeled; tibia in adult specimens (> 50 mm snout-vent length) longer than the distance between the tip of the snout and the posterior margin of the ear. . *Anolis chrysolepis* p. 112
All dorsals small, smooth; tibia in adult specimens shorter than the distance between the tip of the snout and the posterior margin of the ear 3
3. Scales on the upper surface of the snout smooth . . . *Anolis ortonii* p. 136
Scales on the upper surface of the snout keeled 4
4. Snout projecting, strongly in males, slightly in females; adult specimens reaching a snout-vent length of 85 mm; body colour green or purplish with blue spots; scales around midbody 137–181; nuchal and dorsal folds present, ventrals smooth . . . . *Anolis p. punctatus* p. 141
Snout not projecting; adult specimens not exceeding 49 mm snout-vent length; body colour brown; scales around midbody 122–154; nuchal and dorsal folds absent, ventrals keeled . . *Anolis fuscoauratus kugleri*, p. 127

Anolis auratus Daudin

DIAGNOSIS.—A small anolis with at least ten rows of enlarged, keeled, dorsal scales, the keels forming longitudinal ridges; 69–88 scales around midbody. Tibia in adult specimens shorter than the head length. No subdigital expansions, the subdigital lamellae forming a continuous series.

Key to the subspecies.

1. Body colour light golden brown, never grey, males with a greyish blue dewlap, most females without a vertebral band bordered by black; white line on the side of the head distinct in most specimens and continued on the flanks . . . . *Anolis auratus auratus*, p. 99
Body colour grey to greyish brown, only exceptionally light-brown, males with a greenish blue dewlap, most females with a reddish brown to beige wide vertebral band, bordered by black; white line on the side of the head and neck indistinct in most specimens . . . . *Anolis auratus sipaliwinensis* nov. subspec, p. 107

Anolis auratus Daudin (figs. 11, 12; pl. 11; table 8)

Anolis auratus Daudin, 1802b: 89; J. A. Peters & Donoso-Barros, 1970: 48.
Norops auratus: Duméril & Bibron, 1837: 82; Gray, 1845: 207; Duméril & Duméril, 1851: 55; Bocourt, 1873: 108; Bocourt, 1874: pl. XVI fig. 33; Boulenger, 1855b: 95; Van Lidth de Jeude, 1917c: 46; Burt & Burt, 1930: 18; Burt & Burt, 1931: 297.
Draconura 12-striata Berthold, 1840: 899; Berthold, 1842: 18; Berthold, 1843: 62.

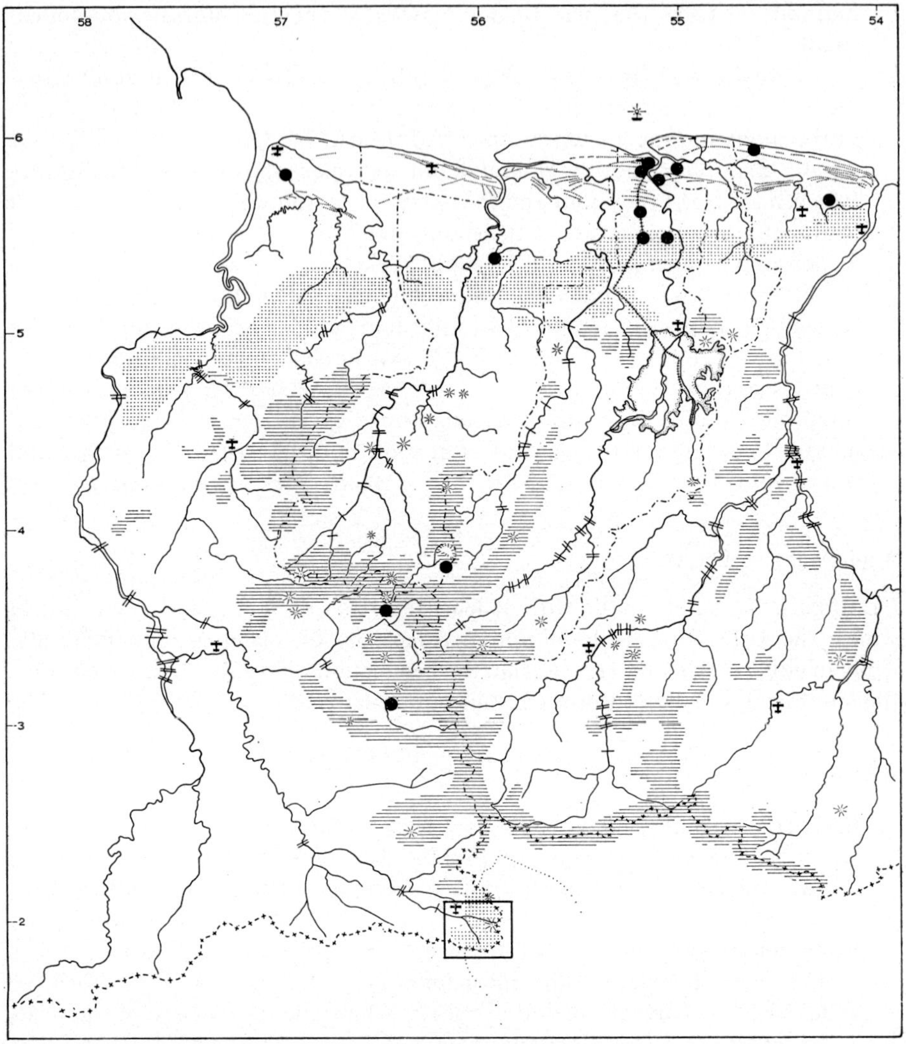

Fig. 11. Distribution of *Anolis a. auratus* Daudin in Surinam. The framed area is represented on a larger scale in fig. 13.

Laemanctus auratus: Fitzinger, 1843: 63.
Dracontura Nitzschi: Fitzinger, 1843: 69 (partly).
Dracontura Bertholdi Fitzinger, 1843: 69.

TYPE-LOCALITY.—Surinam.

MATERIAL. – Surinam.

Surinam: 1♀, RMNH 2799, 1827, leg. H. H. Dieperink; 2♀♀, 1♂, RMNH 2800, leg. H. H. Dieperink; 1♂, RMNH 2997, 1854, leg. Van Es; 1♀, RMNH 11627, leg. H. F. C.

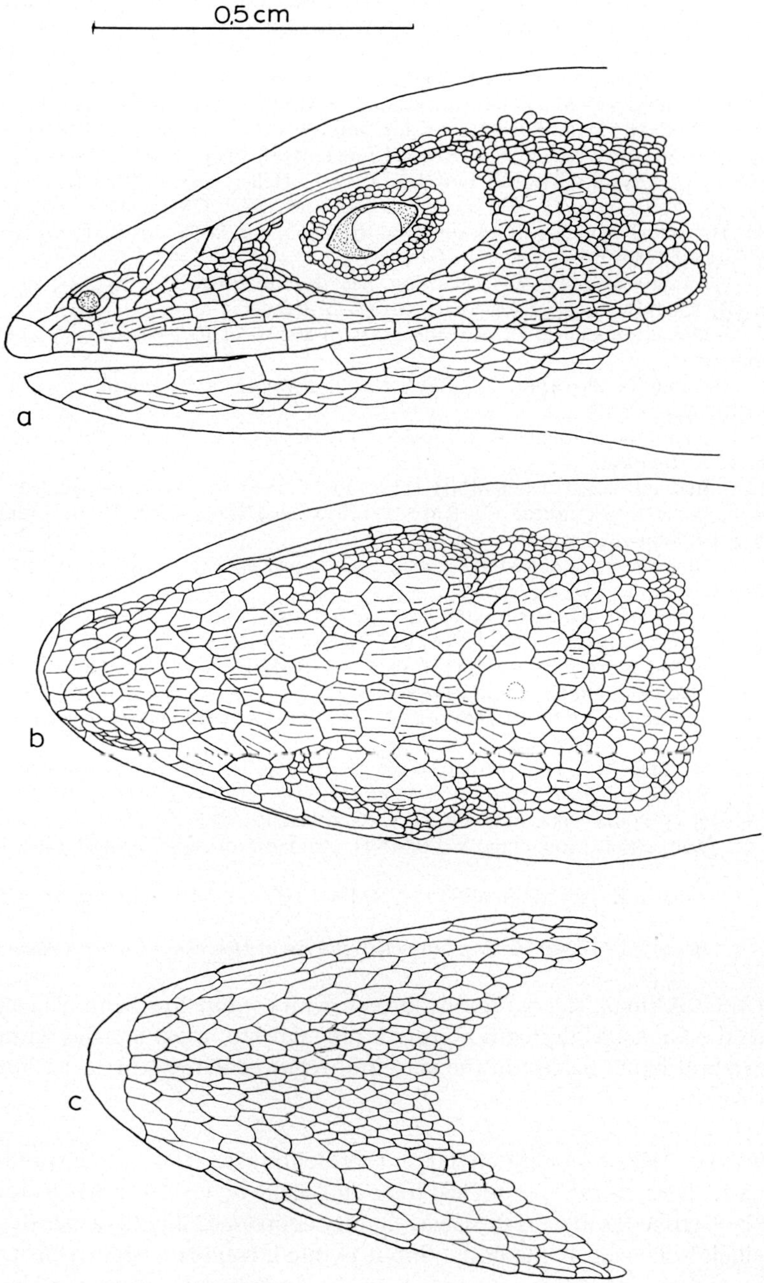

Fig. 12. Anolis a. auratus (Daudin); a, b, c: lateral, dorsal and ventral views of the head of ♂, RMNH 15137, Bigisanti beach.

ten Kate; 1♂, RMNH 16015, leg. G. Witsen Elias; 2 ex., BM 68.11.17.6–7; 1♂, ZMB 4243; 1♂, ZMB 26099, leg. A. O. C. Heller; 1♀, NMW 16448.1.

Paramaribo: 1♂, RMNH 11626, 1911, leg. Jhr. W. C. van Heurn; 1♀, RMNH 15689, leg. W. N. Polder; 1♂, RMNH 15690, 3–V–1968, leg. M. S. Hoogmoed; 1♀, RMNH 15691, 17–VIII–1968, leg. M. S. Hoogmoed; 1♂, RMNH 16014, 30–III–1950, leg. D. C. Geijskes; 1♀, RMNH 16018, August 1911, leg. Jhr. W. C. van Heurn; 1♀, RMNH 16019, 12–XI–1962, leg. G. van Vreden; 2♂♂, RMNH 16020, April 1969, leg. H. M. van Meeuwen; 1 ex., ZMB 25973, 1 ex., ZMB 27104, both leg. A. O. C. Heller; 1 ex., CM 44342, 7–V–1966, 1♀, CM 44346, 27–VI–1966, 1♂, CM 44347, 29–VII–1966, 2♂♂, CM 44349–50, 16–X–1966, all leg. M. H. de la Fuente; 1♂, CM 49528, 4–VII–1968, leg. D. A. Jones; 10 ex., AMNH 8638–40, 8642–48, May 1916, leg. J. A. Samuels.

Nanni Creek: 1♀, RMNH 16012, near Kaaiman Creek, 4–X–1941, leg. D. C. Geijskes; 1♂, RMNH 16494, 18–III–1971, leg. D. G. Reeder.

Plantation Clevia: 1♀, RMNH 15016, 8 km N. of Paramaribo, 7–V–1968, leg. M. S. Hoogmoed.

Plantation Peperpot: 1♂, ZMA 11625, 2 km E. of Paramaribo, 17–XI–1964, leg. T. A. Renssen.

Plantation Marienburg: 1♀, RMNH 15693, 10 km E. of Paramaribo, September 1968, leg. T. A. Renssen.

Bigisanti beach: 1♂, RMNH 15137, 29–IV–1968, leg. M. S. Hoogmoed.

Wane Creek north: 1♀, RMNH 11630, 29–IX–1948, leg. P. H. Creutzberg (Surinam Expedition 1948–49).

Onverwacht: 1♀, RMNH 15694, 27–X–1968, leg. M. S. Hoogmoed; 1♀, CM 49529, 10–VII–1968, leg. D. A. Jones.

Tibiti: 1♂, RMNH 11551, km 1.4, 6–I–1949, 2♀♀, 2♂♂, RMNH 11552–53, 11632–33, km 0.5, 6–I–1949, 1♀, 1♂, RMNH 11554–55, 2nd transect, km 0.5, 14–I–1949, 1♂, RMNH 11556, 2nd transect, km 0.3, 15–I–1949, 1♀, RMNH 11634, 2nd transect, km 0.7, 15–I–1949, all leg. Surinam Expedition 1948–49).

Zanderij: 1♀, RMNH 16011, 13–I–1970, leg. M. S. Hoogmoed & J. J. P. Paats; 1 ex., CM 49527, 4–VII–1968, leg. D. A. Jones.

Powakka: 1♀, 1♂, CM 44343–44, 29–V–1966, leg. M. H. de la Fuente.

Kappel savannah, near Tafel Mountain: 1♂, RMNH 16016, 15–III–1958, 1♂, RMNH 16017, 15/23–III–1958, both leg. D. C. Geijskes.

Wilhelmina Mountains: 1♀, RMNH 16013, south slope, August 1963, leg. S. Ligorie.

Airstrip Kayser Mountains: 2♀♀, RMNH 15692, 1–VIII–1968, leg. M. S. Hoogmoed.

Brazil?: 1♂, ZMB 501, leg. Nitzsch (holotype of *Draconura Nitzschii* Wiegmann).

DIAGNOSIS.—A small, slender anolis, brown with a white line from beneath the eye onto the flanks. Males with a greyish blue dewlap; most females without a wide vertebral band. Scales on the vertebral region enlarged, in 10–11 longitudinal rows.

DESCRIPTION.—Head moderately large, representing 24–26% of the snout-vent length in females, 25–27% in males; long and narrow; 1.7–2.0 times as long as wide; 1.0–1.3 times as long as the tibia; slightly depressed; slightly wider than the neck, which is as wide as the body. Snout pointed, rounded, slightly projecting over the mouth, in profile straight. A slight frontal depression bordered by feeble frontal ridges.

Rostral more or less trapezoid, more than three times as wide as deep, just visible from above; bordered posteriorly by six small scales, the central four bi-

or tricarinate, the outermost two with one keel, and by the first supralabial on each side. Supralabials five to eight, five or six to below the centre of the eye. Loreal region with five to eight horizontal scale rows, decreasing in size from the labials towards the canthus rostralis. Canthal scales three (rarely two or four), the anterior one of which is minute, the posterior two long and slender; the last scale being longest, keeled, forming a conspicuous canthus rostralis. Two rows of suboculars between the supralabials and the minute scales of the lower eyelid, the lower row consisting of very slender scales. No differentiated postoculars, scales on the postocular region decreasing in size from the supralabials towards the temporals and from the eye towards the ear. Temporal region with small, juxtaposed, hexagonal, keeled scales, decreasing in size posteriorly. Ear-opening round to vertically oval, with the longitudinal axis slanting backwards; smaller than the interparietal; behind and above the commissure of the mouth; with a denticulate anterior and a smooth posterior margin; external auditory meatus short, tympanum visible, partly overhung by the ear-margins. Supratemporal arch of scarcely enlarged scales hardly discernable. Scales on the snout large, juxtaposed, hexagonal to polygonal, with two to five keels, in the frontal depression with only one keel. Nostril in a single nasal. Supraciliaries two to seven, mostly five or six, forming a continuous row with the canthal scales, reaching to the posterior corner of the eye; first and second scale long and narrow, keeled; second scale longest, the posterior scales small, slightly enlarged, keeled or smooth; accompanied medially by a second row. Supraorbital semicircles inconspicuous, consisting of keeled scales; a single scale may be enlarged at the point where the distance between both semicircles is smallest; semicircles from both sides separated by one or two rows of small scales. Interparietal large, polygonal, longer than wide, separated from the supraorbital semicircles by one or two rows of small scales. Scales on the parietal region fairly large, juxtaposed, more or less hexagonal, longer than wide, keeled, decreasing in size posteriorly and laterally. Scales on the occipital region juxtaposed, hexagonal, about as long as wide, keeled, smaller than those on the parietal region, larger than those on the nape. Supraocular region with five to eight large scales, forming an oval patch, separated from the supraorbital semicircles by one row of small scales, from the supraciliaries by two or three rows of granules; all supraocular scales distinctly keeled. Eyelids covered with small granules, increasing in size towards the rim, where two rows are neatly arranged. Pupil round.

Two mentals, roughly forming a denticulate crescent, bordered posteriorly by four small scales, two sublabials and two infralabials. Infralabials four to seven, mostly five or six, bordered medially by several rows of sublabials. All scales on the chin keeled, juxtaposed, hexagonal, longer than wide, forming longitudinal rows. Scales on the throat ovate to rhomboidal, keeled, slightly imbricate, smaller than those on the chin in the anterior part, larger in the posterior part. Dewlap in females very small, reaching from below the ears to before the insertion of the forelimbs; in males large, reaching from below the posterior corner of the eye to just posterior of the insertion of the forelimbs; when extended showing eight oblique rows of pointed, rhomboidal, keeled

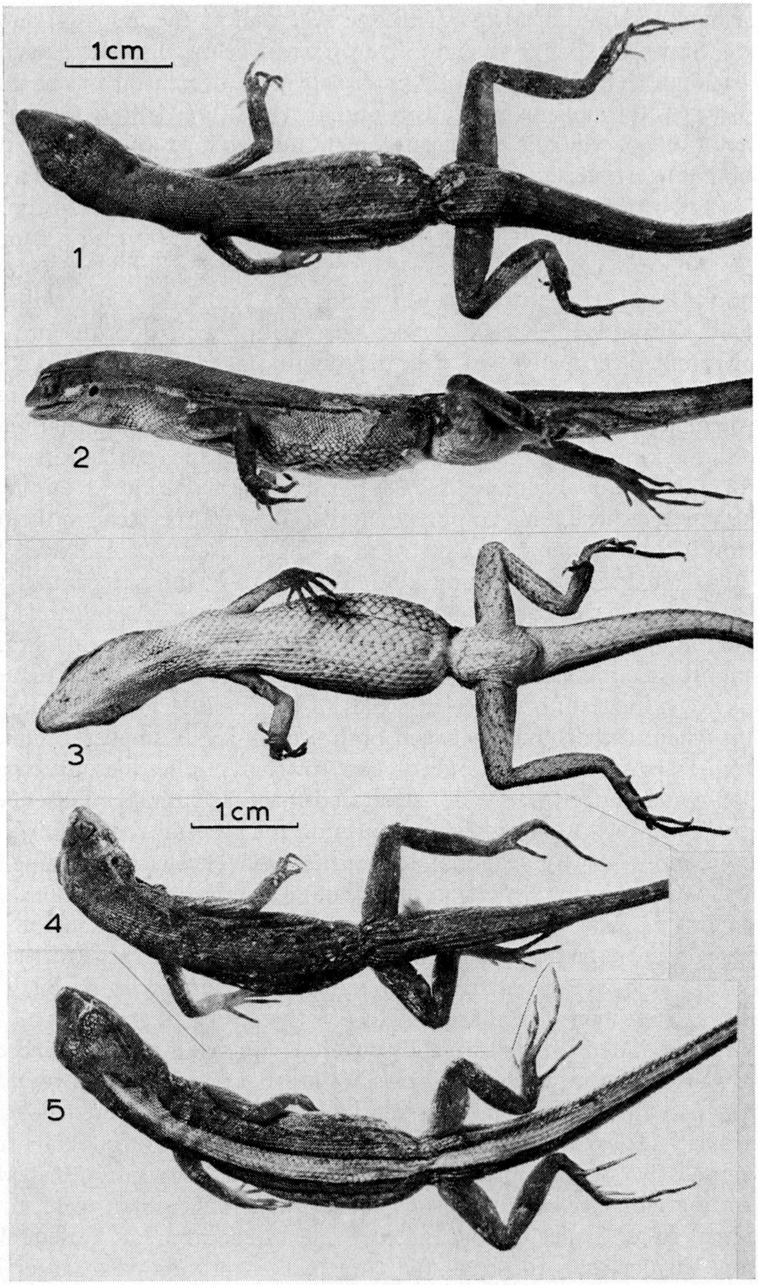

Plate 11. Figs. 1–3. *Anolis a. auratus* Daudin; dorsal, lateral and ventral aspect of ♀, RMNH 16011, Zanderij. Fig. 4. *A. a. sipaliwinensis* nov. subspec.; dorsal aspect of ♂, RMNH 16010a, Sipaliwini. Fig. 5. *A. a. sipaliwinensis* nov. subspec.; dorsal aspect of ♀, RMNH 16000, Sipaliwini.

Table 8. Counts and measurements of *Anolis a. auratus* Daudin. For explanation see table 3.

Reg. no. RMNH	A	B	C	D	E	F	G	H	Q'	J	M	Y	O	P	Q	J'	R'	W	X
11551	♂	49	44(30)	12.2	6.7	6.2			10.2	75	15	10	5-5	5-5	6-6	6-6	3-3	22-?	?
11552	♂	48	73(22)	12.5	6.4	6.1			9.3	72	16	10	5-6	6-6	6-7	6-6	3-3	20-21	31-32
11553	♀	53	137	13.0	6.7	6.1	20	36	10.4	73	18	10	5-5	6-6	2-5	7-6	2-3	19-19	32-32
11554	♂	45	127	12.3	6.6	5.7			9.8	78	19	10	5-6	5-5	6-6	5-5	3-3	22-20	32-35
11555	♀	49	59(33)	?	?	?			?	75	15	10	5-6	6-6	6-6	5-6	3-3	19-18	31-?
11632	♀	48	31(31)	12.2	6.2	5.3	20	35	9.6	69	17	10	5-5	5-5	6-6	6-6	3-3	18-21	33-30
11633	♂	48	71(36)	12.4	6.4	5.7			9.3	81	19	11	5-5	5-5	6-6	6-6	3-3	20-21	33-32
11634	♀	54	28(?)	13.4	7.0	6.1	20	35	10.6	76	17	10	5-5	6-6	5-5	7-6	3-3	19-20	31-31
15016	♀	53	78(38)	12.6	6.7	5.8	20	37	9.8	75	16	11	5-5	7-6	5-6	7-8	3-3	20-20	34-33
15137	♂	42	113	10.4	5.6	5.1	16	30	8.6	77	15	10	5-5	5-6	6-6	7-8	2-3	22-22	?-32
15689	♀	50	127	12.5	7.4	6.3	20	34	10.2	82	13	10	5-6	6-5	3-2	7-7	3-3	18-19	33-33
15690	♂	46	84(35)	11.7	6.3	5.4	21	34	10.2	73	16	10	5-5	7-6	5-4	6-6	3-3	22-24	35-34
15691	♀	54	76(58)	12.8	6.8	6.3	20	34	10.0	86	18	10	6-5	6-6	5-5	7-7	3-3	19-20	35-34
15693	♀	49	77(77)	12.0	6.5	6.2			10.5	69	15	10	5-5	5-5	6-6	6-5	3-3	18-18	31-31
15694	♀	32	85	8.8	5.3	4.6	22	14	7.3	81	14	10	6-6	6-6	5-5	7-6	3-4	21-22	32-35
16011	♀	49	83(76)	12.2	6.6	5.7	18	34	10.4	82	20	10	5-5	4-5	6-6	5-5	3-3	18-17	31-32
16494	♂	45	105	?	?	?			9.7	72	17	10	?	?	?	6-6	3-3	21-21	33-33

scales, smaller than the ventrals; rim scales scarcely smaller than the lateral dewlap scales. Scales on the nape imbricate, rhomboidal, keeled, increasing in size posteriorly, arranged in longitudinal rows; scales on the side of the neck smaller, round, conical, juxtaposed to slightly imbricate. No nuchal fold.

Dorsals distinctly larger than the laterals, slightly smaller than the ventrals; imbricate, rhomboidal anteriorly, hexagonal posteriorly, strongly keeled, in 10–11 longitudinal rows, the keels forming continuous ridges. Laterals small, slightly imbricate, rhomboidal, keeled, slightly enlarged near the dorsals and near the ventrals. Ventrals large, strongly imbricate, rhomboidal to lanceolate, strongly keeled, acuminate, in 13–20 longitudinal rows, the keels forming longitudinal ridges; gradually merging into the laterals. Total number of scales around midbody 69–86. No dorsal fold. Preanal plate with imbricate, keeled, ovate scales, smaller than the ventrals. Enlarged postanal scales present in males, absent in females.

Scales on the tail large, imbricate, rhomboidal, keeled, in longitudinal and oblique series. Scales under the tail also large, imbricate, lanceolate, keeled, in longitudinal series only. Except in the most proximal part, the tail is divided into verticils, three ventral scales corresponding with four dorsal scales. Regenerated part of the tail with irregularly arranged, imbricate, lanceolate, keeled scales, smaller than on the original part of the tail. Breaking points coinciding with the borders between verticils. Tail in cross-section round with ridges, ending in a blunt tip.

Scales on the dorsal and anterior surfaces of the upper arms, on the anterior and lower surfaces of the thighs, on all surfaces of the forearms and lower legs fairly large, imbricate, rhomboidal, keeled, the keels forming ridges; on the ventral and posterior surfaces of the upper arms and on the posterior and dorsal surfaces of the thighs small, granular. Seventeen to 24, mostly 19–22, single, smooth lamellae under the fourth finger; 30–35 under the fourth toe. Palms and soles with small, imbricate, round to rhomboidal, smooth or slightly keeled scales. Fingers and toes subcylindrical.

Colour in life light-brown to golden brown, vertebral band when present beige, lighter than the rest of the dorsum. A white line from beneath the eye, over the ear onto the flanks, where, in its posterior part it may or may not be invaded by dark-brown or black. Belly white, immaculate or with a brown median stripe. Dewlap in males and females greyish blue with white scales. Tongue orange.

Colour in preservative light-brown to beige above. Some females show a wide vertebral band from the occiput onto the tail of a lighter colour than the rest of the back, bordered by black or dark-brown lines. A white lateral stripe from beneath the eye to the groin, but in quite a few specimens it fades on the flanks or is invaded by dark-brown. An indistinct whitish spot may extend backwards from the posterior corner of the eye. The temporal region between the lateral stripe and this spot dark-brown, side of the neck above the lateral stripe slightly darker than the ground-colour. Limbs darker brown than the back or of the same colour, without pattern or with a few lighter, dark-edged ocelli on the back of the thighs. Ventral surface white, immaculate or, rarely in males (three out of eleven), with a vague midventral brown stripe.

A small lizard with a cylindrical to slightly compressed body. Maximum snout-vent length in females 54 mm, in males 49 mm. Limbs moderately long, hind limb 1.6–1.9 times as long as forelimb; when laid along the body there is considerable overlap of forelimb and hind limb, the hind limb reaching to the level of the ear. The undamaged tail in females is 2.5–2.7 times as long as the snout-vent length, in males 2.7–2.8 times. The high number of regenerated tails (13 out of 20 specimens) indicates that the tail of this species is very fragile and that the autotomy mechanism is freely and frequently used.

HABITAT.—This form is restricted to open country where it is present in grass and on low vegetation, never more than 50 cm from the ground. It inhabits the low vegetation on the beaches, the Paramaribo gardens and backyards, as well as the coastal savannah-belt and some isolated small savannahs in the interior. Only one specimen (RMNH 16013) from the Wilhelmina Mountains does not keep to this pattern, as on its label is expressely stated 'in forest'. I am not familiar with this locality, but according to Dr. D. C. GEIJSKES several bare rockslates occur which are partly covered by a type of savannah-forest. Very probably RMNH 16013 was captured in such an area and not in rain-forest and this assumption is strengthened by my never having been able to locate the species in forest, not even in clearings, during my stays in Surinam.

NATURAL HISTORY.—A diurnal lizard which was taken from about 8.00 a.m. to 3.00 p.m., but probably is active until sunset. The species is a heliotherm and frequently was found basking. Several specimens were taken during the evening when they were asleep amongst grass-leaves, clinging to the leaves or to slender twigs. Recently hatched juveniles are not available. Pregnant females, containing two mature eggs, are known from January, May and August, indicating that breeding may occur throughout the year, possibly interrupted by the long dry period.

Five stomachs were examined. They contained items of the following groups: spiders, beetles, crickets, grasshoppers, ants and cicades.

RANGE.—Northern South America, north to Panama: Surinam, Guyana, French Guiana, Venezuela, Colombia, Ecuador, Amazonian Brazil. In Surinam this subspecies is present throughout the country where savannah-type habitats are available, except in the southernmost part where it is replaced by the subspecies described below. It occurs from sea level to at least 200 m (fig. 11).

Anolis auratis sipaliwinensis nov. subspec. (fig. 13; pl. 11; table 9)

MATERIAL.

HOLOTYPE. – 1♂, RMNH 16010a, Sipaliwini, between Base Bivouac and Meyers' farm, ± 8 km E. of airstrip, Surinam, 19–I–1970, leg. M. S. Hoogmoed & J. J. P. Paats.

PARATYPES. – Surinam

Airstrip Sipaliwini, 6 km E.: 1♀, RMNH 15696, 28–VIII–1968, leg. M. S. Hoogmoed; 1♀, RMNH 16006, 18–I–1970, leg. T. A. Renssen.

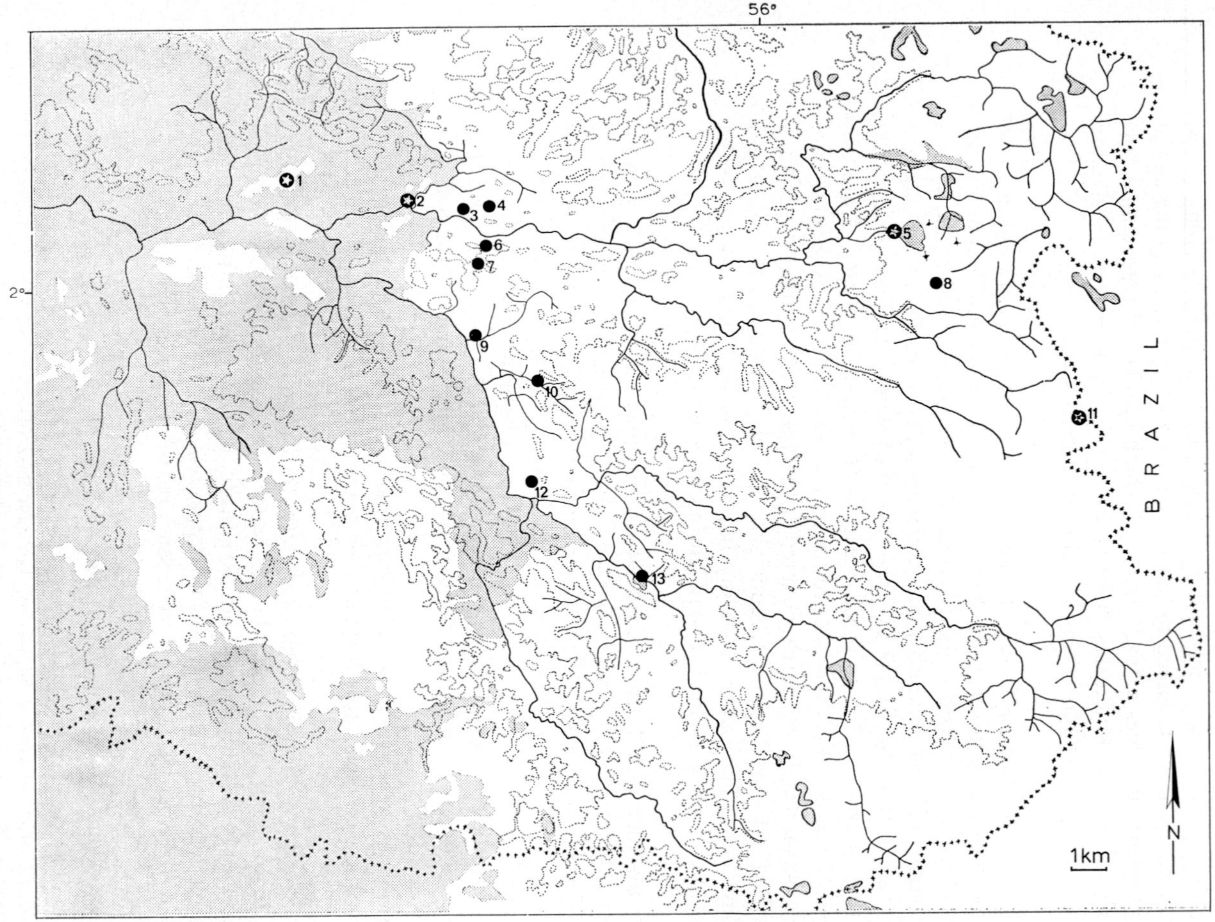

Fig. 13. Detailed map of the Sipaliwini savannah, showing localities where *Anolis auratus sipaliwinensis* nov. subspec. was collected (solid circles). The black-rimmed asterisks indicate important reference points, the grey areas represent forest, the white areas savannah. The dotted line represents the 300 m contour line.

1. airstrip Sipaliwini, 2. Base Bivouac, 3. 2 km SE. of Base Bivouac, 4. 6 km E. of airstrip, 5. Vier Gebroeders Bivouac, 6. between Vier Gebroeders Creek and Meyers' farm, 7. between Base Bivouac and Meyers' farm, 8. Vier Gebroeders Mountain, southern top, 9. Meyers' farm, 10. 2 km SE. of Meyers' farm, 11. airstrip Meyers', 12. between Meyers' farm and South Bi-

Sipaliwini, Meyers' farm: 1♂, RMNH 15698, 31–I–1970, 1♂, RMNH 15699, 6–II–1970, 3♀♀, 2♂♂, RMNH 15700, 22–I–1970, 1♂, RMNH 16001, 30–I–1970, 1♀, RMNH 16002, 31–I–1970, 1♂, RMNH 16003, 2 km SE., 2–II–1970, 1♀, 1♂, RMNH 16004, between Meyers' farm and South Bivouac, 20–I–1970, 1♂, RMNH 16005, between Vier Gebroeders Creek and Meyers' farm, 23–I–1970, 1♀, RMNH 16007, 4–II–1970, 2♂♂, RMNH 16008, 19–I–1970, 1♀, RMNH 16009, 5–II–1970, 1♂, RMNH 16432, 25–I–1970, all leg. M. S. Hoogmoed & J. J. P. Paats.

Sipaliwini, between Base Bivouac and Meyers' farm, ± 6 km E. of airstrip: 1♂, RMNH 15172, 29–VIII–1968, leg. M. S. Hoogmoed; 2♂♂, RMNH 16010b–c, 19–I–1970, leg. M. S. Hoogmoed & J. J. P. Paats.

Sipaliwini, Vier Gebroeders Mountain: 1♂, RMNH 15695, southern top, 1–X–1968, leg. M. S. Hoogmoed.

Sipaliwini, Base Bivouac: 1♂, RMNH 16003, 2 km SE., 27–I–1970, leg. M. S. Hoogmoed & J. J. P. Paats.

Sipaliwini, South Bivouac: 1♀, RMNH 16000, 29–I–1970, leg. F. H. F. Oldenburger.

Diagnosis.—A small, slender anolis, grey with an indistinct white line from the eye to the insertion of the forelimbs. Males with a greenish blue dewlap; females with a very distinct wide vertebral band. Scales on the vertebral region enlarged, in 10–12 longitudinal rows.

Description.—Since in most lepidosis characters this subspecies agrees with *A. a. auratus* Daudin, I refer to the description of the preceeding subspecies. There are, however, some slight differences. The first is that scales in the supraorbital semicircles tend to be slightly larger and in all specimens these supraorbital semicircles are separated only by a single row of scales. Secondly, the number of enlarged dorsal scalerows is more variable, varying from 10–12. Total number of scales around midbody 73–88, that is, in the same range as in *auratus*. The number of supralabials to below the centre of the eye is nearly always five, rarely four or six, so less variable than in *auratus*.

In this subspecies the head is 1.2–1.3 times as long as the tibia, that is, the tibia is relatively shorter than in *auratus*. The snout-vent length in *sipaliwinensis* is slightly less than in *auratus*, females reaching a maximum snout-vent length of 50 mm, males of 48 mm. Also the undamaged tail is slightly shorter, 2.3–2.7 times as long as the snout-vent length in females, 2.0–2.7 times in males.

The most important differences from *auratus* are to be found in the colour pattern. Four colourslides of the holotype, showing dorsal, lateral and ventral aspects and a detail of the extended dewlap are kept in the RMNH together with the specimen. From these slides and from notes in my fieldbook, the following description of the colour pattern of the holotype, while still alive, has been compiled. Dorsal and lateral surfaces of head, flanks, forelimbs and side of tail dark-grey. Back grey with dark-grey chevrons pointing backwards, continued on the anterior part of the tail. Distal part of the tail brownish with dirty white narrow crossbands. Hind limbs dark-grey, marbled with grey. A very indistinct white lateral stripe from beneath the eye to the side of the neck. Ventral parts white, chin, throat and base of tail heavily mottled with grey-brown, a midventral line of the same colour. Dewlap greenish blue.

Table 9. Counts and measurements of *Anolis auratus sipaliwinensis* nov. subspec. For explanation see table 3.

Reg. no. RMNH	A	B	C	D	E	F	G	H	Q'	J	M	Y	O	P	Q	J'	R'	W	X
15172	♂	39	87	11.0	6.2	5.1			8.5	81	16	12	5-5	6-6	6-6	7-7	3-3	20-20	33-31
15696	♀	42	87(87)	11.0	6.3	5.0	16	29	9.1	78	17	11	5-5	6-6	6-6	6-6	3-3	20-19	32-31
15697	♀	47	54(38)	12.3	6.5	5.4			10.1	74	14	10	5-5	6-6	5-5	6-6	2-3	21-?	?
15698	♂	45	118	11.4	6.0	5.3	18	31	9.1	78	15	12	5-5	6-7	6-6	6-6	3-3	19-20	31-28
15699	♂	44	120	11.9	6.4	5.5	18	32	9.5	88	16	10	5-5	6-6	5-5	5-5	2-3	20-21	32-31
15700a	♀	45	115	12.2	7.0	5.5	18	33	9.4	77	17	10	5-5	5-5	6-6	6-6	2-3	19-21	31-31
b	♂	43	114	11.5	6.2	5.5	19	32	10.0	78	15	10	5-5	6-6	6-6	6-5	3-3	21-21	33-33
c	♂	48	50(42)	12.7	7.3	6.2	19	34	10.3	81	16	12	5-5	6-6	6-5	6-6	3-3	20-21	33-32
d	♀	45	120	12.5	6.3	5.5	19	33	9.5	80	16	10	5-5	6-5	7-7	6-6	3-3	20-20	31-32
e	♀	46	117	12.1	6.4	5.5	19	33	10.1	76	16	12	5-6	5-6	6-6	6-6	2-3	21-20	32-32
16000	♀	46	52(28)	11.8	6.2	5.3	18	32	9.2	70	16	12	5-5	6-6	6-5	6-5	2-2	19-19	31-32
16001	♂	45	98(98)	12.1	6.7	5.6	18	32	9.5	79	17	10	5-5	5-5	6-6	6-6	2-2	20-19	31-31
16002	♀	45	69(69)	12.0	6.2	5.4			9.5	76	16	12	5-6	6-6	6-6	6-6	3-3	19-21	31-33
16003	♂	43	98	11.3	6.3	5.3			9.3	73	14	10	5-5	5-5	5-6	6-5	3-3	22-19	?-30
16004a	♀	46	41(41)	12.0	6.3	5.4			9.4	75	17	11	5-5	8-7	6-7	6-6	3-3	19-20	32-?
b	♂	44	59(36)	11.8	6.3	5.3			9.4	78	15	10	5-5	6-5	6-7	6-6	3-3	23-21	36-34
16005	♂	44	27(27)	11.6	6.5	5.4			9.3	83	17	12	5-5	6-6	5-5	6-6	3-3	20-21	33-33
16006	♀	45	99(99)	12.4	6.7	5.5			9.8	77	16	11	5-5	6-6	5-5	6-6	2-2	20-20	30-29
16007	♀	50	117	12.7	7.1	5.9			9.8	76	16	12	5-5	5-5	5-6	6-6	3-3	20-20	31-32
16008a	♂	43	60(60)	11.7	6.4	5.3			9.3	77	19	10	5-5	6-6	4-5	5-6	3-3	21-20	32-32
b	♂	43	117(117)	11.6	6.3	5.4			9.1	78	18	10	4-4	6-7	6-6	5-6	3-3	21-21	31-32
16009	♀	48	20(20)	12.1	7.1	5.7			9.4	76	18	11	5-5	5-5	4-4	6-5	3-3	20-20	?-30
* 16010a	♂	44	101	11.5	6.4	5.3	17	36	8.9	75	19	11	5-6	5-5	5-5	6-6	3-3	20-18	29-30
b	♂	43	88	11.3	6.3	5.2			8.8	79	16	11	5-5	5-6	6-6	6-5	3-3	21-19	33-32
c	♂	46	91(71)	12.2	6.4	5.4			9.2	79	19	10	5-6	6-5	6-4	6-6	2-2	21-20	29-30

In preservative the chevrons are indistinct. All male paratypes agree well with the description of the holotype, the only differences present being caused by variation of colour, which may range from greyish brown to grey. Female paratypes, however, have a pattern very much different from that of the males. In females the dorsal and lateral surfaces of the head, lateral parts of the back and side of tail are dark grey-brown, limbs and flanks light grey-brown, flanks with or without a longitudinal dark-grey stripe. A very distinct reddish brown to beige vertebral band, accompanied by a narrow black line on each side, from occiput onto the tail where it disappears and is continued as a series of spots on the distal part. Both in males and females the lateral white stripe from beneath the eye does not reach beyond the insertion of the forelimbs, furthermore in most specimens it is inconspicuous. Ventral parts white, in many specimens with greyish brown midventral and lateral stripes. Throat with or without greyish brown spots.

HABITAT.—These lizards inhabit extensive savannahs, where they occur on the ground and on rocks. They range from the wet creek-valleys to the dry hill--tops, although there seems to be a preference for the higher country. On hill--tops they live among large boulders; on level pieces of savannah they live in the neighbourhood of burrows of larger animals, in which they take refuge. They were also present on the earthen walls of Meyers' farm.

NATURAL HISTORY.—A diurnal lizard, taken between 7.30 a.m. and 5.00 p.m., when it was frequently observed basking, both on the bare ground between *Bulbostylis* and on rocks. During basking the anterior part of the body is highly elevated from the substratum by stretching the forelimbs. Recently hatched juveniles have not been found. During my visit to Sipaliwini in 1970 (January/February) several pregnant females, each containing one mature egg, were collected. Others did not contain eggs and they may have deposited them already. Only four stomachs were examined. In two stomachs the bulk was taken up by winged ants and termites, the other two contained a beetle, ants, cicadas and caterpillars.

This subspecies is capable of remarkable colourchanges. One specimen (RMNH 16008), which was found in the hollow interior of a partly burnt *Mauritia*-palmstem lying on the ground, was completely black without any pattern discernible. After capture the specimen became lighter and gradually changed to greyish brown, at the same time the pattern becoming more distinct.

RANGE.—Hitherto only recorded from the Sipaliwini savannah in extreme southern Surinam. However, I am convinced that this subspecies inhabits the Brazilian part of this same savannah (Paru savannah) too, because the division in Sipaliwini savannah and Paru savannah along the watershed is artificial, the watershed being so low that it does not constitute a barrier to landdwelling forms. It is present from 250 m to at least 554 m (the top of Vier Gebroeders Mountain) (fig. 13).

REMARKS.—Because the Sipaliwini savannah regularly burns, at least part of the year the ground is covered with greyish ash and the short stems of *Bulbostylis* are black. The colour difference with the nominate form may be an adaptation to this habitat. Also the surface of the rocks is frequently blackish, but this is due to weathering.

Anolis chrysolepis Duméril & Bibron

DIAGNOSIS.—A small ground-dwelling anolis with short, quadrangular head and moderately well developed digital expansions. At least two rows of scales on the vertebral region enlarged. In adult specimens the tibia is longer than the distance between the tip of the snout and the posterior margin of the ear. Infralabials small, usually 13–15.

Key to the subspecies of *Anolis chrysolepis*

1. Scales on the snout small, multicarinate; more than two rows of enlarged vertebral scales; supraorbital semicircles inconspicuous; scales on the upper arm as large as the vertebral scales; dewlap in life blue in males, yellow in females . . . . *A. c. chrysolepis*, p. 112
   Scales on the snout large, with a single keel; only two rows of enlarged vertebral scales; supraorbital semicircles distinct; scales on the upper arm larger than the vertebral scales; dewlap in life scarlet in both sexes . . . . *A. c. planiceps*, p. 125

Anolis chrysolepis chrysolepis Duméril & Bibron (figs. 14, 15; pls. 12, 13)

Anolis chrysolepis Duméril & Bibron, 1837: 94; Bocourt, 1873: 99; Bocourt, 1874: pl. XVI fig. 26; Goeldi, 1902: 529; Van Lidth de Jeude, 1917c: 46; Barbour, 1934: 127; J. A. Peters & Donoso-Barros, 1970: 51.
Draconura chrysolepis: Berthold, 1840: 899; Berthold, 1842: 20; Berthold, 1843: 65.
Dracontura Nitzschii Fitzinger, 1843: 69 (partly).
Anolis lentiginosus O'Shaughnessy, 1875: 279; Boulenger, 1885b: 93; Van Lidth de Jeude, 1904: 88; Van Lidth de Jeude, 1917c: 46; Burt & Burt, 1933: 17; Barbour 1934: 138; D'Audretsch 1967c: 27 (originally cited with questionmark).
Anolis longicrus Roux, 1907: 763; Burt & Burt, 1933: 18; Barbour, 1934: 141.
'55–61. Grondbewoners' Fernandes, 1927: 249 (partly).
Anolis chrysolepis chrysolepis: Vanzolini & Williams 1970: 85.

TYPE-LOCALITY.—Mana, French Guiana.

MATERIAL. – Surinam.

Surinam: 1♀, RMNH 7585, leg. G. M. Versteeg (Tapanahoni Expedition); 1♀, ZMA 10099, 1890, leg. Brussel; 1♀, 1♂, ZMA no number, 1922, leg. J. W. Gonggrijp & G.

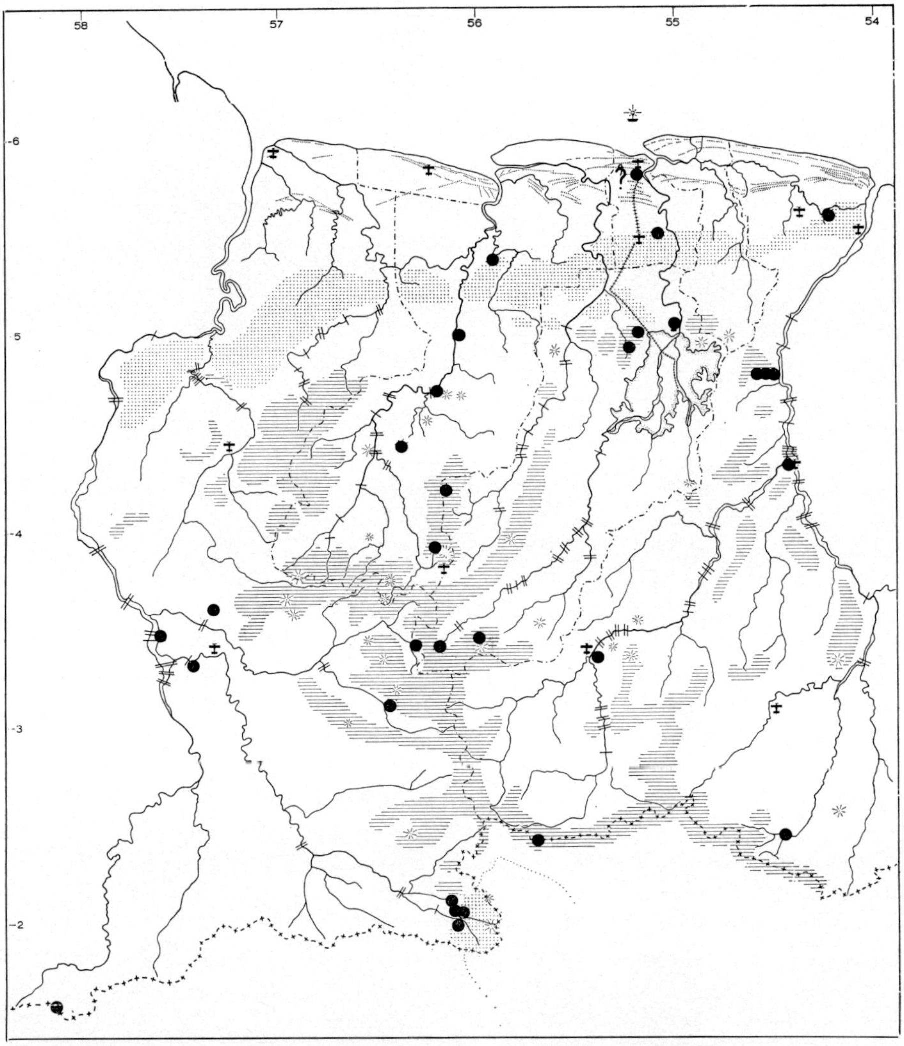

Fig. 14. Distribution of *Anolis c. chrysolepis* Duméril & Bibron in Surinam.

Stahel; 1♂, BM 66.8.14.303, leg. R. Damon; 1♀, 1 juv., BM 68.11.17.8 & 11, leg. R. Damon (locality with questionmark); 1♂, BM 1946.8.8.31, leg. A. Kappler (Holotype of *Anolis lentiginosus* O'Shaughnessy); 1 juv., ZMB 27134, leg. A. O. C. Heller; 3♀♀, 3♂♂, NMW 16822; 1♀, MCZ 65359, leg. Brueghel.

Paramaribo: 2♂♂, ZMB 25952, leg. A. O. C. Heller.

Tibiti: 1♀, RMNH 11638, 2nd transect, km 3.4, 12–I–1949, 1♂, RMNH 11639, 3rd transect, km 2.1, 16–I–1949, both leg. Surinam Expedition 1948–49.

Between Paranam and Kraka: 1♀, RMNH 16476, 15–VIII–1966, leg. Jhr. A. A. van Heurn.

Moengo Tapoe transect: 2♂♂, RMNH 11636, km 5 and 14.6, 21/22–X–1948, 1♀, RMNH 11637, km 3.4, 23–IX–1948, 1♂, RMNH 11665, km 4, 24–IX–1948, all leg. P. H. Creutzberg (Surinam Expedition 1948–49).

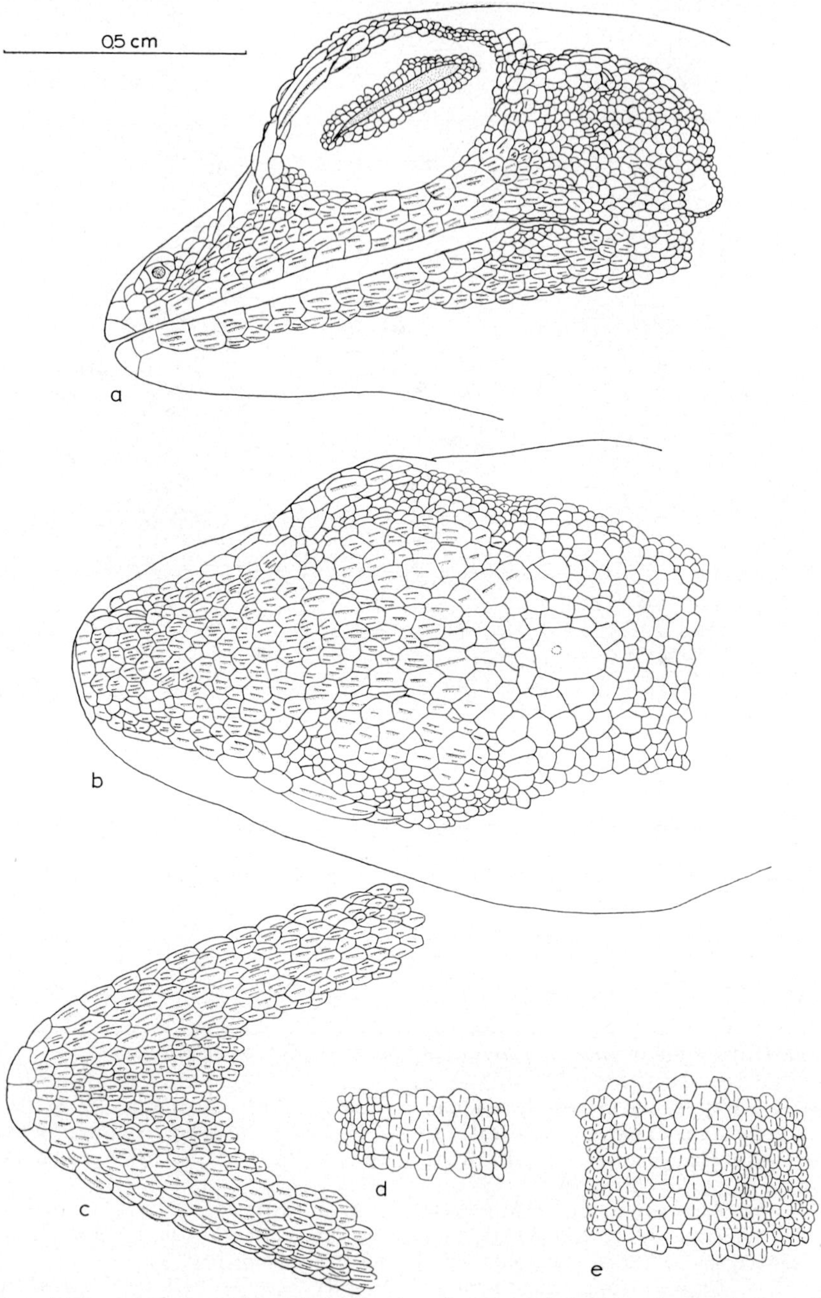

Fig. 15. Anolis c. chrysolepis Duméril & Bibron; a, b, c: lateral, dorsal and ventral views of the head of ♂, RMNH 15251, Airstrip Kayser Mountains; d: middorsal scales of RMNH 15251; e: middorsal scales of ♀, RMNH 15252, Sipaliwini.

Camp 1, Coppename River: 1 juv., BM 1946.4.2.81, 27–VI–1938, leg. I. T. Sanderson.

Brown's Mountain: 1♀, 1♂, RMNH 15680–81, 11–XI–1968, leg. R. Hoogmoed-Verschoor; 1 juv., ♂, RMNH 15682, foot, 13–VII–1968, 1♀, 1 juv. ♂, RMNH 15706–07, 12–XI–1968, both leg. M. S. Hoogmoed; 1♀, RMNH 15724, 25–VIII–1968, leg. W. N. Polder; 1♀, AMNH 108762, 15/16–II–1972, leg. C. W. Myers & J. Daly.

Brokopondo: 1♀, RMNH 11763, 18–VIII–1965, leg. G. F. Mees.

Nassau Mountains: 1♂, RMNH 11640, km 0.5–1, 15–II–1949, 1♂, RMNH 11641, Base Encampment, 16–II–1949, 1 juv. ♂, RMNH 11642, km 0.7, 18–II–1949, 1♂, RMNH 11643, km 0–5, 18–II–1949, 1♂, RMNH 11644, km 5–6, 25–II–1949, 1♀, RMNH 11645, km 2.2, 25–II–1949, 1♀, RMNH 11646, km 1.6, 27–II–1949, 1♂, RMNH 11647, km 5.4, 8–III–1949, 1 juv. ♀, RMNH 11648, S. of km 3.6, 8–III–1949, 1♀, RMNH 11649, S. of km 3.6, 11–III–1949, 1 juv. ♂, RMNH 11650, km 8.6, 12–III–1949, 1♂, RMNH 11651, km 10.5, 12–III–1949, 1 ex., RMNH 11652, km 9, 14–III–1949, 1♀, RMNH 11653, km 13.5, 14–III–1949, 1 juv. ♂, RMNH 11654, km 12.8, 16–III–1949, 1♀, 1♂, RMNH 11655, km 11, 17–III–1949, 1 juv. ♂, RMNH 11656, km 16.2, 22–III–1949, 1♀, 1♂, RMNH 11657, km 16–17, 22/23–III–1949, 1♂, RMNH 11658, km 14.1, 24–III–1949, 1♂, RMNH 11659, N. of km 11.2, 24–III–1949, 2♀♀, RMNH 11661, N. of km 11.2, 17–III–1949, 1♀, RMNH 11662, km 13, 15–III–1949, all leg. Surinam Expedition 1948–49.

Raleigh Cataracts, Coppename River: 1♀, 1♂, RMNH 11765, 8/16–VII–1962, leg. P. H. van Doesburg; 1♂, AMNH 108764, 24–II–1972, leg. J. Daly.

Voltz Mountain reserve: 1♂, RMNH 15679, 5–XI–1968, leg. M. S. Hoogmoed; 1♀, AMNH 108763, trail between Voltz Mountain and Raleigh Cataracts, 23–II–1972, leg. C. W. Myers & J. Daly.

Kroetoe Mountain: 1♂, RMNH 11660, 24–XI–1943, leg. D. C. Geijskes.

Poeloegoedoe: 1♂, RMNH 11663, 30–VIII–1939, leg. D. C. Geijskes.

Hendrik Mountain: 1♂, ZMA no number, 11–III–1922, leg. J. W. Gonggrijp & G. Stahel.

Tafel Mountain: 1♀, RMNH 11664, foot, 5–XI–1943, leg. D. C. Geijskes; 1♀, RMNH 15688, km 26, September 1944, leg. L. Schmidt.

Wilhelmina Mountains: 1 juv., 1♀, 1♂, 1 ex., ZMA 10091–93, 10098, 1926, 1♀, MCZ 65360, 1926, both leg. Wilhelmina Mountains Expedition.

King Frederik Willem IV Cataracts, Corantijn River: 1♀, 1♂, BM 1936.9.3.34–35, leg. C. A. Hudson.

Gonini Encampment, Coeroeni River: 1 juv. ♂, RMNH 15683, 26–VI–1968, leg. M. S. Hoogmoed.

Lucie Camp, Lucie River: 1♂, RMNH 11666, 8–X–1910, leg. J. F. Hulk (Corantijn Expedition).

Upper Gran Rio: 1♀, RMNH 4857, 7–IX–1910, leg. J. F. Hulk (Corantijn Expedition).

Bivouac near Franssen Herderschee Peak: 1♂, ZMA 10090, 13–IX–1908, leg. J. H. A. T. Tresling (Suriname Expedition 1908).

Airstrip Paloemeu: 1 juv. ♀, RMNH 15684, S., 23–V–1968, leg. M. S. Hoogmoed; 1♀, RMNH 15812, May 1969, leg. H. M. van Meeuwen.

Airstrip Kayser Mountains: 1♂, RMNH 15251, NE., 2–VIII–1968, 1♂, RMNH 15666, SW., 5–VIII–1968, 2♀♀, 1♂, RMNH 15667–69, NE., 3–VIII–1968, 1♀, RMNH 15670, NE., 9–VIII–1968, 1 juv. ♂, RMNH 15671, SE., 7–VIII–1968, 2♂♂, RMNH 15672, E., 8–VIII–1968, 2♀♀, RMNH 15673, NE., 2–VIII–1968, 2♀♀, RMNH 15674, SW., 5–VIII–1968, 1♂, RMNH 15675, SE., 1–VIII–1968, all leg. M. S. Hoogmoed; 2♀♀, FMNH 128983–84, 1–II–1960, leg. H. A. Beatty.

Boundary Camp, Itabu Creek Head: 1♀, 1♂, FMNH 30941–42, September/October 1938, leg. E. R. Blake.

Airstrip Sipaliwini: 1♂, RMNH 11764, February 1961, leg. D. C. Geijskes; 1♂, RMNH 15677, 3 km E., 7–IX–1968, leg. M. S. Hoogmoed; 1♂, RMNH 15685, 3 km E., 16–I–1970, 1♂, RMNH 15686, 6 km NW., 9–II–1970, both leg. M. S. Hoogmoed & J. J. P. Paats;

1♂, RMNH 15687, between airstrip and Meyers' farm, 9–II–1970, leg. J. P. Schulz; 1♂, RMNH 16440, 3 km S., 11–II–1970, 1♀, 1 juv. RMNH 16477, 3 km NW., 10–II–1970, both leg. M. S. Hoogmoed & J. J. P. Paats.

Sipaliwini, Base Bivouac: 1♀, RMNH 15252, 2 km N., 4–IX–1968, 1♂, 1 juv. ♂, RMNH 15676, 3.5 km N., 4–IX–1968, 2 juv. ♀♀, RMNH 15678, 0–5 km N., 2–IX–1968, all leg. M. S. Hoogmoed.

Upper Paloemeu, Boundary Mountains, 1st forest bivouac: 1♀, RMNH 11762, 11–IV–1952, leg. D. C. Geijskes.

Litani River: 1♂, RMNH 4470, 7–XII–1903, leg. Gonini Expedition.

Surinam/Guyana.

New River, 750 feet: 1 juv., BM 1939.1.1.69, leg. C. A. Hudson.

Guyana.

Guyana: 1♀, 1♂, RMNH 2811.

Oko River: 1♂, FMNH 35123.

Onora: 1♂, MCZ 65350, leg. R. Snedigar.

Shudikar-wau: 2♀♀, MCZ 65348–49, leg. R. Snedigar.

French Guiana.

Guyane: 1♀, MP 2439, leg. E. Rousseau (paralectotype of *Anolis chrysolepis* Duméril & Bibron).

Cayenne: 1♂, RMNH 2828, 1835, leg. Paris Museum.

Mana: 1♂, MP 2436, leg. Leschenault & Doumerc (lectotype of *Anolis chrysolepis* Duméril & Bibron).

Lower Matarony River: 1♂, 1 ex., MCZ 110388–89, 18–VIII–1968, leg. P. H. Silverstone.

Sophie: 1♂, MCZ 77548, 8–VIII–1963, leg. T. Monath.

DIAGNOSIS.—A small anolis with more than two rows of enlarged dorsals; supraorbital semicircles inconspicuous; scales on the upper arm as large as the vertebral scales; scales on the snout subequal, multicarinate; dewlap in life blue in males, yellow in females.

DESCRIPTION.—Head moderately large to large, representing 27–30% of the snout-vent length in females, 25–27% in males and 26–30% in juveniles; short, about 1.5 times as long as wide; shorter than the tibia (except in juveniles); slightly depressed; wider than the neck, which is as wide as the body. Snout slightly pointed, rounded; not projecting over the mouth. Frontal region concave, frontal ridges absent or slightly indicated. Occipital region slightly bulbous.

Rostral nearly rectangular, two to three times as wide as deep; barely visible from above; bordered posteriorly by four to six small scales and by the first supralabial on each side. Ten to 16, mostly 12–14 supralabials; eight to 13, mostly ten to 12, to below the centre of the eye. Loreal region with five to seven horizontal rows of keeled scales; the upper row not different from the rest. Canthal scales mostly five, rarely four or six, of which the posterior three are enlarged, sharply keeled, forming a conspicuous canthus rostralis. Two rows of enlarged, keeled suboculars between the supralabials and the eyelid, containing a variable number of scales. Postocular scales not enlarged, indistinguishable from the small, granular temporal scales. Ear-opening small; vertically oval or subtriangular, with one tip pointing posteriorly; smaller or larger than

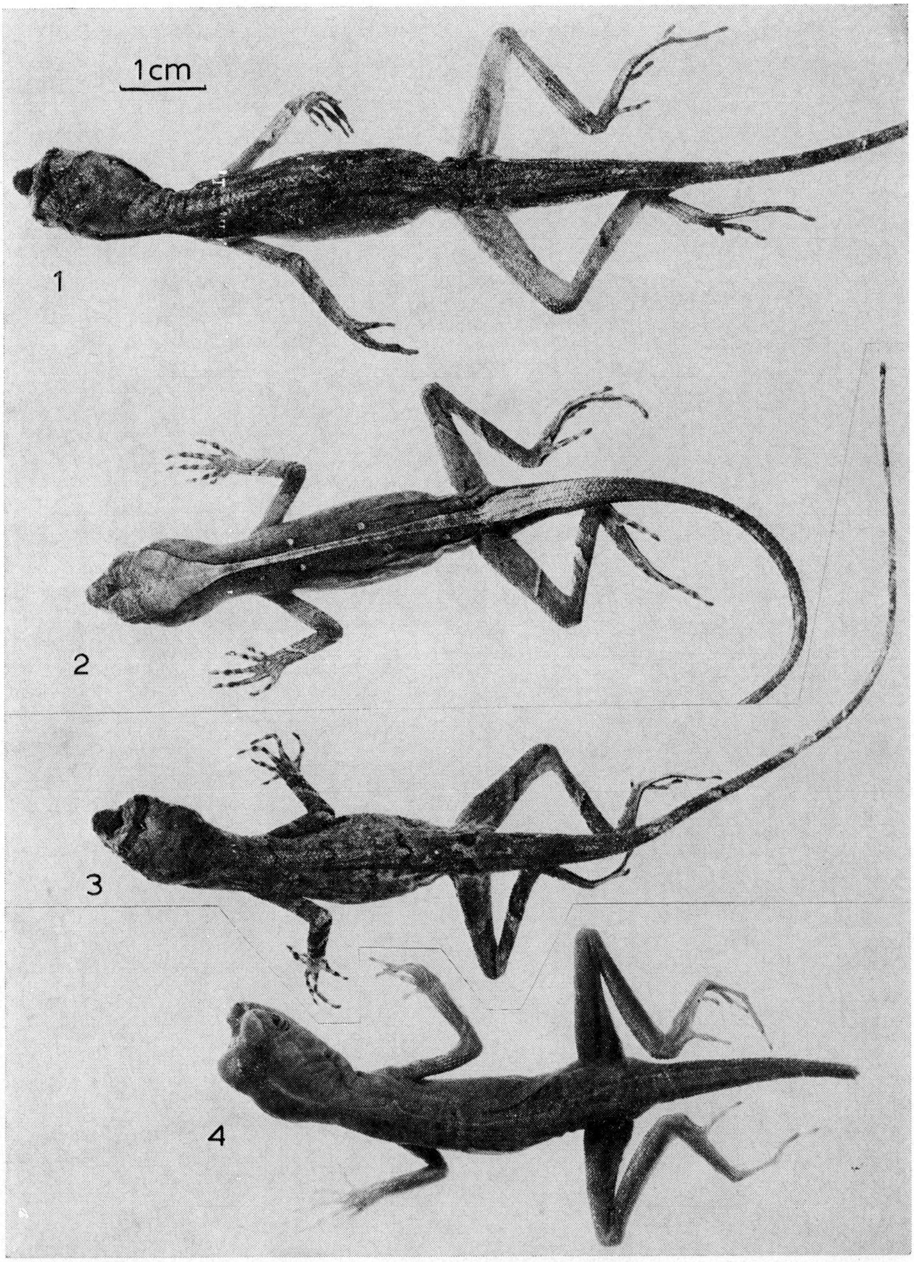

Plate 12. Figs. 1–3. *Anolis c. chrysolepis* Duméril & Bibron. Fig. 1. Dorsal aspect of ♂, RMNH 15687, Sipaliwini. Fig. 2. Dorsal aspect of ♀, RMNH 15670, Airstrip Kayser Mountains. Fig. 3. Dorsal aspect of ♂, RMNH 15251, Airstrip Kayser Mountains. Fig. 4. *Anolis c. planiceps* Troschel; dorsal aspect of ♀, RMNH 2812, Surinam.

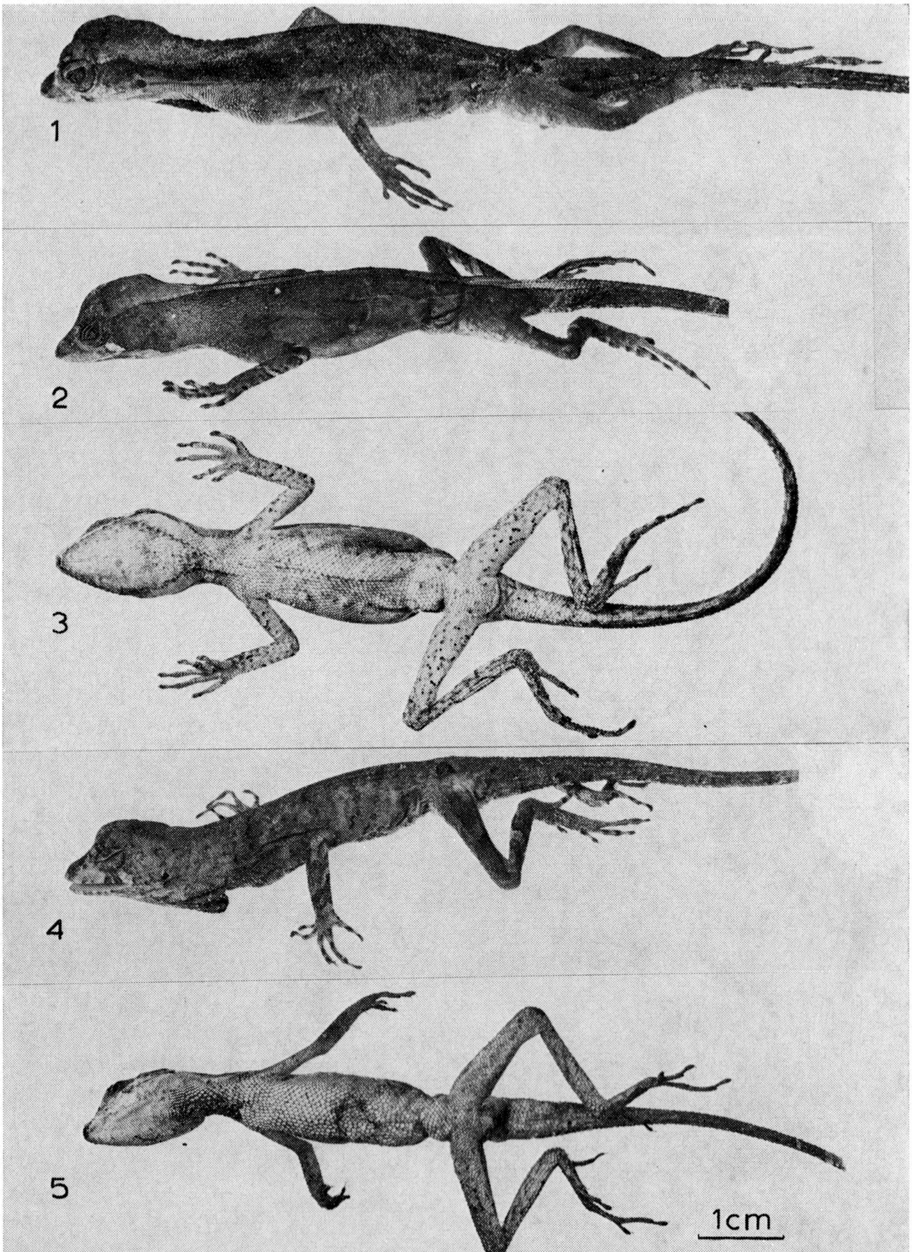

Plate 13. Anolis c. chrysolepis Duméril & Bibron. Fig. 1. lateral aspect of ♂, RMNH 15687, Sipaliwini. Figs. 2–3. Lateral and ventral aspect of ♀, RMNH 15670, Airstrip Kayser Mountains. Fig. 4–5. Lateral and ventral aspect of ♂, RMNH 15251, Airstrip Kayser Mountains.

or equal to the interparietal; behind and slightly above the commissure of the mouth; surrounded by small scales forming a smooth margin, or, in the Itabu Creek and Sipaliwini specimens, some scales on the anterior border slightly enlarged and projecting, forming a denticulate anterior margin; external auditory meatus short, tympanum clearly visible. A supratemporal arch of enlarged scales is absent, though its position is indicated by the arrangement of the temporal scales. Scales on the snout small, mostly hexagonal, with three longitudinal keels. Nostril in a single nasal. Supraciliaries two to six, mostly three to five, keeled; forming a continuous row with the canthal scales; reaching to beyond the centre of the eye; second scale longest; medially the supraciliaries are accompanied by a row of enlarged, elongate scales. Supraorbital semicircles inconspicuous, consisting of slightly enlarged scales in the Itabu Creek and Sipaliwini specimens; in contact or separated by one to three scales; the anterior scales of the semicircles keeled, the posterior ones smooth. Interparietal variable in size and shape; separated from the supraorbital semicircles by two to four scales. Scales on the parietal and occipital regions very large, irregularly shaped, smooth, decreasing in size posteriorly, larger than those on the nape. Supraorbital region in males with a patch of slightly enlarged multicarinate scales, separated from the supraorbital semicircles by one row of smaller scales, separated from the supraciliaries by several rows of granules; in females the scales of this region are as large as the scales on the snout, with a single keel. Eyelids covered with minute granules, not enlarged near the rim. Pupil round.

Two small mentals, roughly forming a semicircle; bordered posteriorly by five or six small scales and two infralabials. Infralabials 11–16, mostly 13–15. Sublabials absent. Scales on the chin hexagonal, elongate; decreasing in size from the infralabials towards the median; near the infralabials with two or three, near the median with one longitudinal keel. Scales on the throat rounded, keeled, smaller than those on the chin. Dewlap both in males and females small, extending from below the ears to in front of the forelimbs, when extended showing four or five oblique rows of rounded, keeled scales, which are slightly smaller than the ventrals; rim scales smaller but larger than the scales on the throat. Scales on the nape small, juxtaposed, polygonal, keeled, larger towards the median, smaller than the dorsals. Scales on the side of the neck smaller than those on the nape, juxtaposed, round. No nuchal fold.

Dorsals much larger than the laterals, slightly smaller than the ventrals, forming a wide longitudinal band in which the scales decrease in size from the median laterally; dorsals juxtaposed, polygonal, keeled. In males the dorsals are larger than in females. Laterals granular, small, juxtaposed, round, gradually merging into the ventrals. Ventrals imbricate, rhomboidal with a rounded posterior margin; large, increasing in size towards the median; keeled, the keels forming indistinct longitudinal ridges. Total number of scales around midbody 118–154 in females, 104–138 in males. No dorsal fold.

Preanal plate with small, round, juxtaposed granules, much smaller than the ventrals. Enlarged postanal scales absent. Everted hemipenis bifurcate, toadstool-shaped with a short stem and a hood divided into two lateral bulbs with reticulate surface and a triangular, wrinkled flap projecting anteriorly. Sulcus spermaticus on the posterior surface of the stem to the saddle between the two bulbs where it bifurcates, sending a branch to the tip of each bulb along the posterior surface.

Scales on the tail large, hexagonal, keeled, the keels forming distinct longitudinal ridges; juxtaposed on the base of the tail, slightly imbricate on the rest of the tail. Scales under the tail identical in shape, imbricate. Tail not divided into verticils. In none of the examined specimens does the tail show a regenerated part. Tail in cross-section round with ridges, ending in a knob with five ridges.

Scales on the anterior and upper surfaces of the forelimbs, on the anterior and lower surfaces of the thighs and on all surfaces of the lower legs large, rhomboidal, juxtaposed, keeled, the keels forming ridges; on the posterior and lower surfaces of the forelimbs and on the posterior and upper surfaces of the thighs small, granular. Eighteen to 27, mostly 20–23, single lamellae under the fourth finger, 10–17, mostly 12–14, smooth ones to the end of the digital expansion, scales under the ultimate phalanx multicarinate; 27–35, mostly 29–33, single lamellae under the fourth toe, 18–25, mostly 20–24, smooth ones to the end of the digital expansion, scales under the ultimate phalanx multicarinate. Palms and soles with round, flat to keeled, juxtaposed scales, smaller than the ventrals; scales on the upper surface of hands and feet identical to those on the upper surface of the lower forelimbs, respectively on the lower legs. Digital expansions on fingers and toes moderately well developed, i.e., the third phalanx is slightly wider than the basal one. Proximal part of fingers and toes depressed, distal part compressed.

In colour and pattern there is a striking sexual dimorphism. In life females have a golden brown back with a more or less distinct lighter vertebral line, which may be greyish brown or even silvery grey. This vertebral line starts at the tip of the snout, from there it first widens and then narrows to form a rhomboid marking on the upper surface of the head. Just anteriorly to the forelimbs it becomes a narrow line, which continues to the sacral region, where it widens again and occupies the entire dorsal surface of the base of the tail. Between forelimbs and hind limbs there are some short side-branches or a few isolated light spots accompanying it. Limbs with creamy cross-bars. Belly orange-yellow. Dewlap lemon yellow with dirty white and brown scales. Under the eye there is a creamy spot which is more distinct in juveniles than in adult females. One female (RMNH 15669) shows the first type of pattern described for males. Males have a wide grey vertebral band (with or without a scattering of red scales) from the tip of the snout to the base of the tail, which is interrupted just behind the tip of the snout by a dark-brown patch, and, between the eyes, by a dark-brown cross-band. A number of dark-brown chevrons, with the tips pointing posteriorly, may or may not be present between the limbs. On the back between the hind limbs there are two semicircles or triangles which do or do not touch. Belly white to light-brown. Dewlap cobalt blue with dirty white and brown scales. Under the eye a white spot may or may not be present. In another type of pattern, which is rare in most localities, but is fairly common in the Nassau Mountains, the grey vertebral zone is bordered laterally by wide dark-brown bands, which start at the posterior corner of the eye and fade between the forelimbs and hind limbs. Between the limbs these bands may encroach upon the grey band and may even fuse across the back, thus interrupting the grey vertebral band. In both male types the flanks are greyish to light-brown. Tongue orange. Iris brown to golden.

In preservative there is not much change of colour. In females the vertebral line becomes light- to dark-grey, the belly becomes whitish pale brown. The vertebral line is bordered laterally by a narrow dark-brown line. There may or may not be a narrow dark-brown line from the corner of the mouth to the base of the forelimbs, separating the brown lateral surface of the neck and the pale brown throat. Belly and throat may be immaculate or slightly mottled with brown. When spotted there are one or more, more or less distinct V-shaped marks on the throat, converging on the dewlap. In some specimens an indistinct brown median line is present. Dewlap white.

Males may have the belly and throat immaculate or densely mottled with blackish brown. V-shaped marks on the throat rarely present. Dewlap blue.

In both sexes the tail is dark-grey with or without narrow lighter cross-bands. Limbs, fingers and toes with dark-brown crossbands. Colour of the tongue pale orange to whitish.

A small lizard, with a cylindrical body. Maximum snout-vent length in females 60 mm, in males 64 mm. Limbs long and slender, hind limb approximately twice as long as forelimb, when laid along the body there is considerable overlap of forelimb and hind limb, the hind limb reaching to between the eye and the tip of the snout, or slightly beyond. The undamaged tail in females is 1.6–2.0 times as long as the snout-vent length, in males 1.8–2.0 times, in juveniles 1.4–1.7 times.

HABITAT.—This subspecies is an inhabitant of the forest-floor and only rarely ventures onto the base of trees and low bushes, never attaining more than 50 cm from the forest-floor. *A. c. chrysolepis* Duméril & Bibron is found only on ground that is covered with a thick layer of dead leaves and twigs; bare ground is avoided by the species. There is no preference for the closeness of creeks, specimens being found everywhere in the forest.

NATURAL HISTORY.—A diurnal form which was taken between 9.00 a.m. and 5.30 p.m. and seemingly does not bask. At least, none of the specimens collected by me was seen in the sunshine and all were in shady places. Five pregnant females are known from January-April, three of them containing two mature or nearly mature eggs, two containing one mature egg each. Recently hatched juveniles with a snout-vent length of 25–35 mm, taillength 35–39 mm, are present in February, March, May, June-September and November. Thus, it may be concluded that breeding takes place throughout the year. This species moves in a peculiar way. The lizards run for short distances over the forest-floor, then stop and remain immobile on the spot. When they are immobile they are virtually invisible because of the disruptive pattern which makes them look like a small stick, the females especially so. They remain stationary until the danger is quite near and then make a short spurt.

The contents of six stomachs were examined. They contained two spiders, one diplopod, six orthoptera (Acrididae, Gryllidae), two small homoptera (Jassidae?), two ants (Formicidae: Attinae, Myrmecinae), one caterpillar, one termite, one mosquito (Tipulidae) and one flymaggot (Brachycera). From this list of food items it might be concluded that orthoptera are preferred to other food, but it may be that due to the habits of this lizard orthoptera are the most frequently encountered prey.

RANGE.—Surinam, French Guiana, southern Guyana, northern Brazil (Pará, Amapá). In Surinam this subspecies is present throughout the country in rain-forest from the northern border of the coastal savannah belt south to the Brazilian border. It is present from sea level to at least 600 m on the Hendrik Mountain (fig. 14).

REMARKS.—All Surinam specimens of *A. chrysolepis* Duméril & Bibron (except RMNH 2812, see under *A. c. planiceps* Troschel) undoubtedly belong to this subspecies. However, some specimens from the southwestern part show differences from specimens from the remainder of Surinam. In some characters they resemble *A. c. planiceps* Troschel, although only slightly so. This southwestern corner seems to be an area in which the two subspecies gradually merge into each other. For instance, the juvenile male (RMNH 15683) from Gonini Encampment had a greenish dewlap instead of the cobalt-blue one present in the remainder of Surinam. Some Sipaliwini specimens and the Itabu Creek specimens have the scales of the supraorbital semicircles slightly enlarged, the interparietal oval and larger than the earopening and the anterior margin of the ear-opening slightly denticulate. However, in all other characters they agree with specimens from the remainder of Surinam.

From my data it is evident that allometric growth occurs. In adult specimens the length of the tibia is always more than the distance between tip of snout and posterior margin of the ear-opening (head length). In juveniles with a snout-vent length of 35 mm or less, the tibia is always shorter than the head. Males have a proportionately faster growth of the tibia than females. In a male of 39 mm snout-vent length the tibia is only slightly longer than the head length; in males with a snout-vent length of 44 mm or more the tibia is already distinctly longer. In a female of 39 mm snout-vent length the tibia is still distinctly shorter than the head length, in females with a snout-vent length of 41–44 mm tibia and head length are equal, whereas in larger females the tibia is longer. The same applies to the tail length as compared to the snout-vent length. The larger the snout-vent length, the longer the proportional tail length both in females and males.

In most populations the second type of pattern in males is rare. For instance, among seven males from the airstrip Kayser Mountains there is only one with the brown lateral bands and among six males from Sipaliwini, there is also one. However, among 14 males from the Nassau Mountains there are five with lateral bands. From the Surinam specimens it might be concluded that there is a cline from east to west in which this character becomes less abundant. Therefore, it would be interesting to have more data on French Guianese specimens and check whether an east to west cline exists.

VANZOLINI & WILLIAMS (1970) extensively dealt with *A. chrysolepis* Duméril & Bibron in a thorough paper. It is regrettable that some of the Surinam localities they mention are wrong. On p. 21 they list two specimens from the Berlin Museum (ZMB 25982) as coming from Paramaribo. I studied the same specimens which were bought by the museum from A. O. C. HELLER, a missionary who stayed in Surinam during the turn of the century (HOLTHUIS, 1959; HUSSON, 1962). Most of HELLER's material is labelled as coming from Paramaribo, but this is not correct. Paramaribo was only the port of shipment. During my collecting trips in Surinam I paid special attention to the environments of Paramaribo, to ascertain whether several inland species occurred there. This proved not to be so. *A. c. chrysolepis* Duméril & Bibron is one of the species that is only found farther inland (see under range). A more probable place of origin is the region near Berg en Dal, some 80 km south of Paramaribo. HELLER stayed there for prolonged periods and the species under discussion occurs in that region.

Another incorrect locality is La Haye (Hague). A locality with this name or a slightly different one (Haque) is indicated on several maps on the place where VANZOLINI & WILLIAMS also place it (Map C, p. 17 and p. 121). The specimen in question, RMNH 2812, is one of the older specimens in our collection and still has the original label attached. This label is written in French as was usual in those days and the complete text on it was:

Anolis refulgens Schlegel
La Haye   Suriname
1828

Here La Haye and 1828 are written in the lower left corner, a position on a label where collectors' or donors' names are mentioned. In the lower right corner is the place of origin, in this case Surinam. Thus, this label does not indicate that the specimen came from La Haye, Surinam, but simply that it came from Surinam and was received from La Haye (= The Hague) in the Netherlands in 1828. Upon searching our files I found some correspondence between TEMMINCK, the first Director of the Rijksmuseum van Natuurlijke Historie and VAN DEN KASTEELE, Director of the Kabinet van Zeldzaamheden (Cabinet of Rarities) in The Hague, now no longer in existence. From this correspondence it is clear that between 1822 and 1835 several shipments of natural history objects were transferred from the Hague to Leiden. Unfortunately this correspondence apparently is no longer complete, but from what remains it is clear that at least part of the material came from Surinam (for example in 1823: 250 shells). These facts make it very probable that RMNH 2812 came from this same source. The locality thus becomes simply Surinam. However, this locality is very likely wrong, as will be explained under *A. c. planiceps* Troschel.

There is one locality mentioned by VAN LIDTH DE JEUDE (1904) and by VANZOLINI & WILLIAMS (1970) that obviously is a misspelling on our labels (RMNH 4470). On all our labels this locality reads as Hanie. The specimen was collected by the Gonini Expedition, but no locality of that name could be found in their report (FRANSSEN HERDERSCHEE, 1905a) or on the maps accompanying it. Most probably the Litani (or Litanie) River is meant. VANZOLINI & WILLIAMS (1970) interpreted the locality on their map C correctly.

The above changes fortunately do not interfere too much with the conclusions by VANZOLINI & WILLIAMS. The most important result is that RMNH 2812 and ZMB 25985 no longer play a part in the quantitative data of VANZOLINI & WILLIAMS. Where they say there is a break in characters between Dunoon and La Haye (p. 38) or between La Haye and Tibiti (p. 44) one should read that it occurs between Dunoon and Tibiti, where the break occurs between Dunoon and Paramaribo (p. 43) one should read 'between Dunoon and Moengo Tapoe', and for a break between Paramaribo and Moengo Tapoe (p. 51) one should read 'between Tibiti and Moengo Tapoe'. In all instances the 'Paramaribo' specimens fall completely within the range of the Surinam specimens. In four out of six instances (tables 14, 49 and 103, graph 16) RMNH 2812 fits the Guyana sample better than the Surinam sample. in one case it fits both samples (table 88) and in another it fits the Surinam sample better than the Guyana sample, although it falls well within the range of the latter (table 67). From this it can be concluded that RMNH 2812 does not belong with this subspecies as its place of origin would indicate, but with the Guyanese subspecies.

VANZOLINI & WILLIAMS (1970) discard the oldest name available for this species (*nitens* Wagler) for the following reasons: the inadequacy of the original description, the type locality ('America') is insufficiently defined, the type is lost and the use of this name has been confused. None of these reasons is sufficient to discard this name, for the original description is completely valid and the name has been in use until recently, alongside the now accepted name *A. chrysolepis* Duméril & Bibron which was more widely used. The only means of discarding *nitens* Wagler and ensuring nomenclatural stability is to ask the International Commission on Zoological Nomenclature to use its plenary powers to suppress *nitens* Wagler. A suggestion to this end has been made by me to VANZOLINI & WILLIAMS. Some of the specimens from Guyana that I assign to this subspecies probably have wrong localities. For the RMNH 2811 specimens an explanation will be given under *A. c. planiceps* Troschel. Of the FMNH 35123 specimen from the Oko River I can only say that it is a typical *A. c. chrysolepis* Duméril & Bibron with a blue dewlap. This makes it very probable that the locality is wrong, because all of the surrounding area is occupied by *A. c. planiceps* Troschel. However, the localities Onora and Shudikarwau are correct beyond doubt.

Anolis chrysolepis planiceps Troschel

Anolis refulgens Duméril & Bibron, 1837: 91; Duméril & Duméril, 1851: 56; Bocourt, 1874: pl. XVI fig. 25.
Draconura refulgens: Berthold, 1840: 899; Berthold, 1842: 20; Berthold, 1843: 65.
Dracontura nitens: Fitzinger, 1843: 69.
Draconura nitens: Gray, 1845: XVII, 207.
Anolis planiceps Troschel, 1848: 649.
Anolis chrysolepis: Boulenger, 1885b: 89.
Anolis nitens: Boulenger, 1885b: 91; Van Lidth de Jeude, 1917c: 46; Barbour, 1934: 143.
Anolis n. nitens: J. A. Peters & Donoso-Barros, 1970: 61.
Anolis chrysolepis planiceps: Vanzolini & Williams, 1970: 85.

TYPE-LOCALITY.—Caracas, Venezuela

MATERIAL.—Surinam.
Surinam: 1♀, RMNH 2812, 1828.
Guyana.
Guyana: 2♂♂, RMNH 11766.
Georgetown: 1 ex., MCZ 8080, leg. T. Barbour.
Oko Mountains: 1♀, FMNH 26740; 1♂, FMNH 27282.
Kaburi: 3 ex., MCZ 81306–07, 81502, 31–V–1964, leg. G. Gorman.
Kamakusa, Mazaruni River: 1♂, MCZ 65352.
Forest Settlement, Mazaruni River: 1♂, MCZ 39690, 23–VIII–1935, leg. N. A. Weber.

DIAGNOSIS.—A small anolis with only two rows of slightly enlarged dorsal scales; supraorbital semicircles with distinctly enlarged scales; scales on the upper arm larger than the vertebral scales; a median row of enlarged scales on the snout, scales on the snout with only a single keel; dewlap in life scarlet in both sexes.

DESCRIPTION.—In most characters this subspecies resembles *A. c. chrysolepis* Duméril & Bibron and I may refer to the description thereof. Only the differences between the two will be mentioned below.

The scales on the snout with only a single keel, larger and fewer in number than in *A. c. chrysolepis* Duméril & Bibron. A median row of enlarged scales from between the nostrils to the frontal depression. Frontal ridges distinct. Scales of supraorbital semicircles distinctly enlarged. Interparietal large, oval to polygonal, larger than the ear-opening. Only two rows of dorsal scales over the vertebral column slightly enlarged, smaller than the upper arm scales.

The colour of the dewlap also differs. From a slide of a living female taken at Echerak Creek, Guyana and made available to me by Mr. A. N. WARREN it is evident that females have a scarlet dewlap. BEEBE (1944b) mentiones this same colour for *A. nitens* (Wagler) from Kartabo. VANZOLINI & WILLIAMS (1970) say that the dewlap-colour in life is uniformly red. In preservative both males and females have whitish dewlaps.

In females there is a narrow light middorsal stripe, starting from the interparietal and continued on the tail. Over the sacrum it widens considerably. Males have a very variable pattern, but mostly resemble males of *A. c. chrysolepis* Duméril & Bibron.

HABITAT.—BEEBE (1944b) gives for *A. chrysolepis* Duméril & Bibron as favourite haunts 'leaves of low shrubs and bushes in the jungle and along streams' and for *A. nitens* (Wagler): 'It is found in the jungle both on tree trunks and on low bushes, but usually on the latter'.

NATURAL HISTORY.—Nothing is known regarding Surinam individuals, but BEEBE (1944b) observed breeding in this species between March and August in Guyana.

RANGE.—Guyana, Venezuela, Trinidad and maybe extreme western Surinam.

REMARKS.—Judging by the label the two specimens RMNH 11766 together with RMNH 2811 originated in 'Guyane'. The two specimens in the original no. 2811 clearly are *A. c. chrysolepis* Duméril & Bibron. RMNH 2812 according to the label comes from Surinam, but clearly is *A. c. planiceps* Troschel which probably does not occur in Surinam. The solution for at least part of this confusion may be that formerly the border between Guyana and Surinam was not the Corantijn River, but between the Corantijn and the Berbice Rivers. Until 1799 the mouth of the Devil's Creek, three miles east of the Berbice River was the border (BENJAMINS, 1898). Moreover, both Guyana and Surinam formerly belonged to the Netherlands. The solution for RMNH 2812 might be that it originates from the region between Devil's Creek and Corantijn River and was collected there before 1799. For RMNH 2811 and 11766 it may be that two specimens (RMNH 2811) were collected in the region that is now known as Surinam and that the other two (RMNH 11766) were collected in present-day Guyana and assembled together under the general locality 'Guyane'.

Anolis refulgens Duméril & Bibron was described as having come from Surinam. DUMÉRIL & BIBRON received their specimen from the Leiden Museum and gave as one of the synonyms '*Anolis refulgens* Schlegel. Mus. de Leyde'. From this it can be concluded that they received their specimen under that name from Leiden. However, SCHLEGEL never published this name and it just represents a label name. In our old collections there is a specimen (RMNH 2812) from Surinam (see above) which bears on its label the name *Anolis refulgens*. In all characters this specimen agrees very well with the description of DUMÉRIL & BIBRON (1837) and with the illustration in DUMÉRIL, BOCOURT & MOCQUARD (1873), the differences being due to individual variation and to sexual dimorphism, the Paris specimen is a male, RMNH 2812 a female. Very probably RMNH 2812 formed one lot with the holotype, the policy of the RMNH in its early years being that duplicates of a species could be used for exchange. So, for the holotype of *A. refulgens* Duméril & Bibron the same applies as has been stated above for RMNH 2812.

Anolis fuscoauratus kugleri Roux (figs. 16, 17; pl. 14; table 10)

Anolis kugleri Roux 1929: 29.

TYPE-LOCALITY.—El Mene, Acosta District, Estado Falcón, Venezuela.

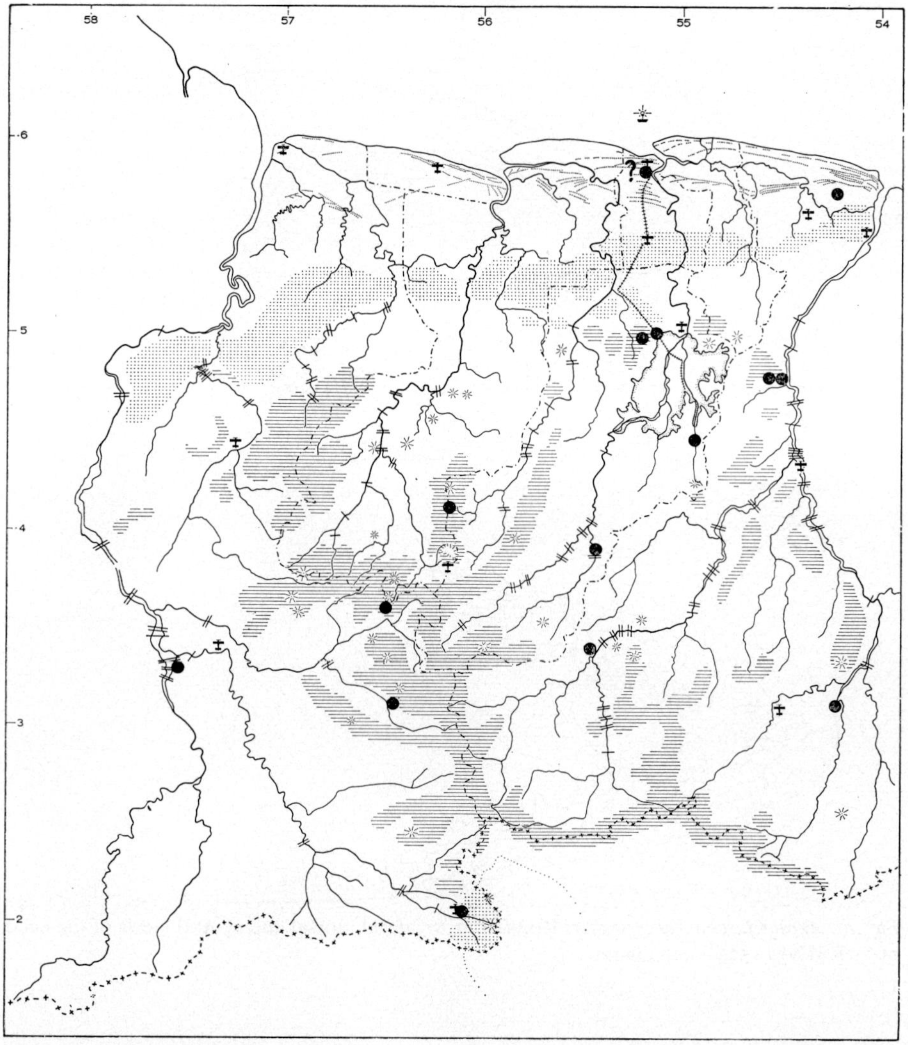

Fig. 16. Distribution of *Anolis fuscoauratus kugleri* Roux in Surinam.

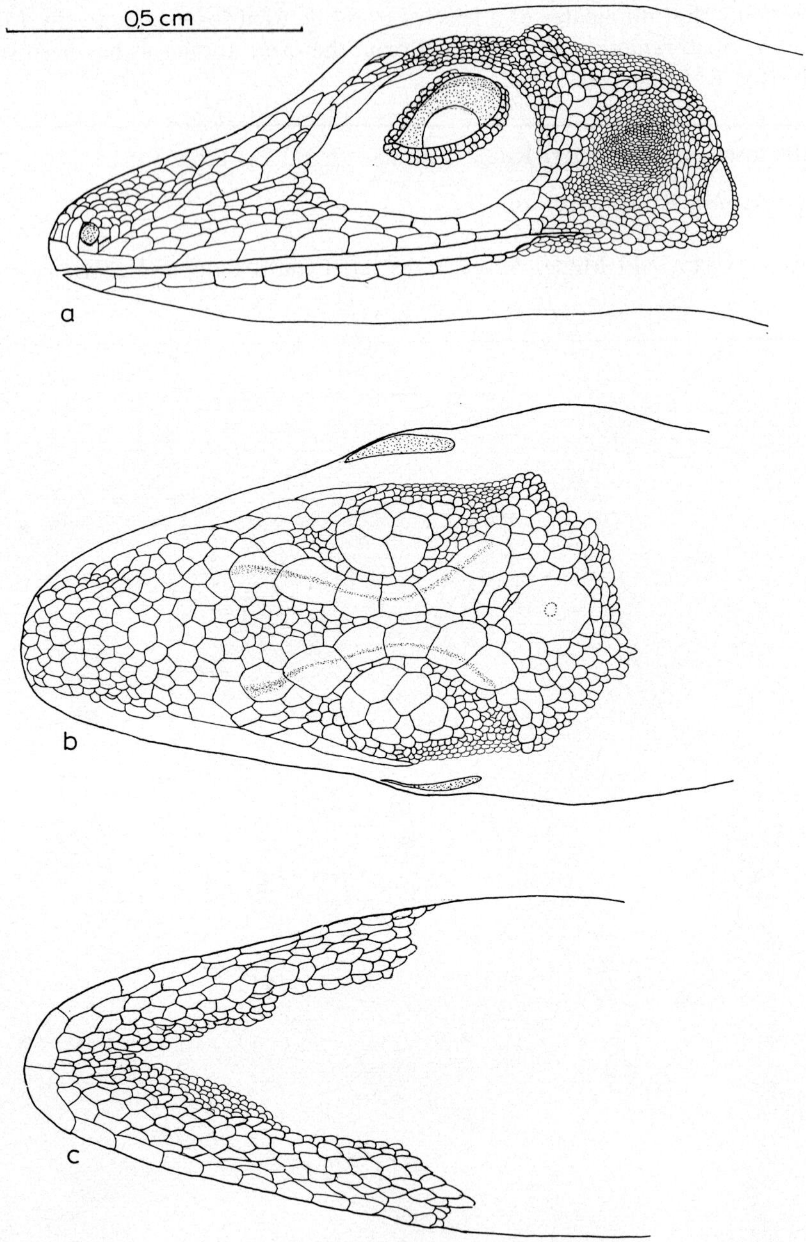

Fig. 17. Anolus fuscoauratus kugleri Roux; a, b, c: lateral, dorsal and ventral views of the head of ♀, RMNH 15253, Sipaliwini.

MATERIAL. – Surinam.

Paramaribo: 1 ex., ZMB 25983, leg. A. O. C. Heller.

Moengo Tapoe transect, Third Encampment: 1♂, RMNH 11631, km 14.9, 25–X–1948, leg. Surinam Expedition 1948–49.

Brown's Mountain: 1♂, RMNH 17169, 30–XII–1971, leg. G. F. Mees.

Railway km 121, Mahami Creek: 1♀, RMNH 11635, 30–XI–1949, leg. C. Bleijs.

Nassau Mountains, Base Encampment: 1♂, RMNH 11550, March 1949, 1♂, RMNH 11557, 25–II–1949, 1♀, 1♂, RMNH 11558, 26–II–1948, all leg. Surinam Expedition 1948–49.

Nassau Mountains: 1♀, RMNH 11560, km 11.2, 21–III–1949, leg. Surinam Expedition 1948–49.

Langatabbetje, Upper Sara Creek: 1♀, RMNH 15662, 12/14–XII–1965, leg. G. F. Mees.

Emma Mountains, Base Bivouac; 1♂, RMNH 15664, 23–VII–1959, leg. F. Jonkers.

Mirahoedoe, Pikien Rio: 1ex., ZMA 12234, 23–VII–1908, leg. J. H. A. T. Tresling (Suriname Expedition).

Wilhelmina Mountains: 1♂, RMNH 15665, Camp 3, km 19, June 1963, leg. H. P. Pijpers (Expedition Maguire).

Tigri Encampment, New River: 1♀, RMNH 15656, 27–VI–1968, leg. M. S. Hoogmoed.

Airstrip Kayser Mountains: 1♂, RMNH 15659, NE., 12–VIII–1968, 1♀, RMNH 15660, NE., 12–VIII–1968, 1♂, RMNH 15661, 2–VIII–1968, all leg. M. S. Hoogmoed; 1♀, 1♂, 1 juv., FMNH 128986, 128988–89, 1–II–1961, leg. H. A. Beatty.

Airstrip Paloemeu: 1♀, USNM 159028, leg. T. H. Lewis.

Feti Creek, Litani River: 1♀, RMNH 15663, 9–VIII–1939, leg. D. C. Geijskes.

Airstrip Sipaliwini, Base Bivouac: 1♀, RMNH 15174, 9–IX–1968, 1♀, RMNH 15253, 4–IX–1968, 1♀, RMNH 15657, N., 2–IX–1968, 1♂, RMNH 15658, 26–VIII–1968, all leg. M. S. Hoogmoed.

Guyana.

Cauli, Aremu (Arimu) River: 1 ex., UMMZ 58545, leg. C. W. Barbe.

Mackreba Falls, Kurupung River: 1 ex., UMMZ 85222, 22–II–1939, leg. A. S. Pinkus.

Kaieteur Falls, Potaro River: 1 ex., MCZ 66935, leg. M. Locke; 1♂, MCZ 77379, 3–XII–1963, leg. G. Underwood.

DIAGNOSIS.—A small semi-arboreal anolis with a cylindrical to slightly compressed body. Dorsal scales small; scales around midbody 122–154; ventrals keeled; scales on the snout keeled. Tibia in adult specimens shorter than the head length. Digital expansions moderately well developed. Females brown above with a lighter vertebral stripe, or with a series of dark-brown spots on the lighter vertebral region, dewlap red with white scales. Males brown above with a wide, dark-brown vertebral zone, a black spot in front of the forelimbs, dewlap purplish red or yellow with white scales.

DESCRIPTION.—Head moderately large to large, representing 24–28% of the snout-vent length in both sexes; long and narrow, 1.6–1.8 times as long as wide; longer than the tibia, which is 80–93% of the head length; slightly depressed; wider than the neck, which is slightly narrower than the body. Snout slightly pointed, rounded, not projecting over the mouth, concave in profile. A slight frontal depression bordered by frontal ridges. Occipital region flat.

Rostral nearly rectangular, 2.5 times as wide as deep; hardly visible from above; bordered posteriorly by six or seven small scales and by the first supralabial on each side. Eight to 12 supralabials, seven to ten, mostly seven or eight, to below the centre of the eye. Loreal region with five to seven, mostly five or six, horizontal scale rows; the upper row consisting of long, slender scales. Canthal scales six or seven, mostly six, of which the posterior three are much larger than the anterior scales; keeled, forming a conspicuous canthus rostralis. Four to five keeled suboculars, forming a semicircle with the distinctly smaller postoculars. Postoculars decreasing in size dorsally. Temporal region with small granules. Ear-opening vertically oval, small; distinctly smaller than the interparietal; behind the commissure of the mouth; surrounded by small scales forming a smooth margin; external auditory meatus short, tympanum well visible. A more or less distinct supratemporal arch of slightly enlarged scales from the upper end of the postocular row to the dorsal tip of ear-opening; scales in the anterior portion largest. Scales on the snout small, juxtaposed, hexagonal to polygonal, keeled; smallest and smooth in the frontal depression. Nostril in a single nasal. Supraciliaries two to six, mostly three or four, forming a continuous row with the canthal scales; reaching to above the centre of the eye or slightly beyond; second scale longest, the anterior two scales slightly keeled. Supraorbital semicircles with seven to ten, mostly seven or eight large scales, the anterior ones keeled, the posterior ones smooth; semicircles from both sides separated by one row of small scales; continued posteriorly by a row of slightly enlarged scales connecting them with the postoculars. Interparietal large, elongate, polygonal, larger than the ear-opening; separated from the supraorbital semicircles by two or three, exceptionally one, rows of small scales. Scales on the parietal region fairly large, juxtaposed, polygonal, smooth, decreasing in size posteriorly. Scales on the occipital region granular, slightly larger than those on the nape. Supraocular region with an oval patch of large, slightly keeled scales, largest near the supraorbital semicircles, smaller towards the supraciliaries; separated from the supraorbital semicircles by one row of small scales; rest of the supraocular region with small granules. Eyelids covered with minute granules, two rows of enlarged scales bordering the rim. Pupil round.

Two mentals, roughly forming a semicircle; bordered posteriorly by four to six small scales, two sublabials and two infralabials. Infralabials 8–12, mostly 10–11, bordered medially by a few rows of slightly enlarged sublabials. The other chin scales granular. Scales on the throat smooth, larger than those on the chin, except along a longitudinal, medial band where they are distinctly smaller. Dewlap in females very small, reaching from below the level of the ears to just before the forelimbs; in adult males very large, reaching from just before the eyes to mid-abdomen, in young males smaller; when extended showing seven to nine approximately horizontal rows of dorso-ventrally flattened, triangular, protruding scales; rim scales smaller, imbricate; scales on the lateral part of the dewlap larger than the ventrals. Scales on the nape juxtaposed, small, round, smooth, granular; identical to the dorsals. Scales on the side of the neck smaller, identical to the laterals. No nuchal fold.

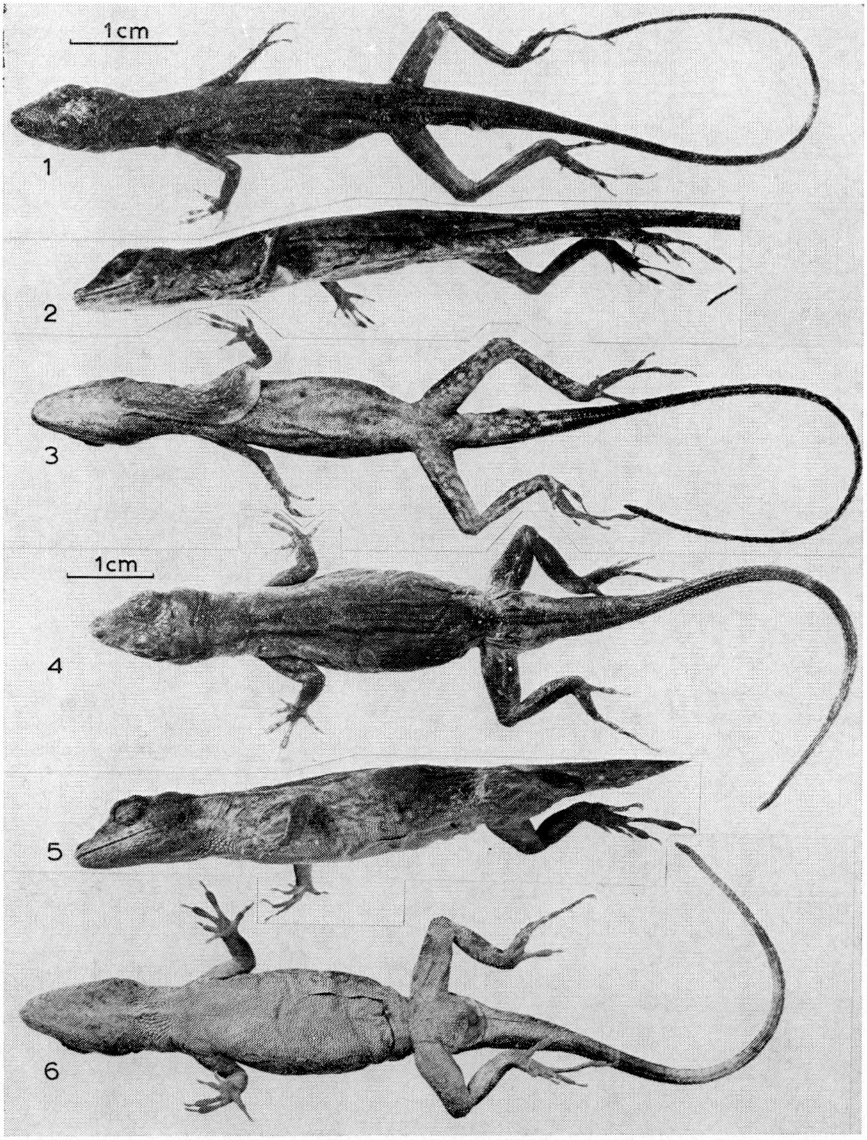

Plate 14. Figs. 1–3. *Anolis fuscoauratus kugleri* Roux; dorsal, lateral and ventral aspect of ♂, RMNH 15661. Figs. 4–6. *Anolis ortonii* Cope; dorsal, lateral and ventral aspect of ♀, RMNH 15655, Tigri Encampment.

Table 10. Counts and measurements of *Anolis fuscoauratus kugleri* Roux. For explanation see table 3.

Reg. no.	A	B	C	D	E	F	G	H	Q'	J	S'	O	P	Q	J'	R'	W	X
RMNH																		
11550	♂	46	43(43)	12.1	6.7	5.5	18	33	10.3	144	8-10	8-8	11-11	4-4	5-5	6-6	22(16)-21(15)	34(27)-34(27)
11557	♂	45	82	12.3	6.7	5.8	18	32	9.9	150	9-9	7-8	10-10	4-4	6-6	6-6	22(14)-21(14	31(23)-31(23)
11558a	♀	47	72(72)	12.4	7.2	5.5	18	35	11.0	132	8-9	8-8	10-10	3-4	6-6	6-6	24(16)- ?	34(26)-34(25)
b	♂	45	64(29)	12.2	6.8	5.7	18	34	11.4	154	7-8	8-9	11-12	4-4	7-7	6-7	24(16)-26(17)	37(28)-34(26)
11560	♀	47	79	12.1	6.9	5.4	20	35	10.5	142	9-9	10-8	11-11	3-3	5-5	6-6	23(15)-24(15)	36(27)-?
11631	♂	33	39(39)	9.0	5.3	5.0	12	24	7.3	143	7-7	8-8	10-10	4-6	5-5	6-6	20(13)-20(14)	28(21)-28(21)
11635	♀	48	70(50)	12.7	7.2	6.0	19	34	10.8	?	8-9	8-7	11-11	4-4	7-7	6-6	22(15)-22(14)	33(24)-?
15174	♀	46	92	?	?	?	?	?	?	?	?	?-8	?-9	?	?-6	?-6	21(14)-22(15)	31(22)-33(24)
15253	♀	41	80	11.3	6.4	5.1			9.5	146	7-7	9-10	8-9	4-3	5-5	6-7	?	35(26)-35(27)
15656	♀	47	92	12.6	7.4	5.9	19	32	10.4	150	8-9	9-8	12-11	4-5	6-6	6-6	24(15)-23(14)	33(24)-35(26)
15657	♀	41	79	11.7	6.6	5.2	17	32	9.7	122	9-10	7-7	8-9	2-4	6-6	6-6	21(14)-20(13)	32(24)-31(24)
15658	♂	45	89	12.3	6.8	5.6			10.5	138	8-9	7-8	10-11	4-4	6-6	7-6	24(16)-23(16)	33(25)-34(25)
15659	♂	38	61	9.3	5.3	4.6	13	26	7.6	142	7-7	7-8	11-10	4-4	5-5	6-6	22(14)-23(15)	33(23)-35(26)
15660	♀	31	51(51)	8.6	5.3	4.5	12	23	6.9	143	8-8	7-7	12-9	3-3	5-5	6-6	21(14)-20(13)	29(21)-30(22)
15661	♂	40	67	10.8	6.2	5.4	16	28	8.7	149	7-8	8-8	10-10	3-3	5-5	6-6	21(14)-21(14)	32(23)-33(24)
15662	♀	45	89	12.0	6.9	5.8	18	33	9.9	140	7-8	8-7	10-9	4-4	6-6	6-6	20(13)-21(14)	34(26)-30(23)
15663	♀	44	88	12.0	6.6	5.6	19	35	10.4	141	7-8	8-7	10-11	4-4	6-6	7-7	23(15)-24(16)	35(26)-34(25)
15664	♂	49	92	13.2	7.2	6.1	19	35	10.9	122	7-7	8-8	10-10	4-5	6-6	7-6	24(16)-24(16)	36(26)-36(27)
15665	♂	44	81	12.2	6.5	5.6			10.2	?	7-7	8-8	10-12	3-3	7-6	6-7	23(16)-20(14)	32(24)-31(23)
FMNH																		
128986	♀	47	75(75)	12.1	7.1	6.2			10.2	147	7-7	8-8	9-9	4-3	5-5	6-6	24-21	31(22)-33(25)
128988	♂	41	73	10.9	6.1	5.4			9.2	?	7-8	7-7	11-10	3-3	6-5	6-6	22(13)-22(15)	33-33
128989	juv.	31	54	8.8	5.1	4.4			6.9	?	8-8	8-9	8-8	4-4	6-6	6-6	23-23	36-35
MCZ																		
77379	♂	45	79	11.8	6.1	5.1			11.1		6-7						25-25	37-37
UMMZ																		
58545	?	45	84	11.6	6.4	5.6	18	34	10.6		6-6						24-22	36-36

Dorsals slightly larger than the laterals, both smaller than the ventrals. Ventrals juxtaposed, oval to round, smooth, in oblique rows on the male's chest. Ventrals and laterals gradually merging into each other. Total number of scales around midbody 122–154. No dorsal fold. Preanal plate with small, round to oval, smooth, juxtaposed scales, smaller than the ventrals. Enlarged postanal scales absent in most specimens, slightly enlarged scales present in two males from the airstrip Kayser Mountains (RMNH 15661 and FMNH 128988).

Scales on the anteriormost part of the tail small, round to hexagonal, smooth, juxtaposed; posteriorly larger, hexagonal, keeled, imbricate, in longitudinal series, the scales on the upper edge slightly larger than the rest. Scales under the anteriormost part of the tail small, round to squarish, smooth, juxtaposed; posteriorly larger, lanceolate to hexagonal, longer than wide, keeled, in longitudinal series. Except in the most proximal and the most distal part, the tail is divided into verticils. The anterior verticils with four keeled scales below, five keeled scales above; the posterior verticils with three keeled scales below, four keeled scales above. In two specimens only (RMNH 11558b, 11635) part of the tail is regenerated. Scales on the regenerated part identical on the upper and lower surface, slightly larger than the scales on the original tail, hexagonal, keeled, imbricate, in longitudinal rows, not arranged in verticils. Breaking points tend to coincide with the border between verticils. Tail slightly compressed, ending in a tip with five serrated ridges.

Scales on the lower and posterior surfaces of the forelimbs and hind limbs small, round, smooth, juxtaposed; scales on the upper and anterior surfaces of the forelimbs and hind limbs larger, rhomboidal to hexagonal, keeled, the keels forming ridges, juxtaposed to slightly imbricate. Twenty to 26 single, smooth lamellae under the fourth finger, 13–17 to the end of the digital expansion; 28–37 single, smooth lamellae under the fourth toe, 21–28 to the end of the digital expansion. Palms and soles with round, flat, smooth, slightly imbricate scales, as large as the ventrals. Digital expansions on fingers and toes moderately well developed, i.e., the third phalanx scarcely wider than the basal one. Proximal part of fingers and toes depressed, distal part compressed.

Colour in life of females (RMNH 15253, 15656, 15660) beige above with a wide, composed vertebral stripe, yellow-brown in the centre, bordered by an orange-brown and a black line; starting with a sharp point just posterior to the interparietal, fading on the tail. Flanks light-brown with small white spots. A light-brown transverse band between the eyes, bordered by a dark-brown one on its posterior side, may or may not be present. Throat and belly, ventral surfaces of the tail and legs vermiculated with dark-brown or grey and white, in different degrees. In RMNH 15656 the belly is nearly completely white with some scattered brown spots, in 15253 there is more brown, whereas in 15660 the belly is brown with white spots. Scales of the dewlap-rim white. Dewlap red with white scales. One other female (RMNH 15663) bears the note 'back dark-brown, belly spotted with grey'. In life, males (RMNH 15658–59, 15661) are brown above with a wide dark-brown vertebral zone, a black spot in front of the forelimbs. Flanks with small yellow spots. Throat, belly and ventral surface of limbs grey or brown with white spots or white with brown spots. The ventral surface of the base of the tail is yellow, posteriorly the ventral surface of the tail has creamish and brown bands. Dewlap in RMNH 15659 and 15661, both from the airstrip Kayser Mountains, purplish red with rows of bluish white scales. Scales of the dewlap-rim white. In RMNH 15658 from Sipaliwini the dewlap is yellow with rows of white scales. Iris golden-brown.

In preservative the colour does not change very much. Two types of females can be recognised, one type (five females) has the back as described above with the vertebral stripe, the other type (seven females) has a series of dark-brown, longitudinally rectangular spots on the vertebral area, the rest of the back of a lighter brown. The back of the thighs shows a number of ill-defined lighter spots. All males have the same type of pattern. Most specimens (females and males) show a light-brown transverse band between the eyes, bordered posteriorly by a dark-brown band. Upper surface of fingers and toes with dark transverse bands in both sexes.

A small lizard, with a cylindrical to slightly compressed body. Maximum snout-vent length in females 48 mm, in males 49 mm. Limbs long and slender; hind limb 1.7–2.0 times as long as forelimb; when laid along the body there is considerable overlap of forelimb and hind limb, the hind limb reaching to between the ear and the eye, or to the eye. The undamaged tail in females is 1.7–2.0 times as long as the snout-vent length, 1.6–2.0 times as long in males.

HABITAT.—This subspecies occurs in the lower layers of the forest, from the forest-floor to about three metres above the ground. It seems to prefer small trees and bushes, that is, slender twigs and stems. Frequently it was encountered near creeks. From my field notes this habitat may be exemplified: RMNH 15174: on tree, ± two metres above the ground, creek-bank. RMNH 15253: forest floor. RMNH 15656: forest floor. RMNH 15657: on tree, ± three metres above the ground. RMNH 15658: creek-bank, on twigs, 30 cm above the ground. RMNH 15659: on slender stem, near ground, creek-bank. RMNH 15660: on liana quite near the ground, creek-bank. RMNH 15661: on bamboo rod near house.

NATURAL HISTORY.—A diurnal form, which seemingly does not bask; all specimens I collected were in the shade. Four females (RMNH 11560, 15174, 15662–3) each contained one mature egg. Breeding thus seems to occur throughout the year (March, August, September, December). BEEBE (1944b) found a pregnant female in Kartabo, containing one well developed and one less developed egg, in May. No recently hatched juveniles are available. These lizards can change their ventral colour rapidly, as was demonstrated by RMNH 15661, which, when collected, had a white belly with brown spots, but shortly afterwards, during a shock-like state, changed it to brown with white spots. Four stomachs that were examined contained three spiders, one diplopod, one bristletail (Thysanura: *Machilis*), two caterpillars, one homopteron, one termite and one beetle. One stomach contained some pieces of shed skin. The eating of the shed skin is a normal process in this genus.

RANGE.—Northern South America: Surinam, Guyana, French Guiana, Brazil (Pará, Amazonas, Amapá), Venezuela (Falcón, Bolivar). The nominate form occurs on the Amazonian slopes of the Andes in Colombia, Ecuador, Peru and Brazil. In Surinam *A. fuscoauratus kugleri* Roux is present in forested country on both sides of the coastal savannah-belt, from sea level to a height of at least 500 m (fig. 16).

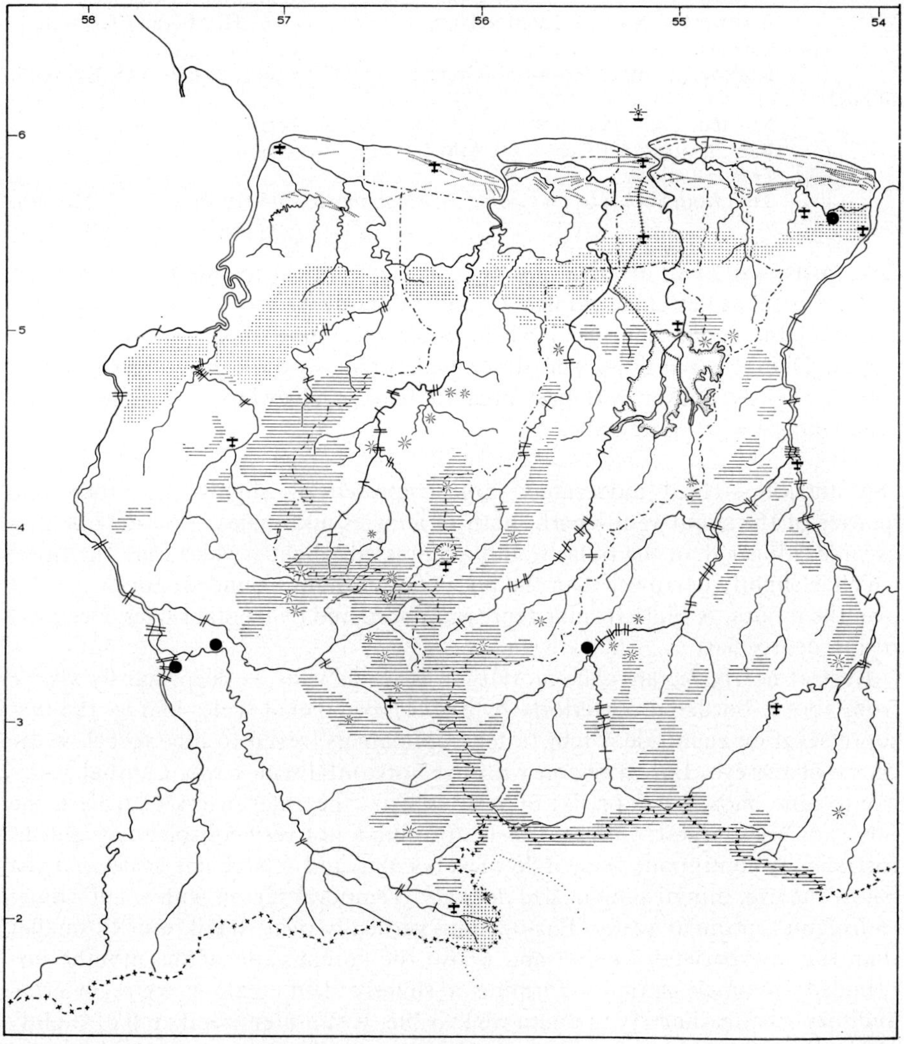

Fig. 18. Distribution of *Anolis ortonii* Cope in Surinam.

Anolis ortonii Cope (figs. 18, 19; pl. 14; table 11)

Anolis ortonii Cope, 1869: 97; Boulenger, 1885b: 51.

TYPE-LOCALITY.—Rio Napo or Upper Rio Marañón, Ecuador or Peru.

MATERIAL. – Surinam.
Surinam: 1♂, BM 68.11.17.9 (locality with question-mark).
Moengo Tapoe transect: 2♂♂, RMNH 11629, km 1, 22–IX–1948, leg. P. H. Creutzberg (Surinam Expedition 1948–49).
Tigri Encampment, New River: 1♀, RMNH 15655, 30–VI–1968, leg. M. S. Hoogmoed.
Airstrip Coeroeni: 1♀, RMNH 15255, August 1959, leg. D. C. Geijskes.
Airstrip Paloemeu: 1♀, RMNH 15711, 5–V–1969, leg. H. M. van Meeuwen.
Guyana.
Kurupung, Upper Mazaruni River: 1♀, UMMZ 83629, 20–III–1938, leg. A. S. Pinkus.
Kaieteur Falls, Potaro River: 1♂, MCZ 66026, April 1961, leg. M. Locke.
Bank Demerara River: 1♂, UMMZ 56013, 22–VIII–1914.
French Guiana.
Trail from Sophie to La Greve: 1♀, MCZ 76680, 10–VIII–1963, leg. T. Monath.

DIAGNOSIS.—A small arboreal anolis with a cylindrical to slightly compressed body. Dorsal scales small; scales around midbody 140–173; ventrals smooth; scales on the snout smooth. Tibia in adult specimens shorter than the head length. Digital expansions well developed. Light to dark grey-brown above, with a slightly darker vertebral area in females. Dewlap in females small, yellow-ochre with white scales.

DESCRIPTION.—Head moderately large, representing slightly less than one quarter of the snout-vent length, both in females and males; 1.6 times as long as wide; longer than the tibia; slightly depressed; slightly wider than the neck, which is slightly narrower than the body. Snout blunt, rounded, not projecting over the mouth. A slight frontal depression bordered by frontal ridges. Occipital region nearly flat.

Rostral nearly rectangular, more than twice as wide as deep; hardly visible from above; bordered posteriorly by five to seven small scales and by the first supralabial on each side. Eight to ten supralabials, seven to nine to below the centre of the eye. Loreal region with five horizontal scale rows. Canthal scales five to nine, mostly five or six; of various sizes, the anterior ones smallest, the posterior ones largest; with low keels, forming a not very conspicuous canthus rostralis. A continuous semicircle of subocular and postocular scales, largest below the eye, diminishing in size dorsally. Temporal region with small, round to irregular granular scales. Ear-opening vertically oval; small, mostly smaller than the interparietal; behind and above the commissure of the mouth; surrounded by small granules forming a slightly denticulate margin; external auditory meatus short, tympanum well visible. A supratemporal arch of slightly enlarged scales from the upper end of the postocular row to the dorsal tip of the ear-opening may or may not be discernible; if present, the scales in its anterior portion largest. Upper surface of the snout with polygonal, flat, smooth scales; smallest on the latero-anterior part of the snout, larger on the posterior and median part and in the frontal depression. Nostril in a single nasal. Supraciliaries two to five, mostly four, forming a continuous row with the canthal scales; reaching to above the centre of the eye; second scale longest, narrowest, the posterior two scales minute, the anterior two scales slightly angular. Supraorbital semicircles with 8–11, mostly eight or nine, scales of which the second is largest; the anterior scales keeled; the semicircles from both sides in contact. Interparietal elongate, polygonal; in most specimens larger than the ear-opening (in one specimen smaller), separated from the supraorbital semicircles by one or two rows of scales. Scales on the parietal region relatively large, polygonal, diminishing in size posteriorly. Scales on the occipital region small, polygonal, diminishing in size posteriorly, the posterior ones indistinguishable from those on the nape. Supraocular region with an oval patch of enlarged, smooth scales, separated from the supraorbital semicircles by one row of small scales; posteriorly and laterally it is surrounded by small granules. Eyelids covered with minute granules, two rows of enlarged scales bordering the rim. Pupil round.

Two mentals forming a semicircle, bordered posteriorly by three or four small scales, two sublabials and two infralabials. Infralabials 9–11, bordered medially by one row of sublabials; the other scales on the chin gradually grading off into granules. Scales on the throat larger than those on the chin, smooth. Dewlap in females small, reaching from a level just anterior of the ear to the inset of the forelimbs; in males large, reaching from a level just anterior to the ear to midabdomen; when extended showing five oblique rows of dorso-ventrally flattened, triangular, protruding scales; rim scales smaller, imbricate; all dewlap scales larger than the ventrals. Scales on the nape juxtaposed, small, round, smooth, identical to the dorsals. No nuchal fold.

Dorsals and laterals juxtaposed, small, round, smooth, smaller than the ventrals. Ventrals juxtaposed, round, smooth. Ventrals and laterals gradually merging into each other. Total number of scales around midbody 140–173. No dorsal fold. Preanal plate with juxtaposed, small, round to oval, smooth scales, slightly smaller than the ventrals. Enlarged postanal scales absent.

Scales on the tail small, hexagonal, keeled, slightly imbricate, in longitudinal rows; in females the median row slightly enlarged; in males the scales in this row are considerably enlarged, forming a low, denticulate crest. Scales under the tail large, hexagonal, keeled, slightly imbricate, longer than wide, in longitudinal rows. Distally the keels increase in height, forming five serrated ridges. Except in the most proximal and the most distal part, the tail is divided into verticils. Each verticil with three keeled scales below, four keeled scales above. In one specimen (RMNH 15655) the tip of the tail is regenerated and in this part no verticils are discernable. Tail round to slightly compressed in females, compressed in males, ending in a five-ridged tip.

Scales on the upper surface of forelimbs and hind limbs small, hexagonal,

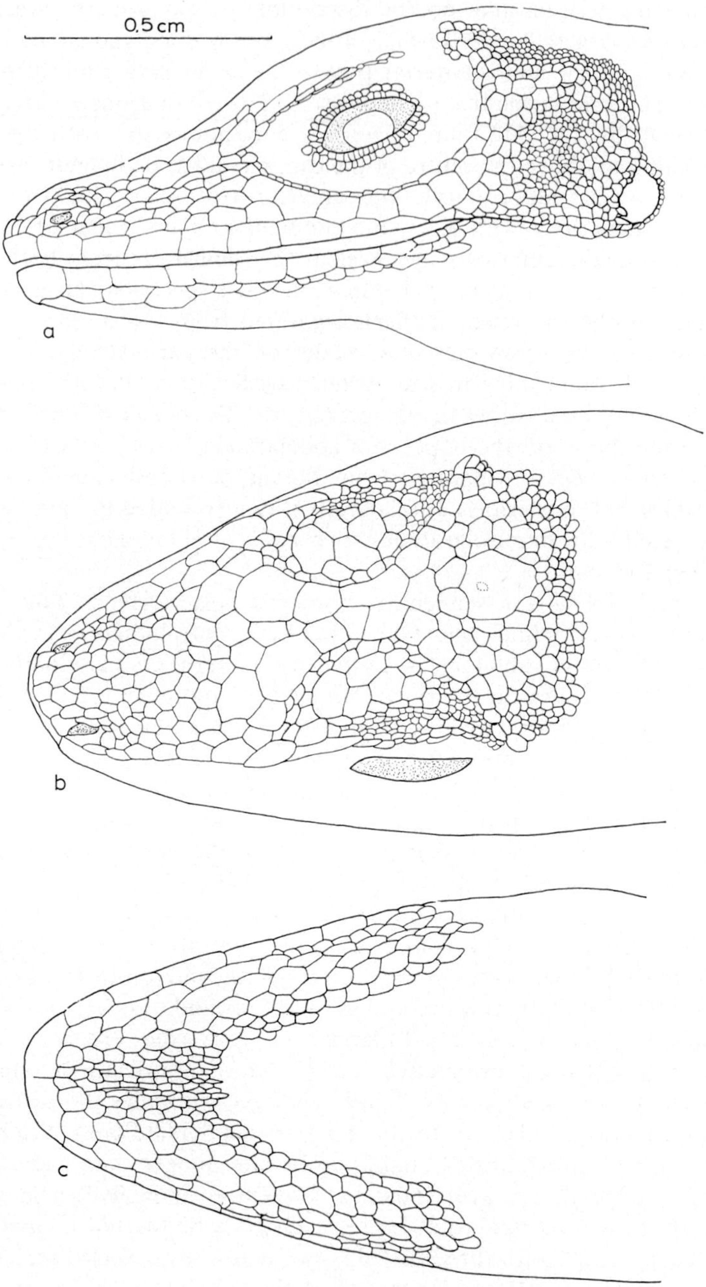

Fig. 19. Anolis ortonii Cope; a, b, c: lateral, dorsal and ventral views of the head of ♀, RMNH 15255, Airstrip Coeroeni.

Table 11. Counts and measurements of *Anolis ortonii* Cope. For explanation see table 3.

Reg. no.	A	B	C	D	E	F	G	H	Q'	J	S'	O	P	Q	J'	R'	W	X
RMNH																		
11629a	♂	49	81	12.8	7.9	6.2	20	33	9.8	173	9-10	7-8	10-10	4-4	5-5	9-9	27(19)-26(17)	38(29)-35(26)
b	♂	51	77	13.2	8.3	6.5	21	35	10.4	158	8-8	8-9	9-9	5-2	5-5	5-6	28(18)-28(19)	42(32)-39(30)
15255	♀	48	72	31.1	7.9	6.5			9.2	140	11-10	8-9	10-10	4-4	5-5	6-6	?(18)-25(19)	31(25)-32(26)
15655	♀	51	64(43)	13.5	8.4	6.3			9.6	158	8-9	8-8	11-11	4-4	5-5	6-6	25(19)-25(18)	37(29)-36(28)
15711	♀	52	81	13.4	8.4	6.2	20	30	9.4	151	8-8	7-7	10-10	5-3	5-5	5-5	25(18)-26(19)	38(31)-37(29)
MCZ																		
66026	♂	38	57	10.8	7.4	6.3			8.4								27-31	37-36
76680	♀	45	62	11.7	7.4	5.7			9.5		6-6						22-26	35-33
UMMZ																		
56013	♂	44	65	12.2	8.0	6.1	19	28	10.1		8-8						22-25	33-34
83629	♀	34	?	10.8	7.3	5.8	13	22	7.8		9-8						27-23	33-35

smooth to indistinctly keeled; on the lower and posterior surfaces small, round, smooth, on the anterior surface of the underarm and on the anterior surface of the lower legs slightly larger, hexagonal, keeled. Twenty-five to 28 single, smooth lamellae under the fourth finger, 17–19 to the end of the digital expansion; 31–42 single, smooth lamellae under the fourth toe, 25–32 to the end of the digital expansion. Palms and soles with small, round, smooth scales. Digital expansions on fingers and toes well developed, i.e., the third phalanx is much wider than the basal one. Proximal part of fingers and toes depressed, distal part compressed.

Colour in life of a female (RMNH 15655): body yellowish grey, black to dark-brown on the vertebral region, a brown line between the eyes, upper surface of the head with a tinge of pink, flanks reticulated with dark-brown, tail grey with brownish transverse bands, belly grey, dewlap yellow-ochre with a few horizontal rows of white scales. Iris golden. Nothing is known about the colour of the living male, probably, however, these will agree with the females in most characteristics, except perhaps in the colour of the dewlap.

In preservative the colour of the back ranges from light grey-brown (nearly cream-coloured) to dark grey-brown. The vertebral area is slightly darker in females, of the same colour as the rest of the back in males. Limbs with ill-defined lighter spots. Chin and throat with brown or grey spots. Belly creamish to dark-brown, immaculate or with a few brown spots. Tail with brown cross-bands. Upper surface of the toes with indistinct darker transverse bands. Tongue creamy white.

A small lizard with a cylindrical to slightly compressed body. Maximum snout-vent length in females 52 mm, in males 51 mm. Limbs short, hind limb one and a half times as long as forelimb in females, 1.7 times as long in males; when laid along the body there is considerable overlap of forelimb and hind limb. The undamaged tail is 1.5–1.6 times as long as the snout-vent length in both sexes.

HABITAT.—Only few observations on the habitat of this species are available, but all indicate that this is a lizard of the rain-forest, where it inhabits the base of trees (RMNH 11629: tree-stub in clearing; RMNH 15655: base of tree, edge of clearing; RMNH 15711: on tree next to recently fallen tree) around clearings in the forest. These data agree with the observations of VANZOLINI & WILLIAMS (1970) who stated that this species 'is a trunk anole but occurs only in more open situations'.

NATURAL HISTORY.—A diurnal lizard. RMNH 15655, collected June 30, contains two mature, oval, eggs. No more data on this subject are available.

RANGE.—Northern South-America, in the Amazonian Basin and adjacent areas: Surinam, Guyana, French Guiana, Brazil (Amazonas), Peru, Bolivia, Equador and Colombia. In Surinam the species probably is present throughout the country, on both sides of the coastal savannah-belt. It occurs from sea level to at least 170 m (fig. 18).

Anolis punctatus punctatus Daudin (figs. 20, 21; pl. 15; table 12)

?Le huitième (*Lacertus viridis*) Fermin, 1769: 209.
?Kleine Struik-Hagedis (*Lacertus viridis*) Fermin, 1770: 185.
Anolis punctatus Daudin, 1802b: 84; Van Lidth de Jeude, 1904: 88; Barbour, 1934: 147.

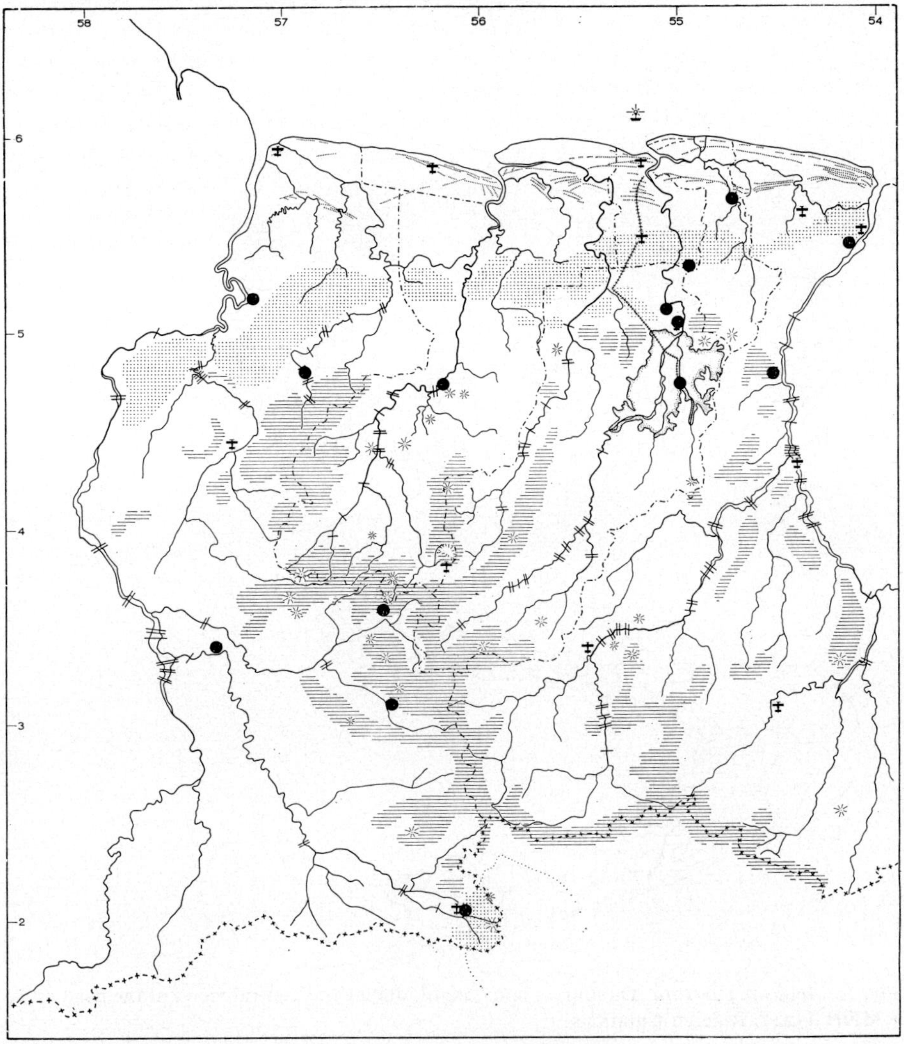

Fig. 20. Distribution of *Anolis p. punctatus* Daudin in Surinam.

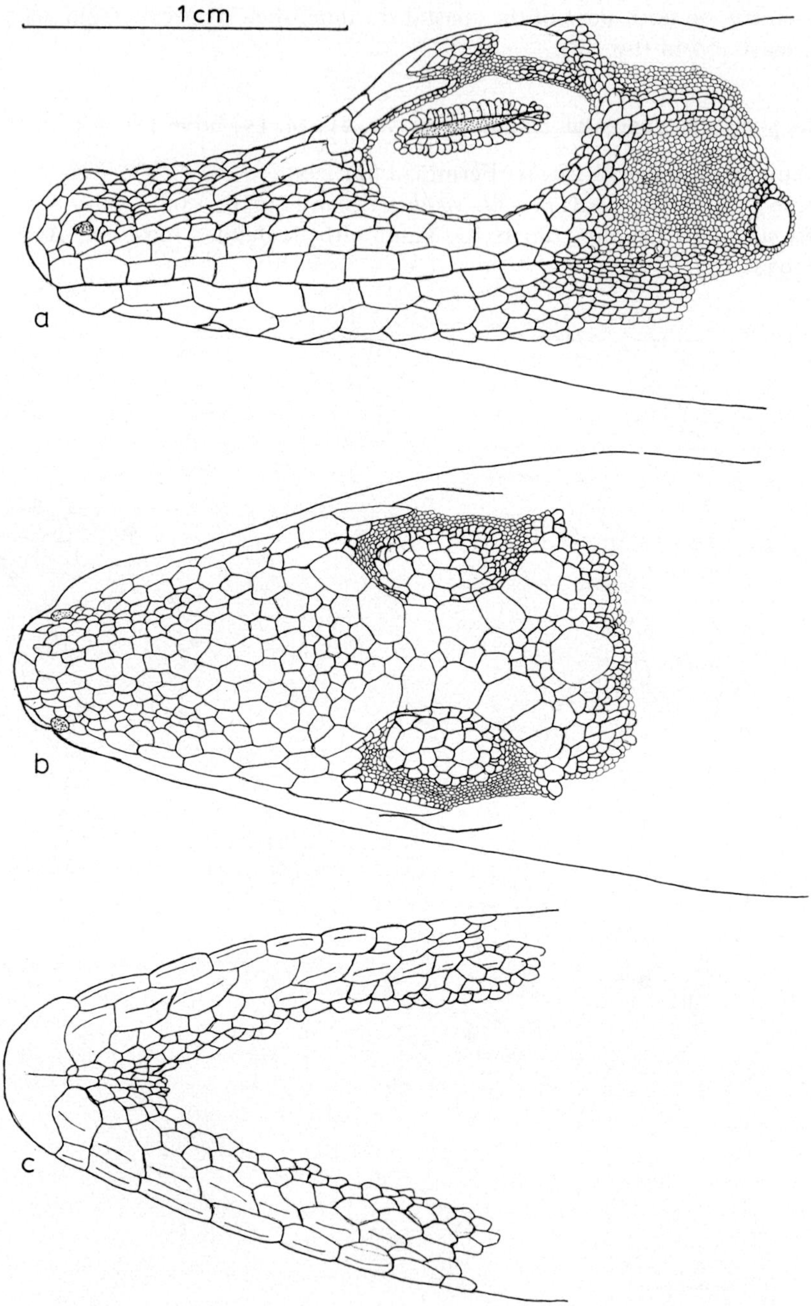

Fig. 21. Anolis p. punctatus Daudin; a, b, c: lateral, dorsal and ventral views of the head of ♂, RMNH 15247, Raleigh Cataracts.

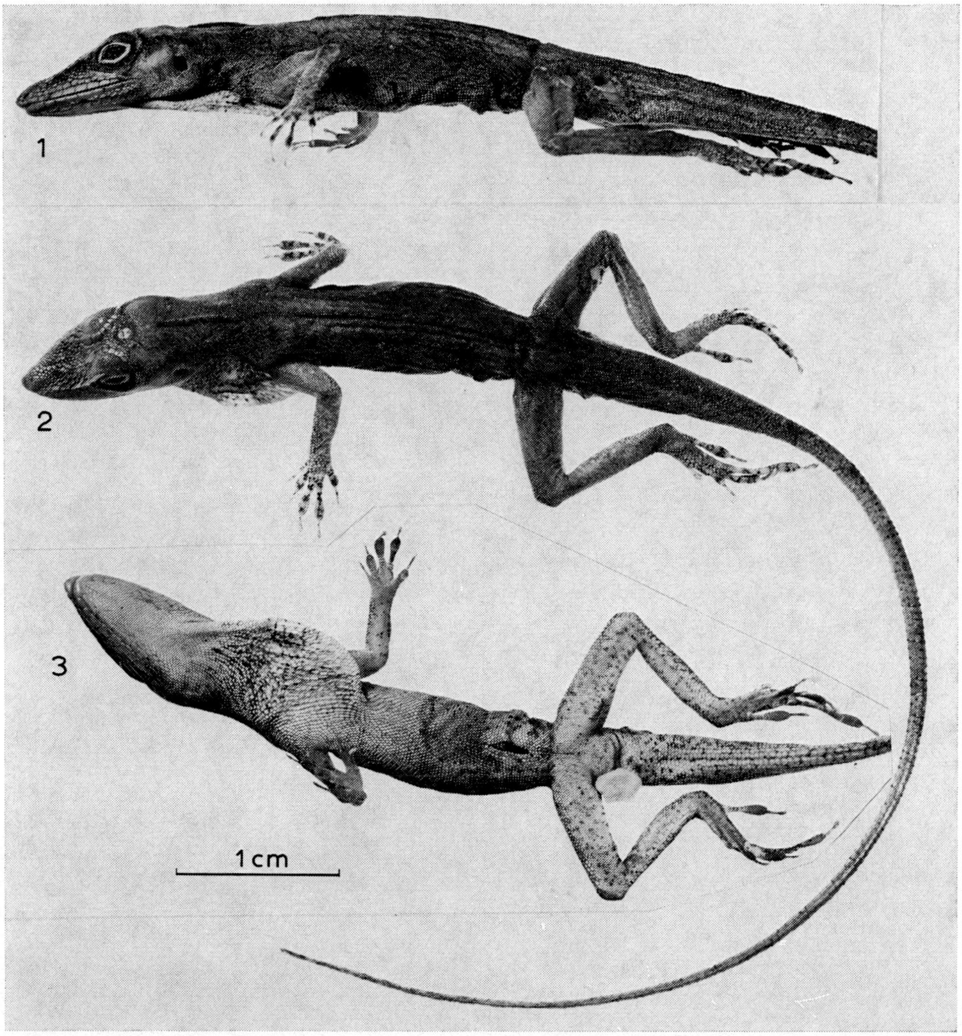

Plate 15. Anolis p. punctatus Daudin. Figs. 1–3. Lateral, dorsal and ventral aspect of ♂, RMNH 15650 a, Sipaliwini.

TYPE-LOCALITY.—South America.

MATERIAL. – Surinam.

Surinam: 1♂, RMNH 2835, leg. H. H. Dieperink; 1♀, RMNH 16435, 1901, leg. Coppename Expedition; 1♀, SMNS no number, leg. A. Kappler; 1♀, 1♂, NMW 16821; 1♀, ANS 7888, leg. Hering.

Kramakka, Commewijne River: 1♀, CM 44348, 21–IX–1966, leg. M. H. de la Fuente.

Washabo, Corantijn River: 1♀, RMNH 11546, August 1949, leg. Radin.

Table 12. Counts and measurements of *Anolis p. punctatus* Daudin. For explanation see table 3.

Reg.no.	A	B	C	D	E	F	G	H	Q'	J	S'	O	P	Q	J'	R'	W	X
RMNH																		
11546	♀	76	199	22.9	12.0	9.8	31	56	17.0	155	10-9	9-8	8-8	2-2	5-6	6-6	32(22)-32(22)	? -45(36)
11559	♀	40	79	12.0	6.8	5.7	14	26	8.0	?	10-9	8-9	9-10	2-2	5-4	7-6	28(21)-30(22)	40(34)-40(34)
11628	♂	85	183	23.4	11.5	9.5	30	?	17.3	158	12-10	7-8	9-10	2-2	5-5	6-7	33(25)-32(24)	45(36)-45(37)
15075	♂	83	196	21.5	10.9	9.8			17.3	168	9-9	9-9	10-10	3-2	5-5	5-6	32(24)-32(23)	44(36)-45(37)
15247	♂	81	155(?)	23.2	11.8	9.8			17.2	172	7-8	9-9	8-7	2-2	6-6	7-7	33-31	46-45
15649	♀	?	136	?	?	?			13.1	181	?	?	?	?	?	?	34(26)-34(26)	46(38)-49(40)
15650a	♂	71	166	21.2	10.5	9.0			16.4	156	10-11	9-9	9-10	2-2	6-6	8-8	31(24)-31(24)	48(39)-43(36)
b	♀	73	166	20.5	11.4	9.3			15.9	146	8-7	8-10	10-9	3-2	7-6	7-7	30(21)-29(21)	44(36)-44(36)
15651	♀	70	178	21.2	12.4	10.5			15.4	160	8-9	7-8	9-?	2-2	7-7	5-6	31(23)-31(23)	42(34)-41(33)
15652	♂	70	165	21.1	10.5	9.9	28	53	16.0	137	9-9	8-9	10-10	2-3	5-5	7-8	30(23)-29(22)	43(36)-44(36)
15653	♂	82	191	24.2	12.1	10.3	31	58	18.4	171	9-10	9-8	9-9	1-1	5-5	7-7	32(24)-30(23)	43(34)-43(36)
15654	♂	84	199	23.2	12.0	9.5	30	60	18.4	161	10-10	?-9	?-9	2-2	6-6	5-5	? -31(23)	43(34)-45(36)
16435	♀	41	87	12.2	7.2	6.1	16	27	8.8	?	10-9	8-8	11-11	2-2	5-5	8-8	29(22)-28(22)	41(36)-44(38)
ZMA																		
12706	♂	67	148	19.2	9.8	9.1			15.6	160	11-9	8-8	9-9	2-2	5-5	7-7	32(24)-30(22)	46(38)-46(38)
ZMB																		
7659	♂	76	176	22.0	11.0	9.7			15.6		10-10	8-9	9-7	2-2		5-4	31-33	46-45
SMNS																		
-	♀	66	149	18.3	9.9	6.8	24	45	14		9-8	9-8	11-9		6-6	6-6		
NMW																		
16821.1	♀	72	112(112)	19.7	11	9.4	26	48	14.3		6-6	8-8	8-8		5-4	7-7		
.2	♂	73	162	20.2	10.5	9.0	28	50	16.9		7-7	8-10	9-9		5-5	7-7		
ANS																		
7888	♀	77	110(110)	20.5	11.9	9.0	30	52	16.6	153	9-8	?-8	6-9	2-2	5-5	5-6	?	?
MCZ																		
57047	?	77	?	22.6	11.8	9.3			17.0		8-8						32-32	44-45
57048	♀	64	153	19.2	10.8	8.6			13.9		6-6						30-32	44-43

Forest Service Encampment 8, Mapane area: 1♂, RMNH 15075, 11–V–1968, leg. M. S. Hoogmoed.

Bigiston, Marowijne River: 1♂, RMNH 15654, September 1963, leg. L. D. Brongersma.

Upper Nickerie River: 1♂, RMNH 11628, leg. H. van Capelle.

Raleigh Cataracts, Coppename River: 1♂, RMNH 15247, 7–XI–1968, 1♀, RMNH 15649, airstrip under construction, 7–XI–1968, both leg. M. S. Hoogmoed.

Berg en Dal: 1♂, ZMB 7659, leg. H. B. Möschler.

Brokopondo: 1♂, ZMA 12706, 8–X–1966, leg. H. Nijssen.

Sikakamp, Sara Creek: 1♀, RMNH 15651, 25–VIII–1965, leg. G. F. Mees.

Nassau Mountains: 1♀, RMNH 11559, N. of km 5.4, 12–III–1949, leg. Surinam Expedition 1948–49.

Airstrip Coeroeni: 1♂, RMNH 15652, 12–VIII–1959, leg. D. C. Geijskes.

Wilhelmina Mountains: 1♂, RMNH 15653, June 1963, leg. H. P. Pijpers (Expedition Maguire).

Airstrip Kayser Mountains: 1♀, 1♂, FMNH 128985, 128987, 1–II–1961, leg. H. A. Beatty.

Sipaliwini, Base Bivouac: 1♀, 1♂, RMNH 15650, 1–IX–1968, leg. M. S. Hoogmoed.

Guyana.

Black Water Creek: 1 ex., MCZ 57047.

Onora: 1♀, MCZ 57048.

DIAGNOSIS.—A medium-sized arboreal anolis with a slightly compressed body. Dorsal scales small; scales around midbody 137–181; ventrals smooth, scales on the snout keeled. Tibia in adult specimens shorter than the head length. Digital expansions well developed. A distinct nuchal and dorsal crest present in males. Green or purplish with small sky-blue dots. Dewlap large, yellow--ochre with white (males) or yellow (females) scales. Males with a knob on the snout.

DESCRIPTION.—Head large, representing 26–30% of the snout-vent length, both in females and males; long, approximately twice as long a wide, longer than the tibia; slightly depressed; slightly wider than the neck, which is as wide as the body. In males the snout is pointed, produced into a knob projecting over the mouth; in females this knob is absent, the snout blunt, only slightly projecting over the mouth. A deep frontal depression. Occipital region with two slightly swollen spheres.

Rostral nearly triangular; just visible from above; bordered posteriorly by eight to ten small scales and by the first supralabial on each side. Eight to 12 supralabials; seven to ten (mostly eight or nine) to below the centre of the eye. Loreal region with four to seven, mostly five, horizontal scale rows. Canthal scales four to eight, the posterior four or five large, sharply keeled, forming a distinctly angular canthus rostralis. A continuous semicircle of preocular, subocular and postocular scales; largest below the eye, tapering off to both ends. Temporal region with small, round, granular scales. Ear-opening large; slightly above the commissure of the mouth; surrounded by small granules forming a finely denticulate margin; external auditory meatus short, causing the tympanum being well visible. A supratemporal arch of slightly enlarged scales from the posterior corner of the eye to above the ear-opening; the scales in its anterior portion largest, gradually tapering off posteriorly. Upper surface of the snout with polygonal scales of different sizes; slightly convex on the anterior part, flat in the posterior part and in the frontal depression. Nostril in a single nasal. Supraciliaries one to three, mostly two, the posterior one very long and narrow, largest, just reaching over the centre of the eye; forming a continuous row with the canthal scales. Supraorbital semicircles with 6–12, mostly eight to ten scales of which the second or third is largest; separated by one row of small scales. Interparietal polygonal to oval or round; in most specimens larger than the ear-opening (two specimens equal, one specimen smaller); separated from the supraorbital semicircles by two or three rows of scales. Scales on the occipital and parietal regions small, polygonal, juxtaposed; slightly larger than the scales on the nape. Supraocular region with a round patch of enlarged scales medially; grading into granules anteriorly, posteriorly and laterally, separated from the supraorbital semicircles by one row of scales. Eyelids covered with minute granules, two rows of enlarged scales bordering the rim. Pupil round.

Mental semicircular, semidivided by a medial longitudinal cleft; bordered posteriorly by two to four scales, two sublabials and two infralabials. Infralabials 6–11, mostly nine or ten, with an elevated ridge along the mouth, bordered medially by one keeled row of enlarged sublabials; the other scales on the chin gradually grading into granules. Scales on the throat granular, smooth. Dewlap very small in females, very large in males, reaching from beneath the eyes to the chest just posterior to the forelimbs; when extended showing a number of more or less horizontal rows of dorso-ventrally flattened, triangular, protruding scales; rim scales smaller; all dewlap scales smaller than the ventrals. Scales on the nape juxtaposed, small, round, granular, identical to the dorsals. A low fold on the nape from the occipital region to above the insertion of the forelimbs in males, absent in females.

Dorsals and laterals juxtaposed, small, round, granular, smooth, much smaller than the ventrals. Ventrals round to oval, smooth, slightly imbricate. No distinct border between the ventrals and the laterals. Total number of scales around midbody 137–181. A low fold on the back from behind the insertion of the forelimbs to above the insertion of the hind limbs in males, absent in females. Preanal plate with small granules, smaller than the ventrals. Enlarged postanal scales are absent in most specimens, in some, however, they are present.

Scales on the tail small, round, keeled, slightly imbricate, irregularly arranged in the proximal part; larger, hexagonal, keeled, slightly imbricate, in longitudinal rows in the distal part. Scales under the tail large, slightly imbricate, hexagonal, wider than long, smooth in the proximal part; hexagonal, longer than wide, keeled, in the distal part; in longitudinal rows. Distally the keels on the scales increase in height, forming four serrated ridges. Tail slightly compressed, ending in a four-ridged knob. Except in the most proximal and the most distal part, the tail is divided into verticils. Each verticil with three keeled scales below, the anterior verticils with five or six dorsal granules above, the posterior ones with four keeled scales above. In none of the specimens is the tail regenerated, which seems to indicate that it is not readily disposed of.

Scales on the forelimbs and on the upper surface of the hind limbs small, granular, smooth to slightly keeled; scales on the lower surface of the hind limbs as large as the ventrals or slightly smaller. Twenty-eight to 34 single, smooth lamellae under the fourth finger, 21–26 to the end of the digital expansion; 38–48 single, smooth lamellae under the fourth toe, 31–40 to the end of the digital expansion. Palms and soles with small, granular scales. Digital expansions on fingers and toes strongly developed, i.e., the third phalanx is much wider than the basal one. Proximal part of toes and fingers depressed, distal part compressed.

Colour in life grass-green above with sky-blue dots, eyelids purplish with a yellow rim. Chin bluish, throat greenish, belly white to purple-green. In bright sunlight the lizards are purple with sky-blue dots. Dewlap in males yellow-ochre with rows of white scales, in females the scales on the small dewlap are yellow. Iris orange-brown.

In preservative the colour of the back ranges from deep purple to beige with or without small white spots. A transverse beige band may or may not be present on the back of the head. The purple animals show no pattern, the beige females have a series of dark-brown blotches on the vertebral region, continued on the tail. Belly from nearly white through brownish white to purplish white, immaculate or with a number of dark-brown to blackish spots. Upper surface of fingers and toes with transverse black bands. Tongue creamy.

A medium-sized lizard of slender built with a slightly compressed body. Maximum snout-vent length in females 77 mm, in males 85 mm. Limbs long, hind limb nearly twice as long as forelimb in both sexes, when laid along the body there is considerable overlap of forelimb and hind limb, the hind limb reaching to between the ear and the eye. The undamaged tail is 2–2.6 times as long as the snout-vent length in females, 1.9–2.4 times in males.

HABITAT.—An arboreal lizard from the rain-forest. Most specimens collected by me were on the trunks or branches of large trees, at least one meter above the ground. RMNH 15649 was captured in the crown of a tree (10 m high) immediately after it was felled. Thus, *Anolis p. punctatus* Daudin ranges from the base of trees into the canopy. One specimen was collected on a leaf of a small tree overhanging a creek (see under natural history).

NATURAL HISTORY.—Specimens of this subspecies are active in daytime, most specimens avoiding direct sunlight, but on one occasion a specimen (♂, RMNH 15650) was observed basking. In contrast to the other specimens collected this specimen was deep purple with blue spots, obviously as a result of the direct sunlight. The other specimens collected in daytime (RMNH 15247, 15649, ♀ 15650) were collected in the shade and were grass-green with blue spots. RAND & HUMPHREY (1968) studying *Anolis punctatus* Daudin near Belém found this species in the same general habitat as described above. According to their observations *A. punctatus* Daudin is a non-heliotherm, which was never observed basking and the specimen observed basking by me may be regarded as an exception. One specimen (RMNH 15075) was collected at 10.30 p.m., sleeping on a large leaf of a small (3 m) tree, overhanging a creek. The leaf was hanging at an angle of about 45°. The anole was situated along the midrib, its head pointing downwards, the hind limbs stretched, parallel to the tail, the distal part of which curved back towards the head. The forelimbs were flexed and pressed against the body. Its colour was grass-green with light-blue spots.

Once a male and a female (RMNH 15650) were captured on neighbouring trees, three quarters of an hour after each other. This might indicate pair--formation in this subspecies, a tendency that also has been established near Belém by RAND & HUMPHREY (1968).

This subspecies is capable of remarkable colour changes in short periods as was demonstrated by a ♂, RMNH 15650, which, a few moments after collecting, from deep purple became grass-green. In the other specimens the colour changes took place immediately after injection with pentabarbitol or shooting with dust shot; they adopted the purple colour, so characteristic of many preserved animals, becoming spotted in a few minutes. The purple colour arose around the point of insertion of the needle or around the shotholes.

Three pregnant females, two containing one, the other containing two mature eggs were collected in August and early September, indicating that breeding may be restricted to the early long dry season.

Only one stomach was examined. It contained one grasshopper, one ant, one fly, one beetle and one homopteron.

RANGE.—Northern South America from Peru through Brazil to Venezuela (Amazonas) and the Guianas, south to Rio de Janeiro. In Surinam the species is present throughout the country, on both sides of the coastal savannah belt. *A. p. punctatus* Daudin occurs from sea level to at least 250 m (fig. 20).

Genus **Iguana** Laurenti

DIAGNOSIS.—Large, green lizards with cylindrical bodies and long compressed tails. A serrated dorsal crest from the head onto the tail. A large gular appendage. Tympanum large and well visible. Femoral pores present. Fourth toe longer than the third. Lateral teeth indistinctly tricuspid, with serrated edges. Two species, one in a large part of Central and South America, e.g., Surinam, the other on some of the Antilles.

Iguana iguana iguana (Linnaeus) (figs. 22, 23; pl. 16)

'Leguanen' Herlein, 1718: 173; Kappler, 1863: 74; Franssen Herderschee, 1905a: 77; Franssen Herderschee, 1905b: 921; De Goeje, 1908: 1018; Kayser, 1912: 453, 455, 467; Geijskes, 1957: 237, 254.

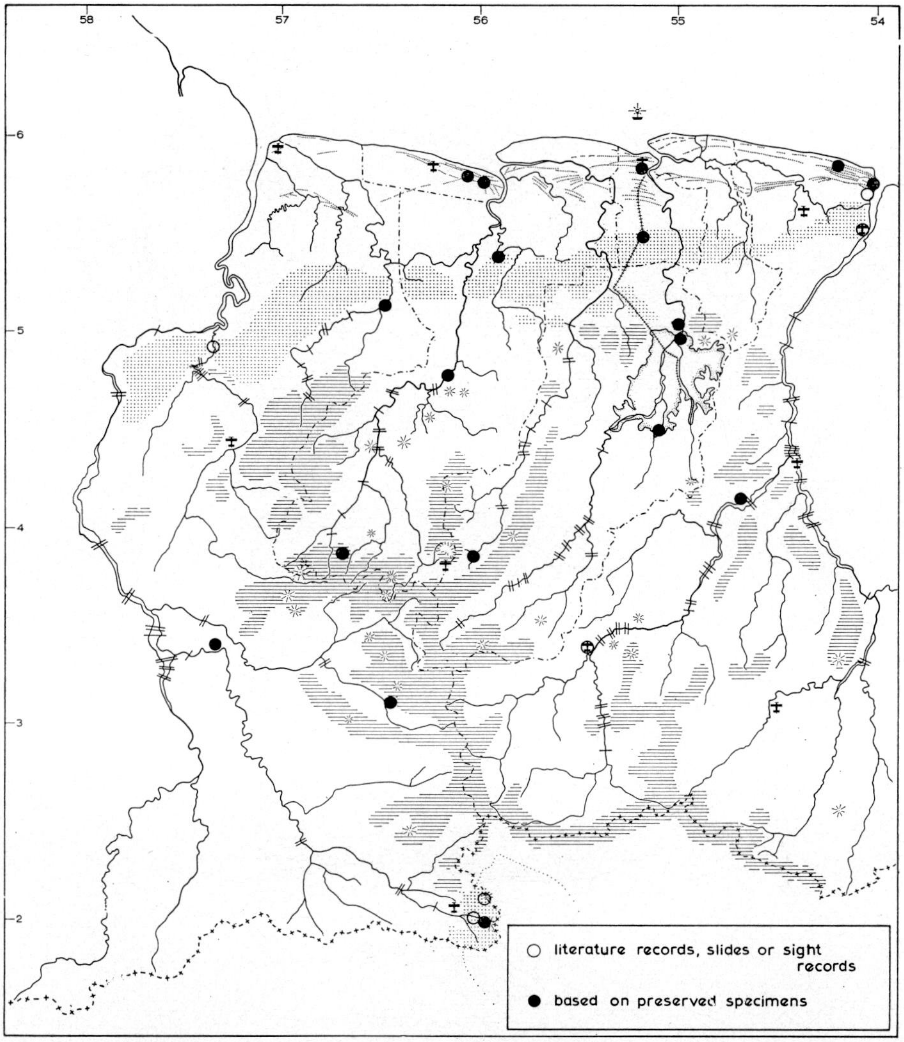

Fig. 22. Distribution of *I. i. iguana* (Linnaeus) in Surinam.

Lacerta, seu Leguana, Surinamensis, pectinata, & strumosa, caerulea, faemina Seba, 1734: 149, pl. 95 fig. 2.

Liguana Senembi, *in N. Hispania,* Tamacolin *dicta, pectinata, & strumosa* Seba, 1734: 152, pl. 97 fig. 3.

Lacertus, Amboinensis, pectinatus, & strumosus, maximus, Senembi & Iguana *dictus, amphibius* Seba, 1734: 153, pl. 98 fig. 1.

Iguana cauda tereti, longa, cylindrica, crista gulae pendula antice dentata, sutura dorsali denticulata Gronovius, 1756: 82.

Lacerta Iguana Linnaeus, 1758: 206; Bancroft, 1769b: 90.

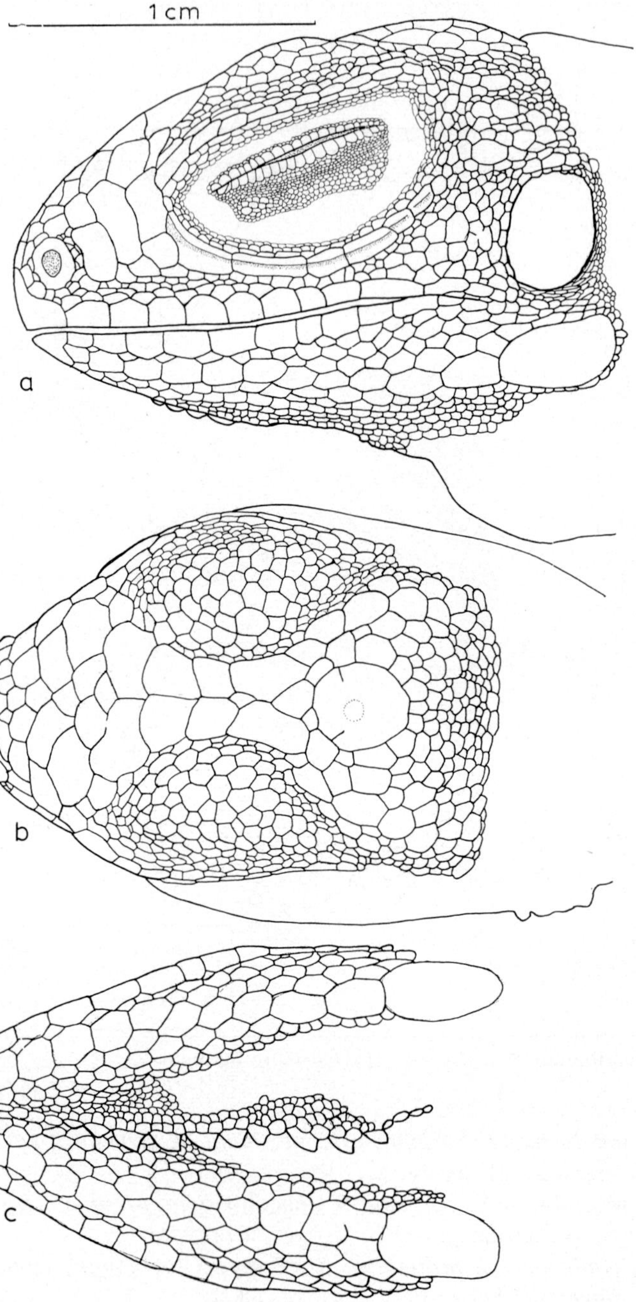

Fig. 23. I. i. iguana (Linnaeus); a, b, c: lateral, dorsal and ventral views of the head of juv. RMNH 15258, Coronie Road.

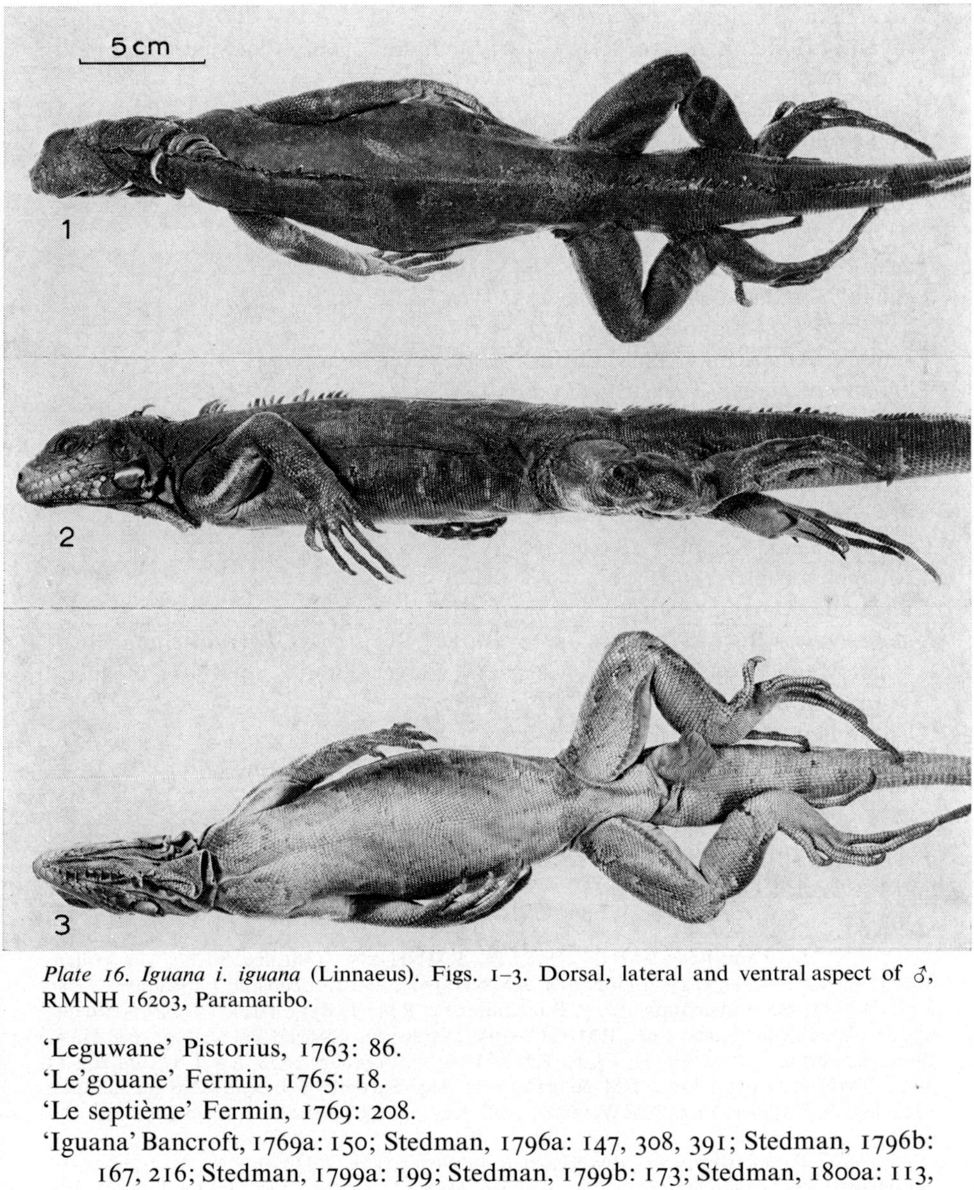

Plate 16. Iguana i. iguana (Linnaeus). Figs. 1–3. Dorsal, lateral and ventral aspect of ♂, RMNH 16203, Paramaribo.

'Leguwane' Pistorius, 1763: 86.
'Le'gouane' Fermin, 1765: 18.
'Le septième' Fermin, 1769: 208.
'Iguana' Bancroft, 1769a: 150; Stedman, 1796a: 147, 308, 391; Stedman, 1796b: 167, 216; Stedman, 1799a: 199; Stedman, 1799b: 173; Stedman, 1800a: 113, 184.
'Leguaan' Hartsinck, 1770: 100; Kappler, 1854a: 120; Kappler, 1863: 75; Bakhuis 1902: 76, 98, 99, 115; Van Capelle, 1903: 56, 85, 150; Menne, 1946: 21; Geijskes, 1957: 232.
'De zevende' Fermin, 1770: 184.
'L'Iguane' Daubenton, 1784: 640; Lacépède, 1788a: 267; Lacépède, 1788b: 322; Lacépède, 1799: 106; Desmarest, 1833: 402; Lacépède, 1847: 179,

'Wayamaka' Stedman, 1796a: 308.

'Wayamaca' Stedman, 1796a: 147, 391; Stedman, 1799a: 199; Stedman, 1799b: 68, 173.

'Iguan' Stedman, 1799b: 68.

'Der Leguan' Bechstein, 1800a: 480; Kappler, 1881: 107.

Iguana delicatissima: Latreille, 1802a: 255; Daudin, 1802a: 263; Anslijn, 1827: 68; Fischer, 1832: 35; Kappler, 1881: 139; Westerouen van Meeteren, 1883: 23.

Iguana coerulea Daudin, 1802a: 286.

'Leguan' Von Sack, 1821a: 70, 91, 225; Von Sack, 1821b: 127, 162; Von Sack, 1821c: 238; Kappler, 1862: 177.

'Common American Iguana' Griffith & Pidgeon, 1830: 223.

'Laguaan of guana' Teenstra, 1835: 437.

Iguana tuberculata: Anonymus, 1841: 87; Duméril & Duméril, 1851: 63; Kappler, 1881: 166; Im Thurn, 1883: 130; Boulenger, 1885a: 189; Kappler, 1885: 816; Kappler, 1887: 124; Boettger, 1893: 63; Van Capelle, 1903: 55; Penard & Penard, 1906: no. 81; Van Lidth de Jeude, 1917f: 372.

Lacerta iguana: Kappler, 1854b: 46.

'Leguane' Kappler 1862: 176.

'Eidechsen' Martin, 1886: 44.

Iguana iguana: Burt & Burt, 1930: 10; Burt & Burt, 1931: 267; Boschma, 1947: 40; Menne, 1952: 10; Goin, 1967: 119; Haverschmidt, 1968: 55; Geijskes, 1970: 279.

'Groene leguanen' Geijskes, 1957: 277.

Iguana i. iguana: D'Audretsch, 1965c: 21; J. A. Peters & Donoso-Barros, 1970: 149.

TYPE-LOCALITY.—'Indiis', here restricted to the confluence of the Cottica River and the Perica Creek, Surinam.

MATERIAL. – Surinam.

Surinam: 1 ex., RMNH 2781; 1 ex., RMNH 2850, 1861, leg. Snellen van Vollenhoven; 1 ex., RMNH 3133; 2 embryo's, RMNH 7507, leg. Coppename Expedition; 1 ex., RMNH 7521, November 1905, leg. P. Buitendijk; 1♀, RMNH 8335, 1946; 1♂, RMNH 9361, leg. Jhr. W. C. van Heurn; 1 ex., RMNH 10405; 2 embryo's, RMNH 16184, 1901, leg. H. A. Boon (Coppename Expedition); 1 juv., RMNH 16189, February 1888, leg. J. H. Spitzly; 13 juvs., RMNH 16191; 2 juvs., BM 66.9.20.10–11, leg. Bartlett; 3 juvs., SMNS no number, 1844, leg. A. Kappler; 1 ex., NMW 13043, 1908, leg. A. O. C. Heller; 1♀, ANS 8097, leg. Hering.

Paramaribo: 1 ex., RMNH 4728, July 1920; 3 ex., RMNH 7517, 1911, leg. W. C. van Heurn; 1 ex., RMNH 7519, 7–III–1939, leg. H. W. C. Cossee; 2 ex., RMNH 7520, 1911, 4 ex., RMNH 7522, 1911, all leg. Jhr. W. C. van Heurn; 2 ex., RMNH 7523, 5–III–1939, leg. H. W. C. Cossee; 1♂, RMNH 16190, 21–IV–1963, leg. A. M. Husson & P. Staffeleu; 2♂♂, RMNH 16201; 1♂, RMNH 16202, 12–III–1967, leg. J. van der Land; 1♂, RMNH 16203; 12–VII–1968, leg. M. S. Hoogmoed; 1 juv., ZMA no number, 13–XII–1922, leg. J. W. Gonggrijp & G. Stahel; 1 juv., BM 1946.4.2.88, 21–I–1938, 2 juvs., BM 1946.4.2.89–90, 25–I–1938, 4 juvs., BM 1946.4.2.91–94, 28–I–1938, 1♂, 11 juvs., BM 1946.4.2.95–99, 1946.4.3.1–7, 2–II–1938, all leg. I. T. Sanderson; 2 juvs., ZMB 25994, 25999, 1♂, NMW 13035, 1908, all leg. A. O. C. Heller; 1 ex., CM 44351, 24–VII–1966, 1♀, CM 44352, August 1966, both leg. M. H.

de la Fuente; 1 ex., AMNH 1696, 1906, leg. Moelaert; 5 ex., AMNH 8150, 8612–15, 1916, leg. J. A. Samuels.

Coronie road: 1 juv., RMNH 15258, picket 187–188, 22–XII–1948, 1 juv., RMNH 16195, picket 197–198, 23–XII–1948, 1♀, RMNH 16199, km 0, picket 216, 20–XII–1948, all leg. Surinam Expedition 1948–49.

Coronie: 1 juv., RMNH 16200, 20 miles E., 29–IV–1966, leg. W. Vogd & J. Meisenbacher.

Wia Wia transect: 1 juv., RMNH 16193, 4th swamp, km S. 1.85, 16–XI–1948, 1 juv., RMNH 16194, 4th swamp, km S. 1.39, 16–XI–1948, both leg. Surinam Expedition 1948–49.

Galibi: 1♀, RMNH 16192, 5–XI–1948, leg. Surinam Expedition 1948–49.

Tibiti: 1 juv., RMNH 16196, 12–I–1949, 3 juvs., RMNH 16197, 15–I–1949, all leg. Surinam Expedition 1948–49.

Zanderij: 1♀, RMNH 16198, 3–VIII–1948, leg. D. C. Geijskes; 1♀, BM 1946.4.3.8, 5–IX–1938, 1♀, BM 1946.4.3.9, 30–VIII–1938, both leg. I. T. Sanderson.

Stondansie Cataracts, Nickerie River: 1♂, RMNH 16471, 3–II–1971, leg. D. G. Reeder.

Brokopondo: 1♀, ZMA 12781, 12–IX–1966, 1♀, ZMA 12782, 27–IX–1966, both leg. H. Nijssen.

Brokopondo artificial lake: 1♀, 1♂, RMNH 16187, 2 km S. of barrage, 1–V–1965, leg. G. F. Mees.

Marowijne Creek: 1 ex., ZMA 12656, 7–VI–1966, leg. H. Nijssen.

Wilhelmina Mountains: 1 juv., RMNH 16178, transect I, August–September 1943, leg. D. C. Geijskes.

Base Bivouac near De Kock Mountain, Upper Saramacca River: 1 juv., RMNH 16186, 12–III–1958, leg D. C. Geijskes (Tafel Mountain Expedition).

Drietabbetje: 2 ex., RMNH 7524, 10–XI–1904, leg. G. M. Versteeg (Tapanahoni Expedition).

Airstrip Coeroeni: 1 juv., RMNH 16185, 12/14–VIII–1959, leg. D. C. Geijskes.

Airstrip Kayser Mountains: 1 juv., RMNH 16205, 22–IV–1969, leg. H. M. van Meeuwen; 2 juvs., FMNH 128981–82, 1–II–1961, leg. H. A. Beatty.

Sipaliwini, Vier Gebroeders Bivouac: 1♂, RMNH 16204, 29–IX–1968, leg. M. S. Hoogmoed.

Diagnosis.—A very large, arboreal lizard with a cylindrical body and a long compressed tail. A serrated dorsal crest from the head onto the tail, a large gular appendage with a very large, round scale just below the ear. Femoral pores present. Number of scales around midbody 183–231. On the neck some longitudinal series of enlarged, conical scales. Body grass-green all over with a distinct white line on the upper arm. Tail banded white and black on a grass-green background.

Description.—Head moderately large, representing 19–22% of the snout-vent length in females, 19–23% in males and 22–25% in juveniles; 0.9–1.0 times as long as the tibia in adults, 1.1–1.3 times in juveniles; very slightly depressed in juveniles, in adults deeper than wide, as wide as the neck which is as wide as the anterior part of the body. Snout blunt, rounded, convex to straight in profile, slightly projecting over the mouth.

Tongue arrowhead-shaped, slightly bifid anteriorly; villose. Teeth indistinctly tricuspid, with serrated edges, laterally compressed; increasing in size posteriorly.

Rostral large; pentagonal to nearly triangular; one and a half to two times as wide as deep; visible from above; bordered posteriorly by 9–13 small scales and the first supralabial on each side. Eight to 11, mostly nine or ten, supralabials, large, roughly quadrangular anteriorly; decreasing in size posteriorly and becoming more elongately rectangular, the sixth to ninth under the centre of the eye. Nostril directed laterally, placed in a single large, bulbous nasal. Nasal separated from the rostral by one row of small scales or in contact with the rostral; separated from the first supralabial by one to three rows of small scales. Loreal region with three to six, mostly four or five horizontal rows of scales, the lower two or three rows consisting of very small scales, the upper rows consisting of large scales. Two or three, mostly three, canthal scales; anterior canthal scale separated from the nasal by one or two rows of small scales, containing three to seven scales. Canthus rostralis indistinct, rounded. Suboculars four to nine, mostly five to seven; distinctly enlarged; the anterior largest, the posterior smallest; slightly keeled, forming an indistinct suborbital rim. Postoculars not different from the temporals. Temporal region with small, irregularly shaped, polygonal, juxtaposed, flat scales; with a triangular patch of larger scales immediately above the posterior supralabials; some scattered larger scales in the supratemporal region, sometimes tending to form a supratemporal band. Ear-opening large, round, with smooth margins; external auditory meatus absent, tympanum superficial, clearly visible. Eyelids covered with minute, granular scales, a row of enlarged scales along the rim. Pupil round, the margin may be undulating in some places. Scales on the snout irregularly shaped, polygonal, small on the anterior part, larger on the posterior part, without transitional zone. Supraorbital semicircles in contact, consisting of approximately five to nine distinctly enlarged scales, on the posterior end of the semicircle the scales merge into the scales of the parietal region and it is difficult to determine where the semicircle ends. Interparietal small, irregularly shaped, mostly longer than wide; much smaller than the ear-opening; separated from the supraorbital semicircles by one row of scales. Parietal region with fairly large polygonal scales, decreasing in size towards the temporal and occipital region. Scales on the occipital region much smaller than those on the temporal region. Supraocular region with small polygonal scales, largest in the centre, decreasing in size towards the margins. Supraciliaries 8–14, mostly 9–11; the anterior one largest, elongate; the rest rectangular to quadrangular; it is difficult to determine the posterior scale of the supraciliary row.

Mental more or less mushroom-shaped, with concave sides; mostly wider than long; bordered posteriorly by one to three, mostly two, small scales. Eight to 14, mostly 9–11, quadrangular infralabials, increasing in size from the first to about halfway the series and then decreasing in size, the posterior scale smallest; anteriorly one, posteriorly two rows of enlarged sublabials, ending in a very large round scale below the ear-opening, which is as large as or larger than the ear-opening; sublabials anteriorly in contact with the infralabials, posteriorly separated from them by one or more rows of small scales. Scales on chin and throat minute, round to hexagonal, juxtaposed. A large gular appendage from just posterior of the postmental scales to the gular fold, anterior margin denticulate, with 11–16 enlarged, triangular, laterally compressed scales. A transverse gular fold is present.

Scales on the nape small, rectangular to hexagonal, convex, juxtaposed, in transverse rows. Scales on the side of the neck identical to those on the nape, decreasing in size towards the throat. Dorsolateral region of neck with a number of enlarged, conical scales, forming irregular longitudinal rows, extending from the back of the head to just anterior of the insertion of the forelimbs. Two antehumeral folds. The gular fold extends obliquely dorsally and posteriorly on the side of the neck. A less distinct fold on each side from the large scale below the ear-opening onto the side of the neck, parallel to the other antehumeral fold.

Dorsals and laterals small, as large as the scales on the nape, rectangular to quadrangular, obtusely and diagonally keeled, slightly imbricate, in transverse rows. A distinct, uninterrupted vertebral crest of enlarged scales from the back of the head to the tip of the tail. In juveniles the vertebral scales are low, blunt and conical, in adults these scales are falciform. The scales on the neck may be longer than those on the back. On the tail the height of the crest decreases until the upper margin of the tail is just serrated. Ventrals larger than the dorsals, lanceolate to rhomboidal, smooth, imbricate, in transverse series only. Total number of scales around midbody 183–231, mostly 191–219. Preanal area with scales identical to the ventrals, more or less arranged in transverse series. Preanal pores absent; femoral pores 12–18, mostly 14–17, surrounded by five or six small scales.

Scales on the anterior part of the tail like the dorsals; posteriorly rhomboidal, larger and keeled, the keels forming longitudinal ridges, in transverse and oblique series. Scales under the anterior part of the tail lanceolate, slightly keeled; under the posterior part of the tail larger, distinctly keeled, the keels forming longitudinal ridges. Tail divided into indistinct verticils with five dorsal and four ventral scales. Scales on the regenerated part of the tail rhomboidal to lanceolate, keeled, smaller than on the original tail, imbricate, irregularly arranged, the keels not forming longitudinal ridges; no trace of the dorsal crest. Tail compressed, oval with a sharp dorsal ridge in cross-section.

Scales on the limbs rhomboidal; keeled, imbricate on the dorsal and anterior surfaces of the forelimbs, on the anterior surface of the thighs and on the ventral surface of the lower legs; on all other surfaces smooth or slightly keeled, smaller and juxtaposed; on the dorsal surfaces as large as the dorsals, on the ventral surfaces slightly smaller than the ventrals. Twenty-three to 29, mostly 25–28, single lamellae with three keels under the fourth finger; 31–41, mostly 34–39, single lamellae with two keels under the fourth toe. Palms and soles with small, lanceolate, imbricate, smooth scales. Fingers and toes distinctly compressed, oval with a flat base in cross-section.

Colour in life grass-green all over, slightly lighter on the ventral parts. In old specimens the overall colour fades to a bluish grey. A distinct white line on the lower anterior surface of the upper arm, from the insertion to near the elbow.

The upper surface of the head in adults may be bluish green or greyish. A number of more or less distinct vertical white stripes, accompanied by a black zone is present on the lower part of the flanks. Gular appendage with black and reddish brown stripes. Infra- and sublabials and anterior margin of gular appendage with reddish brown spots. Tail with white and black transverse bands on a grass--green background. In old specimens the tail shows black bands, alternating with bluish grey bands. Iris brown with a golden rim around the pupil.

Colour in preservative dark bluish green above, yellowish green beneath. A wide, brown vertical band over the eyelids. Four to five short, vertical white--green stripes on the lower flanks, bordered posteriorly by a black stripe which-tapers off into a black zone; very distinct in most, except in the largest specimens. A white-green stripe on the upper arm. Gular appendage dirty white with about four oblique dark-brown stripes, slanting ventrally and posteriorly. In juveniles the back, flanks and limbs are uniform, in adults they may be covered with numerous dirty white round spots. Tail with distinct, narrow, white bands, followed posteriorly by wider black or dark-brown bands. Each set of white and black (dark-brown) bands is separated from the following set by a wide bluish green band. On the posterior part of the tail the white is absent and there only black (dark-brown) and bluish green bands alternate.

A very large lizard with a cylindrical body and long, compressed tail. Maximum snout-vent length (measured by me) in females 322 mm, in males 360 mm. The undamaged tail is 2.9–3.1 times as long as the snout-vent length in females, 2.5–3.1 times in males and 2.4–2.9 times in juveniles. Limbs long, stout, hind limb reaching to the eye. Hind limb 1.5–1.7 times as long as forelimb in juveniles, in a single adult male it is 1.4 times as long.

HABITAT.—This is an arboreal lizard, frequenting forest edges along rivers. It can be found from the ground to the upper layers of the canopy and is present in cultured gardens in Paramaribo, in rain-forest along creeks and rivers, in ridge-forest, in swamps, on the beach and on savannahs. The presence of water seems to be important but not necessary, as a specimen was observed in savannah-country high on the slopes of the Moro Grande do Cimeterio far from the nearest creek. These data agree with those of RAND & HUMPHREY (1968) and others cited by them. RAND & HUMPHREY suggest that the relation with edges of rivers is artificial and caused by the difficulty of observing these lizards in the closed forest.

NATURAL HISTORY.—A diurnal lizard which was observed from 10.30 a.m. to 4.00 p.m. but probably has a longer activity period; these data are based on only four specimens. Basking behaviour was observed several times and most narratives of journeys in the interior mention iguanas basking on branches of trees along rivers (IM THURN, 1883; KAYSER, 1912; FRANSSEN HERDERSCHEE, 1905a, b). RAND & HUMPHREY (1968) did not collect temperature data on this species, but report to have it seen basking frequently. MC-GINNIS & BROWN (1966) made a detailed study of the thermal behaviour of this species and concluded that it was a heliotherm. Pregnant females are known from early August to late September. They contained from 24–57 mature, round to oval eggs. BEEBE (1944b) reports clutches of 4 and 7 eggs from late September and early October respectively, from Kartabo, but all other authors give numbers from 25–50. I only saw a single juvenile that probably had hatched recently, its snout-vent length being 63 mm; it was collected in February. Some larger specimens (snout-vent length 74–75 mm) from March and October still show a distinct umbilical scar and, thus, may be recently hatched. However, other specimens of the same size have the scar completely closed. According to VAN CAPELLE (1903) the iguanas deposit their eggs 'vooral in de maanden September en October' (mainly in the months September and October). HAGMANN (1909) mentions the same period for egg-laying on Mexiana. IM THURN (1883) and KAPPLER (1887) agree in that the eggs are deposited early in the dry season. KAPPLER (1887) says that the eggs are laid in August and that the juveniles will hatch in October, November and early December (KAPPLER, 1862; 1863). From most narratives (BAKHUIS, 1902 (September); VAN CAPELLE, 1903; FRANSSEN HERDERSCHEE, 1905a, b (September/October); IM THURN, 1883) it is evident that sandbanks in and along rivers are favourite egg-laying sites. KAPPLER (1887) also mentions this same place. HAGMANN (1909), IM THURN (1883) and KAPPLER (1887) observed that the eggs were deposited at the end of a long tunnel which is dug by the female. BEEBE (1944b) found the eggs in Kartabo 'under a bank' and 'at the foot of a dead stump'. SWANSON (1950) found quite different data in Panama, where eggs are laid from early February to March and seemingly are laid on the ground and covered with dirt.

RAND (1968) extensively described the nesting process of the iguana on a small islet near Barro Colorado Island in Panama. In contrast to SWANSON he found that the eggs were deposited at the end of a tunnel. Both authors agree about the time of egg-laying.

HIRTH (1963) reports that the eggs on Tortuguero Beach, Costa Rica are laid in the hottest and driest part of the year, in March and early April, in deep holes on the beach. The juveniles in Panama and in Costa Rica take the same time to hatch (three months) as in Surinam. Nothing is known about mating in Surinam; in Central America the mating season seems to be from October to December (HIRTH, 1963; SWANSON, 1950). More details on the reproduction of this species in Central America are given by FITCH (1970). IM THURN (1883) reports that one nest usually contained 30–40 eggs, but that sometimes a larger number was found, which led him to suppose that several females may deposit in the same place. In an unpublished report on the lizards of Bigisanti, HILL and GREEN report the iguana to live on the first sandy ridge. There it is said to 'live in burrows about one foot deep and three or four yards long. They live in colonies of three to five'. As far as I know this behaviour of iguana living in groups in burrows has not previously been reported and it is well worth further investigation. Probably this strange behaviour has been induced by the absence of large trees on the first sandy ridge. A comparable phenomenon has been reported (SCHIFTER, 1972) for *Chamaeleo chamaeleon* (Linnaeus), from North Africa, where specimens on the edges of oases live in burrows in the sand and have adopted a terrestrial way of life, quite in contrast to their relatives in the oasis which are arboreal. The iguana is arboreal, but there are differences between adults and juveniles (RAND & HUMPHREY, 1968; SWANSON, 1950). The juveniles are frequently observed on the ground and on the branches of shrubs (COTT, 1926; KAPPLER, 1887), but the adults do not seem to come to the ground very often. When disturbed, the iguanas dive from branches overhanging rivers or creeks, into the water, where they swim submerged for considerable distances (KAPPLER, 1887; SWANSON, 1950). In the Bigisanti reserve iguanas were occasionally found in the surf at low tides. HILL and GREEN suppose that this may be an aid in thermoregulation. GOIN (1968) observed an iguana during rain accompanied by storm, that intentionally exposed itself in the top of a tree. When the first drops of rain started to fall, the iguana came to the tree and climbed to the top. This agrees well with SWANSON's (1950) observations who, in Panama, noted that iguanas seem to prefer heavily foliaged trees when the sun is shining and, when it is raining, more open trees. However, during prolonged, cool, wet periods the iguanas seem to leave the tree tops and stay on the ground. No such active searching for rain as reported by GOIN has been noted before. It will be interesting to find out whether this behaviour reflects merely a peculiar habit of one individual, or whether more individuals exhibit it.

When cornered, the iguana is well capable of defending itself by biting, clawing and lashing sideways with the long whip-like tail. Adult iguanas probably only are eaten by the large cats (BEEBE, 1944b; IM THURN, 1883) and, maybe, occasionally by boas. VAN CAPELLE (1903) reports that he found the gizzard of a large vulture (*Sarcorhamphus papa* (Linnaeus)) filled with the remains of a large iguana. Juveniles of course have more enemies and will fall victim to most carnivores. BEEBE (1944b) reports a juvenile being dropped by a swallow-tailed kite (*Elanoides forficatus yetapa* (Vieillot)). The greatest predator on adult iguanas and on the eggs doubtlessly is man, who consumes both.

The iguana has been reported from stomachs of the savannah hawk (*Heterospizias m. meridionalis* (Latham)) by HAVERSCHMIDT (1968).

All stomachs examined, both of juveniles and adults, contained only vegetable matter (no fruits or stones) that could not be identified further. HILL and GREEN reported that the stomachs of specimens collected on Bigisanti contained leaves and flowers of *Canavalia maritima*. They also reported caterpillars, but assume that these were taken accidentally during feeding on the vegetation. In contrast to the general opinion (BATES, 1965; SWANSON, 1950) that juvenile iguanas are insectivorous, my finds seem to indicate a purely vegetarian diet. KAPPLER (1887) says that young iguanas feed on (young, recently germinated) plants, which is in accordance with my finds.

RANGE.—From southern Costa Rica and adjacent Caribbean islands to southern Brazil and Paraguay. North of Costa Rica, into Mexico *I. i. iguana* Linnaeus is replaced by another subspecies. In Surinam this lizard is present throughout the country, in savannahs, in the rainforest, in cultivated land, on the beach and on mountains. It occurs from sealevel to at least 500 m. According to DUNN (1944) it reaches 1000 m in Colombia (fig. 22).

VERNACULAR NAME.—VAN LIDTH DE JEUDE (1917f) gives three names for this species. In the Surinam language it is called 'Legoeana', in Arowak 'Joewama' and in Carib 'Wajamakka'. KAYSER (1912) gives as the Trio name for this lizard 'uana'.

Genus **Plica** Gray

DIAGNOSIS.—Arboreal lizards with compressed, cylindrical or depressed bodies. Five to eight subequal suboculars; no enlarged postmentals; supraciliaries forming a crest over the eye; supraoculars enlarged; supraorbital semicircles distinct; interparietal very large, wider than the distance between the orbits, much larger than the ear-opening. Ear-opening with slightly denticulate anterior and smooth posterior margin. A gular pouch is present. Tail without a middorsal crest. Only two species in northern South America; both occur in Surinam and can be distinguished with the following key.

Key to the species of *Plica*

1. Body strongly depressed, tufts of spiny scales on the neck, a strong lateral fold, loreal region with five to eight horizontal scale rows, supralabials mostly five or six, total number of scales around midbody 117–144 . . . . *Plica plica*, p. 159
   Body cylindrical, no tufts on the neck, no lateral fold, loreal region with three or four horizontal scale rows, supralabials mostly four, total number of scales around midbody 52–65 . . . . *Plica umbra*, p. 167

Plica plica (Linnaeus) (figs. 24, 25; pl. 17)

Lacerta plica Linnaeus, 1758: 208.
Agama umbra: Daudin, 1802a: 375.
Hypsibatus agamoides Duméril & Bibron, 1837: 254; Duméril & Duméril, 1851: 69.
Ptychosaurus punctatus Fitzinger, 1843: 59.
Ptychosaurus Plica: Fitzinger, 1843: 59.
Ptychopleura Plica: Fitzinger, 1864: fig. 11; Wretschko, 1867: 79.
Hypselopus plica: Kappler, 1881: 166.
Polychrus marmoratus: Kappler, 1885: 817 (partly); Kappler, 1887: 126 (partly).
Uraniscodon plica: Boettger, 1893: 63; Van Lidth de Jeude, 1898: 20; Van Lidth de Jeude, 1904: 88; Van Lidth de Jeude, 1917g: 438.
Plica plica: Burt & Burt, 1930: 19; Etheridge, 1970: 245; Reed & Borowsky, 1970: 161.

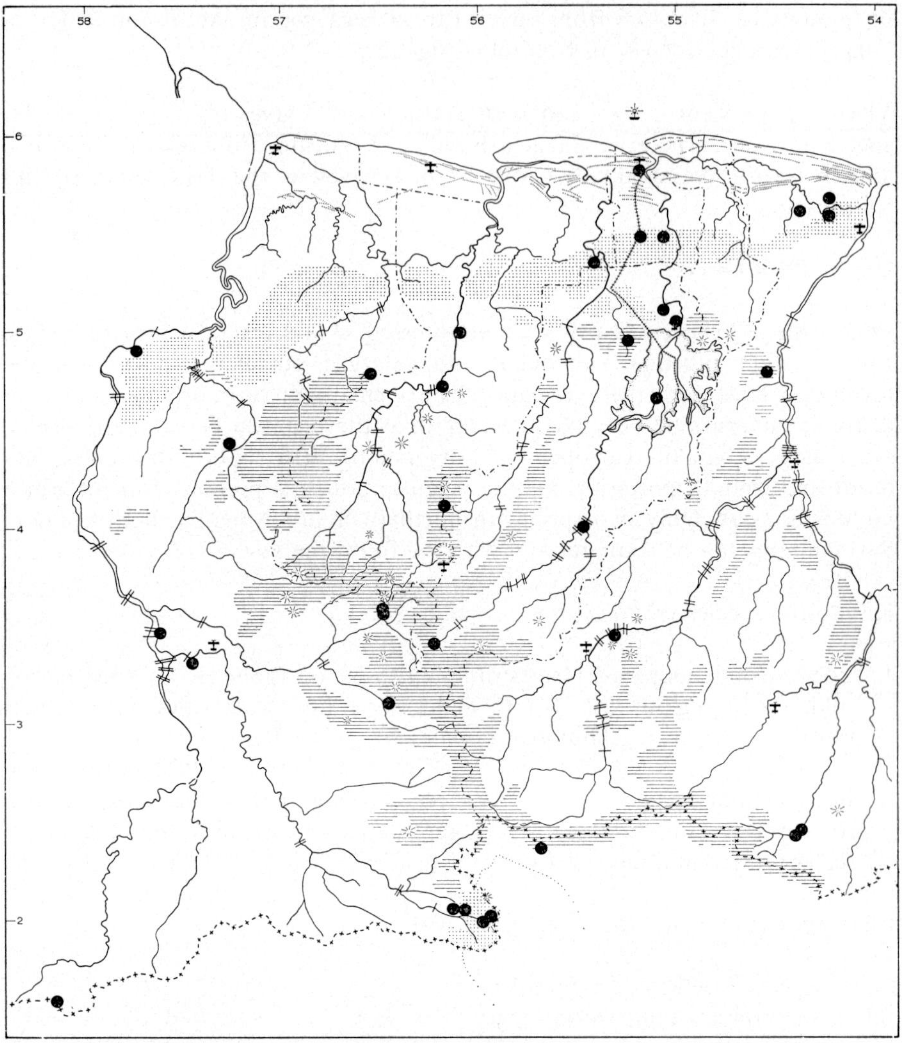

Fig. 24. Distribution of *Plica plica* (Linnaeus) in Surinam.

TYPE-LOCALITY.—'Indiis', restricted by ETHERIDGE (1970) to vicinity of Paramaribo, Surinam. Here restricted to the confluence of the Cottica River and the Perica Creek, Surinam.

MATERIAL. – Surinam.

Surinam: 1 ex., ZMA 11627; 1 ex., ZMA 11629, 1922, leg. J. W. Gonggrijp & G. Stahel; 2♀♀, 1 ex., SMNS 115, 1843, leg. A. Kappler; 1♀, NMW 8191, leg. Werner; 1♀, NMW 13903, 1911.

Paramaribo: 1 ex., RMNH 7530, 1911, leg. Jhr. W. C. van Heurn; 1 juv., ZMB 26378, leg. A. O. C. Heller.

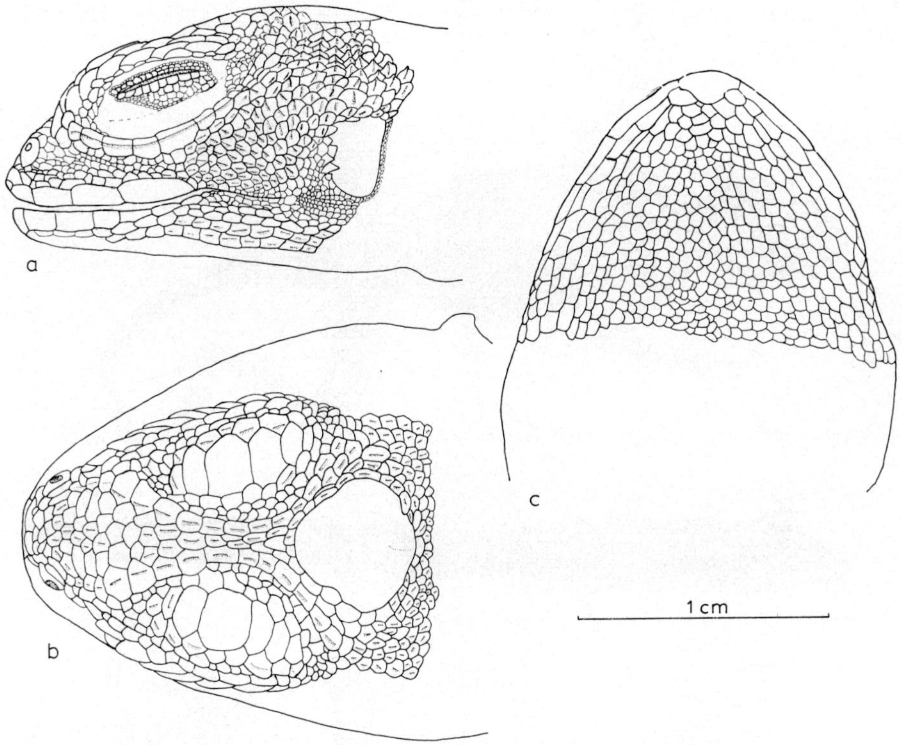

Fig. 25. P. plica (Linnaeus); a, b, c: lateral, dorsal and ventral views of the head of juv., RMNH 15257, King Frederik Willem IV Cataracts.

Encampment Moeroe Creek: 1♂, RMNH 16138, 3 km S., 24–VII–1968. leg. M. S. Hoogmoed.

Zanderij: 1 ex., BM 1946.4.3.43, 14–VIII–1938, leg. I. T. Sanderson.

Powakka: 1 ex., CM 44355, 9–X–1966, leg. M. H. de la Fuente.

Moengo: 3 ex., RMNH 16173, 17–VI–1969, leg. H. M. van Meeuwen.

Moengo Tapoe: 1♂, RMNH 16153, km 0,4, 24–IX–1948, leg. P. H. Creutzberg (Surinam Expedition 1948–49).

Djai Creek: 1♀, RMNH 16154, 7–X–1948, 1♂, RMNH 16155, km 8.6, 10–X–1948, both leg. Surinam Expedition 1948–49.

Mataway: 1 ex., CM 44354, 16–VII–1966, leg. M. H. de la Fuente.

Transect Fallawatra-Coppename River: 1 ex., RMNH 7531, January 1901, leg. H. van Capelle.

Coppename River: 1♀, BM 1946.4.3.44, Camp 1, 8–VII–1938, leg. I, T. Sanderson.

Voltz Mountain Reserve: 1 juv., RMNH 16143, Raleigh Cataracts, 7–XI–1968, leg. M. S. Hoogmoed; 1♀, RMNH 16158, 3–IV–1963, leg. A. M. Husson & P. Staffeleu: 1 juv., RMNH 16162, October 1956, leg. Radin,.

Berg en Dal: 1 ex., ZMB 7660, leg. H. B. Möschler.

Brokopondo: 1♂, RMNH 16167, 3–XI–1965, leg. G. F. Mees,

Brown's Mountain: 1 juv., RMNH 16144, 16–X–1968, 1♀, RMNH 16145, 10–XI–1968, 1♂, RMNH 16147, 12–XI–1968, all leg. M. S. Hoogmoed; 1♂, 1 juv., RMNH 16148, 17–X–1968, leg. J. J. Jansen; 1 juv., RMNH 16152, N. slope, 11–I–1970, leg. M. S.

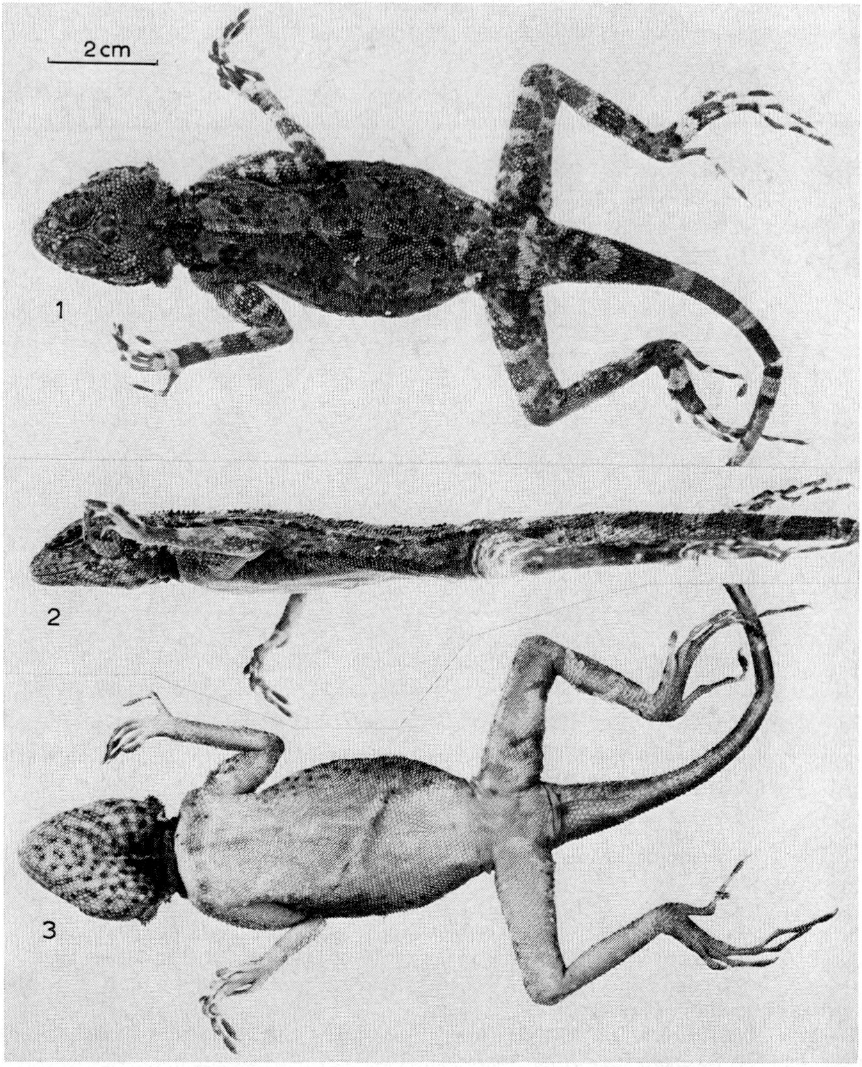

Plate 17. Plica plica (Linnaeus). Figs. 1–3. Dorsal, lateral and ventral aspect of ♂, RMNH 16149, Kayser Mountains.

Hoogmoed & J. J. P. Paats; 1♂, RMNH 16172, 19–IX–1938, leg. D. C. Geijskes; 1♀, RMNH 17268, 21–I–1972, leg. G. F. Mees.

Island near Dessiong, Suriname River: 1♂, RMNH 16161, 14–VIII–1963, leg. D. C. Geijskes.

Nassau Mountains, km 10.5: 1♂, RMNH 16156, 12–III–1949, 1♀, RMNH 16157, 24–III–1949, both leg. Surinam Expedition 1948–49.

King Frederik Willem IV Cataracts, Corantijn River: 1♂, 1 juv., RMNH 15257, 3/8–VIII– 1959, leg. D. C. Geijskes.

Gonini Encampment, Coeroeni River: 1♂, RMNH 16137, 26–VI–1968, leg. M. S. Hoogmoed.

Wilhelmina Mountains: 1♂, RMNH 16163, Lucie Camp, June–October 1963, leg. H. P. Pijpers; 1♂, RMNH 16164, near Camp 2, km 10, 24–VII–1963, leg. S. Ligorie.

Emma Mountains: 1♂, RMNH 16160, 3–VIII–1959, leg. F. Jonker.

Goddo: 1 ex., ZMA 11628, 15–VIII–1908, leg. J. H. A. T. Tresling (Suriname Expedition).

Encampment Upper Gran Rio: 1 ex., RMNH 7532, 3–IX–1910, leg. J. F. Hulk, (Corantijn Expedition).

Tapanahoni River: 1♂, RMNH 16168, near Magneet Rock, 25–VIII–1904, leg. G. M. Versteeg (Tapanahoni Expedition).

Airstrip Kayser Mountains: 1♀, RMNH 16149, NE., 9–VIII–1968, 1♀, 2♂♂, RMNH 16150–51, NE., 12–VIII–1968, all leg. M. S. Hoogmoed; 1♂, RMNH 16166, 26–VI–1963, leg. S. Ligorie; 2♂♂, FMNH 128950–51, 1–II–1961, leg. H. A. Beatty.

Boundary Camp, Itabu Creek Head: 1♂, FMNH 30931, September/October 1938, leg. E. R. Blake (Sewell Avery Expedition).

Sipaliwini: 1♀, 1♂, RMNH 16139, Base Bivouac, 6–IX–1968, 1♂, RMNH 16140, Vier Gebroeders Bivouac, 23–IX–1968, 1♀, 1♂, RMNH 16141, 3 km N. of Vier Gebroeders Mountain, 23–IX–1968, 1♂, RMNH 16142, Base Bivouac, 1–IX–1968, 1♂, RMNH 16146, 4 km E. of airstrip, 6–X–1968, all leg. M. S. Hoogmoed; 1♀, RMNH 16169, 1–II–1966, 1♂, RMNH 16170, 7–II–1966, both leg. G. F. Mees; 1♂, PMYU.R. 5913,airstrip, 11–XII–1961, leg. R. Freund.

Upper Litani River: 1♂, RMNH 16159, 10–XII–1903, 1♂, RMNH 16171, 29–XI–1903, both leg. G. M. Versteeg (Gonini Expedition).

Brazil.

Apisike: 1 juv., RMNH 16165, 20–IV–1952, leg. D. C. Geijskes.

DIAGNOSIS.—A large lizard, with strongly depressed body and head, and with long legs. Neck with tufts of spiny scales, a low vertebral crest from nape onto the tail. A strong, curving lateral fold with spinose scales. Green with rust brown transverse bands.

DESCRIPTION.—Head large, representing 25–28% of the snout-vent length in females, 26–30% in males; 1.1–1.3 times as long as wide in females, 1.1–1.2 times in males, in both sexes, however, mostly 1.2 times; strongly depressed; 1.3–1.6 times as wide as deep in females, 1.5–1.7 times in males; 0.9–1.0 times as long as the tibia in adults, 0.8–1.0 times in juveniles; much wider than the neck, which is much narrower than the body. Snout broadly rounded, convex in profile, not projecting over the mouth.

Tongue arrowhead-shaped with rounded tip, villose. Anterior teeth conical, lateral teeth tricuspid, with blunt central tip. The 'canines' in the upper jaw are enlarged, slightly more so in the males than in the females.

Rostral rectangular; four times as wide as deep; scarcely visible from above; bordered posteriorly by five or six small scales and the first supralabial on each side. Supralabials large, rectangular, five or six (rarely four); the last one under the centre of the eye. Nostril round to oval, directed laterally, in a single bulbous nasal. Nasal separated from the rostral and the supralabials by two or three rows of scales. Loreal region with five to eight horizontal rows of scales, the scales of the upper row largest. One large, keeled canthal scale, separated from the nasal by one or two small scales; canthus rostralis distinct. Five to eight distinctly enlarged, keeled suboculars; separated from the supralabials by three or four rows of small, keeled scales. No enlarged postoculars. Temporal region with rhomboidal, keeled scales with imbricate anterior margins, the keels ending in a sharp tip, pointing anteriorly. Ear-opening oval; large, as large as the interparietal; with a slightly denticulate anterior, and a smooth posterior margin; external auditory meatus short, tympanum well visible. Eyelids covered with small, slightly imbricate scales, a row of larger scales along the rim. Pupil round. Scales on the snout small, increasing in size posteriorly; irregularly shaped, conical to keeled, anterior margins imbricate; without definite arrangement. Supraorbital semicircles distinct, consisting of 8–13, mostly 10–12 distinctly enlarged keeled scales, the first or second largest; semicircles of both sides in most specimens separated by one row of small scales, in some they are in contact. Interparietal large, triangular, slightly wider than long in most specimens, about $\frac{1}{4}$–$\frac{1}{6}$ as wide as the head in adults, about $\frac{1}{3}$ in juveniles; separated from the supraorbital semicircles by one row of scales. Scales on the parietal region like those on the temporal region, but slightly larger, more spinous posteriorly. Occipital region with small, quadrangular to hexagonal, tuberculate, juxtaposed scales. Supraocular region with five, rarely three, four or six, transversely enlarged scales, separated from the supraorbital semicircles and from the supraciliaries by two or three rows of small scales. Supraciliaries elongate, seven to nine directed posteriorly and medially, accompanied by a row of enlarged scales on each side; forming a crest over the eye and a continuous row with the canthal. All dorsal head scales with at least one, but mostly numerous small pustules (asperities).

Mental small, triangular to hexagonal; wider than long; bordered posteriorly by three or four small scales. Four to eight, mostly five to seven, large, rectangular infralabials; the anterior one largest, decreasing in size posteriorly. Chin scales decreasing in size from the infralabials to the midventral line; rhomboidal, smooth, the anterior margins imbricate. Scales on the throat small, spinose, imbricate. One longitudinal and two transverse subgular folds, forming a subgular pouch. The anterior fold halfway between the ears and the insertion of the forelimbs shallow, continued on the sides of the neck; the posterior fold immediately behind it, deep, continued on the body as an antehumeral and lateral fold.

Scales on the nape small, rhomboidal, spinose, imbricate. Scales on the side of the neck granular. Side of the neck with tufts of spine-like scales: two just behind and below the ear, two on the upper part, of which the anterior one is largest. Scales on the antehumeral fold strongly spinose. A crest from the nape onto the tail, highest on the neck.

Dorsals and laterals subequal, small, decreasing in size towards the ventrals, rhomboidal, slightly imbricate, strongly spinose, the spines pointing towards the vertebral crest; in transverse and in oblique series. A low vertebral crest of laterally compressed triangular scales, with 63–76 scales from the beginning on the nape to the posterior border of the hindlimbs. A strong lateral fold which is the continuation of the posterior subgular fold and the antehumeral fold. It passes over the forelimbs, and then down and up again towards the hind limbs. Scales in this lateral fold enlarged, strongly spinose. A second, straight fold with enlarged, spinose scales on the border between flanks and belly from the axilla to the groin. Ventrals small, as large as the dorsals, rhomboidal, smooth, imbricate, in oblique series, strongly mucronate towards the flanks. Total number of scales around midbody 117–144. Preanal plate with small, irregularly shaped, smooth, slightly imbricate scales, larger than the ventrals. Preanal and femoral pores absent. A detailed description of the hemipenis is given by DOWLING, MAJUPURIA & GIBSON (1971a).

Scales on the tail small, rhomboidal to rounded, keeled, strongly mucronate, imbricate, the keels pointing towards the vertebral row, in transverse and in oblique series. On the anterior part of the tail the vertebral scale row is enlarged. Scales under the tail rhomboidal to lanceolate; feebly keeled, only slightly mucronate, imbricate, the keels forming longitudinal ridges, in transverse and longitudinal series. Tail autotomic, divided into verticils; anterior verticils with four ventral and six dorsal scales; posterior verticils with four ventral and seven dorsal scales. Scales on the regenerated part of the tail smaller, elongate, of irregular shape, feebly keeled, irregularly arranged. Tail round in crosssection.

Scales on the dorsal, anterior and posterior surfaces of the forelimbs and lower legs and on the dorsal and posterior surfaces of the thighs rhomboidal, spinose, slightly imbricate, in longitudinal and oblique rows; on the ventral surface of the forelimbs and shanks and on the ventral and anterior surfaces of the thighs rhomboidal, smooth, imbricate, in longitudinal and oblique rows; all scales larger than the dorsals. Twenty to 24 single, keeled, mucronate lamellae under the fourth finger; 27–33 under the fourth toe. Palms and soles with rhomboidal to rounded, feebly keeled, mucronate, imbricate scales, smaller than the ventrals, decreasing in size towards the toes. Fingers and toes distinctly compressed.

Colour in life green above with rust-brown transverse bands. The transverse bands containing black spots. Upper arms also with rust-brown bands with black spots. Forearms and hind limb with black transverse bands. Tail with transverse brown bands. A rust-brown band from the eye to the ear, bordered dorsally and ventrally by black lines. Adult males have a bright orange belly; underside of tail and limbs orange-yellow. Adult females have the belly orange-brown, juveniles pale orange. Throat in males white, in females grey; with brown spots. Gular pouch black in males, dark-brown in females.

Colour in preservative greyish green above with darker transverse bands in which black spots or circles are present. Antehumeral fold black. A few large, black spots with a grey zone between them on the side of the neck. Tail, limbs, fingers and toes with brown to black transverse bands. An indistinct darker stripe from the posterior corner of the eye to the ear, a dark spot may be present below the eye. Belly, underside of limbs and tail immaculate, white to light-brown. Gular pouch black in males, dark-brown in females, throat with brown spots, more or less distinctly arranged in longitudinal series converging on the gular pouch.

A large lizard, with a strongly depressed head and body and a round tail. Maximum snout-vent length in females 135 mm, in males 140 mm. The undamaged tail is 1.6–1.7 times as long as the snout-vent length in females, 1.5–1.9 times in males. Limbs long and slender, hind limb 1.4 times as long as forelimb in both sexes. When laid along the body there is considerable overlap of forelimb and hind limb, the hind limb reaching to between the eye and the snout.

HABITAT.—This lizard inhabits rain- and savannah-forest, where it usually occurs on the trunks of large trees and less frequently on faces of large boulders (BEEBE 1944b; UNDERWOOD, 1962; personal observation, 1968). Only once was this species found on the ground, probably in the process of crossing from one tree to another. On tree trunks it occurs from 1.5 m upwards and probably reaches the larger branches of the crown. This lizard is also present in at least two isolated patches of rain-forest on the Sipaliwini savannah.

NATURAL HISTORY.—A diurnal lizard, taken between 9.10 a.m. and 3.30 p.m. Several individuals were observed basking in the early morning, but the specimens taken in the afternoon were in the shade. Pregnant females, containing two (once) or three (twice) mature, round eggs are available from February, April and August. One recently hatched juvenile (snout-vent length 45 mm, taillength 80 mm), without an exact collecting date, is available. On two occasions a female and a male were collected together on the same tree, indicating that pair-formation occurs in this species. The first pair (RMNH 16151) was collected on August 12, 1968, the second pair (RMNH 16139) on September 6, 1968. BEEBE (1944b) found a pregnant female in May and a recently hatched juvenile in April. From these data it can be concluded that breeding takes place at least from February to September but that it probably does not occur during the long dry period.

On tree trunks this lizard rests with the head downwards, this position may be maintained for hours. The anterior part of the body is raised high from the substratum by the stretched forelimbs. The animal's weight is borne for the greater part by its hind limbs which are stretched or flexed. BEEBE (1944b) gives a good illustration of this attitude. When pursued the lizard may do two things, either it tries to escape notice by pressing itself close to the substratum without leaving its place, whatever happens; or it tries to escape by running up the tree and trying to keep the trunk between itself and the observer. When running up the tree it makes short dashes with pauses in between.

Eleven stomachs were examined. Ten of them contained ants, one of which exclusively, the eleventh contained a millipede and a large beetle. In only six was the bulk of the volume made up by the ants, in the four others there were only few ants. Other food items taken by this species include a spider, 14 small beetles, two wasps, two stingless bees (*Trigona*), two grasshoppers, a bug and a buglarva, two caterpillars, a cockroach and a small cicada. Two stomachs contained vegetable remains, which in one case could not be identified and in the other case turned out to be a pod. Probably this vegetable matter was eaten accidentally while consuming caterpillars (which were present in both stomachs). Thus, *P. plica* (Linnaeus) is most probably strictly insectivorous, with a slight preference for ants, but also taking other arthropods in a large variety. One stomach and one intestine contained nematodes.

RANGE.—Northern South America east of the Andes, including Trinidad and Grenada: Surinam, Guyana, French Guiana, Brazil (Pará, Amazonas), Venezuela (Monagas, Sucre, Delta Amacuro), Ecuador, Bolivia and Colombia east of the Andes. In Surinam the species occurs throughout the country, on both sides of the coastal savannah-belt and is still present in two forest-islands on the Sipaliwini savannah. It occurs from sea level to at least 500 m (fig. 24).

VERNACULAR NAME.—In the Surinam language this lizard is called 'agama', a name also used for *Plica u. umbra* (Linnaeus), *Polychrus marmoratus* (Linnaeus), *Tropidurus torquatus hispidus* (Spix), and *Uranoscodon superciliosa* (Linnaeus).

Plica umbra umbra (Linnaeus) (figs. 26, 27; pl. 18)

Lacerta Umbra Linnaeus, 1758: 207.
?'Devil of the woods' Stedman, 1796b: 19.
?'Bosduivel' Stedman, 1799b: 233.
Ophryessa umbra: Fitzinger, 1826: 47.
Hypselopus umbra: Kappler, 1881: 166.
Uraniscodon umbra: Boettger, 1893: 63; Van Lidth de Jeude, 1904: 88; Van Lidth de Jeude, 1917g: 438.
Hypsibates umbra: Schlegel, 1858: 19.
Tropidurus unicarinatus Werner, 1899: 480.
'50. Rotsbewoner' Fernandes, 1927: 249.
Plica umbra: D'Audretsch, 1967b: 7.
Plica u. umbra: Etheridge, 1970: 252; J. A. Peters & Donoso-Barros, 1970: 231.

TYPE-LOCALITY.—'Meridionalibus', restricted by ETHERIDGE (1970) to vicinity of Paramaribo, Surinam. Here restricted to the confluence of the Cottica River and the Perica Creek, Surinam.

MATERIAL. - Surinam.

Surinam: 1 ex., RMNH 7541, leg. H. F. C. ten Kate; 1♀, 1♂, RMNH 16112; 3♂♂, RMNH 16131, February 1888, leg. J. Spitzly; 1♂, 1 juv., RMNH 16132, 1903, Saramacca Expedition; 1 ex., ZMA 11630, 1887/90, leg. Van Brussel; 1 ex., ZMA 11633; 1 ex., ZMA 12429; 3 ex., ZMA no number, 1922, leg. J. W. Gonggrijp & G. Stahel; 1 ex., ZMB 7888, leg. Ehrhardt; 1♀, 1♂, NMW 13893, 1899, leg. Fruhstorfer; 1♀, 4♂♂, NMW 13899, 1911, leg. Rolle; 2 ex., NMW 18727–28 (lectotype- and paralectotype of *Tropidurus unicarinatus* Werner); 1♀, NMW 18754; 2 ex., ANS 8329–30, leg. C. Hering.

Paramaribo: 1 ex., RMNH 7533, 7–III–1939, leg. H. W. C. Cossee; 3 ex., RMNH 7534, 1911, leg. Jhr. W. C. van Heurn; 1 juv., RMNH 16115, 27–III–1950, 1♀, 1 juv., RMNH 16116–17, 30–III–1950, 1♀, RMNH 16118, 3–V–1950, 1♀, RMNH 16119, 22–V–1950, 1 juv., RMNH 16120, August 1938, all leg. D. C. Geijskes; 2 juvs., ZMB 26381, 2♂♂, NMW 13897,

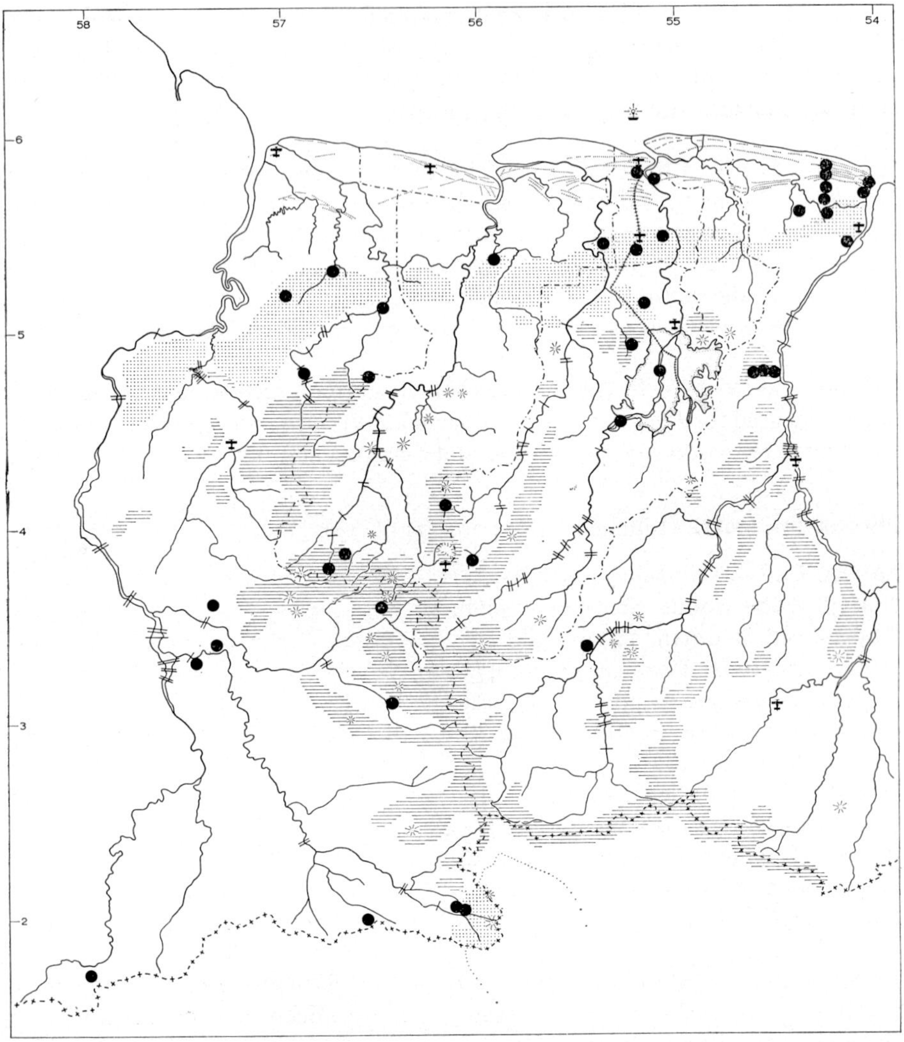

Fig. 26. Distribution of *Plica u. umbra* (Linnaeus) in Surinam.

1910, all leg. A. O. C. Heller; 2♀♀, 2♂♂, NMW 13898, 1915; 1♀, FMNH 121084, 20–IX–1960, leg. H. A. Beatty.

Road Zanderij-Paramaribo: 1♂, RMNH 16127, 21–VIII–1955, leg. D. C. Geijskes.

Plantation Peperpot: 1 ex., ZMA 11624, 2 km E. of Paramaribo, 15–XI–1964, leg. T. A. Renssen.

Matta: 1♂, RMNH 16114, 14–IV–1963, leg. A. M. Husson & P. Staffeleu.

Forest Bivouac Zanderij: 1 ex., ZMA 12299, 10–V–1956, leg. J. van der Kamp.

Powakka: 2♂♂, RMNH 16128, 10–IX–1963, leg. M. H. de la Fuente.

Moengo: 2♀♀, RMNH 16136, 17–VI–1969, leg. H. M. van Meeuwen.

Wia Wia transect: 1♂, RMNH 16096, km S. 11.2, 26–XI–1948, 1♂, RMNH 16097, km S. 9.2, 28–XI–1948, both leg. Surinam Expedition 1948–49.

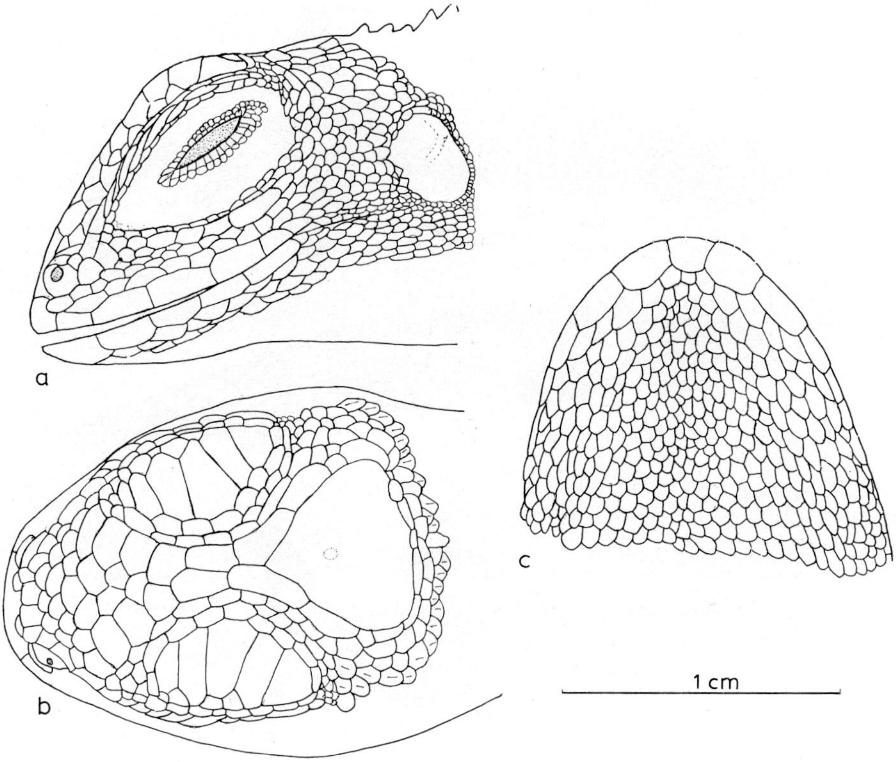

Fig. 27. Plica u. umbra (Linnaeus); a, b, c: lateral, dorsal and ventral views of the head of ♀, RMNH 15250, Brown's Mountain.

Swamp Camp, km S. 3.8, 8th ridge from sea: 1♀, 3♂♂, RMNH 16090, 16–XI–1948, 1 juv., RMNH 16091, 19–XI–1948, 2♀♀, 1♂, RMNH 16092, 20–XI–1948, 2♀♀, 1♂, RMNH 16093–94, 21–XI–1948, 1♀, RMNH 16095, 17–XI–1948, all leg. Surinam Expedition 1948–49.

Third Encampment: 1♀, RMNH 16086, km 14.7, 14–X–1948, 1♀, RMNH 16087, km 16.6, 15–X–1948, 1♂, RMNH 16088, km 14.9, 21–X–1948, all leg. Surinam Expedition 1948–49.

Djai Creek, km 8.6: 1♀, RMNH 16084, 8–X–1948, 1♀, RMNH 16085, 10–X–1948, both leg. P. H. Creutzberg (Surinam Expedition 1948–49).

Wane Creek N.: 1♀, RMNH 16083, km 8.6, 1–X–1948, leg. P. H. Creutzberg (Surinam Expedition 1948–49).

Moengo Tapoe: 1♀, RMNH 16082, km 2.5, 23–IX–1948, leg. P. H. Creutzberg (Surinam Expedition 1948–49).

Galibi: 1♂, RMNH 16089, 6–XI–1948, leg. Surinam Expedition 1948–49.

Langamankondre: 1♂, RMNH 15121, 13–VI–1968, leg. M. S. Hoogmoed.

Bigiston: 1♀, RMNH 16133, September 1963, leg. L. D. Brongersma.

Maratakka River: 1♀, RMNH 16451, 2nd savannah, 6–III–1971, leg. D. G. Reeder.

Tibiti: 1♀, RMNH 16098, Kabo Creek, 17–I–1949, leg. Surinam Expedition 1948–49.

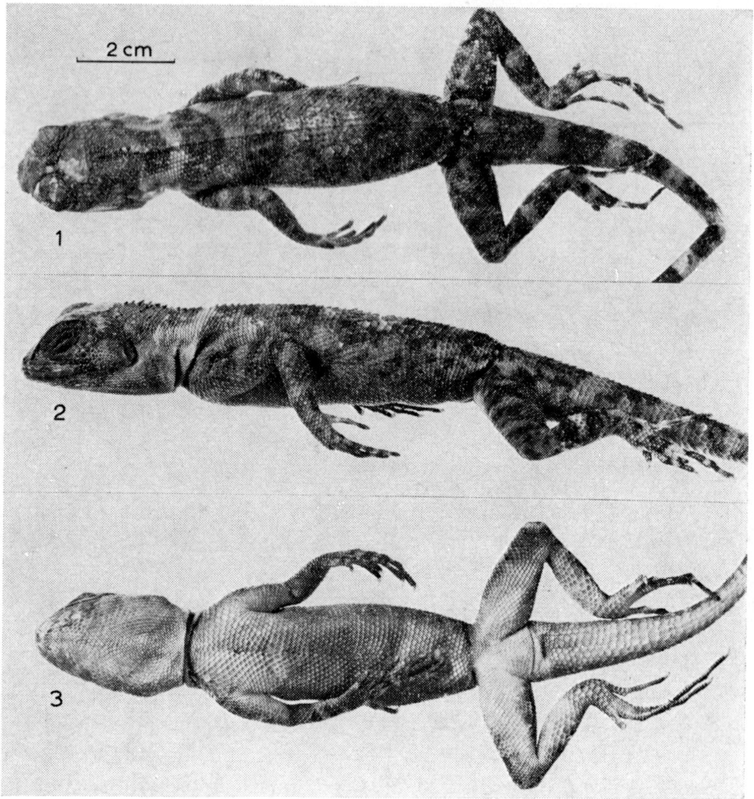

Plate 18. Plica u. umbra (Linnaeus). Figs. 1–3. Dorsal, lateral and ventral aspect of ♂, RMNH 16079, Encampment Gonini.

Second Bivouac, Kaboeri Creek, Corantijn River: 1♀, RMNH 16492, 27–III–1971, leg. P. Staffeleu & D. G. Reeder.

Stondansie Cataracts, Nickerie River: 1♀, RMNH 16462, 30–I–1971, leg. P. Staffeleu.

Upper Nickerie River: 1 juv., RMNH 16134, leg. H. van Capelle (Nickerie Expedition).

Transect Fallawatra-Coppename River: 1♂, RMNH 16113, leg. Nickerie Expedition.

Railway km 91.5: 1♂, BM 1946.4.3.42, 4–XI–1938, leg. I. T. Sanderson.

Brown's Mountain: 1♀, RMNH 15250, 12–XI–1968, leg. M. S. Hoogmoed; 1♂, SM 21, 24–X–1970, leg. D. G. Reeder.

Bedotti: 1 ex., RMNH 16177, 6–VII–1965, leg. G. F. Mees (from stomach of *Ciccaba virgata* (Cassin) (Strigidae)).

Mamadam, Suriname River: 1♀, RMNH 16130, 22–VII–1965, leg. G. F. Mees.

Nassau Mountains: 1 juv., RMNH 16099, km 0.2, 15–II–1949, 1♀, RMNH 16100, Base Encampment, 17–II–1949, 1 juv., RMNH 16101, Base Encampment, 18–II–1949, 1♂, RMNH 16102, km 0.7, 22–II–1949, 1♀, RMNH 16103, km 1.2, 22–II–1949, 1♀, 1♂, RMNH 16104, Base Encampment, 23–II–1949, 1♀, RMNH 16105, Base Encampment, 25–II–1949, 1♀, RMNH 16106, km 4, 13–III–1949, 1♀, RMNH 16107, km 12.5, 18–III–1949, 1♂,

RMNH 16108, km 8.9, 21/22–III–1949, 1 juv., RMNH 16109, km 10.5, 25–III–1949, 1♀, 1♂, RMNH 16110, Base Encampment, March 1949, all leg. Suriname Expedition 1948–49.

Linker Coppename River: 1 ex., ZMA 13225, 9–V–1967, leg. H. Nijssen.

Wilhelmina Mountains: 1 juv., RMNH 16111, transect I, 5–IX–1944, leg. D. C. Geijskes; 1♂, RMNH 16124, 26–VIII–1963, 1♂, RMNH 16126, 14–VIII–1963, both Camp 3, km 19, leg. H. P. Pijpers (Expedition Maguire); 1♂, RMNH 16129, 5–VIII–1963, leg. S. Ligorie (Expedition Maguire); 1 ex., 1 juv., ZMA 11631–32, 1926, leg. Wilhelmina Mountains Expedition.

Emma Mountains: 1♂, RMNH 16135, Base Bivouac, 23–VII–1959, leg. F. Jonker (Expedition Emma Mountains).

Base Bivouac near De Kock Mountain: 1 juv., RMNH 16121, Upper Saramacca River, 12–III–1958, leg. D. C. Geijskes (Tafel Mountain Expedition).

Encampment Gonini: 1♂, RMNH 16079, 26–VI–1968, leg. M. S. Hoogmoed.

Airstrip Coeroeni: 3♀♀, 3♂♂, RMNH 16122, 12–VIII–1959, 1♂, RMNH 16123, 12/14–VIII–1959, both leg. D. C. Geijskes.

Airstrip Kayser Mountains: 1♀, FMNH 128952, 1–II–1961, leg. H. A. Beatty.

Airstrip Paloemeu: 1 juv., RMNH 15815, May 1969, leg. H. M. van Meeuwen.

Itabu Creek: 1♂, FMNH 30928, Middle Dump, September/October 1938, leg. E. R. Blake (Sewell Avery Expedition).

Airstrip Sipaliwini: 1♂, RMNH 16080, 3 km E., 15–I–1970, 1♀, RMNH 16081, 11–II–1970, both leg. M. S. Hoogmoed & J. J. P. Paats; 1♂, RMNH 16126, February 1961, leg. D. C. Geijskes.

Surinam/Guyana.

New River, 750 feet: 1♀, 1 ex., BM 1939.1.1.72–73, leg. C. A. Hudson.

Diagnosis.—A medium-sized lizard with a cylindrical body and long, tapering tail. Neck with a deep, black, antehumeral fold on each side, no tufts of spinose scales. A low vertebral crest on the nape and anterior part of the body. No lateral folds with spinose scales. Interparietal very large, about half as wide as the head. Light-green with transverse black bands.

Description.—Head moderately large, representing 23–27% of the snout-vent length; 1.2–1.5 times as long as wide; 1.0–1.1, rarely 0.9 times as long as the tibia in adults, 1.3 times in a juvenile; slightly depressed; wider than the neck, which is narrower than the body. Snout broadly rounded, straight in profile, not projecting over the mouth.

Tongue triangular with rounded tip, villose. Anterior teeth of upper jaw conical, those on the premaxilla very small, on the anterior part of the maxilla very long, longer in males than in females. Teeth on the posterior part of the maxilla tricuspid, laterally compressed (only slightly so), shorter than the teeth on the anterior part. Teeth on the anterior part of the lower jaw very small, conical; larger, conical 'canines'; lateral teeth tricuspid, laterally compressed, shorter than the 'canines'. A marked sexual dimorphism is present in the size of the 'canines' both of upper and lower jaw; these teeth being longer in the male.

Rostral visible from above; more or less T-shaped with a short vertical bar pointing posteriorly; horizontal bar three to four times as wide as deep; bordered posteriorly by four small scales and the first supralabial on each side. Four, rarely three or five, large, rectangular supralabials, the posterior one largest and under the centre of the eye. Nostril directed dorsally and laterally; in a single, bulbous nasal. Nasal separated from the rostral and the supralabials by one row of scales, or in contact with the rostral. Loreal region with three or four horizontal rows of keeled scales, scales of the upper row largest. One large canthal scale, in contact with the nasal or separated from it; canthus rostralis distinct. Suboculars slightly enlarged, keeled; forming a continuous row with the (not enlarged) postoculars; separated from the supralabials by one or two rows of scales. Temporal region with small rhomboidal to hexagonal, smooth, imbricate, slightly conical scales, smaller near the ear, larger towards the parietal region. Ear-opening round, large, although smaller than the interparietal, with a slightly denticulate anterior and a smooth posterior margin; external auditory meatus absent, tympanum superficial, clearly visible. Eyelids covered with minute granular scales, two rows of enlarged scales along the rim. Pupil round. Scales on the snout roughly arranged in concentric semicircles with their centre on the interparietal; irregularly shaped, slightly convex, increasing in size posteriorly, juxtaposed to slightly imbricate. Supraorbital semicircles distinct, consisting of seven to ten, mostly eight or nine, distinctly enlarged scales; semicircles of both sides in contact. Interparietal enormous, triangular, wider than long, about half as wide as the head; in contact with the supraorbital semicircles. Scales on the parietal region like those on the temporal region but larger. No occipital scales. Supraocular region with five, rarely six, transversely enlarged scales; separated from the supraorbital semicircles by two, rarely three, rows of small scales; from the supraciliaries by two rows of scales, of which one is slightly enlarged, anteriorly, by one row posteriorly. Supraciliaries 8–11, mostly nine or ten; forming a continuous row with the canthal; forming a crest over the eye; elongate, the anterior five or six extending obliquely posteriorly and medially, the sixth or seventh extending parallel to the longitudinal axis, the posterior three or four extending obliquely posteriorly and laterally. All dorsal head scales with numerous small pustules (asperities).

Mental small, triangular to trapezoid; wider than long; bordered posteriorly by three or four, rarely five, small scales. Five to six large, rectangular infralabials, the anterior one largest, decreasing in size posteriorly. Several rows of enlarged, hexagonal, smooth, slightly imbricate sublabials. Scales on the chin small, more or less rhomboidal, smooth, imbricate, decreasing in size towards the median. Scales on the throat identical to those on the chin but towards the chest decreasing in size and mucronate. Two transverse subgular folds, a very deep anterior one halfway between the ear and the insertion of the forelimbs and an indistinct prepectoral fold just anterior of the forelimbs. The anterior fold is continued for a short distance on the sides of the neck; the posterior fold forms distinct antehumeral folds, extending obliquely dorsally and posteriorly, and a gular pouch.

Scales on the nape and on the side of the neck similar to the dorsals but smaller. Scales in the antehumeral fold minute, spinose. A low, serrated nuchal crest, continued on the anterior part of the back, of triangular, laterally compressed scales.

Dorsals and laterals small, rhomboidal, keeled, mucronate, imbricate, the tips of the scales pointing dorsally; in oblique series. Anterior part of the back with a low crest which disappears shortly behind the forelimbs. Vertebral scales on the posterior part of the back slightly larger than the other dorsals. Number of vertebral scales from the beginning of the crest on the neck to the posterior border of the hind limbs 56–58, about 17–28 of these form the short serrated nucho-dorsal crest. Ventrals as large as the dorsals, in oblique series, rhomboidal, faintly keeled, slightly mucronate, imbricate, the apices of the scales directed laterally. Dorsals, laterals and ventrals merge into each other. The laterals tend to be bluntly keeled. Total number of scales around midbody 52–65. Preanal plate with small rhomboidal to irregularly shaped, smooth, imbricate scales, more or less arranged in transverse rows. Preanal and femoral pores absent.

Scales on the anterior part of the tail identical to the dorsals, on the rest of the tail rectangular, obliquely keeled, slightly mucronate, imbricate, in oblique series. Scales under the anterior part of the tail rhomboidal, feebly keeled, strongly mucronate, imbricate; under the rest of the tail elongate, rectangular to lanceolate, strongly keeled, less mucronate distally. Tail autotomic, divided into indistinct verticils, anterior verticils with three ventral and four dorsal scales; posterior verticils with three ventral and five dorsal scales. Scales on the regenerated part of the tail very small, rhomboidal, keeled, not mucronate, imbricate, irregularly arranged. Tail round in cross-section.

Scales on the limbs rhomboidal, keeled, mucronate, imbricate, in oblique and in longitudinal series; on the dorsal and anterior surfaces larger than the dorsals; on the posterior surfaces (except of the lower legs) smaller; on the ventral surfaces as large as the dorsals. Twenty to 23 single, mucronate lamellae under the fourth finger; 23–29, mostly 26–29, single, sharply keeled, mucronate lamellae under the fourth toe. Palms and soles with small, rhomboidal, bluntly spinose, imbricate scales, decreasing in size towards the toes. Fingers and toes slightly compressed, subcylindrical with a serrated lower edge.

Colour in life light-green with black bands above, males with a bluish grey head. Belly, underside of legs and underside of tail purplish brown with whitish spots. Throat yellow-ochre to orange.

Colour of the back in preservative light- to dark-brown with transverse bands of black spots. A single black spot on the anterior nuchal crest-scales, a short transverse band behind it, starting from the ears, and a crescent-shaped band starting in the antehumeral folds, on the anterior part of the back. Between it and the hindlimbs there are four or five transverse bands on the body. Limbs, fingers and toes with transverse bands of black spots. Tail brown above with black transverse bands. Belly, underside of limbs, underside of tail greyish to light-brown with or without dirty white spots. Throat yellow, immaculate in the centre, with ill-defined greyish spots surrounded by brown or grey along the sides. The head, both in males and in females, may be grey. In some males, the side of the neck anterior of the antehumeral fold and the nape in front of the crescent-shaped band are yellowish.

A medium-sized lizard with a cylindrical body and a long, tapering conical tail. Maximum snout-vent length in females 87 mm, in males 96 mm. The undamaged tail is 1.8–2.2 times as long as the snout-vent length in both sexes. Limbs long, stout, hind limb in both sexes 1.5–1.6 times as long as forelimb. When laid along the body there is considerable overlap of forelimb and hind limb, the hind limb reaching to the posterior corner of the eye.

HABITAT.—All available data indicate that this is an arboreal lizard, living on tree trunks and branches from about 2 m upwards. It is present in a wide variety of habitats: rain-forest, ridge-forest, savannah-forest and forest edges. According to RAND & HUMPHREY (1968) the habitat of this species near Belém agrees with that in Surinam. Some Surinam specimens (RMNH 15121, 16081), however, were taken on isolated trees, the first one on a tree in a Carib Indian village, the second one on a small tree about 30 m from the edge of the forest. In Belém they do not seem to be present on isolated trees.

NATURAL HISTORY.—A diurnal lizard, which was collected by me between 11.30 a.m. and 4.30 p.m. None of these specimens was observed basking; all were taken on shady parts of treetrunks. RAND & HUMPHREY (1968) also found that this species near Belém is a nonheliotherm. Pregnant females containing two large, elongate, rounded eggs are available from January to March, May, September to November, indicating that breeding probably takes place throughout the year, perhaps with a short break from June to August. Recently hatched juveniles are not known for a certainty. Only three small specimens from March (two) and August (one) may have hatched recently, though the umbilical scar is already closed. Their snout-vent length is 41–48 mm, the taillength of the smallest specimen is 81 mm. The few notes on breeding of this species supplied by BEEBE (1944b) for specimens from Kartabo indicate the same breeding season as in Surinam. BEEBE (1944b) states that this species is about three times as abundant as *P. plica* (Linnaeus). DA CUNHA (1961) simply states that it seems to be more common than *P. plica* (Linnaeus). Judging from the Surinam material available *P. u. umbra* (Linnaeus) is about twice as abundant as *P. plica* (Linnaeus). This, however, is not the complete picture. The Surinam Expedition 1948–49, working in the northern part of the country, collected 37 *P. u. umbra* (Linnaeus) and only five *P. plica* (Linnaeus). During two trips, which took me through a large part of Surinam, I collected five *P. u. umbra* (Linnaeus) and 20 *P. plica* (Linnaeus). Whether this inversion in abundance is due to differences between the places visited, to a change in abundance in time or to the alertness of the collector is not known. It does, however, suggest that BEEBE's and DA CUNHA's remarks should be handled with care and that the relative abundance of these species has to be investigated more thoroughly.

The contents of ten stomachs were examined. The greater part was made up by ants, belonging to at least four different species. Three stomachs contained exclusively ants, five contained a single other item (spider, cockroach, beetle (three)) and two contained more other items namely three (centipede, beetle, caterpillar) and seven (spider, wasp, five beetles). This agrees well with BEEBE's

(1944b) data, who found ants in all stomachs investigated; many of them in three, only two in another stomach; all stomachs also contained other food items. From this it is clear that *P. u. umbra* (Linnaeus) is a food-specialist with a high preference for ants, taking other arthropods only in small numbers and in very limited variety (spiders, wasps, cockroaches, beetles). This is an important difference with *P. plica* (Linnaeus) which is less fastiduous in its diet and in addition to fair quantities of ants takes considerable quantities of a large variety of other arthropods. LYDEKKER (1896) reports this species to be herbivorous, but this clearly was a mistake.

Five stomachs contained nematodes.

RMNH 16177 was found in the stomach of an owl (*Ciccaba virgata* (Cassin)). Another specimen (RMNH 16127) was taken from the stomach of a snake (*Bothrops* spec.).

RANGE.—Northern South America: Surinam, Guyana, French Guiana, northeastern Brazil. In Surinam the species is present throughout the country on both sides of the coastal savannah belt. It occurs from sea level to at least 500 m (fig. 26).

REMARKS.—Rand & Humphrey (1968) during their studies near Belém found that there may be some sort of pair bond in this species. The same seems to occur in Surinam, where one pair (RMNH 16094) was taken on the same tree. Also in the related *Plica plica* (Linnaeus) pair formation was observed (see above). LYDEKKER (1896) reported pair formation in this species.

VERNACULAR NAME.—In the Surinam language this is one of the iguanid lizards called 'agama'.

Genus **Polychrus** Cuvier

DIAGNOSIS.—Large lizards with strongly compressed bodies without dorsal crest, very long cylindrical tails and well developed pentadactylous limbs. Third finger as long as the fourth finger, third toe as long as the fourth toe. Femoral pores present. Eyes covered partly by the eyelids which are grown together. Head with large, polygonal, flat, subequal scales. A small dewlap; no transverse gular fold. Anterior teeth conical, lateral teeth tricuspid, laterally compressed. Five species in southern Central America, north to Nicaragua, and in South America, south to Argentina. Only one of these species is present in Surinam.

Polychrus marmoratus (Linnaeus) (figs. 28, 29; pl. 19)

Iguana cauda longissima tereti conica, crista dorsali vel gulae nulla Gronovius, 1756: 83.
Lacerta marmorata Linnaeus, 1758: 208.
'Surinaamse Gemarmerde Hagedis' Houttuyn, 1764: 178.

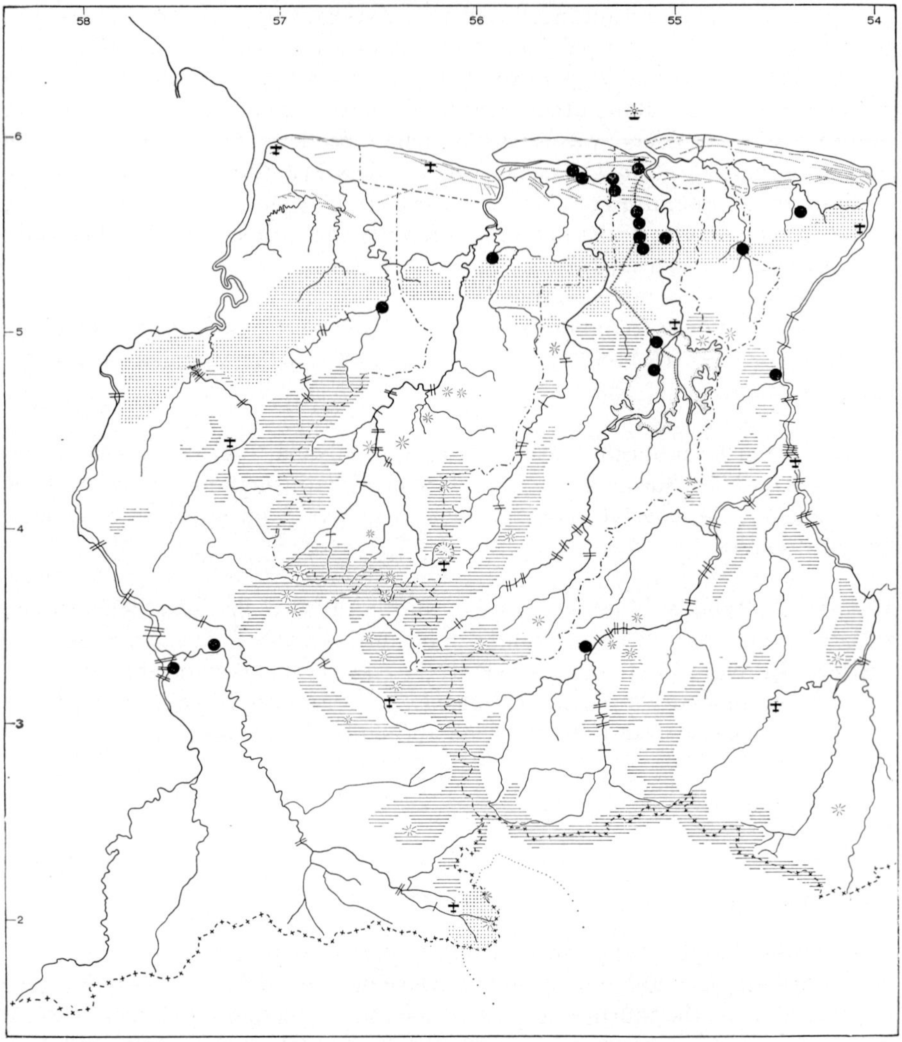

Fig. 28. Distribution of *Polychrus marmoratus* (Linnaeus) in Surinam.

'Le sixième' Fermin, 1769: 207.
'Le onzième' Fermin, 1769: 210.
?'Mexican Cameleon' Bancroft, 1769a: 151.
?'Mexicanische Chamaleon' Bancroft, 1769b: 90.
'De zesde' Fermin, 1770: 184.
'De elfde' Fermin, 1770: 186.
'Agama' Stedman, 1796b: 19; Stedman, 1799b: 233.
'Mexican cameleon' Stedman, 1796b: 19.
'Mexicaansche Kameleon' Stedman, 1799b: 233.

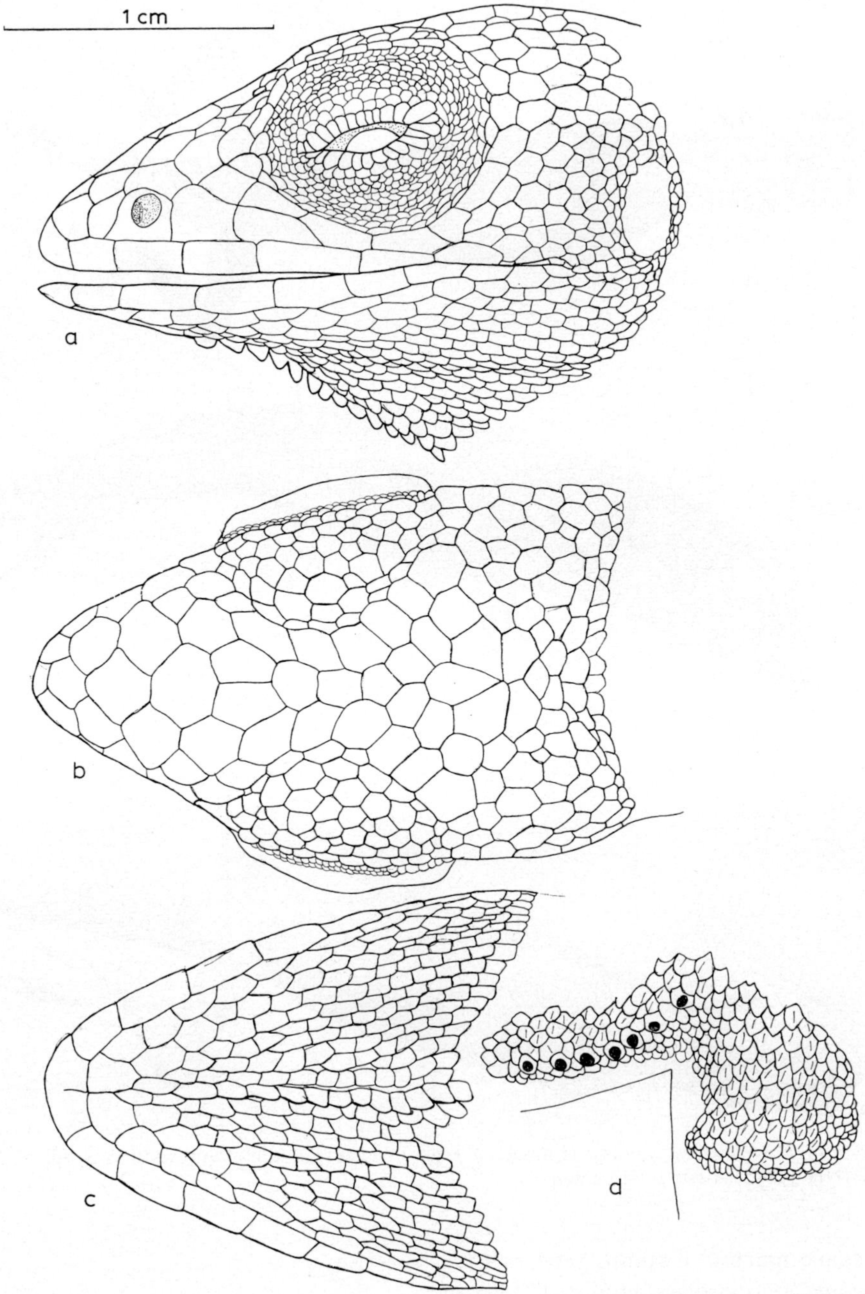

Fig. 29. Polychrus marmoratus (Linnaeus); a, b, c: lateral, dorsal and ventral views of the head of ♂, RMNH 15152, Airstrip Paloemeu; d: ventral surface of right thigh, showing the femoral pores and anal plate of RMNH 15152.

Plate 19. Polychrus marmoratus (Linnaeus). Figs. 1–3. Dorsal, lateral and ventral aspect of ♂, RMNH 15152, Airstrip Paloemeu.

'Iguane marbré' Daudin, 1802a: 261.
Agama marmorata: Daudin, 1802a: 433.
'Amerikanischen Chamäleons' Von Sack, 1821a: 122.
'Agamma' Von Sack, 1821a: 123, 124, 125, 126; Von Sack, 1821b: 219, 221, 223, 224, 225.

'Chamaeleons' Von Sack, 1821a: 124, 125.
'Amerikaansche kameleon' Von Sack, 1821b: 218, 219, 222, 223.
'Kameleon' Von Sack, 1821b: 221.
Polychrus marmoratus: Duméril & Bibron, 1837: 65; Fitzinger, 1843: 61; Duméril & Duméril, 1851: 54; Van der Hoeven, 1855: 528; Schlegel, 1858: 21; Kappler, 1885: 817 (partly); Kappler, 1887: 126 (partly); Boettger, 1893: 60; Van Lidth de Jeude, 1898: 20; Van Lidth de Jeude, 1904: 88; Van Lidth de Jeude, 1917g: 437; Burt & Burt, 1930: 21; Burt & Burt, 1931: 284.
Polychrus marmorata: Penard & Penard, 1906: no. 81.

TYPE-LOCALITY.—'Hispania', here restricted to the vicinity of Paramaribo, Surinam.

MATERIAL. – Surinam.

Surinam: 4 ex., RMNH 2839, 1827, leg. H. H. Dieperink; 3 ex., RMNH 2841, 1846; 1 ex., RMNH 2959, 1862, leg. Bressaer; 1 ex., RMNH 4503, November 1905, leg. P. Buitendijk; 1 ex., RMNH 5742; 1 ex., RMNH 7501, 1883, leg. J. J. Halfhide; 1 ex., RMNH 7502, leg. H. F. C. ten Kate; 1 ex., RMNH 7504 (locality with question-mark); 1 ex., RMNH 7506, June 1910, leg. D. Bolten; 2 ex., RMNH 7508, 1883, leg. C. J. Hering; 1 ex., RMNH 7550 (locality with question-mark); 1♀, RMNH 16031; 3 ex., RMNH 16032, leg. G. Witsen Elias; 1 ex., RMNH 16044, 1883, leg. C. J. Hering; 2 ex., ZMA 10365, 10367; 2 ex., ZMA 12236–37, leg. G. Stahel; 6 ex., ZMA 12251–56; 3 ex., ZMA 12257–59, 1887/90, leg. Van Brussel; 2 ex., ZMA 12334–35; 4 ex., ZMA no number, 1922, leg. J. W. Gonggrijp & G. Stahel; 1♀, NMW 12903, 1897; 2♂♂, NMW 12904, 1899; 4♀♀, 1♂, NMW 12905, 1911; 1♀, NMW 12909; 1♀, NMW 12910, 1902; 1♀, ANS 8141, leg. Keller; 1♀, 1♂, ANS 8144–45, leg. Hering.

Paramaribo: 3 ex., RMNH 7505, 1911, leg. Jhr. W. C. van Heurn; 1♀, RMNH 16030; 1♂, RMNH 16034, 11–VI–1941, 1♀, 1 juv., RMNH 16036, August 1938, both leg. D. C. Geijskes; 2 ex., ZMA 10773, 12250, 12–VIII–1947, leg. G. Stahel; 12 ex., ZMA 12238–49, leg. Périn; 1♀, BM 1946.4.2.86, 14–II–1938, 1♀, BM 1946.4.2.87, 25–I–1938, both leg. I. T. Sanderson; 5 ex., ZMB 25972, 25974, 26398, 9♀♀, 4♂, 1 ex., NMW 12906–08, 1910, all leg. A. O. C. Heller; 1♂, CM 44362, 3–VII–1966, 1♀, CM 44364, 10–VIII–1966, 1♂, CM 44365, 15–VIII–1966, 1♂, CM 44366, 24–VIII–1966, 1♀, 1♂, CM 44367–68, 8–IX–1966, all leg. M. H. de la Fuente; 1♀, CM 49531, 9–VII–1968, leg. D. McDonald; 2 ex., AMNH 8141, 8652, 1916, leg. J. A. Samuels.

Tambahredjo: 1♀, RMNH 16023, 15–XI–1968, leg. M. S. Hoogmoed.

Groningen: 2 ex., AMNH 8617–18, May 1916, 3 ex., AMNH 8654–55, 8673, 1916, all leg. J. A. Samuels.

Uitkijk: 1♀, CM 44363, 28–VII–1966, leg. M. H. de la Fuente.

Santigron: 1♀, RMNH 15103, 1967, leg. W. N. Polder.

Onverwacht: 1♂, RMNH 16022, leg. W. N. Polder; 1♀, RMNH 16024, 13–I–1970, leg. M. S. Hoogmoed & J. J. P. Paats; 1♀, CM 49532, 10–VII–1968, leg. D. A. Jones.

Republiek: 1♀, RMNH 16029, May 1954, leg. D. C. Geijskes.

Zanderij: 1♀, RMNH 15102, 1967, 1♂, RMNH 16045, April 1967, both leg. W. N. Polder; 1 ex., ZMA 12333, 29–IV–1956, leg. J. van der Kamp; 1♂, BM 1946.4.2.82, 19–IX–1938, 1♂, BM 1946.4.2.83, 5–X–1938, 1♂, BM 1946.4.2.84, 9–VIII–1938, 1♂, BM 1946.4.2.85, 5–IX–1938, all leg. I. T. Sanderson.

Powakka: 1♂, CM 44369, 18–IX–1966, leg. M. H. de la Fuente.

Moengo: 1♀, RMNH 16025, 1969, leg. H. M. van Meeuwen.

Tibiti: 2♀♀, 1♂, RMNH 16037–39, 16–I–1949, 1♀, RMNH 16040, 17–I–1949, 1♀, RMNH 16041, 19–I–1949, all leg. Surinam Expedition 1948–49.

Pininica, near Java: 1♀, RMNH 16033, 9–V–1963, leg. A. M. Husson & P. Staffeleu.

Stondansie Cataracts, Nickerie River: 1♀, RMNH 16460, 30–I–1971, leg. D. G. Reeder.

Kabel: 1♀, RMNH 16026, W., 8–VII–1964, leg. P. Leentvaar.

Gansee: 2♂♂, RMNH 16027, 6–XI–1951, leg. H. Heyde; 1♀, RMNH 16035, 21–VIII–1942, leg. L. Schmidt.

Nassau Mountains, Base Encampment: 1♀, RMNH 16042, 20–II–1949, 1♀, RMNH 16043, 24–II–1949, both leg. Surinam Expedition 1948–49.

Tigri Encampment, New River: 1 juv., RMNH 16021, 1–VII–1968, leg. M. S. Hoogmoed.

Airstrip Coeroeni: 1♀, RMNH 16028, 12–VIII–1959, leg. D. C. Geijskes.

Airstrip Paloemeu: 1♂, RMNH 15152, 26–V–1968, leg. M. S. Hoogmoed.

DIAGNOSIS.—A large lizard with a strongly compressed body and a very long cylindrical tail. Dorsals larger than the laterals. A low crest from the mental to the dewlap. Ventrals keeled. Total number of scales around midbody 68–80. Green with five or six oblique yellow bands on the flanks, vertebral region brown. Three black lines on the side of the head, radiating from the eye.

DESCRIPTION.—Head large, bulky, but representing only 23–28% of the snout-vent length in females, 24–27% in males; 1.5–1.7 times as long as wide in females 1.6–1.7 times in males; 1.5–1.7 times as long as the tibia in females, 1.4–1.6 times in males; slightly depressed; distinctly wider than the neck, which is as wide as the body. Snout broadly rounded, not projecting over the mouth. Frontal depression absent or just indicated, frontal ridges absent.

Tongue elongate, arrowhead-shaped, villose; with a rounded, slightly nicked tip and two small vertical flaps near the glottis. Anterior teeth and anterior lateral teeth conical, posterior lateral teeth tricuspid, more or less laterally compressed, subequal.

Rostral pentagonal; more than twice as wide as deep; clearly visible from above; bordered posteriorly by two (rarely three) scales on the dorsal surface of the snout, sometimes two on the lateral surface of the snout, and the first supralabial on each side. Five to seven large and three smaller supralabials; fifth (rarely fourth or sixth) under the centre of the eye. Nostril in a single nasal, which is in contact with the supralabials. Loreal region with two scales between the nasal and the suboculars and one or two scales between them and the canthals. Two or three canthals between the nasal and the supraciliaries. Suboculars and postoculars forming a semicircle of enlarged scales. Temporal region with small, flat, irregularly shaped, juxtaposed scales, decreasing in size posteriorly. Ear-opening large, vertically oval; surrounded by small scales forming a smooth margin; external auditory meatus absent, tympanum superficial, visible. A straight series of slightly enlarged supratemporals that mark the edge between dorsal and lateral surface of the skull. Eyelids partly grown together; covered with granular scales and a row of enlarged scales around the rim. Pupil round. Scales on the snout large, irregularly shaped; arranged in three or four more or less transverse rows. Supraorbital semicircles scarcely enlarged, consisting of four to eight, mostly six or seven, scales; separated by one (exceptionally none) row of small scales. Enlarged interparietal absent. Parietal and occipital region with irregularly shaped scales of different sizes; smaller than the scales on the snout. Supraocular region with polygonal scales, smaller than those on the snout, decreasing in size towards the supraciliaries. Supraciliaries 9–13, mostly 11–12, all small, forming a continuous row with the canthals.

Mental trapezoid, posteriorly with a median cleft; bordered posteriorly by zero to two, mostly one small scale, two sublabials and the anterior infralabial on each side. Infralabials seven to nine, of which five to seven, mostly six, are large, elongate, the remainder small. Several rows of sublabials. Scales on the rest of the chin small, elongate, a median row of raised, triangular scales, forming a serrated crest from the mental to the dewlap. Scales on the throat rhomboidal to lanceolate, imbricate. Both in males and females a small dewlap is present, reaching from below the eye to the insertion of the forelimbs; scales on the lateral surfaces of the dewlap in three or four horizontal rows of rhomboidal, smooth, imbricate scales, the rows separated by minute granules; rim scales lanceolate, smooth, imbricate.

All head scales, except those on the chin show small pustules, which are arranged in many longitudinal, wavy rows on the scales anterior to the eyes, on the suboculars, on the supralabials and on the sublabials.

Scales on the nape forming a convex margin with those on the occipital region of the head; small, rhomboidal to lanceolate, keeled, imbricate. Scales on the side of the neck smaller, squarish to round, juxtaposed, smaller than the dorsals. No nuchal fold.

Dorsals larger than the laterals, of about the same size as the ventrals; rhomboidal, keeled, imbricate, arranged in oblique rows. Laterals rhomboidal to rectangular, keeled, juxtaposed to slightly imbricate, in oblique rows. Ventrals lanceolate, keeled, slightly mucronate, strongly imbricate, in oblique rows. Laterals gradually merging into the dorsals on one side, into the ventrals on the other. Total number of scales around midbody 68–80. No dorsal fold. Preanal plate with scales slightly smaller than the ventrals but otherwise identical. Extended hemipenis bifurcate; smooth with a deep sulcus spermaticus on the posterior surface, which bifurcates just below the bifurcation of the hemipenis, from where it turns to the inner surface of each lobe and is continued to near the tip. Tip of each lobe smooth or with small warts. Enlarged postanal scales absent. Preanal pores absent; 8–13, mostly 9–11, femoral pores; well developed in males, indicated in females.

Scales on the tail large, larger than the dorsals; rhomboidal, keeled, slightly imbricate, in oblique rows; the keels forming longitudinal ridges. Scales under the tail identical to those on the tail in females; in males under the base of the tail larger than those on the dorsal surface. None of the specimens examined had a regenerated tail, only a few specimens lacked part of the tail. Thus, the tail seems to lack autotomy.

Scales on the dorsal, anterior and ventral surfaces of the upper arms and thighs and on all surfaces of the lower legs and forearms small, rhomboidal, keeled, imbricate; on the posterior surfaces of the forelimbs and thighs smaller, granular. Twenty seven to 35, mostly 29–32, single lamellae under the fourth finger;

35–44, mostly 37–41 single lamellae under the fourth toe; lamellae under the proximal part of the fingers and toes with several short, longitudinal rows of sharp-pointed spinules, under the distal part longitudinally striated. Lamellae under the proximal part small, hardly enlarged; distally increasing in size, wider than long, the last one very much larger than the other lamellae, smooth. Palms and soles with small, rhomboidal, keeled, slightly imbricate scales. Fingers and toes compressed, third finger as long as the fourth, third toe as long as the fourth.

Colour in life green with five or six oblique, wide yellow bands, bordered posteriorly by narrow black bands. In a juvenile specimen (RMNH 16021) the entire back is green, in adults only the flanks, the vertebral region is light brown. Dorsal surface of the head green, lateral surfaces yellowish green to bluish green with black stripes. Ventral parts dirty white. Limbs green. Tail greyish with brown cross-bands. Iris golden. The oral cavity is black, the tongue pale orange. Because this species shows remarkable and fast colour changes, many specimens do not fit this description. The colour may change to nearly completely brown and all intermediate stages are possible. An extensive description of a living specimen is given by BREDER (1946).

In preservative the dorsal colour is very variable, ranging from bluish green to purplish brown. On the flanks there are five or six oblique, more or less distinct, black or brown stripes, which may or may not be in connection on the vertebral region. Vertebral region with black to brown cross-bands or spots arranged in transverse series. Ventral parts yellowish to greyish blue with spots or areas of different colours (e.g., brown, grey, bluish). The side of the head with three curved, black lines radiating from the eye, one to the dorsal tip of the ear, one to the corner of the mouth and one to the lips; this last line may be continued for a short distance on the lower lips and the chin. Tail of the same general colour as the body, with dark-brown cross-bands. Limbs without markings or with oblique brown bands.

A large lizard, with strongly compressed body. Maximum snout-vent length in females 137 mm, in males 114 mm. Limbs short and stout, hind limb 1.3–1.5 times as long as forelimb; when laid along the body there is considerable overlap of forelimb and hind limb, the hind limb reaching the axilla. The undamaged tail in females is 2.5–3.0 times as long as the snout-vent length, in males 2.6–2.9 (one exception with 2.1) times and in the single available juvenile 2.4 times.

HABITAT.—Most of the specimens for which habitat data are available were captured in the neighbourhood of water, mostly in bushes along rivers. Thus this species seems to prefer the edge of the forest where it borders rivers. It is generally present in the lower bushes, but it may be found on the ground and in treetops as well (cf. RAND & HUMPHREY, 1968).

NATURAL HISTORY.—A diurnal species which was observed basking several times. The active specimens I collected (RMNH 15152, 16021, 16023) were taken between 7.00 a.m. and 4.00 p.m. One specimen (RMNH 16024) was collected during frog-hunting at about 8.30 p.m., sleeping on a slender, nearly horizontal, branch of a felled tree, lying in a mokko-mokko (*Montrichardia*) swamp. It was about 50 cm above the surface of the water, its head towards the end of the branch which it clasped with hands and feet, its tail stretched along the branch. Pregnant females containing four or six round, immature, eggs are known from August and September. One recently hatched juvenile (RMNH 16021) with a snout-vent length of 42 mm, tail length 102 mm, from July is available. BEEBE (1944b) found pregnant females in Kartabo in January and August, containing a maximum of seven eggs. These data indicate a breeding season lasting at least from January to September and probably extending throughout the year. This lizard is capable of remarkable colour changes. Within a few minutes the colour can change from bright green to a dull dark--brown and all intermediate stages with more or less green and brown can be assumed. According to BEEBE (1944b) these lizards try to escape notice by resting immobile. In this way they show strong resemblance to leaves and are hardly detectable. They may even maintain the most uncomfortable attitudes for several hours. When, however, they are annoyed they show a fierce defence display, extending the dewlap, the mouth wide opened and trying to bite, which can be rather annoying for the attacker. The oral cavity is lined with bluish black, against which the pale-orange tongue with two posterior red flaps shows very well. This probably also helps to discourage would-be predators. *Polychrus* is very agile in climbing, which is indicated by the sharp, recurved claws and the rugose subdigital lamellae. According to COTT (1926) it jumps from one branch to the other. The same author describes the tail as prehensile, but I think this is incorrect and that the tail cannot support the animal's weight as is true for really prehensile tails. I have the impression that its main function is to serve as a counterpoise to the body when the animal moves between branches.

An examination of the stomachcontents of four specimens as well as the intestinal contents of two of these specimens indicates that the species is omnivorous, as one specimen contained some small fruits and four large seeds as well as a number of insects. The bulk, however, was formed by insects and a spider. Arthropods eaten by this species include a spider, beetles, three bees, ants, three locusts, two cicadas, a phasmid, a cricket and a tabanid fly. BEEBE (1944b) also found that the species is omnivorous, two of the six specimens he examined containing vegetable remains as well as insect remains.

RANGE.—South America east of the Andes, south to Rio de Janeiro. In Surinam the species is present throughout the country from sea level to at least 170 m. Reported to occur at 925 m in Venezuela (TEST, SEXTON & HEATWOLE, 1966) (fig. 28).

VERNACULAR NAME.—In the Surinam language this species is called 'agama' and this name also has been adopted by Trio and Oayana Indians. This name, however, is not specific and is used for several species of iguanids.

Genus **Tropidurus** Wied

DIAGNOSIS.—Medium-sized lizards with depressed bodies, well developed legs and moderately long, round tails. Dorso-nuchal crests absent or present. Only one or two suboculars, when two, the posterior one very large; supraciliaries forming a crest over the eye; one or two pairs of enlarged postmentals. Interparietal large, slightly wider than the distance between the orbits, smaller than the ear-opening. Ear-opening with a strongly denticulate anterior margin. No gular pouch. Tail with a low middorsal crest. Nineteen species in South America

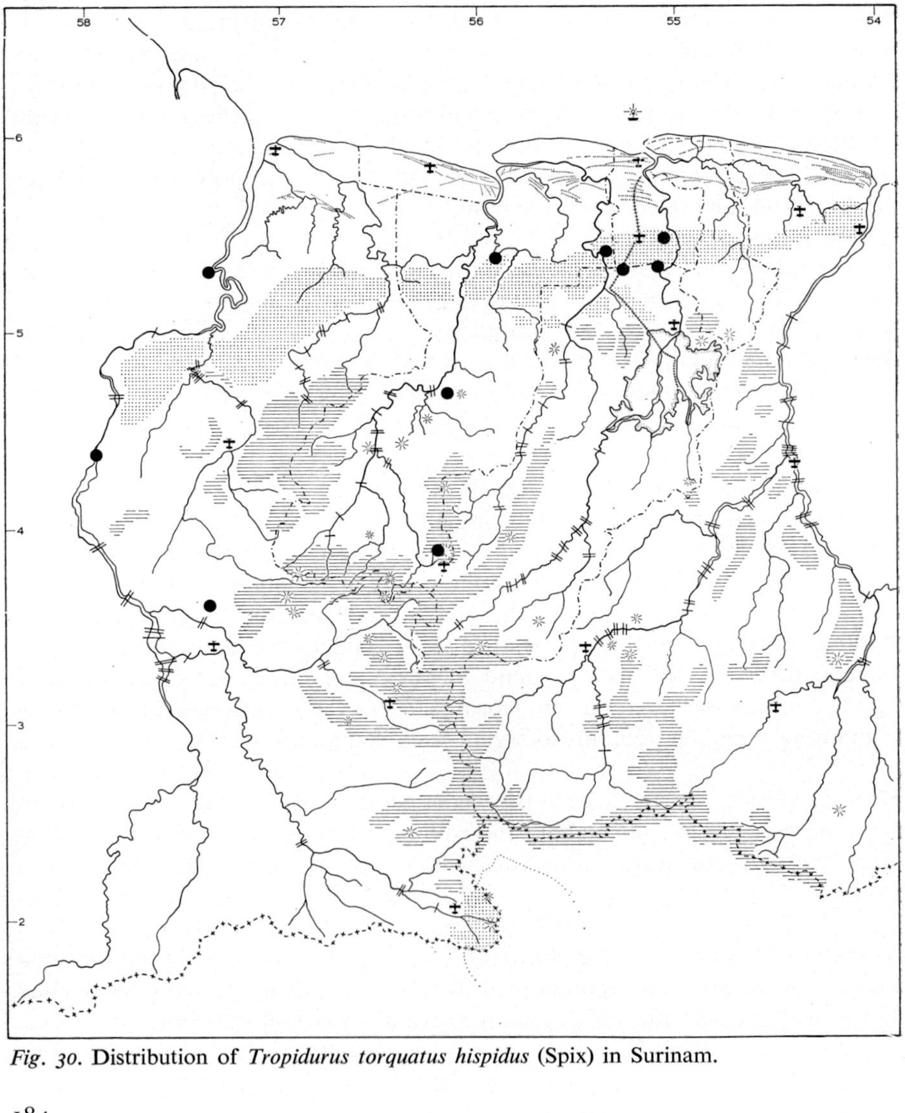

Fig. 30. Distribution of *Tropidurus torquatus hispidus* (Spix) in Surinam.

and on the Galapagos Islands. Twelve species in South-America, of which only one occurs in Surinam.

Tropidurus torquatus hispidus (Spix) (figs. 30, 31; pl. 20)

Agama hispida Spix, 1825: 12.
Tropidurus torquatus: Van Lidth de Jeude, 1917g: 438.
'54. Rotsbewoners' Fernandes, 1927: 249.
'Steen-mopskopleguaan' D'Audretsch, 1967b: 7.

TYPE-LOCALITY.—Rio de Janeiro and Bahía, Brazil.

MATERIAL. – Surinam.

Surinam: 1♂, RMNH 2904, leg. H. H. Dieperink.

Tibiti: 1♀, RMNH 16174, km 3.7, 18–I–1949, leg. Surinam Expedition 1948–49.

Troeli Creek: 1♀, RMNH 15135, 8–V–1968, leg. M. S. Hoogmoed; 1♂, RMNH 16181, 20–VI–1968, leg. H. A. M. de Kruyf.

Powakka: 2 ex., CM 44356–57, 23–V–1966, 1 ex., CM 44358, 26–VI–1966, 1 ex., CM 44360, 21–IX–1966, 1 ex., CM 44361, 23–X–1966, all leg. M. H. de la Fuente.

Railway km 62: 1♀, 4♂♂, RMNH 16175, 26/IV–16/V–1949, leg. A. Brouwer (Surinam Expedition 1948–49).

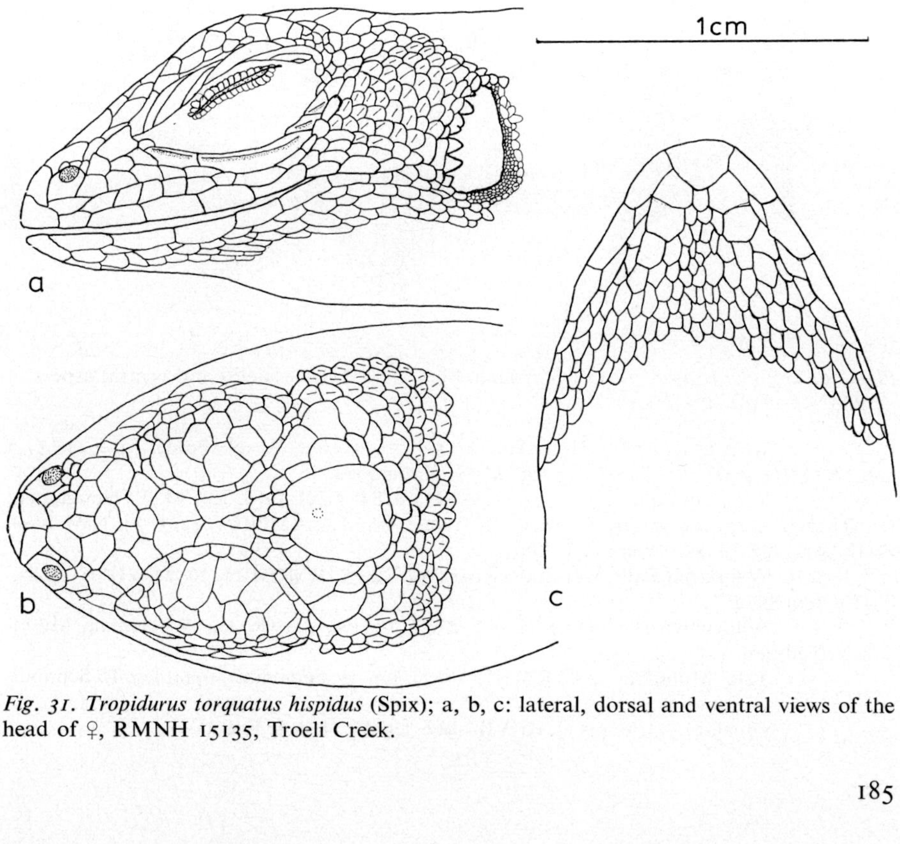

Fig. 31. Tropidurus torquatus hispidus (Spix); a, b, c: lateral, dorsal and ventral views of the head of ♀, RMNH 15135, Troeli Creek.

Plate 20. Tropidurus torquatus hispidus Spix. Figs. 1–3. Dorsal, lateral and ventral aspect of ♂, RMNH 16182 d, Kraka.

Kraka: 1♂, RMNH 16180, 14/15–IV–1968, leg. W. N. Polder; 3♀♀, 4♂♂, 3 juvs., RMNH 16182, 10–VII–1968, leg. M. S. Hoogmoed.

Voltz Mountain: 3♀♀, 1♂, RMNH 16183, 5–XI–1968, leg. M. S. Hoogmoed; 1♂, RMNH 17275, 24–III–1972, leg. G. F. Mees; 5♀♀, 1♂, AMNH 108765–70, NW. base, 23–II–1972, leg. C. W. Myers & J. Daly.

Wonotobo Falls, Corantijn River: 3♀♀, 4♂♂, RMNH 16176, 22–VII–1959, leg. P. H. Creutzberg.

Wilhelmina Mountains: 1♀, 1♂, ZMA 12349–50, 1926, leg. Wilhelmina Mountains Expedition.

Tafel Mountain: 4♀♀, RMNH 16179, km 26, September 1944, leg. L. Schmidt.

Guyana.

Oreala: 1♂, CM 44359, 10–VII–1966, leg. M. H. de la Fuente.

DIAGNOSIS.—A medium-sized lizard with depressed body, well developed legs and a moderately long, round tail. No dorso-nuchal crest. Two suboculars, the posterior one very large; supraciliaries forming a crest over the eye; supraorbital semicircle distinct; interparietal large, smaller than the ear-opening, slightly wider than the distance between the orbits; two large postmentals. Number of scales around the body 71–94. Grey above with two longitudinal rows of black spots. An interrupted black collar in front of the forelimbs.

DESCRIPTION.—Head large, representing 25–31% of the snout-vent length in females, 26–29% in males and 26–32% in juveniles; 1.1–1.4 times as long as the tibia in adults, 1.2–1.4 times in juveniles; distinctly depressed; slightly wider than the neck, which is narrower than the body. Snout obtusely pointed, straight to slightly convex in profile, not projecting over the mouth.

Tongue arrowhead-shaped with rounded tip, villose. Anterior teeth conical, lateral teeth tricuspid, increasing in size posteriorly.

Rostral pentagonal; two to three times as wide as deep; visible from above; bordered posteriorly by four to six scales, of which two to four are central and two lateral; between the central and lateral scales, the nasal may or may not touch the rostral. Seven to nine, mostly seven, rectangular supralabials; the anterior five large, the posterior two to four small; the fifth under the centre of the eye. Nostril directed dorsally and laterally; in a single, bulbous nasal. Nasal separated from the supralabials by a single scale; in contact with or narrowly separated from the rostral. Loreal region with three or four horizontal rows of scales; those of the lower row largest. Two canthal scales between the nasal and the first supraciliary. Canthus rostralis distinct. Two, rarely three, sharply keeled suboculars; the posterior one very large, in contact with or separated from the supralabials by one row of scales. Postoculars not differentiated. Temporal region with irregularly hexagonal, keeled, slightly imbricate scales. Ear-opening vertically oval, large with a strongly denticulate anterior margin and a smooth posterior margin; external auditory meatus short, tympanum very well visible. Eyelids covered with minute, irregularly shaped granules, arranged in horizontal rows, two rows of slightly enlarged scales along the rim. Pupil round with a projecting dorsal spine, a ventral projection may or may not be present. Scales on the snout of irregular arrangement, shape and size; convex, juxtaposed. Supraorbital semicircles distinct, consisting of six to eleven, mostly seven to nine, enlarged scales, semicircles of both sides in contact. Interparietal large, roughly trapezoid, wider posteriorly than anteriorly; from wider than long to longer than wide; in juveniles slightly less than half, in adults a quarter or less, the width of the head, smaller than the ear-opening, separated from the supraorbital semicircles by one row of scales. Parietal region with few, enlarged, irregularly shaped scales. Scales on the occipital region like those on the nape. Supraocular region with five to seven, mostly five, transversely enlarged scales; separated from the supraorbital semicircles by one row of small scales; from the supraciliaries by two or three rows of scales; decreasing in size towards the supraciliaries. Supraciliaries five to nine, mostly seven or eight, elongate; the anterior four or five extending obliquely posteriorly and medially; the fifth or sixth directed parallel to the longitudinal axis; the posterior scales extending obliquely posteriorly and laterally; the posteriormost scale usually very small; forming a continuous row with the canthal scales and a crest over the eye. All dorsal head scales with numerous small pustules (asperities).

Mental small, triangular; slightly longer than wide; bordered posteriorly by one very small central and two large lateral scales. Six or seven, mostly seven, infralabials, gradually decreasing in size posteriorly. Several rows of sublabials, of which the innermost has the largest scales. Scales on the chin small, rhomboidal, imbricate. Scales on the throat identical to those on the chin. No transverse gular folds, no collar, no gular pouch.

Scales on the nape small, rhomboidal, imbricate, keeled, strongly mucronate, in longitudinal and oblique rows; scales on the side of the neck smaller, in vertical and oblique rows. Side of the neck with two oblique folds, directed from anteroventral to posterodorsal, the anterior fold very deep, forming a deep pocket, the second fold shallow. Scales in the pocket minute, tubercular, juxtaposed.

Dorsals and laterals identical to the scales on the nape, in longitudinal and oblique rows, the keels forming numerous slanting ridges. Number of vertebral scales between the interparietal and the posterior border of the hindlimbs 85–102. Ventrals smaller than the dorsals, rhomboidal, imbricate, smooth, in curved transverse rows. Dorsals, laterals and ventrals smoothly merge into each other. Total number of scales around midbody 71–94. Preanal plate with small, rhomboidal, imbricate, smooth scales, as large as the ventrals. Preanal and femoral pores absent.

Scales on the tail rhomboidal, imbricate, keeled, strongly mucronate, in transverse rows; in the anterior part larger than the dorsals. Scales under the anterior part of the tail rhomboidal, imbricate, smooth, mucronate, irregularly arranged; under the posterior part of the tail lanceolate, keeled, in transverse rows. Vertebral scale row slightly enlarged, forming a low serrated crest. Tail not divided into verticils. Base of the tail in males strongly swollen. Scales on the regenerated part of the tail smaller than on the original tail, rhomboidal to lanceolate, imbricate, keeled, slightly mucronate on the dorsal, not mucronate on the ventral part; in oblique rows or irregularly arranged. Tail round in cross-section in females, in males the anterior part is round or horizontally oval, the posterior part vertically oval.

Scales on the dorsal and anterior surfaces of the forelimbs, on the posterior and dorsal surfaces of the thighs and on the posterior, dorsal and anterior surfaces of the lower legs as large as the dorsals, rhomboidal, imbricate, keeled, mucronate in oblique and longitudinal rows; scales on the ventral and posterior surfaces of the forelimbs, on the anterior and ventral surfaces of the thighs and on the ventral surface of the lower legs as large as the ventrals, rhomboidal, imbricate, smooth, in oblique and longitudinal rows. Fifteen to 20, mostly 16–18, single, sharply keeled, mucronate lamellae under the fourth finger; 22–30, mostly 24–27, single lamellae under the fourth toe; distinctly keeled, strongly mucronate in the proximal part, smooth or keeled, with one mucrone and two denticles or only two denticles, in the distal part. Palms and soles covered with scales like those on the underside of the limbs, but differing from them in being mucronate. Fingers and toes slightly compressed, subcylindrical with a serrated lower edge.

Colour in life grey with two rows of squarish black spots on the back. The region behind the second prehumeral fold is black and with its opposite number forms an interrupted collar. One female from Kraka (RMNH 16182 i) lacked any indication of pattern and was completely grey. Throat both in males and females black. In juveniles the throat is bluish with lighter dots. Belly dirty white to white. In adult males the preanal plate and the ventral surface of the thighs are black. 'Iris very pale bronze to very pale metallic greenish with a suffusion of black on each side of the pupil. Tongue unpigmented, slightly pinkish' (field notes of MYERS & DALY).

Colour in preservative, above light-grey to brownish grey with an indistinct pattern of black spots in adults. In juveniles the black spots are more distinct and are more or less arranged in transverse and longitudinal bands. Each spot accompanied posteriorly by a white line. A black antehumeral bar, starting in the second antehumeral fold and extending to the vertebral area where it is narrowly separated from its opposite number. Anteriorly and posteriorly these antehumeral bars are accompanied by white lines in juveniles, by lines slightly lighter than the rest of the back in adults. In juveniles the upper surface of the tail shows a distinct pattern of black and white spots. Limbs of juveniles with lighter spots above, a white line on the posterior surface of the thighs. Limbs of adults uniform above, of the same colour as the back. Toes with greyish cross-bars. Throat black to dark-grey. Chin with white spots on a brown background or dirty white. Belly and underside of the tail immaculate, white to light-brown. Preanal region and ventral surface of the thighs black in males.

A medium-sized lizard with a depressed body and a moderately long, round tail, of which, in adult males the posterior part is compressed. Maximum snout-vent length in females 86 mm, in males 122 mm. Recently hatched juveniles range from 31–42 mm. The undamaged tail is 1.4–1.8 times as long as the snout-vent length in females, 1.5–1.7 times in males and 1.7 times in juveniles. Limbs moderately long, stout, hind limb 1.5 times as long as forelimb in females, 1.5–1.7 times in males, 1.4–1.6 times in juveniles. When laid along the body there is considerable overlap of forelimb and hind limb, the hind limb reaching to between the posterior margin of the ear and the centre of the eye.

HABITAT.—All available Surinam data indicate that this is a ground- and rock-dwelling lizard. It seems to be restricted to dry, sunny habitats and is found both in the coastal sandy savannahs and on rockslates and boulders in the interior. It was never found in the forest itself, only in open places, usually surrounded by forest. It is absent from the Sipaliwini savannah, although suitable places seem to be available there.

NATURAL HISTORY.—A diurnal lizard which was taken between 9.45 a.m. and 4.00 p.m. Basking was frequently observed. MYERS and DALY noted that this species is 'not active at first light, but unlike *Ameiva* they stay out until dusk and many were active (although seemingly a bit sluggish) during a late afternoon rain'. Only one (small, snout-vent length 59 mm) pregnant female, containing three large, mature eggs is known from January. Recently hatched juveniles were taken in July; of these one had hatched very recently, having a large umbilical scar and a snout-vent length of 31 mm; two others have the umbilical scar smaller and are larger (41–42 mm), thus probably being a few weeks old.

The species is very abundant, very shy, and it can be approached only with difficulty.

MYERS and DALY collected a series of this species from under a flake of granite, where a colony of two to three dozen was sleeping. They were caught by prodding under the flake. According to these collectors these lizards are as fast at night as during the day and they seem to start running as soon as they awake.

Ten stomachs were examined. All stomachs contained at least one ant. In five stomachs the bulk was taken up by ants, in one by stingless bees (*Trigona*) and in the remaining three by other arthropods. Among the ants, *Atta*'s were most numerous. Other arthropods taken include: a spider, beetles and a beetle grub (present in eight stomachs), six caterpillars and a butterfly, wasps, bees, bugs, two cicadas, a fly, a grass-hopper, a dragonfly, an ant-lion and a bunch of eggs of an orthopterous insect. Two stomachs contained vegetable matter, namely seeds with plumes. RAND & RAND (1966) also noticed that these objects were taken. They did not examine stomachs, but only faecal pellets and found that the commonest item present was ants. In addition these pellets contained 'a large variety of orders of insects'.

LEON *et al.* (1970) found similar data in Venezuela where 22.95% of the total volume of preys taken consisted of Hymenoptera, mainly vespid wasps and ants. Other important groups taken include Orthoptera (19.45%), Coleoptera (14.00%) and Lepidoptera (13.22%). From their data they conclude that this species mainly relies on vision to obtain its prey.

RAND & RAND (1966) made an extensive study of this species in Belém, Brazil. They found it to be strongly heliotherm, the specimens gaining their high body temperature by basking or, when the sun disappeared, by pressing themselves against previously heated surfaces. The species was active from about 7.00 a.m. to just after 6.00 p.m. They also found that *T. torquatus* (Wied) is restricted to open areas and suggest that the high eccritic temperature of the species is a major factor in its restriction to this habitat. They consider it to be arboreal, because most of its time is spent above the level of the ground, either on the lower part of treetrunks or on other surface irregularities. In Surinam I got the impression that this was a terrestrial lizard in the coastal savannah, but this may have been caused by the fact that I had no opportunity to study living specimens for prolonged periods and that the specimens observed by me had been chased. Most of the observations by RAND & RAND probably also hold true for the Surinam specimens.

RANGE.—Surinam, Guyana, Brazil (Pará, Goiás), Venezuela (Barinas, Sucre, Nueva Esparta) and Bolivia. In Surinam the subspecies is present throughout the country in suitable places, although no records south of the Wilhelmina Mountains are available. This will be due partly to lack of exploration of that area, partly to its complete absence from areas such as the Sipaliwini savannah, where, although the suitable sites were extensively searched for this subspecies, no specimens were seen during two and a half months of intensive collecting. It occurs at least to 250 m above sea level on top of the Voltz Mountain. From Venezuela it is reported to occur at 1050 m (TEST, SEXTON & HEATWOLE, 1966) (fig. 30).

REMARKS.—BURT & BURT (1930) were the first to treat *Tropidurus hispidus* (Spix) as a subspecies of *T. torquatus* (Wied). At that time they did not discuss their decision, but they gave their arguments one year later (BURT & BURT, 1931) when treating the South American lizards in the AMNH. According to them the specimens from Bolivia show an intermediate character between *T. torquatus* (Wied) and *T. hispidus* (Spix) and thus the two forms have to be united into one species. From the key for the species of *Tropidurus* occurring east of the Andes it is clear that the most conspicuous character separating the two subspecies is the size of the dorsal scales, a character already used by BOULENGER (1885b) to separate the two forms. Further differences between the two forms are given by DA CUNHA (1961). From his data it is clear that the forms show an overlap in most characters and moreover that they are geographically separated, *T. t. torquatus* (Wied) occurring in Argentina, Paraguay, Bolivia, southern central and northeastern Brazil, *T. torquatus hispidus* (Spix) occurring in Venezuela, the Guianas, Amazonian Brazil and Bolivia. I agree with Da Cunha and therefore I named the Surinam form *T. torquatus hispidus* (Spix). J. A. PETERS & DONOSO-BARROS (1970) still treat *torquatus* and *hispidus* as different species. According to their distributional data both species occur in the Guianas, a mistake that probably was caused by BOULENGER's (1885b) misstatement that *T. torquatus* (Wied) should occur in 'Guianas: Brazil'. Probably this erroneous statement was a result of BOULENGER's copying the distributional data which DUMÉRIL & BIBRON (1837) gave for their *Ecphymotes torquatus* (Wied), which in reality is *T. torquatus hispidus* (Spix). The specimens from the Guianas show a uniform appearance and all belong to *T. torquatus hispidus* (Spix). I am not aware of any specimen of *T. t. torquatus* originating from that region.

VERNACULAR NAME.—This is another species which, in the Surinam language, is called 'agama'.

Genus **Uracentron** Kaup

DIAGNOSIS.—Medium-sized lizards with depressed bodies, well developed legs and short, depressed, spinose tails that lack autotomy. Interparietal large, wider than the interorbital distance, larger than the ear-opening. One large, keeled subocular; a row of loreolabials. Ear-opening with smooth margins. Two transverse gular folds, no gular pouch. This genus is purely Amazonian and is composed of four species, one of which occurs in Surinam.

Uracentron azureum (Linnaeus) (figs. 32, 33; pl. 21)

Lacerta azurea Linnaeus, 1758: 202.
'Lezard à queuë d'Epy' Fermin, 1765: 19.

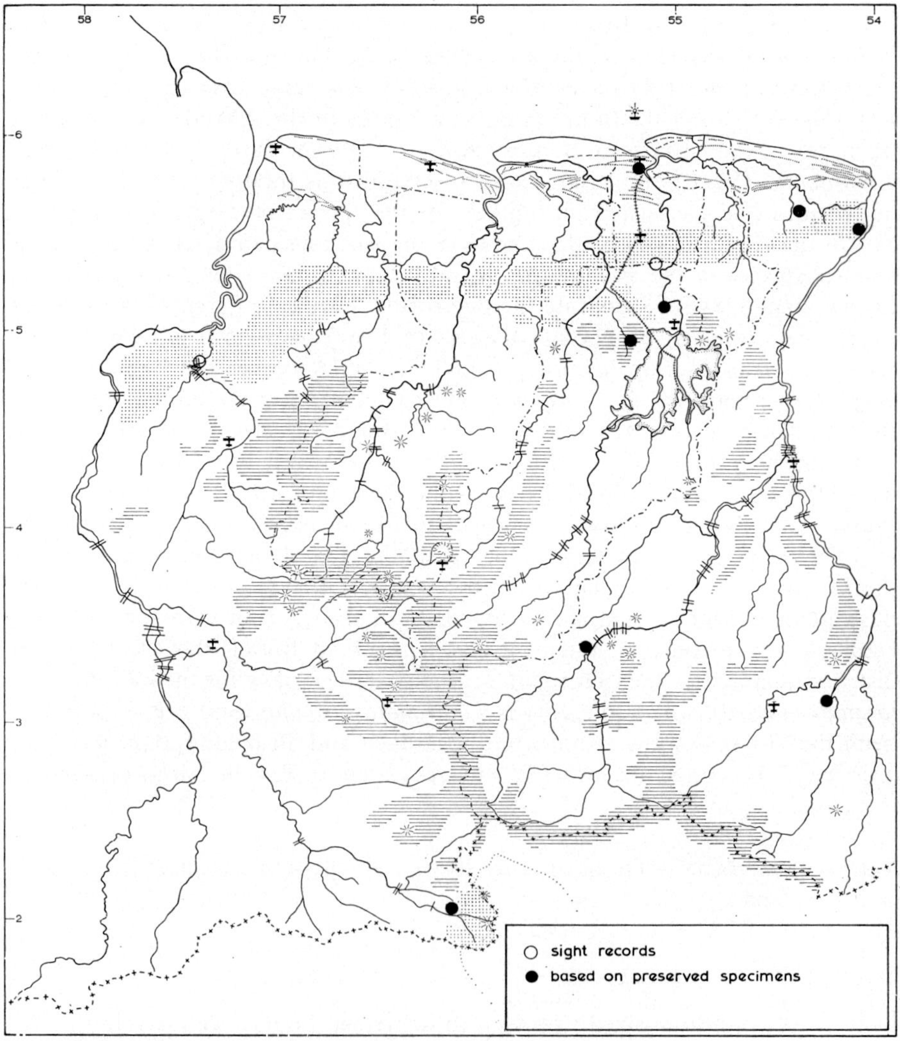

Fig. 32. Distribution of *Uracentron azureum* (Linnaeus) in Surinam.

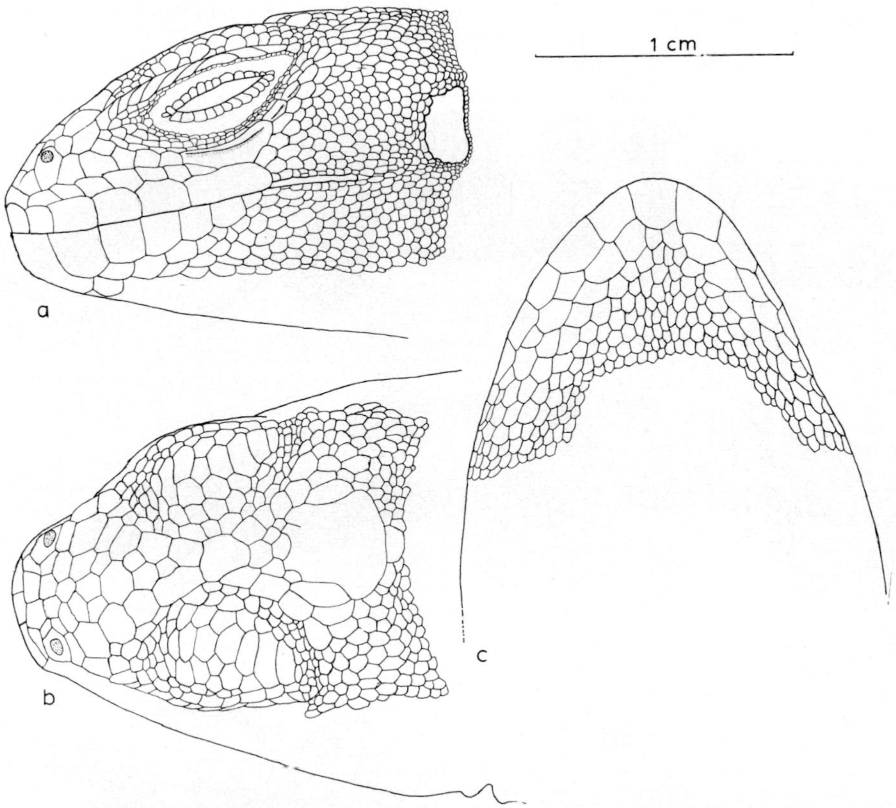

Fig. 33. Uracentron azureum (Linnaeus); a, b, c: lateral, dorsal and ventral views of the head of ♀, RMNH 15249, Brown's Mountain.

'Le dixième' Fermin, 1769: 210.
'De tiende' Fermin, 1770: 185.
'Mananiroe of Serpent' Hartsinck, 1770: 99.
Stellio brevicauda Latreille, 1802b: 29.
Stellio azureus: Latreille, 1802b: 34.
Stellio brevicaudatus Daudin, 1802b: 40.
Uromastix caeruleus Merrem, 1820: 56.
'Die buntgestreifte oder geringelte Eidechse' Von Sack, 1821a: 225.
'Bontgestreepte of geringde Hagedis' Von Sack, 1821c: 238.
Caudiverbera caerulea: Schinz, 1822: 51.
Doryphorus azureus: Schinz, 1833: 92; Duméril & Bibron, 1837: 371; Duméril & Duméril, 1851: 85; Desmarest, 1856: 76, pl. 11 fig. 3; Kappler, 1881: 166; Kappler, 1885: 817; Kappler, 1887: 127; Van Lidth de Jeude, 1917e: 274.
Urocentron azureum: Fitzinger 1843: 77; Van Lidth de Jeude, 1904: 89.
Hoplurus azureus: Schlegel, 1858: 17, 544, pl. 2 fig. 33.

Plate 21. Uracentron azureum (Linnaeus). Figs. 1–3. Dorsal, lateral and ventral aspect of ♀, RMNH 15249, Brown's Mountain.

Uracentron azureum: Etheridge, 1968: 50; Hoogmoed, 1968: 611; J. A. Peters & Donoso-Barros, 1970: 273.

TYPE-LOCALITY.—'Africa', restricted by ETHERIDGE (1968) to vicinity of Paramaribo, Surinam. Here restricted to the confluence of the Cottica River and the Perica Creek, Surinam.

MATERIAL. – Surinam.

Surinam: 2 ex., RMNH 2908, 1862, leg. Krauss; 2 ex., RMNH 2909; 1 ex., ZMA 10708; 1♀, ZMA 12230, 1922, leg. Emma Mountains Expedition; 1 ex., ZMB 675, leg. Von

Borke; 1 ex., ZMB 676, leg. Von Sack; 2 ex., SMNS 135, 1844, leg. A. Kappler; 1 ex., MCZ 4500; 2 ex., ANS 8605–06.

Paramaribo: 3 ex., RMNH 7525, surroundings, 1901, leg. H. van Capelle.
Moengo: 1 ex., CM 6669, August 1932, leg. G. Dambach.
Marowijne River: 1 ex., ZMB 6762, leg. A. Kappler.
Berg en Dal: 2 ex., NMW 13926.1–2, 1910, leg. A. O. C. Heller.
Brown's Mountain: 1♀, RMNH 15249, 16–X–1968, leg. M. S. Hoogmoed.
Airstrip Paloemeu: 1 ex., RMNH 13423, 29–XI–1965, leg. G. F. Mees.
Feti Creek, Litani River: 1 juv., RMNH 13425, 7–VIII–1939, leg. D. C. Geijskes.
Airstrip Sipaliwini: 1 ex., RMNH 13424, February 1961, leg. D. C. Geijskes.

DIAGNOSIS.—A medium-sized lizard with a depressed body, well developed legs and a short, depressed, spinose tail. Tail not autotomic. Scales on the tail in 20–22 whorls. One very large subocular. Green with black transverse bands.

DESCRIPTION.—Head large, representing 24–29% of the snout-vent length, 1.1–1.7, mostly 1.3–1.5 times as long as wide; distinctly depressed; slightly narrower than the neck, which is wider than the adjoining part of the body. Snout broadly rounded, convex in profile, not projecting over the mouth.

Tongue arrowhead-shaped, villose; with a pointed, slightly nicked tip. Teeth conical in adults, tricuspid in a juvenile, decreasing in size posteriorly.

Rostral rectangular; three times as wide as deep; hardly visible from above; bordered posteriorly by four small scales and the first supralabial on each side. Five to seven, mostly five, large supralabials; the posterior one largest and under the centre of the eye. Rostral and supralabials separated from the rest of the head scales by a row of small scales. Nostril directed dorsally, in a single nasal. Loreal region with two or three scales. One large canthal; separated from the nasal by a short suture between a loreal scale and one from the dorsal surface of the head. Canthus rostralis rounded. Three suboculars of which the posterior one is beneath the centre of the eye and very large, in contact with the posterior supralabial. An oblique row of slightly enlarged postoculars. Temporal region with small, hexagonal, convex, juxtaposed scales. Ear-opening large; more or less oval; surrounded by small scales forming a smooth margin; external auditory meatus very short, tympanum nearly superficial, very well visible. Enlarged supratemporals absent. Eyelids covered with small scales along their periphery; lower eyelid with a central patch of larger, polygonal scales; a row of enlarged, rectangular scales along the rims. Pupil round. Scales on the snout polygonal, large, convex or flat. Supraorbital semicircles with slightly enlarged scales, the left and the right one in contact. Interparietal very large; as wide as long or longer than wide, triangular to trapezoid; widest posteriorly; approximately representing $\frac{1}{4}$ of the head-length in adults, $\frac{1}{3}$ in one juvenile specimen; in contact with or separated from the supraorbital semicircles by one row of small scales. Parietal and occipital regions with small, hexagonal to polygonal, slightly convex scales, decreasing in size posteriorly; distinctly smaller than the scales on the snout. Supraocular region with transversely enlarged scales, separated from the supraorbital semicircles and from the supraciliaries by one to three rows of small scales. Supraciliaries 8–11, elongate, forming a continuous row with the canthal; the anterior six or seven extending obliquely posteriorly and dorsally; the posterior three extending obliquely posteriorly and ventrally.

Mental small; rectangular, trapezoid or mushroom-shaped; as long as wide or longer than wide; bordered posteriorly by two to four small scales. Enlarged infralabials four to eight, mostly six. Several rows of sublabials. Scales on the chin round to hexagonal, small. Scales on the throat small, hexagonal to round; in the anterior part juxtaposed, in the posterior part slightly imbricate. Two transverse gular folds, the anterior one just posterior of the ears, the posterior one continued on the sides of the neck just anterior of the insertion of the forelimbs.

All head scales, except those on the chin and throat show numerous small, round pustules, dispersed irregularly over the scale surface.

Scales on the nape and the anterior part of the back small, round to quadrangular, juxtaposed, convex, in oblique series. Scales on the side of the neck identical, in vertical and oblique series. An antehumeral fold on the side of the neck just anterior of the insertion of the forelimbs (continuation of the second gular fold) turning backwards over the forelimbs and continued on the flanks, where it disappears halfway between forelimbs and hind limbs.

Scales on the posterior part of the back larger, rhomboidal, slightly imbricate, obtusely keeled, increasing in size towards the tail, in oblique and in transverse series. Laterals smaller than the dorsals, round to oval, juxtaposed, obtusely keeled to convex, in oblique and in vertical series, with minute granules between the scales. Ventrals larger than the dorsals, rhomboidal, imbricate, smooth, in oblique and in transverse series. Dorsals, laterals and ventrals smoothly merge into each other. Total number of scales around midbody 104–120. Preanal plate with polygonal, slightly imbricate, smooth scales, larger than the ventrals. Preanal and femoral pores absent. The umbilical scar persists in adults.

Scales on the tail large, rectangular to rhomboidal, strongly mucronate, arranged in 20–22 whorls; under the tail lanceolate. Regenerated tails absent. Tail horizontally oval in cross-section.

Scales on the limbs small, rhomboidal, slightly imbricate, smooth, except on the dorsal and anterior surfaces of the lower legs, where they are obtusely keeled, in longitudinal and transverse rows. Twenty seven to 29 single lamellae under the fourth finger; 30–32 under the fourth toe, under the proximal part multicarinate, under the distal part smooth. Palms and soles with the same type of scales as on the rest of the limbs. Fingers and toes compressed.

Colour in life of RMNH 15249, above bright grass-green with black cross--bands. Belly, throat and ventral surface of the forelimbs yellow-green, chin green. Ventral surface of the hind limbs chrome-yellow. Scales on the dorsal surface of the tail green, the outer and next to outer row with the tips of the spines yellow; scales under the tail yellow-ochre and green. Iris golden brown. The juvenile specimen (RMNH 13425) has some notes concerning the colour in life on a label which, in translation, read: 'Head and back yellow-green mottled, tail with yellow spots, throat yellow-green; belly, ventral surface of legs grey-blue'. Thus the juveniles are less gaily coloured than the adults. Coloured plates of living lizards are given by COTT (1926) and HOOGMOED (1968).

In preservative the colour of the back varies from greyish blue and bright blue to greenish blue. Dorsal and lateral surfaces of head and body with transverse black bands. The first band W-shaped, with a small blue patch on the posterior part of the interparietal, on the back of the head, starting at the suboculars. A second crescent-shaped band from the corners of the mouth via the ears over the nape. Third one of the same shape on the neck. The fourth band just posterior to the antehumeral fold, crescent-shaped, the convexity directed bakcwards. Four bands on the body, the posterior one between the hind limbs. In most specimens these bands are complete, but in some the posterior bands may be connected along the vertebral region by longitudinal lines or the left and right half of a band staggered. In adults the posterior bands may be slightly invaded by blue, in juveniles all bands are solid jet-black. Medial part of the supra-ocular region black, snout with irregular black spots. Ventral parts immaculate, greyish blue. Scales on the dorsal surface of the tail greyish blue, the lateral ones with dirty white tips, some black along the sutures. Scales under the tail bluish or dirty white without regularities in pattern. Dorsal surface of limbs reticulated with black lines with blue spots between them. This reticulation is very distinct in juveniles, but may be indistinct in adults. Fingers and toes with black cross-bands.

A medium-sized lizard with a depressed, subcylindrical body, and a short, depressed spinose tail. Maximum snout-vent length 80 mm. Tail 0.5–0.6 times as long as the snout-vent length in adults, in juveniles (snout-vent length 31–32 mm) 0.6–0.7 times; 1.5–1.8 times as wide as deep in adults, 1.7–2.1 times in juveniles. Tail with serrated lateral edges, not autotomic. Limbs short and stout, hind limb 1.2–1.3 times as long as forelimb. When laid along the body there is considerable overlap of forelimb and hind limb, the hind limb reaching the ear.

HABITAT.—The available data indicate that this is an inhabitant of forest. Most specimens were collected in rain-forest. Three specimens (RMNH 7525) from the surroundings of Paramaribo were collected there 70 years ago, when this now highly cultivated area (HOOGMOED, 1969a) still may have shown patches or even larger complexes of undisturbed rain-forest. Another possibility, however, is that they were collected in the belt of mangrove bordering the Suriname River near Paramaribo. COTT (1926) reports them as occurring in this habitat on the island of Marajó in Brazil. All specimens from Surinam of which habitat data are available (RMNH 13423, 13424, 13425, 15249 and two observations by Dr. D. C. GEIJSKES) were collected on tree-trunks from two to five metres above the forest floor, indicating an arboreal way of life. This agrees well with COTT's (1926) observations on Marajó and those of KAPPLER (1885) in Surinam. However, the single specimen collected by BEEBE (1944b) in Kartabo was found 'among some rocks near the shore', a completely different habitat.

I have difficulties in interpreting DA CUNHA's (1961) remark on the habitat of this species 'Eles parecem ser muito comuns nos lugares arenosos e secos da ilha de Marajó' (They seem to be very common in sandy and dry places of the island of Marajó). If this remark is correct, the species shows a remarkable ecological variation, occurring on tree trunks in the Surinam rain-forest and on dry, sandy places on Marajó. I have no idea whether these places are open and sunny or whether there is a substantial growth of trees. If the latter, the habitat may correspond to that found in Kraka, Surinam, where a specimen was observed, by Mr. W. N. POLDER, living on the trunks of trees in an open savannah-forest. Nevertheless, DA CUNHA's remark contradicts COTT's observation and more habitat data on this species will be needed to be able to make more definite statements concerning its habitat. Summarizing the available data, we may tentatively conclude that this is an arboreal lizard, occurring in rain-forest, savannah-forest and mangrove trees along tidal waters.

NATURAL HISTORY.—A diurnal lizard. Only two pregnant females are available. One (ZMA 12230) contains one large, elongate, mature egg (25.4 × 10.5 mm), the female itself having a snout-vent length of 72 mm. Unfortunately no collecting date is available for this specimen. The other specimen (RMNH 15249), collected on October 16th, contains 2 small, round, immature eggs. Two recently hatched juveniles with snout-vent lengths of 31 and 32 mm, taillength of 18 and 22.5 mm, respectively, are available. Only one of them (RMNH 13425) has an exact collecting date, having been collected on July 7th. The other (SMNS 135b), is without date. From these scanty data nothing about the breeding season can be concluded.

Little is known about the feeding habits of this species. KAPPLER (1885) reports it to live in hollow trees, and presumes it to feed on insect larvae living in the mould. COTT (1926) observed a specimen 25 feet up in a tree, while it was busily catching ants. Four stomachs of Surinam specimens examined indicate that this lizard is a food-specialist. Two stomachs contained only ants (Formicinae), belonging to at least three species. A third one also contained three species of ants (Formicinae and Ponerinae), and moreover, one Curculionid beetle and one spider. The fourth stomach contained remains of three species of ants and one small coleopteron. In all stomachs the ants dominated, both in volume and in individuals. Although it can be ascertained that *Uracentron azureum* (Linnaeus) has a preference for ants, it is not very fastidious regarding the ants, as the four stomachs contained at least nine species of ants. Sofar it has not been possible to establish the exact identity of the ants.

RANGE.—Northern South America from Guyana and Manaus in the west to the mouth of the Amazon River in the east. South to the Manjuru River in Brazil (ETHERIDGE, 1968). In Surinam the species occurs throughout the country from sea level to at least 500 m (fig. 32).

REMARKS.—ETHERIDGE (1968) states that the supraoculars always are separated from the supraciliaries by at least two rows of small scales. In Surinam specimens this is true only for the anterior part of the supraocular region, where the number of scale rows may be three. In the posterior part, however, there remains only one scale row between the last enlarged supraocular and the supraciliaries.

Genus **Uranoscodon** Kaup

DIAGNOSIS.—See diagnosis of species. A single species in northeastern South America.

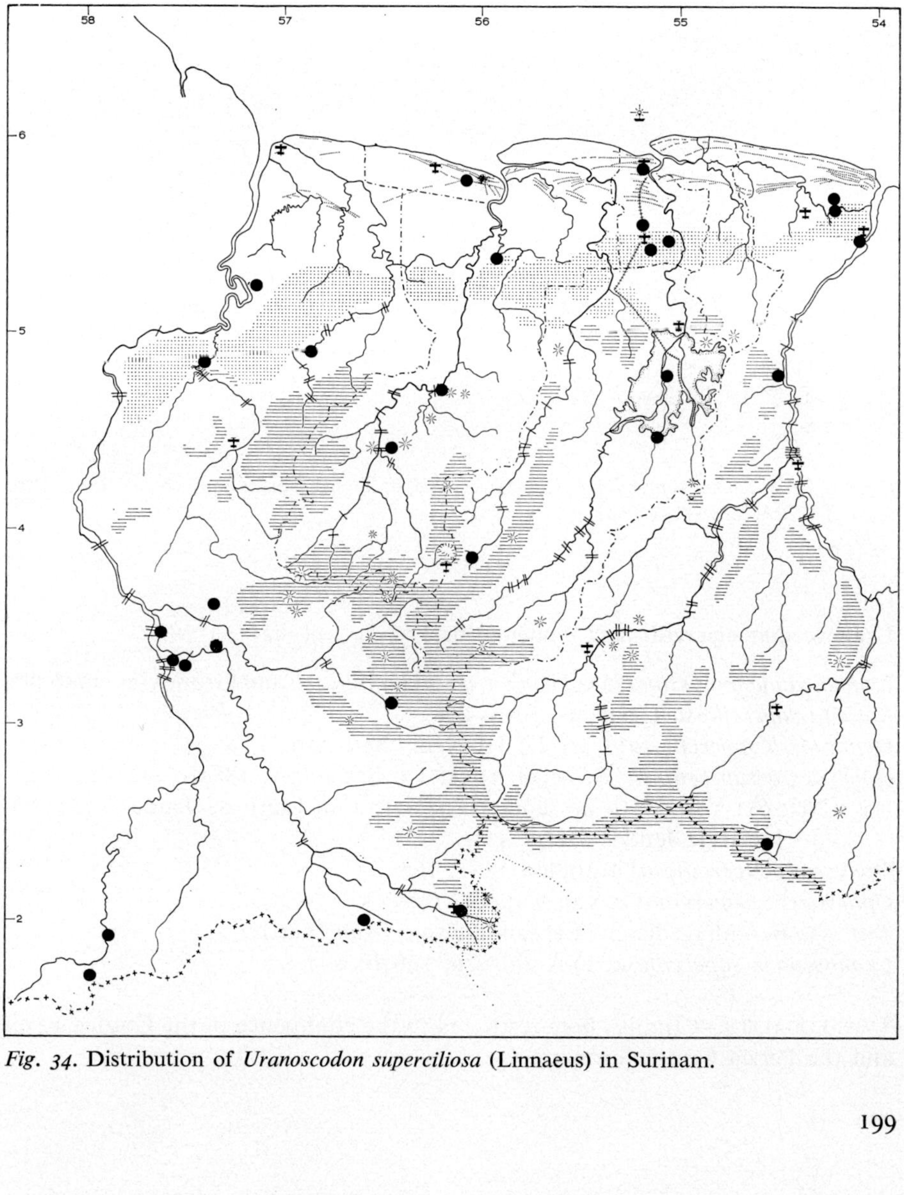

Fig. 34. Distribution of *Uranoscodon superciliosa* (Linnaeus) in Surinam.

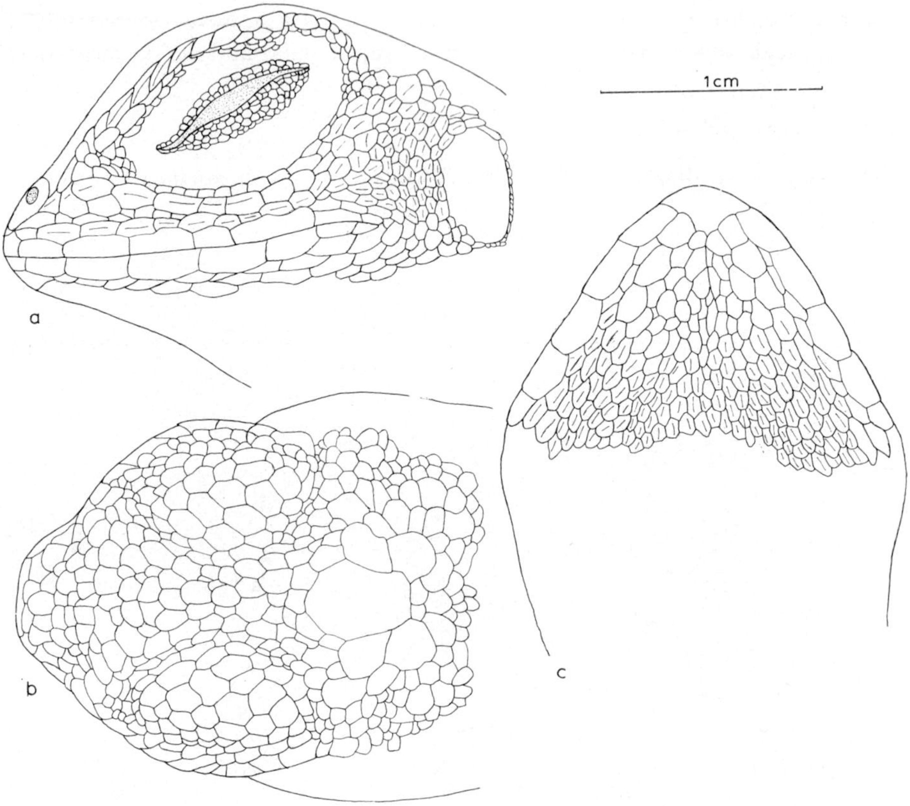

Fig. 35. Uranoscodon superciliosa (Linnaeus); a, b, c: lateral, dorsal and ventral views of the head of ♀, RMNH 15254, Raleigh Cataracts.

Uranoscodon superciliosa (Linnaeus) (figs. 34, 35; pl. 22)

Iguana cauda longissima tereti conica, dorso tantum cristato Gronovius, 1756: 83.
Lacerta superciliosa Linnaeus, 1758: 200.
Ophryoëssa superciliosa: Van der Hoeven, 1855: 529.
Ophryoessa superciliosa: Schlegel, 1858: 19; Boulenger, 1885b: 111; Boettger, 1893: 61; Van Lidth de Jeude, 1898: 20; Van Lidth de Jeude, 1904: 88; Van Lidth de Jeude, 1917g: 437.
Ophycessa superciliosa: Kappler, 1881: 166.
Ophryoessa Superciliosa: Van Capelle, 1903: 51.
?'51–53. Boomhagedissen' Fernandes, 1927: 249 (partly).
Uranoscodon superciliosa: D'Audretsch, 1967b: 6, 7.

TYPE-LOCALITY.—'Indiis', here restricted to the confluence of the Cottica River and the Perica Creek, Surinam.

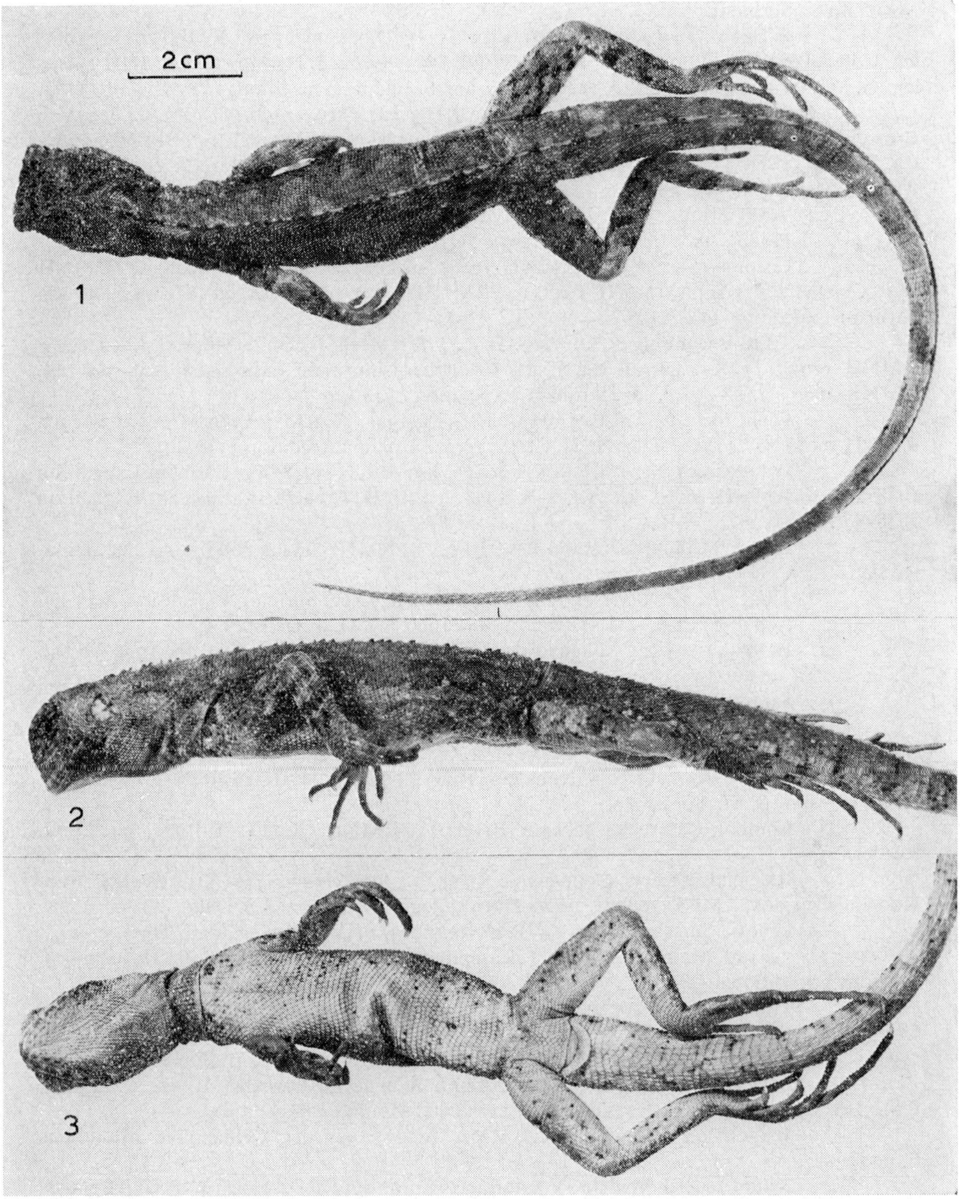

Plate 22. Uranoscodon superciliosa (Linnaeus). Figs. 1–3. Dorsal, lateral and ventral aspect of ♀, RMNH 15254, Raleigh Cataracts.

MATERIAL. – Surinam.

Surinam: 1 ex., RMNH 2914, leg. H. H. Dieperink; 1 ex., RMNH 7534, 1901, leg. Coppename Expedition; 1 ex., RMNH 7536, 1883, leg. C. J. Hering; 2♀♀, RMNH 16064, leg. G. Witsen Elias; 1 ex., ZMA 12290, 1908, leg. Suriname Expedition; 1♀, 1♂, 1 ex., ZMA 12291–92, 12295, leg. G. Stahel; 1 ex., ZMA 12294; 3 ex., ZMA 12296–98, 1887–90, leg. Van Brussel; 1 ex., ZMA no number, 1922, leg. J. W. Gonggrijp & G. Stahel; 1♂, BM 66.9.20.12, leg. Bartlett; 1♀, 3♂♂, NMW 12934–35, 1910, 2♂♂, NMW 12936, all leg. A. O. C. Heller; 1 juv., NMW 12938, leg. Werner; 1♂, NMW 12939, 1899; 1♀, NMW 12940, leg. A. O. C. Heller; 1 ex., ANS 8317.

Paramaribo: 1 ex., RMNH 7535, 7–III–1939, leg. H. W. C. Cossee.

Coronie road: 1 juv., RMNH 16065, picket 216. December 1948, 1♀, RMNH 16074, picket 208, km 1.4, 16–XII–1948, 1♂, RMNH 16075, picket 210, 23–XII–1948, all leg. Surinam Expedition 1948–49.

Third Encampment, km 14.9: 2♀♀, 4♂♂, RMNH 16069, 1st ridge, 6–X–1948, 1♀, RMNH 16072, 13–X–1948, all leg. P. H. Creutzberg (Surinam Expedition 1948–49); 1♂, RMNH 16073, 24–X–1948, leg. Surinam Expedition 1948–49.

Djai Creek: 1♀, 1♂, RMNH 16068, 5–X–1948, 1♀, RMNH 16070, 7–X–1948, 1♂, RMNH 16071, 8–X–1948, all leg. P. H. Creutzberg (Surinam Expedition 1948–49).

Wane Creek N.: 1♀, RMNH 16066, km 6, 2–X–1948, leg. Surinam Expedition 1948–49; 1♀, RMNH 16067, km 7.9, 2–X–1948, leg. P. H. Creutzberg (Surinam Expedition 1948–49).

Kaboeri Creek, Corantijn River: 1♀, 1♂, RMNH 16490, 2–IV–1971, leg. D. G. Reeder.

Tibiti: 1♀, 1♂, RMNH 16076–77, 18–I–1949, leg. Surinam Expedition 1948–49.

Republiek: 1♂, RMNH 16053, May 1960, leg. D. C. Geijskes.

Powakka: 2♂♂, CM 44370–71, 19–V–1966, 1♀, CM 44372, 11–IX–1966, all leg. M. H. de la Fuente.

Saramacca Creek: 1♀, RMNH 16062, 27–IV–1965, leg. G. F. Mees.

Papatam, 4 km S. of Albina: 1♀, RMNH 16046, 19–VI–1968, leg. M. W. H. de Boer.

Avanavero Cataracts, Kabalebo River: 1 juv., RMNH 16491, 9–IV–1971, leg. D. C. Geijskes & M. Boeseman.

Lombok Cataracts, Nickerie River: 1♀, RMNH 16472, 7–II–1971, leg. D. G. Reeder.

Raleigh Cataracts, Coppename River: 1♀, RMNH 15254, 7–XI–1968, leg. M. S. Hoogmoed; 1 juv., AMNH 108771, 19/20–II–1972, leg. C. W. Myers & J. Daly.

Bedotti, Surinam River: 1♂, RMNH 16063, 29–V–1965, leg. G. F. Mees.

Nassau Mountains, Base Encampment: 1♀, RMNH 16078, 14–II–1949, leg. Surinam Expedition 1948–49.

Transect IV, Coppename River: 1♀, RMNH 16060, 29–XI–1943, leg. D. C. Geijskes.

Middle Gran Creek: 1♀, RMNH 16058, 30–VII–1964, leg. M. Boeseman.

Base Bivouac near De Kock Mountain, Upper Saramacca River: 1♀, 1 juv., RMNH 16059, 12–III–1958, leg. D. C. Geijskes (Tafel Mountain Expedition).

Wilhelmina Mountains: 1♂, ZMA 12293, 1926, leg. Wilhelmina Mountains Expedition.

King Frederik Willem IV Cataracts, Corantijn River: 3♀♀, 1♂, RMNH 16055, 3/8–VIII–1959, 1♀, RMNH 16056, all leg. D. C. Geijskes.

Barrington Brown Rapids, New River: 1♀, FMNH 30930, August-October 1938, leg. E. R. Blake.

Tigri Encampment, New River: 1♂, RMNH 16047, E., 28–VI–1968, leg. M. S. Hoogmoed.

Airstrip Coeroeni: 1♀, RMNH 16054, 12–VIII–1959, 2♀♀, RMNH 16057, 12/14–VIII–1959, all leg. D. C. Geijskes.

Airstrip Kayser Mountains: 1♀, RMNH 16050, NE., 2–VIII–1968, 1♀, RMNH

16051, NE., 10–VIII–1968, 1♀, RMNH 16052, NE., 12–VIII–1968, all leg. M. S. Hoogmoed.

Middle Base Camp, Itabu Creek: 1♀, 1♂, FMNH 30927, 30929, September/October 1938, leg. E. R. Blake.

Sipaliwini, Base Bivouac: 1♂, RMNH 16048, 3–IX–1968, 1♂, RMNH 16049, 28–VIII–1968, both leg. M. S. Hoogmoed.

Waremapan Creek, Upper Litani River: 1♀, RMNH 16061, 20–VII–1939, leg. D. C. Geijskes.

Surinam/Guyana.

Upper New River: 1 juv., BM 1936.4.2.55, 10–XII–1935, leg. J. S. Myers.

New River, 750 feet: 1♂, 1 juv., BM 1939.1.1.70–71, leg. C. A. Hudson.

DIAGNOSIS.—A large lizard with compressed body and tail. Head cuboid, short. Upper head scales small, subequal, pyramidical; supraciliaries forming a strong, projecting crest over the eyes; several rows of small, subequal suboculars. A dorsal crest from the neck onto the tail. Dull green to dark-brown above with a dirty white, undulating lateral band.

DESCRIPTION.—Head small, representing 22–24% of the snout-vent length in females, 21–24% in males, 25–26% in juveniles; cuboid, very short; 1.3–1.4 times as long as wide in females and juveniles, 1.2–1.5 times in males; slightly wider than deep; mostly slightly longer than the tibia; wider than the neck, which is as wide as the adjoining part of the body. Snout broadly rounded, not projecting over the mouth, straight in profile.

Tongue arrowhead-shaped, villose, with a rounded tip. Anterior teeth conical, lateral teeth tricuspid, only slightly compressed laterally, subequal.

Rostral rectangular; 3.5–4 times as wide as deep; visible from above; bordered posteriorly by five to seven small scales and by the first supralabial on each side. Supralabials rectangular, four (rarely five or six), the posterior one largest; four to below the centre of the eye. Nostril directed upwards and laterally; in a single bulbous nasal. Nasal separated from the supralabials by one or two rows of small scales; situated medially of the canthus rostralis. Loreal region with four or five horizontal rows of small scales; scales in the lowest row, which is continued to the end of the fourth supralabial, largest. Several rows of subocular scales. Postocular scales in several rows, not enlarged. Temporal scales small, hexagonal to oval, juxtaposed. Ear-opening large, larger than the interparietal; surrounded by small scales forming a smooth margin; external auditory meatus absent, tympanum superficial, clearly visible. Eyelids covered by small granules, scales along the rim slightly larger. Pupil round. Scales on the snout small, subequal, irregularly shaped. Supraorbital semicircles not defined. Interparietal large, longer than wide. A few enlarged scales bordering the interparietal laterally. Scales on the parietal region small; irregularly shaped, diminishing in size laterally. Scales on the occipital region like those on the parietal region, with two enlarged scales, one on each side of the median. Supraocular region elevated, with enlarged, hexagonal, slightly imbricate scales. Interorbital region concave. One or two canthals, forming a distinct canthus rostralis. Seven to eleven, mostly eight or nine, supraciliaries, forming a crest over the eye; the anterior five directed medially and posteriorly; the posterior scales directed laterally and posteriorly; forming a continuous row with the canthals.

All dorsal head scales strongly keeled, except the rostral and the supralabials. Scales on the loreal, temporal, parietal, occipital and supraocular regions with strong longitudinal keels. Scales on the snout partly longitudinally keeled, partly pyramidal. The enlarged scales on the occipital region forming two small, pyramidal horns with furrows along all slopes.

Mental small; more or less triangular; bordered posteriorly by one or two small, median scales, one large scale and the first infralabial on each side. Five or six, rarely seven, rectangular infralabials. Scales on the chin decreasing in size towards the median; hexagonal to rectangular, strongly keeled, juxtaposed.

Scales on the nape, on the side of the neck and on the posterior part of the throat small, rhomboidal, keeled, slightly imbricate; mucronate on the throat, not on the side of the neck and the nape. Anterior part of the throat with small, oval to rhomboidal, keeled, mucronate scales. Two gular folds; the anterior one halfway between the ears and the insertion of the forelimbs; the posterior one just before the insertion of the forelimbs and continued on the sides of the neck as an antehumeral fold that extends obliquely upwards and backwards. Nuchal crest present, continuous with the dorsal crest, but the anterior two or three scales may be separated from the rest by a small scale.

Dorsal and lateral scales small, rhomboidal to rectangular, keeled, the keels pointing to the dorsal crest, mucronate, slightly imbricate, in transverse and oblique series. A serrated dorsal crest, continuous with the nuchal crest, of triangular, erect scales; 51–64 from the beginning to the posterior surface of the thighs. One crest-scale corresponds to two dorsals. Ventrals slightly larger than the dorsals, rectangular to trapezoid, keeled, the keels forming oblique ridges, mucronate, imbricate, in oblique and in 55–64 transverse rows. Total number of scales around midbody 96–112. Scales on the preanal plate like the ventrals. Femoral and preanal pores absent. The everted hemipenis is long and slender, bifurcate, the two tips also long and slender with a reticulated surface; lateral flaps from the base to halfway the tips; sulcus spermaticus on the posterior surface, deep.

Vertebral crest continued onto the tail. Scales on the rest of the tail small, rectangular, obliquely keeled, mucronate, slightly imbricate, in transverse series. Scales under the tail larger, with longitudinal keels, forming longitudinal ridges. Scales on the regenerated part of the tail small, irregularly shaped, slightly keeled, slightly imbricate, arranged irregularly. No crest on the regenerated part. Tail compressed, in cross-section oval with one dorsal and several (± six) ventral keels.

Scales on the limbs small, rhomboidal, keeled, imbricate. Twenty to 25 single, sharply keeled lamellae under the fourth finger; 28–34 under the fourth toe. Third, fourth and fifth finger with a row of keeled, not enlarged scales along their lateral margins; third, fourth and fifth toe with a serrated row of enlarged, keeled scales along their lateral margins, thus increasing the ventral surface of the toes considerably. Fingers subcylindrical, toes triangular in cross section. Palms and soles with small, rhomboidal, keeled, juxtaposed scales.

Colour of the back in life ranging from dull green to dark-brown with a dirty white lateral band. Belly pale yellow. Tail with more or less distinct transverse black bands.

In preservative the colour of the back ranges from pale grey-brown with white spots to dark-brown. The upper eyelids show four radiating dark-brown bands which are connected over the interorbital region with their counterparts of the other side. Lateral band creamy, with undulating dorsal margin, ventrally just narrowly separated from the ventral ground colour by an indistinct brown band. Ventral parts dirty white to light-brown with scattered dark-brown spots. Dorsal surface of limbs and toes of the same colour as the back, with large, transverse black spots. Tail with or without transverse black lines or bands, which may be continuous on the dorsal and ventral surface.

A large lizard with compressed body and tail. Maximum snout-vent length in females 135 mm, in males 138 mm. The undamaged tail is 2.0–2.3 times as long as the snout-vent length in females, 2.1–2.5 times in males and 1.8 times in a single juvenile. Limbs long and stout, when laid along the body there is considerable overlap of forelimb and hind limb, the hind limb reaching to between the eye and the tip of the snout in juveniles and halfgrowns, to between the insertion of the forelimb and the ear in adults. Hind limb 1.6–1.9 times as long as forelimb.

HABITAT.—An inhabitant of forested banks of creeks and rivers, where it is present on the lower part (not higher than two metres) of trees near the water's edge or overhanging the water. It also may be found on the forest floor. The species is restricted to banks of creeks and rivers that have an open vegetation, as in the interior of the forest, and are not closed off by dense bushes and vines. This same type of habitat was found in Belém by RAND & HUMPHREY (1968).

NATURAL HISTORY.—A diurnal species, taken between 11.00 a.m. and 3.45 p.m. Basking was not observed, all lizards taken by me were in shady spots. This agrees with the observations of RAND & HUMPHREY (1968) who found that this species is a nonheliotherm. These lizards try to escape notice by resting immobile on the spot, when on trees by clenching the stem, when on the forest-floor by pressing themselves against it. Frequently it is possible to lift the specimen without trouble (BEEBE 1944b), but some specimens I collected myself tried to escape by jumping into the water and diving to the bottom (cf. RAND & HUMPHREY, 1968). RMNH 16047 was spotted on the bank of a creek. When approached it jumped into the creek, and headed for a log that protruded out of the water. It reached it by running over the watersurface on the hind limbs for a short distance (two metres) and, when the animal slowly sank deeper into the water, by making undulating movements with the tail. During this entire proces the anterior part of the body was raised out of the water. Another specimen that could not be collected first was seen on the base of a tree bordering the inundated bank of the Coeroeni River. When approached it dived into the water where it submerged. It emerged again on the stem of a tree completely surrounded by water, after having covered about five metres under water. When approached again it silently slid into the water and was not seen afterwards. For these activities it seems that the fringed toes probably render good services.

Pregnant females containing three to eight large, elongate, mature eggs are known from July, August, September and October. Females containing small, round, immature eggs are known from March, June and August. BEEBE (1944b) reports a clutch of 11 eggs and another of six eggs, both found in August, the last clutch taken from a hollow 12 feet up a tree. A female from November (Transect IV, Coppename River, RMNH 16060) according to the label had recently deposited her eggs. Another female (Powakka, CM 44372) from November was collected on the ground in front of a burrow in which she was to deposit her eggs. Thus, there seems to be a wide diversity in breeding sites. My Surinam data suggest that there is a direct connection between the number of eggs and the size of the female, the larger the female, the more eggs she produces. Recently hatched juveniles with snout-vent lengths of 47–50 mm, taillength 84 mm (one specimen) are available from March and December. These data indicate that breeding probably takes place throughout the year. Six specimens (RMNH 16069) were collected on the same day along the margin of a small pool. They all are of about equal size and this may indicate that the members of the same egg-clutch stay together for extended periods.

The stomach and intestinal contents of seven specimens were examined. Three specimens contained large, hairy caterpillars. Another specimen had a large, hairy caterpillar in its mouth. Moreover, the stomachs and intestines contained an earthworm, a diplopod, a millipede, three beetles, a beetle grub, four ants, a bug and a grasshopper. Three further stomachs (intestines) contained small twigs and a piece of bark, one intestine contained clay (probably indicating that an earthworm had been digested). RMNH 16046 contained some cocoons of wasp parasites. BEEBE (1944b) kept a specimen in captivity for some time. After fasting for six weeks it started to eat orthopterons, caterpillars and grubs but it did not like termites. Another specimen, when captured, contained a millipede and three caterpillars. These data are quite in accordance with those found for Surinam examples.

RANGE.—Northeastern South America: Surinam, Guyana, French Guiana, Brazil (Pará, Amazonas, Amapá) and Venezuela (Bolivar, Amazonas). In Surinam the species is present throughout the country, from sea level to at least 250 m (fig. 34).

VERNACULAR NAME.—This is another iguanid called 'agama' in the Surinam language. According to the label of RMNH 16061 this lizard is called 'hoepoeloeloe' by the Oayana Indians.

Family SCINCIDAE

This family shows considerable variation in the shape of the body, the size of the limbs and the eye covering. All skinks have scales that are underlain by bony plates. Most species have cycloid scales, but some have other types of scales. In many genera there is a marked tendency towards reduction of the limbs, which may result in species without limbs. The eye may be covered by normal, scaly, movable eyelids, by movable eyelids with a transparent disc in the lower eyelid or by a brille, i.e., a transparent lower eyelid that has grown over the eye. There is also some variation in the number of scales covering the head, caused

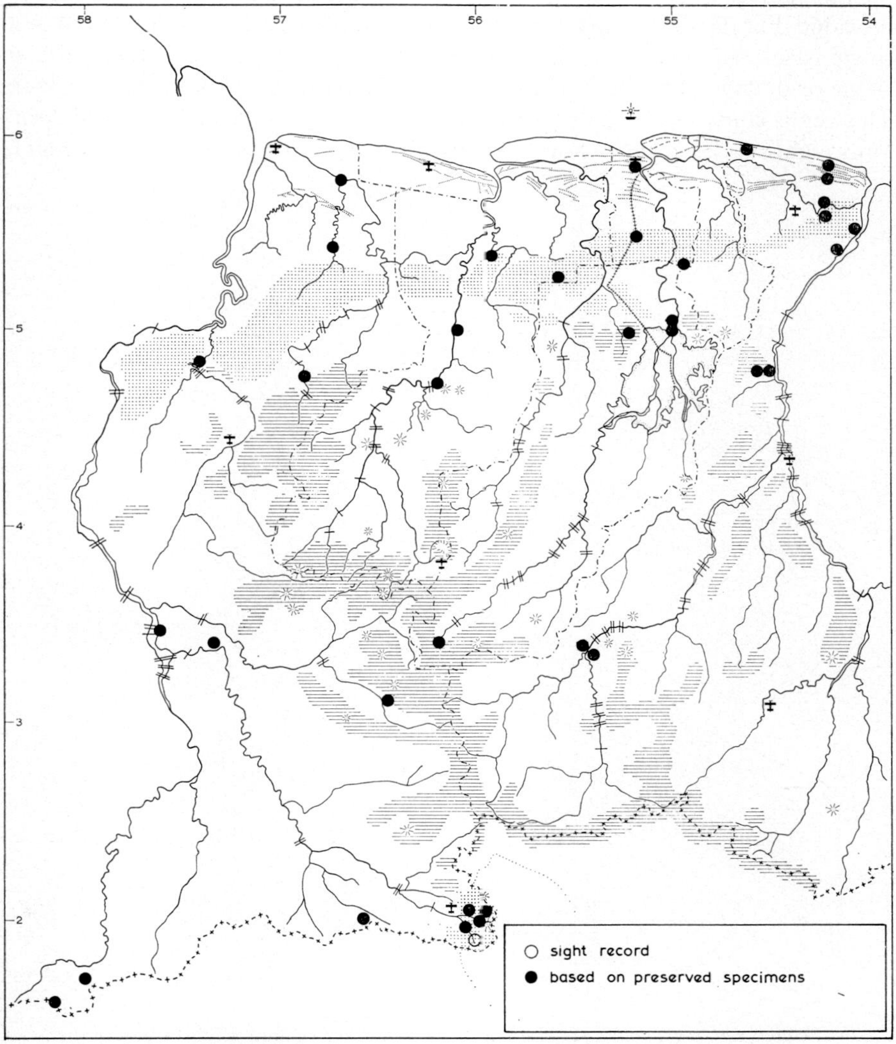

Fig. 36. Distribution of *Mabuya m. mabouya* (Lacépède) in Surinam.

by fusion. Only a single species of this cosmopolitan family is present in Surinam. It can be readily distinguished from other Surinam lizards by the presence of large, cycloid scales on the body, each of which is underlain by a bony plate. The Surinam representative is present both in clearings in the forest and on open savannahs, it is terrestrial to semi-arboreal.

Genus **Mabuya** Fitzinger

Diagnosis.—Medium-sized to large lizards with cylindrical bodies, short legs and short tails. Palatine bones in contact in the median; palatal notch separating the pterygoids, extending forwards to between the centres of the eyes. Eyelids movable. Ear distinct, tympanum more or less deeply sunk. Nostril pierced in a single nasal; supranasals, prefrontals and frontoparietals present. Interparietal single or double, sometimes fused with the parietals (after Boulenger, 1887a). This genus contains about 85 species, dispersed over America, Africa and Asia. Only eight species occur in South America, of which one is present in Surinam.

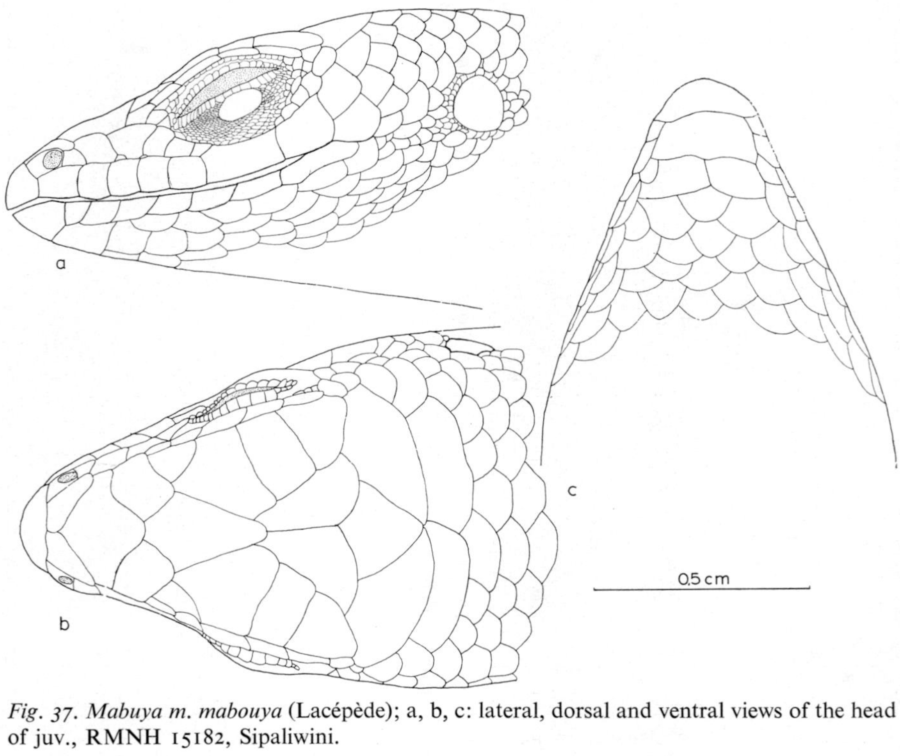

Fig. 37. Mabuya m. mabouya (Lacépède); a, b, c: lateral, dorsal and ventral views of the head of juv., RMNH 15182, Sipaliwini.

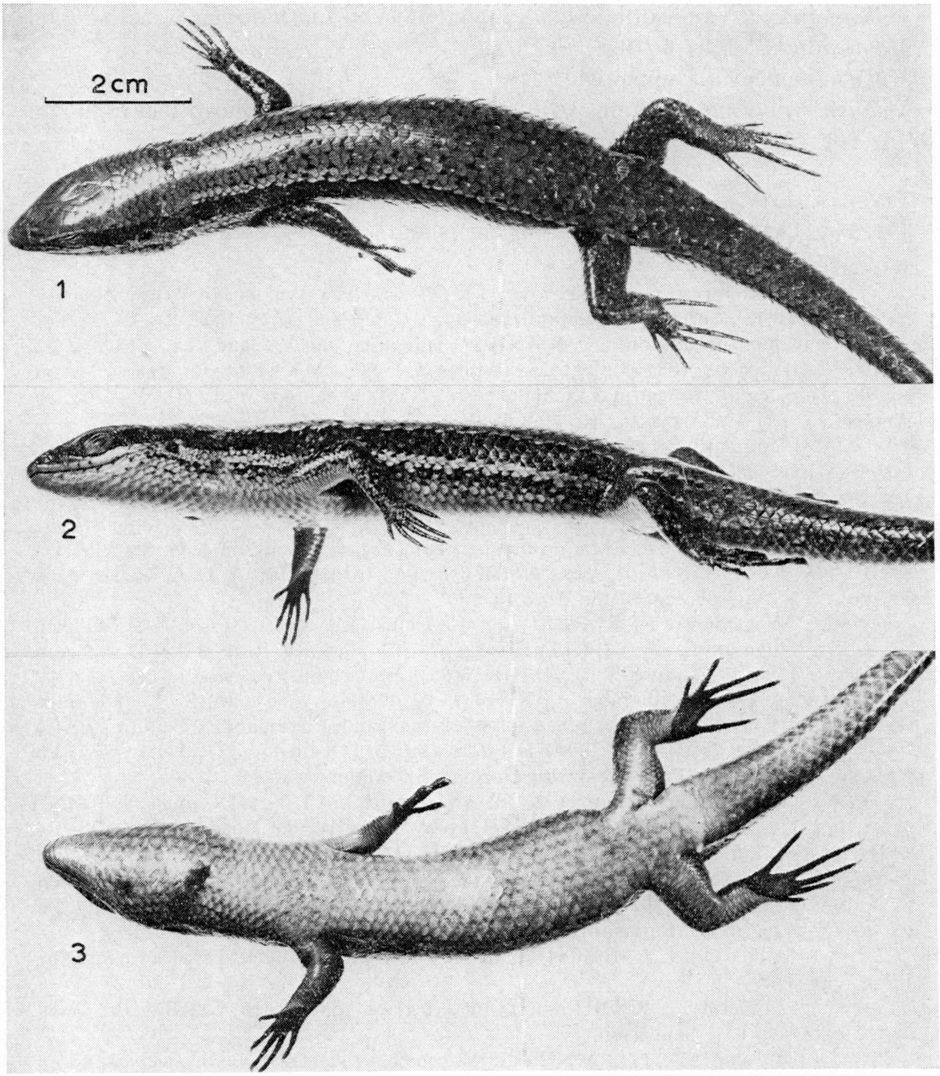

Plate 23. Mabuya m. mabouya (Lacépède). Figs. 1–3. Dorsal, lateral and ventral aspect of ♂, RMNH 15645, Sipaliwini.

Mabuya mabouya mabouya (Lacépède) (figs. 36, 37; pl. 23)

Sincus pedibus pentadactylis, ungiculatis, digitis teretibus Gronovius, 1756: 75.
Lacertus Mabouya Lacépède, 1788a: Synopsis methodica Quadrupedum Oviparorum.
Sincus Aurata Schneider, 1810: 176.
Euprepris surinamensis Hallowell, 1856: 154.
Mabouya agilis: Kappler, 1881: 166.

Mabuia aurata: Van Lidth de Jeude, 1904: 92; Van Lidth de Jeude, 1917g: 438.
Mabuya agilis: Burt & Burt, 1930: 42.
Mabuya mabouya: Dunn, 1936: 537.
Mabuya m. mabouya: Dunn, 1936: 544; J. A. Peters & Donoso-Barros, 1970: 199.

TYPE-LOCALITY.—Antilles, restricted by DUNN (1936) to Lesser Antilles, by SMITH & TAYLOR (1950) to St. Vincent Island.

MATERIAL. – Surinam.

Surinam: 3 ex., RMNH 2510; 1♀, 1♂, RMNH 15594, leg. G. Witsen Elias; 1♀, RMNH 15622, leg. Surinam Expedition 1948–49; 1♂, RMNH 15625, 1904, leg. G. M. Versteeg (Tapanahoni Expedition); 1♀, RMNH 15644, leg. G. van Vreden; 1 ex., ZMA 12302; 2 ex., ZMA 12303–04, 1887–90, leg. C. van Brussel; 6 ex., ZMA 12307–12; 1 ex., ZMA no number, 1922, leg. F. Gonggrijp & G. Stahel; 1 ex., NMW 9550, 1901; 1 ex., NMW 16194, leg. Werner; 1♀, 1♂, ANS 9532–33, leg. Hallowell.

Paramaribo: 3 ex., RMNH 7581, 1911, leg. Jhr. W. C. van Heurn; 1 juv., RMNH 15595, 12–IX–1948, leg. Surinam Expedition 1948–49; 1♀, RMNH 15624, 1911, leg. Jhr. W. C. van Heurn; 1♂, RMNH 15633, 17–VII–1968, leg. M. S. Hoogmoed; 1♂, RMNH 15641, March 1966, 1♀, RMNH 15643, both leg. G. van Vreden; 2 ex., ZMA 12305–06, August 1907, leg. C. A. van Brussel; 1 ex., ZMA no number, 1922, leg. F. Gonggrijp & G. Stahel; 1 ex., ZMB 25969, 1 ex., ZMB 26402, 3 ex., NMW 9564.1–3, 1910, all leg. A. O. C. Heller; 1 ex., CM 44373, 31–V–1966, leg. M. H. de la Fuente.

Wageningen: 1♂, RMNH 15626, 9–VI–1963, leg. A. M. Husson & P. Staffeleu.

Bigisanti beach: 1♂, RMNH 16441, 1–VIII–1969, leg. R. L. Hill & D. J. Green.

Wia Wia transect: 1♀, RMNH 15604, 4th swamp from sea, 13–XI–1948, 1♀, RMNH 15605, km S.3, 7th ridge, 15–XI–1948, 1♀, RMNH 15607, km S. 4.3, 8th ridge, 18–XI–1948, 1♀, RMNH 15608, km S. 9.2, 28–XI–1948, all leg. Surinam Expedition 1948–49.

Wane Creek N: 1♀, RMNH 15602, km 6.6, 1–X–1948, 1♂, RMNH 15603, km 5.7, 4–X–1948, both leg. P. H. Creutzberg (Surinam Expedition 1948–49).

Moengo Tapoe transect: 1♂, RMNH 15597, km 1.5, 21–IX–1948, 1♀, RMNH 15598, km 0.8, 22–IX–1948, 1 juv., RMNH 15599, along Weijne Road, 24–IX–1948, 1♂, RMNH 15600, km 0.15, 24–IX–1948, 1♂, RMNH 15601, Base Encampment, 27–IX–1948, all leg. P. H. Creutzberg (Surinam Expedition 1948–49); 1♀, RMNH 15606, Base Encampment 30–X–1948, 1♀, RMNH 15621, Base Encampment, 25–IX–1948, 1 juv., RMNH 15623, km 4.4, 27–IX–1948, all leg. Surinam Expedition 1948–49.

Albina: 1 juv., RMNH 15596, 19–IX–1948, leg. P. H. Creutzberg (Surinam Expedition 1948–49).

Zanderij: 1♀, RMNH 15642, 1966, leg. G. van Vreden; 1♂, BM 1946.4.4.5, 22–IX–1938, leg. I. T. Sanderson.

Bigiston: 2♀♀, 1♂, RMNH 15627, September 1963, leg. L. D. Brongersma.

Awarra savannah, Maratakka River: 1♂, RMNH 16453, 28–II–1971, leg. D. G. Reeder.

Tibiti: 1♀, RMNH 15609, km 0.2, 10–I–1949, 1♂, RMNH 15610, 3rd transect, km 2.2, 15–I–1949, 1♀, RMNH 15611, 3rd transect, km 2.3, 18–I–1949, all leg. Surinam Expedition 1948–49.

Encampment Goliath Creek: 1 ex., CM 49530, 25–VII–1968, leg. M. S. Hoogmoed.

Forest Service Encampment 8, Mapane area: 1♀, RMNH 15065, 13–V–1968, 1♂, RMNH 15066, 12–V–1968, both leg. M. S. Hoogmoed.

Camp 1, Coppename River: 1♀, BM 1946.4.4.2, 2–VII–1938, 1♀, 1♂, BM 1946.4.4.3–4, 8–VII–1938, all leg. I. T. Sanderson.

Brown's Mountain, Mazaroni top: 1♀, RMNH 15632, 13–VII–1968, 1♀, RMNH 15640, 11–XI–1968, both leg. M. S. Hoogmoed.

Brokopondo: 1♀, RMNH 15592, 25–VII–1964, leg. M. Boeseman.

Afobaka: 1♀, RMNH 15593, 15–IV–1965, leg. G. F. Mees.

Avanavero Cataracts, Kabalebo River: 1♂, RMNH 16493, 6–IV–1971, leg. D. G. Reeder.

Blanche Marie Cataracts, Nickerie River: 1♀, RMNH 16452, 16–II–1971, 1♀, RMNH 16454, 13–II–1971, 1♀, RMNH 16469, 14–II–1971, all leg. P. Staffeleu; 1♂, RMNH 16468, 11–II–1971, leg. D. G. Reeder.

Voltz Mountain Reserve, Raleigh Cataracts: 1♂, RMNH 15639, 6–XI–1968, leg. M. S. Hoogmoed.

Nassau Mountains: 3♀♀, RMNH 15612, km 0.4–0.5. 15–II–1949, 1♂, RMNH 15613, km 0.2, 20–II–1949, 1♀, RMNH 15614, km 2.7, 25–II–1949, 1♀, RMNH 15615, Base Encampment, 28–II–1949, 2♀♀, 1♂, RMNH 15616, km 10.4, 11–III–1949, 3♀♀, RMNH 15617, km 10.5, 12–III–1949, 1♀, RMNH 15618, N. of km 14.5, 16–III–1949, 1♀, RMNH 15619, km 10.7, 18–III–1949, 1♀, RMNH 15620, km 14.5, 23–III–1949, all leg. Surinam Expedition 1948–49.

King Frederik Willem IV Falls, Corantijn River: 1♂, FMNH 30949, August 1938, leg. E. R. Blake (Sewell Avery Expedition).

Airstrip Coeroeni: 1♀, RMNH 15631, 22–VI–1968, leg. M. S. Hoogmoed.

Upper Gran Rio: 1 ex., RMNH 5696, 12–IX–1910, leg. J. F. Hulk (Corantijn Expedition).

Airstrip Paloemeu: 1♀, RMNH 15051, S., 28–V–1968, 1♀, RMNH 15628, 26–V–1968, 1♀, RMNH 15629, 27–V–1968, 1♂, RMNH 15630, N., 30–V–1968, all leg. M. S. Hoogmoed.

Airstrip Kayser Mountains: 1♂, RMNH 15634, NE., 3–VIII–1968, leg. M. S. Hoogmoed; 1 ex., FMNH 128949, 1–II–1961, leg. H. A. Beatty.

Camp Kayser Mountains: 5 ex., FMNH 121830–34, 1–II–1961, leg. H. A. Beatty.

Boundary Camp, Itabu Creek Head: 1 juv., FMNH 30945, September/October 1938, leg. E. R. Blake (Sewell Avery Expedition).

Middle Base Camp, Itabu Creek: 1♀, 3 juvs., FMNH 30946–48, 30950, September 1938, all leg. E. R. Blake (Sewell Avery Expedition).

Sipaliwini: 1 juv., RMNH 15182, Vier Gebroeders Bivouac, 24–IX–1968, 1♀, 1♂, RMNH 15635, 6 km E. of airstrip, 28–VIII–1968, 1♀, RMNH 15636, 8 km E. of airstrip, 29–VIII–1968, 1 juv., RMNH 15637, 3 km N. of Vier Gebroeders Mountain, 25–IX–1968, 1♀, RMNH 15638, Vier Gebroeders Bivouac, 30–IX–1968, all leg. M. S. Hoogmoed; 1♂, RMNH 15645, 6 km E. of airstrip, 15–I–1970, 1♂, RMNH 15646, near Meyers' farm, 22–I–1970, 1♀, RMNH 15647, 3 km E. of Base Bivouac, 24–I–1970, 1♂, RMNH 15648, Vier Gebroeders Bivouac, 6–II–1970, all leg. M. S. Hoogmoed & J. J. P. Paats.

Surinam/Guyana:

New River, 750 feet: 2♀♀, BM 1939.1.1.81–82, leg. C. A. Hudson.

Diagnosis.—A medium-sized lizard with cylindrical body, short legs and short tail. Body covered with large, imbricate cycloid scales, each horny scale underlain by a bony one. Head scales imbricate; chinshields hardly distinguishable from the gulars. Glossy brown with a dark-brown to black lateral stripe.

Description.—Head short and wide, distinctly depressed; as wide as the neck, which is as wide as the adjoining part of the body. Snout rounded. Tongue lanceolate, covered with irregularly shaped, imbricate, scale-like papillae; tip slightly notched. Teeth conical.

Rostral hexagonal; visible from above; twice as wide as deep. A pair of supranasals forming a short median suture or just touching each other behind the rostral. A large, irregularly hexagonal frontonasal; about twice as wide as long.

A pair of quadrangular prefrontals, separated from each other by the frontal and frontonasal forming a suture. Frontal elongate, rhomboid, nearly twice as long as wide. A pair of frontoparietals, irregularly pentagonal, forming a long median suture. Interparietal more or less rhomboidal; as wide as long or slightly longer than wide. Parietals trapezoid with rounded posterior margins; wider than long; forming a short suture behind the interparietal in few specimens. A pair of nuchals, more than twice as wide as long. Four supraoculars; the second one largest; the first one smallest; in direct contact with the supraciliaries. Nostril in the posterior part of an undivided nasal. A small postnasal, which is deeper than long. An anterior and a posterior loreal, the anterior one quadrangular, smallest; the posterior one trapezoid, largest. One preocular; one, rarely two, suboculars. A variable number of small postoculars. Supraciliaries four to seven, mostly five, no noticeable regularity in size. Temporal region with large imbricate scales, identical with those on the nape. Ear-opening moderate, round; bordered by small scales, forming a smooth margin; external auditory meatus long, the tympanum concealed externally. Supralabials seven or eight, mostly eight; sixth largest, under the eye when eight; fifth largest, under the eye when seven. Lower eyelid with an undivided semitransparent disc. Pupil round.

Mental trapezoid with convex anterior margin, approximately 2.5–3 times as wide as long. A moderately large, trapezoid postmental, twice as wide as deep. Two pairs of chinshields, hardly distinguishable from the other scales on the throat. Infralabials five to nine, mostly seven or eight.

All head scales smooth, on the anterior part of the head slightly imbricate; on the posterior part of the head (from behind the eyes) strongly imbricate.

Gulars and ventrals identical, smooth, cycloid, imbricate, in longitudinal and oblique rows. Collar absent. Scales on the nape and on the side of the neck slightly smaller than the dorsals, for the rest identical.

Dorsal and lateral scales identical; varying from smooth to slightly tricarinate; cycloid, imbricate, in longitudinal and oblique series. Forty nine to 56 dorsals between the nuchals and the posterior surface of the thighs. Ventrals smooth, otherwise identical to the dorsals. No distinct boundary between the laterals and the ventrals. Total number of scales around midbody 27–32, most specimens having an even number. Preanal plate with scales identical to the ventrals, however, of the eight scales bordering the vent anteriorly the two centrals are enlarged.

Scales on the tail striated, cycloid to hexagonal, strongly imbricate, in longitudinal and oblique rows; under the tail smooth; otherwise identical. Tip of the undamaged tail with a sharp point. Scales on the regenerated part of the tail smooth, imbricate, tending to be differentiated into transversely enlarged, hexagonal dorsal and ventral scales and cycloid lateral scales. Tail round in cross-section. The tail is very fragile and judging from the high number of individuals with regenerated tails, this defense mechanism is often and freely used.

Scales on the limbs smooth, imbricate, cycloid, smaller than the dorsals, in longitudinal and oblique rows. Scales on the back of the thighs slightly smaller. Twelve to 16, mostly 13–14, single, smooth lamellae under the fourth finger;

15–21, mostly 16–18, single, smooth lamellae under the fourth toe. Palms and soles with small, juxtaposed, tubercular scales. Fingers cylindrical, toes slightly compressed.

Preanal and femoral pores absent in both sexes.

Colour in life glossy brown (in different shades) above, with a dark-brown to black lateral stripe; belly and ventral surface of the tail orange-yellow, chin and throat greyish white.

In preservative the back and the dorsal surface of the head vary from olive brown through light-brown to dark-brown. The dorsum may be without spots, show two longitudinal rows of dark-brown spots, or a number of haphazardly distributed dark-brown spots. A light dorsolateral stripe, continued on the tail, may be present in some specimens, but is absent in most specimens. A broad, dark-brown to black lateral band from the posterior corner of the eye on to the tail; bordered ventrally by a narrow white to creamy stripe from the upper lip through the ear to the groin, which may or may not be continued on the proximal part of the tail as a series of light spots. Lower parts immaculate, white to yellow. Upper surfaces of the limbs of the same colour as the back, with a number of small dark-brown dots.

A medium-sized lizard, stoutly built, with short legs. Maximum snout-vent length in females 103 mm, in males 100 mm. Snout-vent length in recently born juveniles 37–50 mm. Hind limb 1.1–1.5 (mostly 1.3) times as long as fore-limb. When the limbs are laid along the body the fingers and toes do overlap. The undamaged tail is 1.3–1.6 times as long as the snout-vent length in females, 1.5–1.6 times as long in males and 1.2–1.3 times in juveniles.

HABITAT.—An inhabitant of the forest floor and trunks of fallen trees in places where sufficient sunlight is available. Thus, in the forest it can be encountered along trails and roads, on the edges of clearings and along creeks. On the Sipaliwini savannah the species is also present. There it inhabits piles of loosely stacked boulders with many crevices, both in and among the boulders, and some of the rock slates present in several places. These boulders are the essential factor for the occurrence of this species on the Sipaliwini savannah and when suitable piles are present *Mabuya m. mabouya* (Lacépède) occurs, irrespective of the place of the piles, whether near a creek or on top of a hill. This species is frequently seen basking on tree stumps (in the forest) or on boulders (Sipaliwini savannah). FITCH (1968) also noted basking.

NATURAL HISTORY.—According to observations by BEEBE (1925, 1945) this species is (ovo)viviparous. A female collected August 22, 1920 gave birth to four young lizards one week later. Among the Surinam material there are ten juveniles in which the umbilical scar is still visible, all collected in September (six with precise data between 12 and 27 September). This agrees well with BEEBE's observation in Guyana and seems to indicate that this species has only one breeding season, resulting in the birth of young from approximately late August to late September. Among the Surinam material there is one specimen

(RMNH 15624) containing five developing young. Thus the number of young seems to be four to five.

The contents of nine stomachs were examined. Eight contained spiders of different sizes, five contained grasshoppers. Other items present include a scorpion, crickets, cockroaches, beetles, a mosquito, a lepidopterous pupa, a fly and a maggot. The spiders never made up the bulk of the stomach contents. It can be concluded that the diet consists mainly of terrestrial arthropods.

One specimen was taken from the stomach of a small owl (*Glaucidium* spec.), indicating that this species of owl hunts by day. HAVERSCHMIDT (1970) noted hunting in daytime for the barn owl (*Tyto alba hellmayri* Griscom & Greenway) in Surinam.

RANGE.—From southern Mexico (Colim, Veracruz) south through Central America to northern South America where it occurs south to Pallatanga in Ecuador on the Pacific side of the Andes, to Bolivia and southern Brazil on the Atlantic side of the Andes. It also is present in the Lesser Antilles south of St. Martin. In Surinam *Mabuya m. mabouya* (Lacépède) is present throughout the country, both in forested and savannah areas, from sea level to at least 500 m. In Venezuela it occurs to 1830 m (DUNN 1936) (fig. 36).

VERNACULAR NAME.—In the Surinam language this species sometimes is called 'kamra-wientje' or 'kamerawientje', a name that is not specific, because it is also used for *Cnemidophorus l. lemniscatus* (Linnaeus) and merely indicates that it may be encountered in or near houses, 'kamra-wientje' meaning chambermaid. Indians at the airstrip Paloemeu called it 'moeloegoembe', but unfortunately it was not recorded whether they belonged to the Trio or the Oayana tribe.

Family TEIIDAE

The representatives of this family in Surinam show a great diversity in appearance, ranging from the nearly legless *Bachia* to the long-legged genera *Ameiva*, *Cnemidophorus*, *Kentropyx* and *Tupinambis*. There is also a great variation in form, size and number of scales, which range from large to small, from smooth to keeled, from imbricate to juxtaposed, from identical scales to several types of scales. There is a strong tendency to scale-reduction, which is very well illustrated when, for instance, *Ameiva* with numerous, granular, juxtaposed scales on the back and *Tretioscincus* with few, large, imbricate scales are compared. Members of this family can be readily distinguished from members of other families by the presence of large, symmetrical juxtaposed scales on the head which are not fused with the bones of the skull. Scales on body, tail and legs lack underlying osteoderms.

Most habitats are occupied by representatives of this family, however, they are essentially ground- or subsurface-dwellers and none of them is arboreal, though some species do climb tree-trunks to a certain degree. No special adaptations have been developed for this last type of habitat. For several other habitats, such as creeks, leaf-litter and upper soil-layer, adaptations are present.

Recently several species of the families Gekkonidae (KLUGE & ECKHARDT, 1969), Lacertidae (DAREVSKY, 1966) and Teiidae (GRANT, 1958; MASLIN, 1962; THOMAS, 1965; VANZOLINI, 1970b) have been shown to be parthenogenetic. Some of those parthenogenetic teiids are present in Surinam. *Cnemidophorus l. lemniscatus* (Linnaeus) has bisexual populations in part of its range, parthenogenetic communities in other parts. *Gymnophthalmus underwoodi* Grant, *Kentropyx borckianus* Peters and *Leposoma percarinatum* (L. Müller) are completely parthenogenetic.

Both in number of species (20) and in the abundance of specimens this family exceeds the other families of Surinam lizards.

Key to the genera of Surinam Teiidae.

1. Limbs well developed, complete with five fingers and toes, all bearing claws 2
   Limbs rudimentary or well developed, but in this last instance, the inner finger, if distinct, is clawless 12
2. Anterior nasal plates not separated by a frontonasal 3
   Nasal plates widely separated by one or two frontonasals 6
3. Ventral scales smooth 4
   Ventral scales keeled . . . . *Kentropyx*, p. 285
4. Ventral scales large, forming eight to ten longitudinal series only 5
   Ventral scales small, forming more than 20 longitudinal series . . . . *Tupinambis*, p. 362
5. Scaly portion of the tongue not much widened and not or but feebly emarginate posteriorly, retractile into a basal sheath; ventrals in ten longitudinal rows . . . . *Ameiva*, p. 225
   Scaly portion of the tongue arrowhead-shaped and not retractile into a basal sheath; ventrals in eight longitudinal rows . . . . *Cnemidophorus*, p. 262
6. Tail round or cyclo-tetragonal 7
   Tail compressed, with a double, denticulate crest . . *Neusticurus*, p. 330
7. Dorsal scales rhomboidal, in oblique and in transverse rows 8
   Dorsal scales hexagonal or rectangular, in transverse rows only or in transverse and in longitudinal rows 9
8. Interparietal longer than the parietals, head scales with longitudinal striations . . . . *Leposoma*, p. 315
   Interparietal equal in length to the parietals or shorter, head scales smooth or with longitudinal striations on the interparietal and parietals . . . . *Alopoglossus*, p. 216
9. Dorsal scales forming transverse rows only, hexagonal 10
   Dorsal scales forming transverse and longitudinal rows, much larger than the lateral scales, keeled, rectangular . . . . *Cercosaura*, p. 255
10. A transverse row of occipitals; dorsal and lateral scales subequal . . 11

One occipital, accompanied by a postparietal on each side; dorsal scales much larger than the lateral scales . . . . *Prionodactylus*, p. 346

11. Dorsal scales keeled, ventral scales rounded and imbricate posteriorly; enlarged gulars in pairs . . . . *Arthrosaura*, p. 235

Dorsal scales smooth, ventral scales squared off and juxtaposed; gulars subequal, in transverse rows . . . . *Ptychoglossus*, p. 351

12. Limbs well developed, only inner finger (when present) clawless; ear-opening present 13

Limbs rudimentary, digits clawless; ear-opening absent . . *Bachia*, p. 249

13. Lower eyelid with semitransparent disc, movable; inner finger present but clawless 14

Lower eyelid transparent, grown over the eye, immovable; inner finger absent . . . . *Gymnophthalmus*, p. 273

14. Sixteen scales around the body; undamaged tail slightly longer than the snout-vent length, scales on the tail smooth . . . . . *Tretioscincus*, p. 355

Twelve scales around the body; undamaged tail twice as long as the snout-vent length, scales on the tail keeled. . . . . *Iphisa*, p. 279

Genus **Alopoglossus** Boulenger

DIAGNOSIS.—Small lizards with cylindrical bodies, well developed legs with five digits, and round tails with ridges. Tongue covered with oblique plicae. Frontonasal present, separating the nasals; interparietal equal in length to the parietals or shorter, four supraoculars, four or five supraciliaries; three suboculars, the second of which is very long, three times as long as the first; one postmental; three pairs of chinshields, the posterior margins of the last pair rounded. Dorsals rhomboidal, keeled, mucronate, imbricate, in oblique and transverse rows. Ventrals keeled or smooth, with a pointed or a truncate posterior border, in longitudinal and transverse rows. Eyelids movable, lower one with semi-transparent disc. Tympanum visible. Pupil round. A South American genus with 5 species in the northern part of the continent. Only one species occurs in Surinam.

Alopoglossus angulatus (Linnaeus) (figs. 38, 39; pl. 24; table 13)

Lacerta angulata Linnaeus, 1758: 204; Linnaeus, 1760: 204; Linnaeus, 1766: 364; Statius Müller, 1774: 94; Gmelin, 1789: 1061; Merrem, 1820: 55.

'Hoekige Haagdis' Houttuyn, 1764: 130.

'L'Exagonal' Daubenton, 1784: 625.

Lacertus Hexagonus Lacépède, 1788a: Synopsis methodica Quadrupedum Oviparorum; Lacépède 1788c: 450.

'L'Hexagone' Lacépède, 1788a: 327; Lacépède, 1788c: 37; Lacépède, 1799: 130; Desmarest, 1833: 421; Lacépède, 1847: 194.

'Die sechseckige Eidechse' Bechstein, 1800b: 40.

Agama angulata: Daudin, 1802a: 389.

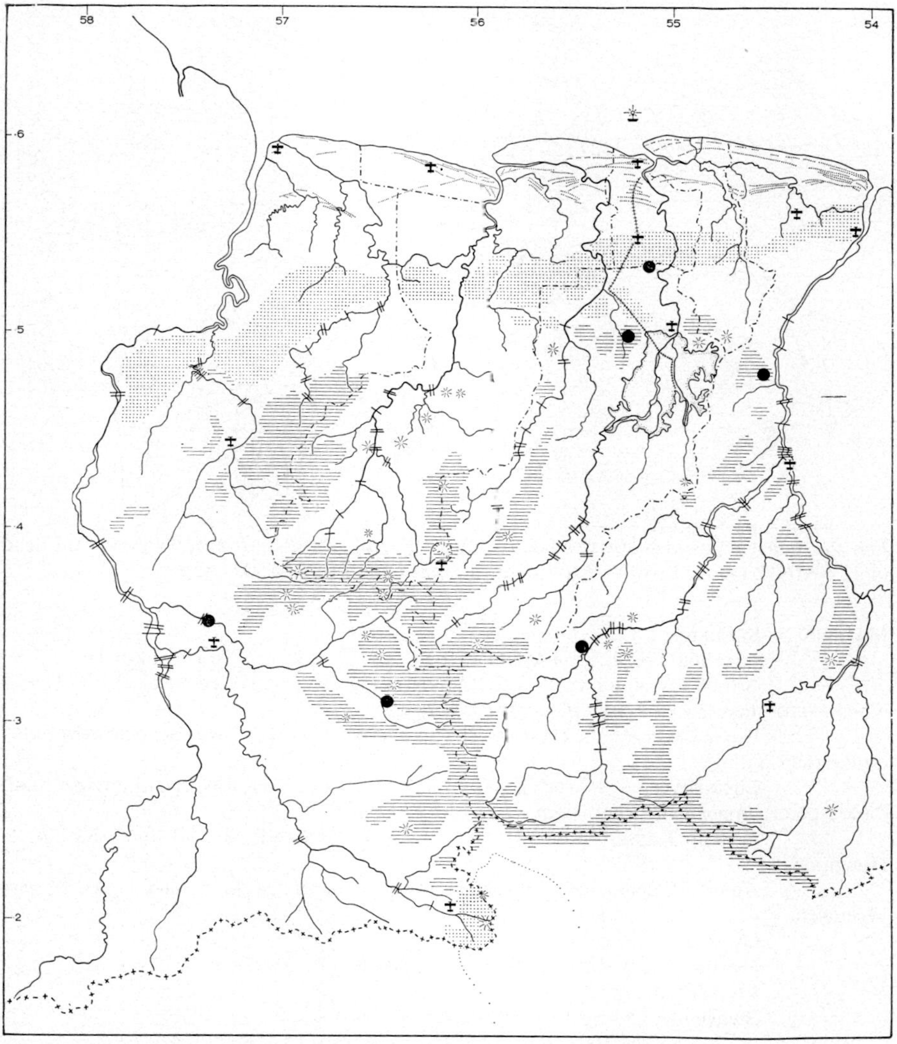

Fig. 38. Distribution of *Alopoglossus angulatus* (Linnaeus) in Surinam.

Stellio hexagonus: Latreille, 1802b: 32.
Alopoglossus copii surinamensis Brongersma, 1946: 231.
Alopoglossus carinicaudatus: Ruibal, 1952: 508; J. A. Peters & Donoso-Barros, 1970: 15.

TYPE-LOCALITY.—'America', here restricted by neotype selection to Brown's Mountain, Surinam.

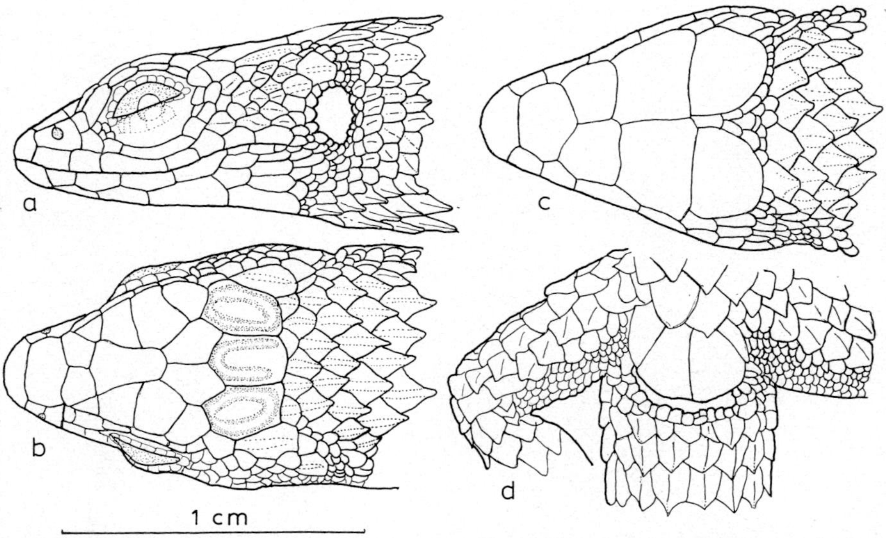

Fig. 39. Alopoglossus angulatus (Linnaeus); a, b, c: lateral, dorsal and ventral views of the head of ♀, RMNH 15200, Brown's Mountain; d: anal region of RMNH 15200.

MATERIAL. – Surinam.

Carolina Creek: 1♀, FMNH 134725, 27–XI–1961, leg. P. Hershkovitz.

Brown's Mountain: 1♀, 1♂, RMNH 15200–01, 20–VIII–1968, leg. M. S. Hoogmoed (15200 neotype of *Lacerta angulata* Linnaeus).

Nassau Mountains: 1♀, RMNH 13446, km 9.3, 11–III–1949, leg. Surinam Expedition 1948–49.

Lucie River: 1♀, RMNH 4858, 2–XII–1910, leg. J. F. Hulk (Corantijn Expedition) (holotype of *Alopoglossus copii surinamensis* Brongersma).

Airstrip Kayser Mountains: 1♀, RMNH 15379, NE., 2–VIII–1968, leg. M. S. Hoogmoed.

Airstrip Paloemeu: 1♀, RMNH 15215, surroundings, May 1969, leg H. M. van Meeuwen.

Guyana.

Marudi: 1♀, AMNH 61381, April 1938, leg. R. Snedigar.

French Guiana.

Bienvenue, valley of Canopi River: 1♀, MP 50–22.

DIAGNOSIS.—A small, brown lizard, with a cylindrical body, well developed legs and a round tail with 12 ridges. Dorsals and laterals rhomboid, imbricate, strongly keeled. Interparietal and parietals with or without longitudinal striations. Third pair of chinshields rounded posteriorly. Tongue with oblique plicae.

DESCRIPTION.—Head short, slightly depressed, snout rounded. Head and neck of equal width, slightly narrower than the body.

Tongue lanceolate, covered with oblique plicae; tip bifid. Anterior teeth both in upper and lower jaw conical; posterior teeth in upper jaw tricuspid, those in lower jaw bicuspid.

Rostral trapezoid, twice as wide as deep, clearly visible from above. Fronto-

Plate 24. Figs. 1–3. *Alopoglossus angulatus* (Linnaeus); dorsal, lateral and ventral aspect of ♀, RMNH 15200, Brown's Mountain. Figs. 4–6. *Bachia cophias* (Schneider); lateral, dorsal and ventral aspect of ♀, RMNH 15183, Sipaliwini.

Table 13. Counts and measurements of *Alopoglossus angulatus* (Linnaeus). For explanation see table 3.

Reg. no.	A	B	C	D	E	F	G	H	I	J	K	L	M	N	O	P	Q	R	S	T	U	V	W	X
RMNH																								
* 4858	♀	28	19(17.5)	7.6	5.1	3.6	9	13.5	0	22	28	19	4	7	7-7	6-6	5-5	4-4	3-3	4+2	8	3	17-18	23-21
13446	♀	32	30(30)	8.5	5.3	4.0	9	14	0	21	29	20	6	8	8-7	6-6	6-5	4-4	3-3	4+3	8	3	18-17	21-21
15200	♀	48	20(20)	11.5	7.8	5.4	13	19	0	21	27	18	6	8	7-7	6-6	5-5	4-4	3-3	4+2	7	3	14-14	21-20
15201	♂	25	37	7.0	4.4	3.3	7	11	10/0-9/0	19	28	20	6	8	8-7	6-6	5-5	4-4	3-3	4+3	7	3	13-13	20-19
15215	♀	27	38	7.2	4.6	3.0	8	11	0	19	27	20	6	7	7-8	5-6	5-5	4-4	3-3	4+2	7	3	15-14	19-21
15379	♀	39	65	9.7	6.7	4.6	11	17	0	22	28	18	6	8	7-7	6-6	5-5	4-4	3-3	4+2	7	3	16-16	20-21
FMNH																								
134725	♀	47	81	12.3	7.9	6.0	12	21	0	21	28	18	6	8	7-7	6-6	5-5	4-4	3-3	4+2	6	3	15-14	19-?
MP																								
50-22	♀	61	81(52)				14	25	0	21	28	?	4	9	?-7	6-6	5-5	4-4	2-3	4+2		3	15-15	? -22
AMNH																								
61381	♀	54	66(39)	13.0	8.5	6.3	15	22	0	20	28	17	4		7-7	6-6	5-5	4-4	?	4+2	6	3	16-16	22-22

nasal single, pentagonal, wider than long, separating the nasals. Two prefrontals, as wide as long, with a short median suture. Frontal hexagonal, longer than wide. Two frontoparietals, pentagonal, longer than wide, with a long median suture. Interparietal pentagonal with nearly parallel longitudinal sides; longer than wide. Parietals separated by the interparietal; slightly longer than the latter or as long; hexagonal; longer than wide. Four supraoculars, the anterior three in contact with the frontal; the first smallest, the third largest. Five supraciliaries. In RMNH 15201, a small hexagonal scale is present between the frontal and the frontoparietals. Nostril in a semidivided nasal. Nasal widely separated from its fellow by the frontonasal; in contact with the first and second supralabials, or sometimes separated from the second supralabial by a small scale. Loreal deeper than long, separated from the supralabials by a frenoocular. Three suboculars, the second of which is very long, three times as long as the first. Postoculars two. Temporals rhomboidal, keeled, juxtaposed, of unequal size; two large supratemporals bordering on the parietal. Ear-opening large, vertically oval; the anterior margin denticulate, the posterior margin nearly smooth; external auditory meatus short, tympanum clearly visible. Lower eyelid with a semitransparent disc, consisting of four or five palpebrals. Supralabials seven or eight, mostly seven; third longest, about twice as long as the second. Pupil round.

Mental trapezoid with rounded anterior margin, about once and a half as wide as long. Postmental heptagonal, wider than long. Three pairs of chinshields, the posterior pair with rounded posterior margins; members of the anterior two pairs in contact with each other and with the infralabials; members of the posterior pair separated from each other by small scales and from the infralabials by larger scales. Infralabials six, rarely five. Gulars in distinct oblique and in indistinct transverse rows, rhomboidal with a low, broad keel, imbricate, seven or eight rows between the posterior pair of chinshields and the anterior margin of the forelimbs. Collar absent. Scales on the side of the neck rhomboidal, keeled, imbricate.

Posterior head scales with a median depression and raised marginal parts.

Dorsals rhomboidal, strongly mucronate, with a distinct, narrow keel, imbricate; in oblique and in 27–29 transverse rows. Ventrals rhomboidal, slightly mucronate, with low, broad keels, imbricate, the keels forming longitudinal ridges; in 18–20 transverse and in six longitudinal rows. In FMNH 134725 the ventrals are pentagonal, without a keel and only slightly mucronate. Dorsals, laterals and ventrals smoothly merge into each other. Preanal plate with four scales, two small laterals and two large medians.

Preanal pores absent, femoral pores absent in females, nine or ten on one side in the single available male.

Scales on the tail rhomboidal, mucronate, keeled, imbricate, in longitudinal and in transverse rows; scales under the tail identical but less distinctly keeled. The keels forming 12 longitudinal, serrated ridges, four dorsally, four ventrally and two on each side.

Scales on the limbs rhomboidal, keeled, imbricate; only the ventral surface of the upper arms and the posterior surface of the thighs are covered with small, granular scales. Lamellae under the fourth finger 13–18, smooth, single; proximally with an indistinct median groove, distally with an indistinct ridge. Lamellae under the fourth toe 19–23, smooth; in the distal part single, in the proximal part in pairs.

In life dark-brown above with faint orange-brown dorsolateral stripes, tail with light-brown and dark-red spots. Ventral parts dirty white with two dark-brown spots on the preanal plate. Iris green. Tongue bluish black.

In preservative brown above with two faint dorsolateral lines on the neck. A dark line from the nostril to the eye, through the upper eyelid to the shoulder. Behind the shoulder it widens to a wide lateral band. Back mottled with dark- and light-brown. Dorsal surface of the head uniformly brown. Ventral parts white, without dark spots in most specimens, except in two adult females which have a dark-brown spot on the lateral part of each ventral. An indistinct light line on the border of flanks and belly. Dorsal caudal scales light-brown, intermingled with dark-brown ones, forming alternating light- and dark-brown transverse bands. Lateral caudal scales dark-brown. Subcaudals white, some of them with large brown spots; near the tip of the tail most subcaudals are brown with small white spots. A black or dark-brown spot on the postero--lateral part of the medial preanals, which may be continued on the adjacent part of the lateral preanals. Palms and soles brown. Dorsal parts of the limbs brown, ventral parts white. Posterior surface of the hind limbs with round, white spots of several sizes. Tongue blue anteriorly, white posteriorly.

A small lizard with a cylindrical body and a round tail with ridges. The maximum snout-vent length in females is 48 mm, the (juvenile) male is 25 mm. Limbs short and slender, hind limb 1.4–1.7 times as long as forelimb in females, 1.6 times in the male; when laid along the body there is considerable overlap, the hind limb not quite reaching the axilla. The undamaged tail is 1.4–1.7 times as long as the snout-vent length in females, 1.5 times in the male.

HABITAT.—All specimens on which data are available were collected near creeks on the rain-forest floor, where it is covered with leaf-litter, in shadowy places.

NATURAL HISTORY.—All specimens were taken in daytime, the three specimens collected by me were taken between 1.00 and 4.50 p.m. Judging by the scanty data available this is a nonheliotherm, no basking behaviour having been seen. Nothing is known about breeding season. Too few specimens were available to examine stomach contents.

RANGE.—Surinam, Guyana, French Guiana, Brazil (Pará, Amapá), Ecuador and Peru. In Surinam the species is present in the forested country south of the coastal savannah belt. It reaches at least to 300 m above sea level (fig. 38).

REMARKS.—Linnaeus described in 1758 a lizard under the name *Lacerta angulata*. Since that time this name has only been cited in the subsequent editions of the Systema Naturae, by HOUTTUYN (1764), by DAUBENTON (1784), by LACÉPÈDE (1788a, c, 1799, 1847), by STATIUS MÜLLER (1774), by BECHSTEIN (1800b), by DAUDIN (1802a), by LATREILLE (1802b), by MERREM (1820) and by DESMAREST (1833). HOUTTUYN did not add new data but just gave an integral translation of LINNAEUS' Latin text. STATIUS MÜLLER changed the sequence slightly and added 'da nun die Schuppen reihenweise stehen, so macht dieses solche Erhöhungen, und zwischen den Reihen wiederum solche Furchen, das sich der Körper im ganze Umfange, (ausgenommen am Unterleibe) vieleckigt oder vielseitig zeigt'.

The description by LINNAEUS is good and enables identification of the species. As *Lacerta angulata* Linnaeus described the species that about 120 years later was again described, now by COPE (1876), under the name *Lepidosoma carinicaudatum* = *Alopoglossus carinicaudatus*. My arguments for this decision are the following:

1. LINNAEUS says that the specimen on which he based his description was collected by ROLANDER in America. As has been shown by HOLTHUIS (1958, 1959), ROLANDER visited only Surinam and St. Eustatius in the Antilles. As on St. Eustatius no lizard is present that answers the description, the specimen must have come from Surinam, where indeed *Alopoglossus carinicaudatus* (Cope) is present.

2. All characters mentioned by LINNAEUS match only one Surinam lizard, namely *Alopoglossus carinicaudatus* (Cope): colour brown; all scales, except the ventrals, strongly keeled and mucronate; head smooth with several wrinkles; under the throat two large, rounded scales; tail 1.5 times as long as the body, hexagonal in cross section and very angular. The large scales under the throat undoubtedly form the third pair of chin-shields, which, together with the wrinkled head and the angular tail, are sufficient to identify the species.

The synonymising of *Lacerta angulata* Linnaeus, 1758 and *Lepidosoma carinicaudatum* Cope, 1876 has the following consequences:

1. As article 23b of the International Code of Zoological Nomenclature now has been declared invalid, *Lacerta angulata* Linnaeus, 1758 has priority over *Lepidosoma carinicaudatus* Cope, 1876 and the name of the taxon under discussion consequently has to be changed into *Alopoglossus angulatus* (Linnaeus, 1758). The consequences of this change are not large, because there are only few publications in which *Alopoglossus carinicaudatus* (Cope, 1876) has been used. Moreover, the use of a Linnaean name ensures the best stability.

2. The type-locality of the species, indicated by LINNAEUS as being 'America', can here be restricted to Surinam. According to HOLTHUIS (1959) ROLANDER only visited the neighbourhood of Paramaribo and the Commewijne River. As ROLANDER travelled in the company of DAHLBERG it seems reasonable to suppose that the visited DAHLBERG's plantations Brouwershaven on the Perica Creek and Carlsburg on the Cottica River. On the other hand, from Paramaribo ROLANDER could have reached without problems the environs of Brokopondo. From the Perica-Commewijne area no recent material is available, from the region south of Paramaribo there are two specimens from Brown's Mountain

(RMNH 15200–01) and one from Carolina Creek (FMNH 134725). As the type specimen seems to be lost (it cannot be found in Stockholm, Uppsala or London (ANDERSSON, 1900; HOLM, 1957)) it seems desirable to indicate a neotype. For this I would like to select RMNH 15200, from Brown's Mountain, Surinam. This specimen is well preserved and shows all characters mentioned by LINNAEUS. The only character that cannot be checked is the length of the tail, the tip is missing. The specimen is a female with a snout-vent length of 48 mm, the length of the remaining part of the tail is 20 mm, the number of scales around midbody is 21, there are 27 transverse rows of dorsals to the posterior margin of the thighs, 18 transverse rows of ventrals, supralabials 7–7,

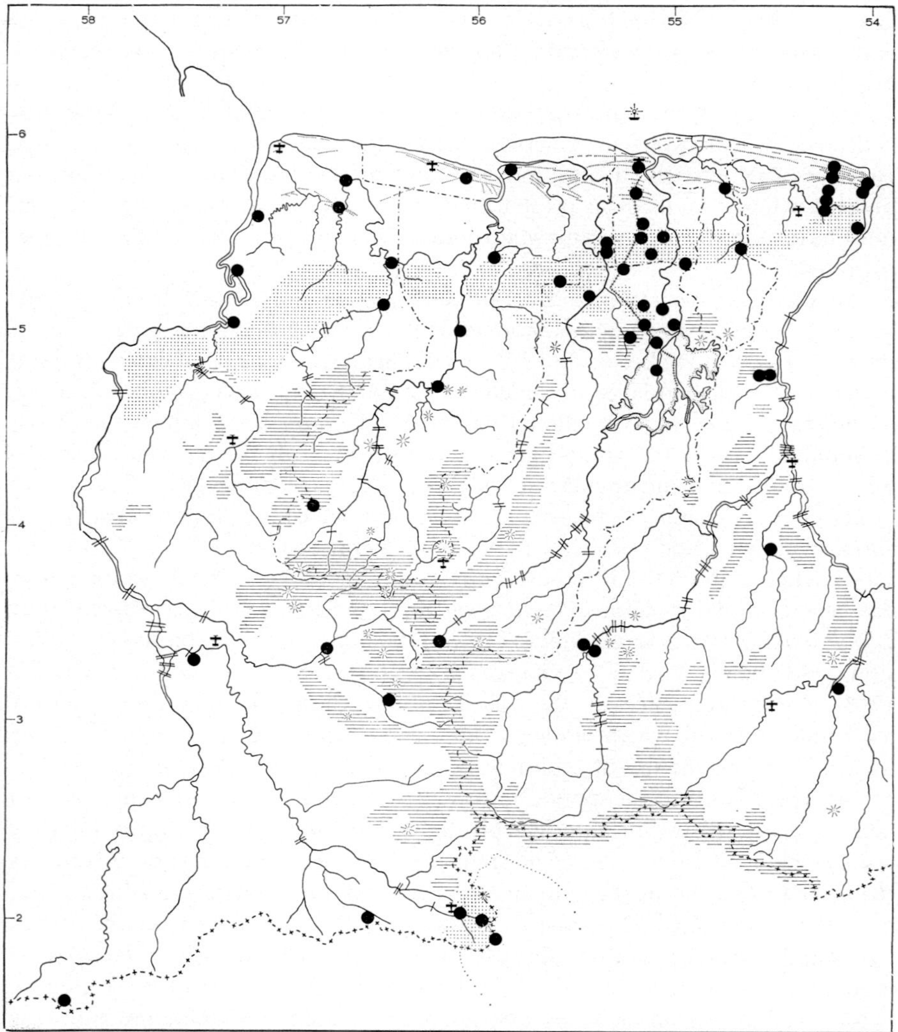

Fig. 40. Distribution of *A. a. ameiva* (Linnaeus) in Surinam.

infralabials 6–6, gulars 7, subdigital lamellae fourth finger 14–14, subdigital lamellae fourth toe 21–20.

Genus **Ameiva** Meyer

DIAGNOSIS.—Medium-sized to large lizards with cylindrical bodies, well developed legs with five digits and long tails. Tongue lanceolate, retractile into a basal sheath; covered with rhomboidal, imbricate, scale-like papillae in oblique rows. Nostril in a divided nasal; the anterior part of the nasal forming a suture with its fellow from the other side. Dorsal and lateral scales granular; ventrals smooth, in six to 16 longitudinal rows. Eyelids movable, lower one with a semitransparent disc. Tympanum visible. Pupil kidney-shaped. The genus ranges from Mexico through Central America and the Antilles to South America. About 15 species, of which seven are South American. Only one species in Surinam.

Ameiva ameiva ameiva (Linnaeus) (figs. 40, 41; pl. 25)

'Hagedissen' Herlein, 1718: 173.
Lacerta, Surinamensis, dorso dilute caeruleo, cauda tenui longiore Seba, 1734: 139, pl. 88 fig. 1.
Lacerta, Surinamensis, major, Ameira *dicta* Seba, 1734: 140, pl. 88 fig. 2.
Lacerta cauda tereti, corpore triplo longiore, squamis laevissimus, abdominalibus oblongo-quadratis Gronovius, 1756: 80.
Lacerta Ameiva Linnaeus, 1758: 202; Anonymus, 1775: 246; Latreille 1802a: 224.
'Lezard (gros)' Fermin, 1765: 20.

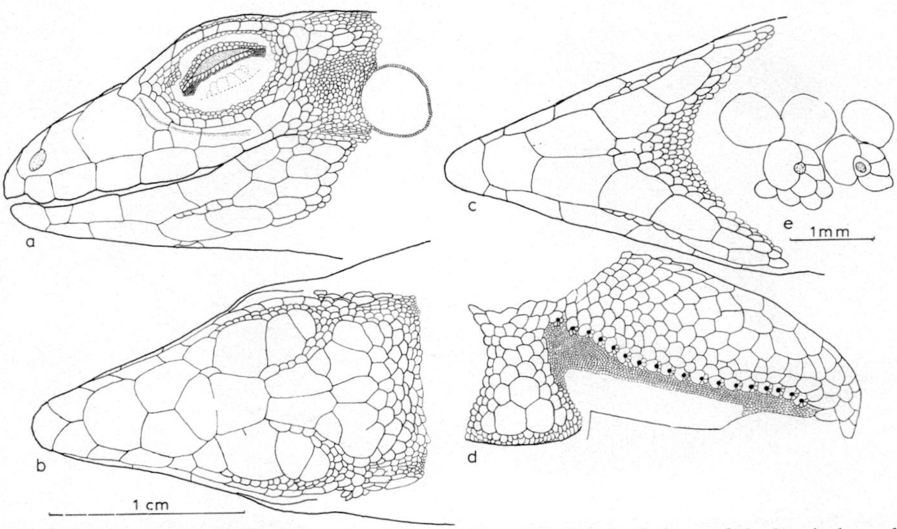

Fig. 41. A. a. ameiva (Linnaeus); a, b, c: lateral, dorsal and ventral views of the head, d: anal region, e: detail of femoral pores of ♀, RMNH 15119, Paramaribo.

Plate 25. Ameiva a. ameiva (Linnaeus). Figs. 1–3. Dorsal, lateral and ventral aspect of ♂, RMNH 15482, Airstrip Kayser Mountains. Figs. 4–6. Dorsal, lateral and ventral aspect of ♀, RMNH 15497, Powakka.

'Le troisième' Fermin, 1769: 206.
'Le quatrième' Fermin, 1769: 207.
'Le cinquième' Fermin, 1769: 207.
'De derde' Fermin, 1770: 182.
'De vierde' Fermin, 1770: 183.
'De vyfde' Fermin, 1770: 183.
'L'Améiva' Daubenton, 1784: 588; Lacépède, 1788a: 328; Lacépède, 1788c: 38; Lacépède, 1799: 130; Desmarest, 1833: 421; Lacépède, 1847: 195.
'Die marmorirte Eidechse' Bechstein, 1800b: 42.
'Lézards voisins de l'Ameiva 1° & 2°' Latreille, 1802c: 251.
Lacerta ameiva: Daudin, 1802a: 98; Cloquet, 1827: 526; Kappler, 1881: 140.
Lacerta litterata Daudin, 1802a: 106.
Lacerta graphica Daudin, 1802a: 112.
'Die gemeinste Eidechse' Von Sack, 1821a: 225.
'De gemeenste Hagedis' Von Sack, 1821c: 239.
Ameiva litterata: Fitzinger, 1826: 51.
'Ameiva' Griffith & Pidgeon, 1830: 205.
Ameiva vulgaris: Duméril & Bibron, 1839: 100; Duméril & Duméril, 1851: 113; Schlegel, 1858: 27; Penard & Penard, 1906: no. 81.
Ameiva surinamensis: Cope, 1863: 67; Kappler, 1881: 166; Boulenger, 1885a: 352; Kappler, 1885: 817; Kappler, 1887: 126; Boettger, 1893: 74; Van Lidth de Jeude, 1904: 89; Van Lidth de Jeude, 1917a: 43; D'Audretsch, 1965: 22.
Ameiva a. ameiva: Barbour & Noble, 1915: 462, 463; Burt & Burt, 1930: 28; Burt & Burt, 1931: 304; Marcuzzi, 1950: 101; Marcuzzi, 1954: 248; Wijffels, 1961: 10; Hoogmoed, 1967: 82; J. A. Peters & Donoso-Barros, 1970: 19.
Ameiva ameiva: Menne, 1952: 10; Menne, 1954a: 16; Menne, 1954b: 64; Haverschmidt, 1968: 55, 57, 59.
'Ameiva's' Menne, 1954b: 63.
'Joi' Geijskes, 1957: 276.

TYPE-LOCALITY.—'America', here restricted to the confluence of the Cottica River and the Perica Creek, Surinam.

MATERIAL. – Surinam.

Surinam: 1 ex., RMNH 3388; 2 ex., RMNH 7512, 1904, leg. G. M. Versteeg (Tapanahoni Expedition); 3 ex., RMNH 7513, November 1905, leg. P. Buitendijk; 1 ex., RMNH 7537, June 1910, leg. D. Bolten; 1♀, RMNH 15496; 3♀♀, 1♂, RMNH 15500, leg. G. Witsen Elias; 1♂, RMNH 15513; 1 ex., RMNH 15517, February 1888, leg. J. H. Spitzly; 1 ex., RMNH 15524; 1 ex., RMNH 15525, leg. H. F. C. ten Kate; 1♂, RMNH 15579, 1910, leg. D. Bolten; 1♀, RMNH 16498, 1971, leg. Surinam Expedition 1971; 1 ex., ZMA 10364; 1 ex., ZMA 10652; 1 ex., ZMA 12273, 1891; 1 ex., ZMA 12274; 2 ex., ZMA 12275–76; 2 ex., ZMA 12277–78, 1887, leg. Van Brussel; 4 ex., ZMA 12279–82, October 1892, leg. Van Brussel; 1♀, BM 5.7.17.74; 1 ex., ZMB 861, leg. Steitz; 1 ex., ZMB 865, leg. Bloch; 1 ex., ZMB 36143; 1♂, NMW 18739, 1911; 1 ex., ANS 9146.

Paramaribo: 4 ex., RMNH 7509, 7–III–1939, 1 ex., RMNH 7514, 2–III–1939, both leg. H. W. C. Cossee; 13 ex., RMNH 7518, 1911, leg. Jhr. W. C. van Heurn; 1♀, RMNH 15119, 5–V–1968, leg. M. S. Hoogmoed; 1♂, RMNH 15467, 11–II–1967, leg. W. N. Polder; 1♀, RMNH 15480, 3–V–1968, leg. M. S. Hoogmoed; 2♀♀, RMNH 15484, 30–VIII–1950, leg. D. C. Geijskes; 1 juv., RMNH 15485, 22-V-1966, 1♀, 4♂♂, RMNH 15486, 20-V-1966, 1♂, RMNH 15487, 23-V-1966, 1 juv., RMNH 15488, 1966, 1♂, RMNH 15489, 19-VI-1966, all leg. W. Vervoort & C. Cornet; 1 juv., RMNH 15493, 12-III-1967, leg. J. van der Land; 2♀♀, RMNH 15494, 14-III-1961, leg. S. Ligorie; 1♀, 1♂, RMNH 15495, 4-IX-1964, leg. L. Werner; 1♂, RMNH 15499, 4-XI-1955, leg. D. C. Geijskes; 1♀, RMNH 15515, May 1965, leg. R. & F. Deuss; 1 juv., RMNH 15518, 27-XI-1954, leg. P. Bolwerk; 1♂, RMNH 15526, 12-IX-1948, leg. P. H. Creutzberg (Surinam Expedition 1948-49); 6 ex., RMNH 15591, February 1966, leg. G. van Vreden; 1♂, RMNH 16411, 14-I-1970, leg. M. S. Hoogmoed & J. J. P. Paats; 1♀, 2♂♂, RMNH 16412, April 1969, leg. H. M. van Meeuwen; 1 ex., ZMA 10663, leg. D. Piet; 2 ex., ZMA 12286-87, leg. Menne; 1 ex., ZMA 12652-54, 22-V-1966, 2 ex., ZMA 12665-66, 28-VI-1966, both leg. H. Nijssen; 1 ex., ZMA no number, leg. J. W. Gonggrijp & G. Stahel; 2 ex., BM 1946.4.3.20, 1946.4.3.30, 2-II-1938, 2♀♀, 1 juv., BM 1946.4.3.21-22, 1946.4.3.26, 29-I-1938, 2♀♀, 1 juv., BM 1946.4.3.23-24, 1946.4.3.28, 21-I-1938, 1♀, 6♂♂, BM 1946.4.3.25, 1946.4.3.29, 1946.4.3.32-34, 1946.4.3.36, 1946.4.3.38, 21-I-1938, 1♀, 1♂, BM 1946.4.3.27, 1946.4.3.31, 25-I-1938, 1♀, 1♂, BM 1946.4.3.35, 1946.4.3.37, 28-I-1938, all leg. I. T. Sanderson; 1 ex., ZMB 26379, 1 ex., ZMB 26424, both leg. A. O. C. Heller; 10♀♀, 8♂♂, NMW 18740, 1910, leg. A. O. C. Heller; 3 ex., CM 44374-76, 5-V-1966, 2 ex., CM 44377-78, 7-V-1966, 1 ex., CM 44379, 5-X-1966, 1 ex., CM 44380, 23-X-1966, 1 ex., CM 44393, 4-V-1966, all leg. M. H. de la Fuente; 8 ex., AMNH 8142-43, 8598, 8601-02, 8605, 8608-09, 1916, leg. J. A. Samuels; 1 ex., MCZ 6077; 1 ex., MCZ 66844; 1♂, SM 17, 10-II-1965, leg. D. C. Geijskes.

Wageningen: 1♂, RMNH 15510, 9-VI-1963, leg. A. M. Husson & P. Staffeleu; 1♀, 1♂, RMNH 16475, 13-III-1971, leg. P. Staffeleu.

Coronie Road: 1♀, RMNH 15567, picket 208, km 0.2, 23-XII-1948, 1♀, RMNH 15568, picket 216, encampment, 23-XII-1948, both leg. Surinam Expedition 1948-49.

Forest Service Encampment MacClemen, Corantijn River: 2♀♀, RMNH 16484, 22-III-1971, leg. P. Staffeleu.

Cupido, Maratakka River: 4♂♂, 1♀, RMNH 16474, 8-III-1971, leg. D. G. Reeder.

Forest Service Encampment Wakay, Corantijn River: 3♀♀, 4♂♂, 1 juv., RMNH 16485, 23-III-1971, 1♀, 3♂♂, RMNH 16791, 24-III-1971, all leg. D. G. Reeder.

Forest Service Encampment Arawarra: 1♀, RMNH 16486, 1971, leg. P. Staffeleu.

Tibiti: 2♀♀, RMNH 15569-70, 12-I-1949, 1 juv., RMNH 15571, 17-I-1949, both leg. Surinam Expedition 1948-49.

Plantation Caledonia, Saramacca River: 2 ex., ZMA 12283-84, 20/25-XII-1929, leg. Van Genderen; 1 ex., ZMA 12285, January 1930.

Lelydorp: 4 ex., ZMA 10722, 12411-13, 2-IV-1963, leg. Van Heusden.

Potribo: 1♂, RMNH 15490, 4-V-1966, leg. W. Vervoort & C. Cornet.

Wia Wia transect: 1♀, 1♂, RMNH 15557, 2nd ridge from sea, 13-XI-1948, 1♂, 1 juv., RMNH 15558, 4th ridge from sea, 13-XI-1948, 2♀♀, RMNH 15559, 15561, 4th swamp from sea, 13-XI-1948, 1♀, 1♂, 1 juv., RMNH 15560, 4th ridge from sea, 13-XI-1948, 1 juv., RMNH 15562, 8th ridge, km S. 3.8, 15-XI-1948, 3♂♂, RMNH 15563, 7th ridge km S. 3.0-3.1, 17-XI-1948, 1♂, RMNH 15565, 8th ridge, km S. 4.1, 20-XI-1948, 2♀♀, RMNH 15566, November 1948, 1 juv., RMNH 15577, 1st encampment, km 3.8, 29-XI-1948, all leg. Surinam Expedition 1948-49.

Third Encampment: 1♂, RMNH 15546, km 14.4, 14-X-1948, 1♀, RMNH 15547, km 15.7, 15-X-1948, 1♀, RMNH 15548, km 15.8, 15-X-1948, 1♂, RMNH 15549, km 16.4, 15-X-1948, 1♂, RMNH 15550, km 16.1, 23-X-1948, 2♀♀, RMNH 15551-52, km 15.8, side transect E., 24-X-1948, all leg. Surinam Expedition 1948-49.

Djai Creek: 1 juv., RMNH 15545, km 8.8, 9-X-1948, leg. P. H. Creutzberg (Surinam Expedition 1948-49).

Wane Creek N.: 1♀, 1♂, RMNH 15536, km 5.8, 29-IX-1948, 1♂, 2 juv., RMNH 15537, km 6.5, 29-IX-1948, 2♀♀, 1♂, RMNH 15538, km 5.8, 30-IX-1948, 1 juv., RMNH 15539, km 6, 30-IX-1948, 1♀, RMNH 15540, km 5.5, 2-X-1948, 1♂, RMNH 15541, km 5.8, 2-X-1948, 2♀♀, 1♂, 2 juvs., RMNH 15542, km 5.8-6.3, 3-X-1948, 2♀♀, RMNH 15543, km 5.8,

1♀, 1♂, 1 juv., RMNH 15544, km 6.2, 4–X–1948, all leg. P. H. Creutzberg (Surinam Expedition 1948–49).

Mooi Wane Creek: 3♂♂, 1 juv., RMNH 15533, N. of km 12, 25–IX–1948, 1♀, 1♂, RMNH 15534, 25–IX–1948, all leg. P. H. Creutzberg (Surinam Expedition 1948–49).

Moengo Tapoe transect: 1♀, RMNH 15529, km 1.71, 22–IX–1948, 1♂, RMNH 15530, km 0.15, 23–IX–1948, 2♂♂, RMNH 15531, base bivouac, 24–IX–1948, 1♀, RMNH 15532, km 0.02, 24–IX–1948, 2♂♂, RMNH 15535, base bivouac, 26–IX–1948, all leg. P. H. Creutzberg (Surinam Expedition 1948–49); 1 juv., RMNH 15553, base bivouac, 31–X–1948, leg. Surinam Expedition 1948–49.

Galibi: 1♀, 1♂, RMNH 15555, 4–XI–1948, 4♀♀, RMNH 15556, 6–XI–1948, both leg. Surinam Expedition 1948–49.

Langamankondre: 1♀, RMNH 15479, 13–VI–1968, leg. M. S. Hoogmoed; 1♀, RMNH 15528, 19–IX–1948, leg. P. H. Creutzberg (Surinam Expedition 1948–49).

Albina: 1♀, 2♂♂, RMNH 15527, 19–IX–1948, leg. P. H. Creutzberg (Surinam Expedition 1948–49); 4♀♀, 2♂♂, 5 juvs., RMNH 15554, trail N., 1–XI–1948, leg. Surinam Expedition 1948–49.

Forest Service Encampment Matapi, Corantijn River: 1♀, 1 juv., RMNH 16497, 13–IV–1971, leg. D. G. Reeder.

Stondansie Cataracts, Nickerie River: 6 hgrns., RMNH 16461, 30–I–1971, leg. D. G. Reeder; 1♀, RMNH 16476, 3–II–1971, leg. P. Staffeleu.

Encampment Goliath, Coesewijne area: 1♂, RMNH 15466, 25–VII–1968, leg. M. S. Hoogmoed.

Matta: 6♂♂, RMNH 15502–06, 15509, 14–I–1963, leg. A. M. Husson & P. Staffeleu.

Troeli Creek: 1♂, RMNH 15236, 20–VI–1968, leg. H. A. M. de Kruijf.

Republiek: 1 juv., RMNH 15523, 29–X–1949, leg. D. C. Geijskes.

Zanderij: 2♀♀, 1♂, BM 1946.4.3.10–12, 7–VIII–1938, 1♀, BM 1946.4.3.13, 5–X–1938, 1 juv., BM 1946.4.3.14, 17–IX–1938, 2♂♂, BM 1946.4.3.15–16, 22–IX–1938, all leg. I. T. Sanderson; 2 ex., FMNH 121080, 121082, 18–IX–1960, leg. H. A. Beatty.

Powakka: 1♀, RMNH 15497, 4–IV–1964, leg. M. H. de la Fuente.

Railway km 62: 4♀♀, 1♂, RMNH 15576, 26/IV–16/V–1949, leg. A. Brouwer (Surinam Expedition 1948–49).

Carolina Creek: 2 ex., FMNH 134726–27, November 1961, leg. P. Hershkovitz.

Forest Service Encampment 8, Mapane area: 1 juv., RMNH 15474, 13–V–1968, 2♀♀, RMNH 15481, 12–V–1968, both leg. M. S. Hoogmoed.

Pininica, near Java: 1 juv., RMNH 15501, 2–V–1963, leg. A. M. Husson & P. Staffeleu.

Camp 1, Coppename River: 2♀♀, BM 1946.4.3.17–18, 8–VII–1938, leg. I. T. Sanderson.

Loksihatti/Brokolonko: 1 ex., FMNH 134729, 8–II–1962, leg. P. Hershkovitz.

Railway km 91.5: 1♂, BM 1946.4.3.19, 4–XI–1938, leg. I. T. Sanderson.

Berg en Dal: 1♀, RMNH 15498, 14–IX–1963, leg. D. C. Geijskes.

Brown's Mountain: 1♀, RMNH 15238, 13–VII–1968, leg. M. S. Hoogmoed; 1♀, RMNH 17278, 25–XII–1971, leg. G. F. Mees.

Brown's Creek: 1 ex., ZMA 12288, 6–XI–1960, leg. H. P. Pijpers.

Railway km 121: 2♂♂, RMNH 15578, 30–XI–1949, leg. C. Bleijs.

Brokopondo: 1 juv., RMNH 15491, 21–IV–1965, 1 ex., RMNH 15492, 22–IV–1965, 1♂, RMNH 15514, 6–VI–1965, 1♀, RMNH 15520, 12–XI–1965, 1♀, RMNH 15521, 14–X–1965, all leg. G. F. Mees; 1♂, RMNH 15519, 20–VII–1964, leg. M. Boeseman; 1 ex., ZMA 12655, 27–V–1966, 1 ex., ZMA 12657, 15–VIII–1966, 1 juv., ZMA 12695, 12–IX–1966, 2 ex., ZMA 13208, 27–III–1967, 1 ex., ZMA 13226, 8–VI–1967, all leg. H. Nijssen.

Kabel: 1♂, RMNH 15516, 19–VI–1964, leg. P. Leentvaar.

Raleigh Cataracts, Coppename River: 1♀, RMNH 15469, 6–XI–1968, leg. M. S. Hoogmoed; 1 juv., RMNH 15508, 7–IV–1963, leg. A. M. Husson & P. Staffeleu; 1♂, RMNH 17276, 26–III–1972, leg. G. F. Mees.

Gansee: 1♂, RMNH 15507, 28–I–1963, leg. A. M. Husson & P. Staffeleu.

Nassau Mountains: 1♂, RMNH 15572, km 0.4, 15–II–1949, 1♂, RMNH 15573, km 0.2, 20–II–1949, 1 juv., RMNH 15574, km 3.8, 28–II–1949, 1♀, RMNH 15575, N. of km 11.2, 24–III–1949, all leg. Surinam Expedition 1948–49.

Gonini Encampment, Coeroeni River: 1♀, RMNH 15237, 25–VI–1968, leg. M. S. Hoogmoed.

Wilhelmina Mountains: 1♂, RMNH 15512, line II, September 1943, leg. D. C. Geijskes.

Airstrip Kayser Mountains: 2♀♀, 1♂, RMNH 15482, 31–VII–1968, 1♂, RMNH 15483, SE., 1–VIII–1968, both leg. M. S. Hoogmoed; 2 ex., 4 juvs., FMNH 128957, 128959–60, 128962–64, 1–II–1961, leg. H. A. Beatty.

Upper Gran Rio: 1 ex., RMNH 5655, 10–IX–1910, 1 ex., RMNH 7516, 9–XI–1910, both leg. J. F. Hulk (Corantijn Expedition).

Airstrip Paloemeu: 1♀, RMNH 15154, S., 23–V–1968, 1♀, 1♂, RMNH 15475, 20–V–1968, 4♀♀, 1♂, RMNH 15476, 25-V-1968, 1♀, RMNH 15477, S., 26–V–1968, 4♀♀, RMNH 15478, 27–V–1968, all leg. M. S. Hoogmoed.

Gonini River, confluence of Emma and Wilhelmina Rivers: 1♂, RMNH 15522, 10–X–1903, leg. G. M. Versteeg (Gonini Expedition).

Janamalékondre, Litani River: 1♀, 1♂, RMNH 15511, 18–VIII–1939, leg. D. C. Geijskes.

Itabu Creek Head: 1 ex., FMNH 30951, Boundary Camp, 1938, leg. E. R. Blake (Sewell Avery Expedition).

Sipaliwini, airstrip: 1♀, 1♂, RMNH 16389, 10–II–1970, 1♀, RMNH 16390, 23–I–1970, 1♂, RMNH 16391, 16–I–1970, 1♀, RMNH 16395, 13–II–1970, 1♂, RMNH 16397, 2 km S., 11–II–1970, 1♀, RMNH 16399, 2 km NW., 12–II–1970, 2♂♂, RMNH 16402, 11–II–1970, 1♂, RMNH 16403, 9–II–1970, 2♀♀, RMNH 16406, 8–II–1970, all leg. M. S. Hoogmoed & J. J. P. Paats.

Sipaliwini, Meyers' farm: 1♂, RMNH 16392, 28–I–1970, 1♂, RMNH 16393, 22–I–1970, 1♂, RMNH 16394, 25–I–1970, 1♂, RMNH 16396, 3 km SE., 2-II-1970, 1♂, RMNH 16400, 28–I–1970, 1♂, RMNH 16401, 31–I–1970, 1♂, RMNH 16405, 19–I–1970, 2♀♀, 2♂♂, 1 juv., RMNH 16407, 23–I–1970, 1♀, RMNH 16408, 22–I–1970, 2 juvs., RMNH 16410, 25–I–1970, all leg. M. S. Hoogmoed & J. J. P. Paats.

Sipaliwini, Vier Gebroeders: 1 juv., RMNH 15184, near mountain, 12–IX–1970, 1♂, RMNH 15225, mountain, 20–IX–1968, 2♀♀, RMNH 15470, bivouac, 12–IX–1968, 1♀, RMNH 15471, bivouac, 18–IX–1968, 1♀, RMNH 15472, 3 km N. of mountain, 20–IX–1968, 1 juv., RMNH 15473, 3 km N. of mountain, 25–IX–1968, all leg. M. S. Hoogmoed; 1♂, RMNH 16398, 10 km W. of bivouac, 6–II–1970, leg. M. S. Hoogmoed & J. J. P. Paats.

Sipaliwini: 1 juv., 1♂, RMNH 17271, May-June 1972, leg. G. F. Mees.

Surinam/Guyana.

New River: 1♂, BM 1939.1.1.74, 750 ft, leg. C. A. Hudson.

Surinam/Brazil.

Sipaliwini, Meyers' airstrip: 1♂, RMNH 16409, 5 km S., 26–I–1970, leg. M. S. Hoogmoed & J. J. P. Paats.

Brazil.

Sipaliwini, Lange Dijk: 1♀, RMNH 16404, 27–I–1970, leg. M. S. Hoogmoed & J. J. P. Paats.

DIAGNOSIS.—A large lizard with a cylindrical body, well developed legs and a long tail. Dorsal and lateral scales granular, ventrals smooth, in ten longitudinal and in 30–34 transverse rows. Nasal plates not separated by a fronto-nasal. In nearly all specimens at least part of the back is grass-green, flanks with dark--brown longitudinal bands and greenish white lines, adult males with creamy spots.

DESCRIPTION.—Head pyramidal; slightly depressed; about twice as long as wide; canthus rostralis very distinct, angulate, rounded; large, posterior part swollen, wider than the neck in old males; small, slender, as wide as the neck in females and young males. Snout pointed. Neck as wide as the body. Tongue lanceolate, without emargination, retractile into a basal sheath; covered with rhomboidal, imbricate, scale-like papillae in oblique rows; tip bifid, smooth with a longitudinal furrow on the dorsal surface of each tip. All lateral teeth tricuspid, laterally compressed; anterior teeth conical.

Rostral pentagonal; distinctly visible from above; slightly deeper than wide. Nostril oval, in an obliquely divided nasal, the anterior part of the nasal forming a short median suture behind the rostral with its fellow from the other side. A hexagonal frontonasal, slightly longer than wide. A pair of irregularly trapezoid or pentagonal prefrontals, forming a short median suture. Frontal hexagonal, longer than wide, anteriorly wider than posteriorly. A pair of irregularly trapezoid frontoparietals, forming a long median suture. A transverse series of irregularly shaped parietals, of which the central one is smallest and the two scales bordering it are largest. Occipital region covered with small, irregularly shaped, juxtaposed scales of different sizes. Supraoculars four; first and fourth subequal, smallest; second largest; second, third and fourth separated from the supraciliaries by a row of granules; third and fourth also separated from the frontoparietals and parietals by a row of granules. Loreal very large, trapezoid, longer than deep. One narrow preocular, about twice as deep as long. Three, rarely two, suboculars; third smallest, second largest. Preocular and anterior two suboculars forming a suborbital ridge. A varying number of enlarged postocular scales. Supraciliaries five to eight, mostly five or six; the second and the third elongate, largest; the rest more or less quadrangular. A row of enlarged supratemporals, decreasing in size posteriorly. A band of slightly enlarged scales in front of the ear-opening; central region of temple covered with small, round to hexagonal, juxtaposed scales. Ear-opening large, surrounded by small scales, forming a smooth margin; external auditory meatus short, tympanum clearly visible. Six to eight, mostly seven, large supralabials, forming a lobed upper lip, followed by a variable number of small scales; sixth labial below the centre of the eye; anterior five labials distinctly larger than the posterior one to three; third and fourth subequal, largest. Lower eyelid with a semitransparent disc of five to six palpebrals. Pupil kidney-shaped.

Mental approximately triangular, with rounded tip; about as wide as long. A pentagonal or heptagonal postmental, slightly longer than wide. Four or five pairs of chinshields; the anterior pair forming a long median suture, largest; the fourth and fifth pair smallest; from the second pair on separated from the lower labials. Five to seven, mostly six, infralabials followed by a variable number of small scales; third infralabial largest. Gulars small, flat, hexagonal, juxtaposed; with a wide transverse band of enlarged scales from ear to ear. Band of scales in front of prepectoral fold enlarged. Scales between the two bands of enlarged scales small.

Scales on the nape and on the side of the neck identical to the dorsals and laterals.

Dorsals and laterals small, granular, in indistinct transverse and oblique rows. Ventrals large, rectangular, wider than long, imbricate, smooth, in ten longitudinal and in 30–34 transverse rows. The anterior rows medially interrupted by a small triangle of smaller, irregularly shaped scales. Each transverse row of ventrals tapers off into the lateral scales by two or three marginals. Total number of scales around midbody 155–167. Preanal plate with a variable number of smooth, irregularly shaped, juxtaposed, enlarged scales, forming a narrow triangle, the apex pointing forwards, surrounded by many small scales.

Femoral pores 17–25 on each side, both in females and males. Pores in males slightly larger than in females; surrounded by three to four scales, of which the one on the median side is largest. Preanal pores absent. Hemipenis a cylinder with two horns on top. Base smooth, central part with transverse plicae, interrupted on the posterior surface only, distal part with plicae on both sides, interrupted on the anterior surface by a smooth oval area and on the posterior surface by the sulcus spermaticus. Sulcus spermaticus ending on top of the hemipenis between the two laterally compressed, triangular horns.

Scales on the tail large, rectangular, keeled (slightly oblique on the lateral part), slightly imbricate along their posterior margins; in transverse rows. The keels do not form longitudinal ridges. Scales under the tail identical, but in the proximal part keels are absent. Scales on the regenerated part of the tail smaller, irregularly shaped to quadrangular, keeled, in frequently interrupted transverse rows. Tail round in cross-section.

Scales on the anterior surface of the forelimbs, on the anterior and lower surfaces of the thighs and on the lower surface of the lower legs large, smooth, more or less hexagonal, juxtaposed; on the lower, posterior and upper surfaces of the forelimbs, on the posterior and upper surfaces of the thighs and on the posterior, upper and anterior surfaces of the lower legs small, smooth, granular. Fifteen to 19 single, keeled lamellae under the fourth finger; 32–43 single, sharply tubercular lamellae under the fourth toe. Under the proximal part of the third toe a number of triangular, strongly enlarged and tubercular scales. Fingers and toes compressed. Palms and soles with small, irregularly shaped, juxtaposed, flat scales. Upper surface of hands and feet with large, imbricate, flat scales.

In life the back is grass-green, with or without a brown part, depending on age. In juveniles the head and anterior part of the back are grass-green, the posterior part of the back brown. In halfgrown specimens the entire back is grass-green. In adults the head and anterior part of the back are brown, posterior part grass-green. The green thus shifts from the head towards the sacrum and confines itself to the sacrum during the process of aging. Throat white. Belly orange in juveniles; in older specimens blue or white, with an orange area near the anus. Ventral surface of the tail orange in juveniles, halfgrowns and females, bluish in adult males. Flanks with dark-brown longitudinal bands and greenish-white lines; in adult males with creamy spots. Forelimbs brown in juveniles, halfgrowns and adults. Hind limbs brown in juveniles; partly green, partly brown in halfgrowns; green in adults. Some juvenile and halfgrown specimens from the Sipaliwini savannah are completely brown above, without any trace of green.

In preservative the back of juveniles and most females is bluish to brown, with or without two longitudinal series of dark-brown spots, with or without a light dorsolateral stripe on the anterior part. A broad, dark-brown band from the eye over the ear to the groin, with some small whitish spots in it, bordered ventrally by a more or less distinct white line, which is most conspicuous in its anterior and posterior parts and continues on the side of the tail. Flanks ventrally of the white line brown with irregular, dark-brown spots. In some females and in males the anterior part of the back is olive, with vermiculate black spots and an indistinct dorsolateral stripe; the posterior part of the back is bluish; the flanks show vertical black bands in which white spots are enclosed. Belly yellowish to bluish, immaculate, except the two or three outermost rows of ventrals which may bear a black spot on each scale. Upper surface of hind limbs brown to olive with a sprinkle of black spots. Tail brown to bluish with black spots. Ventral surface of tail grey to bluish.

A large lizard, maximum snout-vent length in females 131 mm, in males 168 mm. Hind limb more than twice as long as the forelimb. When laid along the body there is a considerable overlap of forelimb and hind limb, the tip of the fourth toe reaching the eye, the forelimb reaching more than halfway to the hind limb. The undamaged tail is 1.9–2.3 times as long as the snout-vent length in females, 2.0–2.3 times in males.

HABITAT.—Sunny places where cover is present, e.g., gardens, roadsides, open spots in the forest, along trails through the forest, creek banks, savannahs, ridges. On the Sipaliwini savannah the species is present in a special type of vegetation along creeks and on the edge of Mauritia swamps, consisting of clumps of a grass (*Trypsacum*) separated by bare areas of soil. Also it may be encountered on the rock outcrops present on mountain slopes and in valleys. Large numbers thrive in the borderzone between savannah and forest-islands. In the forest-islands they are fewer in number.

NATURAL HISTORY.—This sun loving, ground-dwelling lizard is easily disturbed and quickly seeks refuge in the undergrowth or in a burrow (GOELDI & HAGMANN, 1901), causing quite some noise when it makes its long spurts over the fallen leaves. It never ventures far from the ground, the highest points reached being the trunks of fallen trees, lying on the ground. The reproductive season seems to be split into two periods, pregnant females being found in May and from September to November. The normal number of eggs is four although once three and once five eggs were found. Recently hatched juveniles are present from January to May and from September to November. BEEBE (1945) concluded from his data that breeding in Kartabo took place throughout the year. GOELDI & HAGMANN (1901) noticed that this species in the neighbourhood of Belém breeds between December and April and that the normal number of eggs seems to be three to five. However, despite extensive series collected throughout the year, in Surinam pregnant females were found only in the periods mentioned above.

Ameivas have been taken from the stomachs of several species of birds, e.g., the savannah hawk (*Heterospizias meridionalis* (Latham)), the roadside hawk (*Buteo magnirostris* (Gmelin)) and the grey hawk (*Buteo nitidus* (Latham)) (HAVERSCHMIDT, 1968). One specimen was found in the stomach of *Dryadophis boddaertii* (Sentzen), a colubrid snake. BEEBE (1925; 1945) reports *Ameiva* from the stomachs of four other snakes: *Boa constrictor* Linnaeus, *Clelia*, *Xenodon severus* (Linnaeus) and *Drymarchon c. corais* (Boie) and also from the stomach of the lizard *Tupinambis*. The ameiva stomachs investigated contained a large variety of insects, spiders, one scorpion and even some vegetable matter, suggesting that this species eats most things it encounters. BEEBE (1925, 1945) found similar data.

On the airstrip Kayser Mountains males of this species were observed chasing each other and, when they reached a certain speed they elevated themselves on the hind limbs and maintained this bipedal locomotion for several metres.

During the 1970 field-trip reptiles were investigated for the presence of Salmonellae. Two specimens of *A. a. ameiva* (Linnaeus) proved to be infected, *viz.* RMNH 16407 with *Salmonella poona* and RMNH 16409 with *S. caracas*. A paper on this matter is in preparation.

RANGE.—Northern South America, south to southern Brazil, west to the Andes and north to Panama. In Surinam the species occurs throughout the country, from the coastal ridges to the mountaintops in the interior, from rain-forest to savannahs. It reaches at least 500 m above sea level. Other subspecies occur in Brazil, northwestern South America and in Central America (fig. 40).

REMARKS.—*Lacerta litterata* Daudin was listed by J. A. PETERS & DONOSO-BARROS (1970) under 'Incertae sedis'. They founded this decision on the fact that BARBOUR & NOBLE (1915) did not mention this name as a synonym of one of their subspecies of *A. ameiva* and they themselves do not know where to place it. I agree with BOULENGER (1885b) who listed this name as a synonym of *Ameiva surinamensis* (Laurenti). I decided to list *L. litterata* Daudin as a synonym of *A. a. ameiva* (Linnaeus) for the following reasons: the number of longitudinal rows of ventrals in *L. litterata* Daudin was stated to be ten, the same number being found in Surinam *A. a. ameiva* (Linnaeus). Other scale characters falling within the range of *A. a. ameiva* (Linnaeus) are the number of femoral pores (21–22) and the number of transverse rows of ventrals (30). Another reason in favour of synonymising *L. litterata* Daudin with *A. a. ameiva* (Linnaeus) is that DAUDIN himself listed as synonyms lizards depicted by SEBA (1734) that undoubtedly belong to *A. a. ameiva* (Linnaeus).

VERNACULAR NAME.—Apart from lakkadisja (= lizard) there is no special name for this species in the Surinam language. Both Trio Indians and Oayana Indians call it 'joi' (yoi). The Trio's use this same name for *Kentropyx calcaratus* Spix, which by the Oayana's is distinguished by a separate name.

Genus **Arthrosaura** Boulenger

DIAGNOSIS.—Small lizards with a cylindrical body and tail and well developed legs. Tongue lanceolate, without posterior emargination; covered with rhomboidal, scale-like, imbricate papillae in oblique rows; tip bifid, smooth. Anterior teeth conical, the rest tricuspid, laterally compressed. Dorsals hexagonal, in transverse rows only; ventrals rounded and imbricate posteriorly; enlarged gulars in pairs. Postparietals absent. Frontonasal separating the nasals. The genus is restricted to the Amazonian basin. Only two species, both of which are present in Surinam.

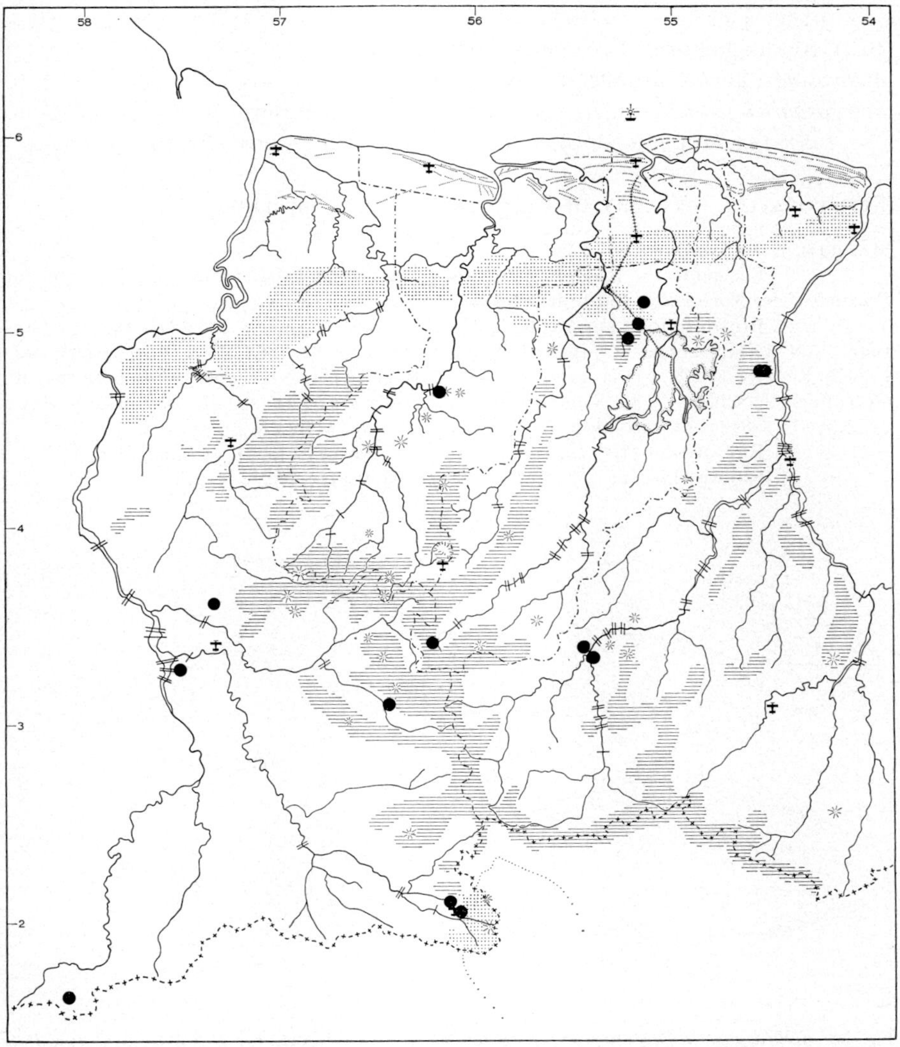

Fig. 42. Distribution of *Arthrosaura kockii* (Van Lidth de Jeude) in Surinam.

Key to the Surinam species of *Arthrosaura*.

1. Supraoculars four; distance between forelimbs and hind limbs at most 1.5 times the length of the forelimb; a white vertebral stripe . *A. kockii*, p. 236
   Supraoculars three; distance between forelimbs and hind limbs 2.0–2.5 times the length of the forelimb; vertebral stripe absent . . . . *A. reticulata versteegii*, p. 242

Arthrosaura kockii (Van Lidth de Jeude); (figs. 42, 43; pl. 26; table 14)

Prionodactylus Kockii Van Lidth de Jeude, 1904: 91; Van Lidth de Jeude, 1917g: 438.
'61. Grondhagedissen' Fernandes, 1927: 249.
Arthrosaura kocki: Brongersma, 1928: 332.
Arthrosaura kockii: Brongersma, 1932a: 77, 81; Burt & Burt, 1933: 55; Da Cunha, 1967: 155; Hoogmoed, 1968: 612; J. A. Peters & Donoso-Barros, 1970: 75.

TYPE-LOCALITY.—Vicinity of Coppename River, Surinam.

MATERIAL. – Surinam.
Surinam: 1♀, RMNH 4464, 1901, leg. Coppename Expedition (holotype of *Prionodactylus Kockii* Van Lidth de Jeude).
Brown's Mountain: 1♂, RMNH 13883, 20–VIII–1968, 1♀, RMNH 15212, 11-XI-1968, 1♂, RMNH 15213, 11–XI–1968, 1♀, RMNH 15214, 12–XI–1968, all leg. M. S. Hoogmoed; 1♀, 1♂, RMNH 15708, 10–XI–1968, 1♀, 1♂, RMNH 16373, November 1968, both leg. R. Hoogmoed-Verschoor & M. S. Hoogmoed; 2♀♀, UF 27024–25, 20–VII–1966, leg. S. Bass; 1♀, AMNH 108783, Mazaronitop, 16–II–1972, leg. C. W. Myers & J. Daly.
Railway km 91.5: 1♀, BM 1946.4.4.1, 4–XI–1938, leg. I. T. Sanderson.
Railway km 121: 1♂, RMNH 13438, May 1949, leg. C. Bleys.

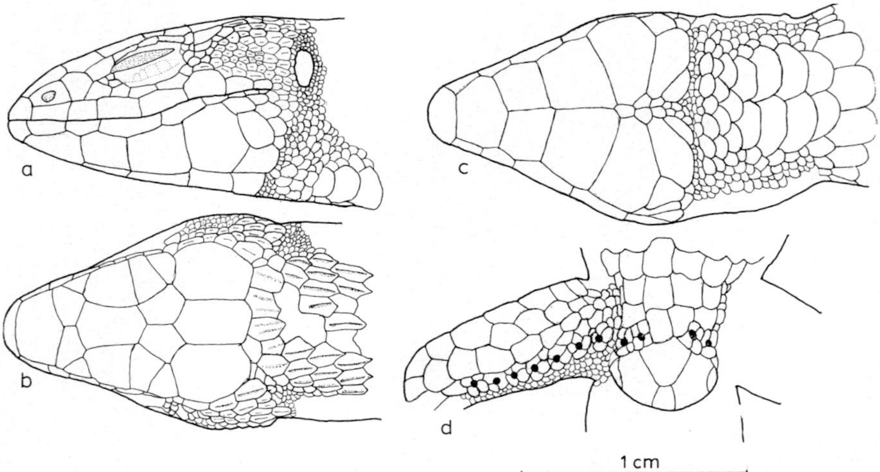

Fig. 43. Arthrosaura kockii (Van Lidth de Jeude); a, b, c: lateral, dorsal and ventral views of the head, d: anal region of ♂, RMNH 15210 c, Voltz Mountain Reserve.

Plate 26. Arthrosaura kockii (Van Lidth de Jeude). Figs. 1–3. Dorsal, lateral and ventral aspect of ♀, RMNH 15210 b, Voltz Mountain Reserve. Fig. 4. Lateral aspect of ♂, RMNH 15210 c, Voltz Mountain Reserve.

Nassau Mountains: 1♂, RMNH 13426, S. of km 1.7–1.8, 22–II–1949, 1♀, RMNH 13427, km 1, 28–II–1949, 1♀, RMNH 13428, km 6.2, 2–III–1949, 1♀, RMNH 13429, S. of km 7, 4–III–1949, 1♂, RMNH 13430, km 10.4, 9–III–1949, 4♀♀, 4♂♂, RMNH 13431, km 8.4–10.4, 11–III–1949, 1♀, RMNH 13432, km 11, 10–III–1949, 1♂, RMNH 13433, km 9, 14–III–1949, 1♀, RMNH 13434, km 13, 15–III–1949, 1♀, RMNH 13435, km 14.5, 16–III–1949, 1♀, RMNH 13436, km 14.7, 21–III–1949, 1♂, RMNH 13437, N. of km 11.2, 29–III–1949, 3♀♀, 1♂, RMNH 13439–42, km 8.6–10, 12–III–1949, all leg. Surinam Expedition 1948–49.

Voltz Mountain Reserve: 2♀♀, 1♂, RMNH 15210, 5–XI–1968, leg. M. S. Hoogmoed; 1♀, RMNH 15226, 1967, leg. W. N. Polder; 1♀, NW. base of Voltz Mountain, 22/23–II–1972, leg. C. W. Myers & J. Daly.

Tigri Encampment, New River: 1♀, RMNH 15227, E., 28–VI–1968, leg. M. S. Hoogmoed.

Wilhelmina Mountains: 1♀, ZMA 12232, 1926, leg. Wilhelmina Mountains Expedition.

Airstrip Kayser Mountains: 1♀, RMNH 13884, NE., 9–VIII–1968, 1♂, RMNH 15380, SE., 1–VIII–1968, both leg. M. S. Hoogmoed.

Suriname River, bivouac 5: 1♂, ZMA 12233, October 1908, leg. J. H. A. T. Tresling (Suriname Expedition).

Airstrip Paloemeu: 2♀♀, RMNH 15052, N., 30–V–1968, 1♂, RMNH 15148, S., 28–V–1968, both leg M. S. Hoogmoed.

Itabu Creek Head: 1♂, FMNH 30922, Boundary Camp, 1938, leg. E. R. Blake (Sewell Avery Expedition).

Sipaliwini: 1♀, RMNH 15193, 1 km E. of airstrip, 5–IX–1968, 2♂♂, RMNH 15194, 2 km N. of Base Bivouac, 4–IX–1968, both leg. M. S. Hoogmoed; 1♂, RMNH 16375, 3 km NW. of airstrip, 9–II–1970, leg. M. S. Hoogmoed & J. J. P. Paats.

French Guiana.

Maripasoula: 1 ex., MCZ 77509, E. of airport, 21–VII–1963, leg. T. Monath.

Tumuc Humac Mountains: 1♀, 1♂, MP 99–71, 72, Upper Lunier River, 1899, leg. Geay.

DIAGNOSIS.—A small lizard with cylindrical body and tail. Distance between forelimbs and hind limbs 1.5 times the length of a forelimb. Supraoculars four. Median suture of the second pair of chinshields 1.5 times as long as that of the first pair. A light vertebral stripe, flanks brown in females, black with creamy spots in males, ventral parts orange-red.

DESCRIPTION.—Head more or less conical, depressed; wider than the neck which is as wide as the body; snout rounded.

Rostral hexagonal; visible from above; more than twice as wide as deep. A large heptagonal frontonasal, separating the nasals. A pair of irregularly hexagonal prefrontals, forming a short median suture. Frontal hexagonal, once and a half as long as wide. A pair of pentagonal frontoparietals, forming a long median suture. Interparietal pentagonal, twice as long as wide, truncate posteriorly. Parietals two, roughly heptagonal, with the posterior lateral part emarginate, as long as the interparietal. A variable (maximum six) number of occipitals in a transverse row; indistinguishable from the scales on the neck. Four supraoculars; second largest, first smallest, in direct contact with the supraciliaries. Nostril in a semidivided nasal. Loreal large, always in contact with the supralabials and the smaller frenoocular. One to three small, wedge-shaped preoculars in a cluster. Three to four suboculars; when three, third largest; when four variable. Two rows of postoculars, the anterior row with a variable number of small scales, the posterior row with four large scales, decreasing in size from the parietals to the labials. Supraciliaries three to six, mostly four to five; the anterior one very large; the third (and fourth) smallest; second and fourth (or fifth) subequal. Temporal region with rather large, hexagonal, keeled scales, increasing in size from antero-ventrally to postero-dorsally;

three to four scales bordering the parietals. Ear-opening large; surrounded by small scales, forming a nearly smooth margin; external auditory meatus short, tympanum clearly visible. Seven to nine (mostly eight) supralabials; the anterior five larger than the posterior two to four; third and fourth largest; fourth under the eye. Lower eyelid with a semitransparent disc of four to five palpebrals. Pupil round. All dorsal and lateral head scales distinctly pitted all over.

Mental trapezoid with convex anterior margin. A large heptagonal postmental. Four pairs of chinshields, the anterior two pairs in contact; the second pair with a distinctly longer median suture (1.5 ×) than the first pair. Seven to nine infralabials, the anterior five distinctly larger; third largest. Anterior three pairs of chinshields in contact with the infralabials, fourth pair separated from them by small scales.

All head scales with a row of small pits along their margin.

Four to five (rarely six) transversely enlarged rows of gulars, smooth, imbricate. Collar with 11–12 scales, the central five much larger than the rest, forming a lobed posterior margin. Scales on the nape hexagonal, keeled, as long as wide near the head, longer than wide near the forelimbs, imbricate, in transverse rows. Scales on the side of the neck round, small, juxtaposed, not arranged in vertical rows.

Dorsals hexagonal, keeled, mucronate, imbricate, in 26–27 (rarely 25 or 28) transverse rows. Laterals rectangular, keeled, not mucronate, imbricate, decreasing in size towards the ventrals; open spaces between the rows with small, round, smooth, imbricate to juxtaposed scales. Ventrals large, square on the central part of the belly, hexagonal towards the sides; smooth; posterior margin rounded, imbricate; in ten longitudinal and in 17–19 transverse rows. Ventrals and laterals distinctly defined by the abutting of the small lateral scales on the larger ventrals. Total number of scales around midbody 35–43. In most specimens five preanal scales in a semicircle, enclosing a sixth one; anterior central scale much larger than the posterior central scale; in some specimens, however, there are three, four or six peripheral scales enclosing an additional one.

Preanal pores absent to two in females, two in males; femoral pores absent to two in females, six to nine on each side in males. Both preanal and femoral pores are in the centre of a group of four scales.

Scales on the tail hexagonal, keeled, mucronate, imbricate, in transverse rows only. Under the anteriormost part of the tail wider than on the dorsal part, without keels, not mucronate; under the rest of the tail identical to the scales on the dorsal part. Scales on the regenerated part of the tail smaller than on the original part of the tail, not mucronate, keels less distinct, only slightly imbricate, in transverse rows. Tail, including the regenerated portion, round in cross-section.

Scales on the upper and anterior surfaces of the forelimbs and on the upper surface of the hind limbs large, rhomboidal, keeled, imbricate; on the anterior and lower surfaces of the hind limbs large, more or less hexagonal, smooth, imbricate; on the lower and posterior surfaces of the forelimbs and on the posterior surface of the hind limbs small, granular. Twelve to 18 (mostly 14–16) single lamellae under the fourth finger, the proximal ones with a shallow median depression; 18–24 lamellae under the fourth toe, divided proximally, with a shallow median depression distally. Palms and soles with small, flat, granular scales. Fingers and toes subcylindrical.

Colour in life brown above with a wide vertebral stripe, greyish on the head, creamy on the anterior part of the back, becoming orange near the sacral region. Flanks brown, in adult males black, with cream-coloured spots. Lateral parts of the belly orange-red, central parts pinkish, throat and chin white. In adult males belly, side of neck and underside of tail bright orange-red, with faint traces of the same colour on the upper labials; chin and throat white. Tail orange dorsally and ventrally. Iris golden brown.

In preservative the back is brown, the side of the head dark-brown. Vertebral stripe from the snout to the tip of the tail, covering the entire dorsal surface of the head from the tip of the snout to the posterior supraocular, from there it narrows to just above the forelimbs and thence is continued as a wide band of equal width to the sacral region where in juveniles it widens again to cover the entire dorsal surface of the tail; in adults it is continued on the tail as a series of white spots alternating with brown ones; greyish to brown on the head, faintly creamy posteriorly, on both sides bordered by a narrow black line. Flanks brown in females and juveniles, with a single row of white, black-edged ocelli from the ear over the forelimb to the groin and on the ventro-lateral part of the tail; a larger ocellus just anterior of the forelimb, coinciding with the outermost, enlarged scales of the collar. In adult males this pattern is the same, except in the region between the forelimbs and the hind limbs, which is jet black and in which most scales are white-tipped. Back of the thighs with one or two white ocelli. Chin, throat and belly immaculately white, except for a few scales on the lateral parts which show some small irregular black spots, more distinct in males than in females. The underside of the tail may be immaculately white, but in older males it becomes dark-grey towards the tip. Anterior part of the tongue grey, gradually changing into white posteriorly.

A small lizard, maximum snout-vent length in females 53 mm, in males 54 mm. Hind limb in females 1.6 times as long as forelimb, in males 1.5–1.7 times and in juveniles 1.3–1.7 times. When laid along the body there is a considerable overlap of forelimb and hind limb. Distance between forelimbs and hind limbs at most 1.5 times the length of a forelimb. The undamaged tail is 1.3–1.6 times as long as the snout-vent length in females and juveniles, 1.6–1.8 times in males.

HABITAT.—All specimens were taken in rain-forest, where they were present in rather open, dry places, where sunlight could penetrate to the forest floor. Most specimens were taken on the ground, a few, however, on the trunks of fallen trees, at most 1 m above the ground. The ground was always covered with a layer of litter, mainly consisting of leaves, a few centimetres thick.

NATURAL HISTORY.—The lizards were frequently seen basking on the forest floor and on fallen logs. When pursued they took refuge in the leaf-litter and, preferably, in the large heaps of dead leaves at the base of certain species of palms. Pregnant females, containing two mature eggs, were found in November and February. Recently hatched specimens with a snout-vent length of 21–30 mm are present in September, November, February and March, indicating that breeding starts in the long wet season (probably early August) and lasts until at least February.

The four stomachs examined all contained orthopterons, also spiders, a mosquito and an earwig.

RANGE.—Surinam, French Guiana, Brazil (Pará, Amapá) and very probably also in Guyana. In Surinam it is present in forested, more or less hilly areas south of the coastal savannah-belt. It has been recorded up to 500 m above sea level (fig. 42).

REMARKS.—Brongersma (1932a) compared *A. kockii* (Van Lidth de Jeude) and *A. reticulata versteegii* Van Lidth de Jeude and I agree with most of the differences he found. However, the differences he mentions under point three are due to lack of material. As may be seen from the preceding description, in *A. kockii* (Van Lidth de Jeude) there are always two pairs of chinshields in contact, in *A. reticulata versteegii* Van Lidth de Jeude there may be two or three (only in three out of ten specimens) pairs in contact, so this character is less reliable than BRONGERSMA thought. A difference between the two species which he does not mention is the relative length of the body. In *A. kockii* (Van Lidth de Jeude) the distance between forelimbs and hind limbs is at most 1.5 times the length of a forelimb, whereas in *A. reticulata versteegii* Van Lidth de Jeude this distance is 2.0–2.5 times the length of a forelimb. This means that of specimens of equal snout-vent length those of *A. kockii* (Van Lidth de Jeude) do have longer legs than those of *A. reticulata versteegii* Van Lidth de Jeude.

DA CUNHA's (1961, 1967) data on this species do not agree very well, despite being based on the same material. Judging by the specimens I examined I am inclined to regard the 1967 data as more reliable than the 1961 ones. In 1961 DA CUNHA said that all females investigated by him from Pará and Amapá lacked preanal and femoral pores. In 1967 he says that indistinct preanal pores may be present. The number of transverse rows of dorsals, stated to vary between 27 and 28 in the 1961 paper, is said to be 26 for all specimens in 1967. The number of scales around the body in 1961 was 34, in 1967 varies from 34–42, is, however, mostly 38–41. In 1961 DA CUNHA stated that there were 26 scales 'between the gulars and the preanals', in 1967 he says that the number of transverse ventrals varies from 17–20, mostly 18–19, the number of pairs of gulars is said to be four. Although the 1961 paper suggested there might be some geographical variation, from the more reliable data in the 1967 paper it is clear that the specimens from Amapá and Pará agree completely with those from Surinam and French Guiana.

Arthrosaura reticulata versteegii Van Lidth de Jeude (figs. 44, 45; pl. 27; table 15)

Arthrosaura Versteegii Van Lidth de Jeude, 1904: 89; Van Lidth de Jeude, 1917g: 438.
Arthrosaura versteegii: Brongersma, 1932a: 77, 78, 79, 81; Burt & Burt, 1933: 56; Da Cunha, 1967: 150; J. A. Peters & Donoso-Barros, 1970: 75.
Arthrosaura reticulata versteegii: Brongersma, 1935: 264.

TYPE-LOCALITY.—Cottica Mountains, French Guiana.

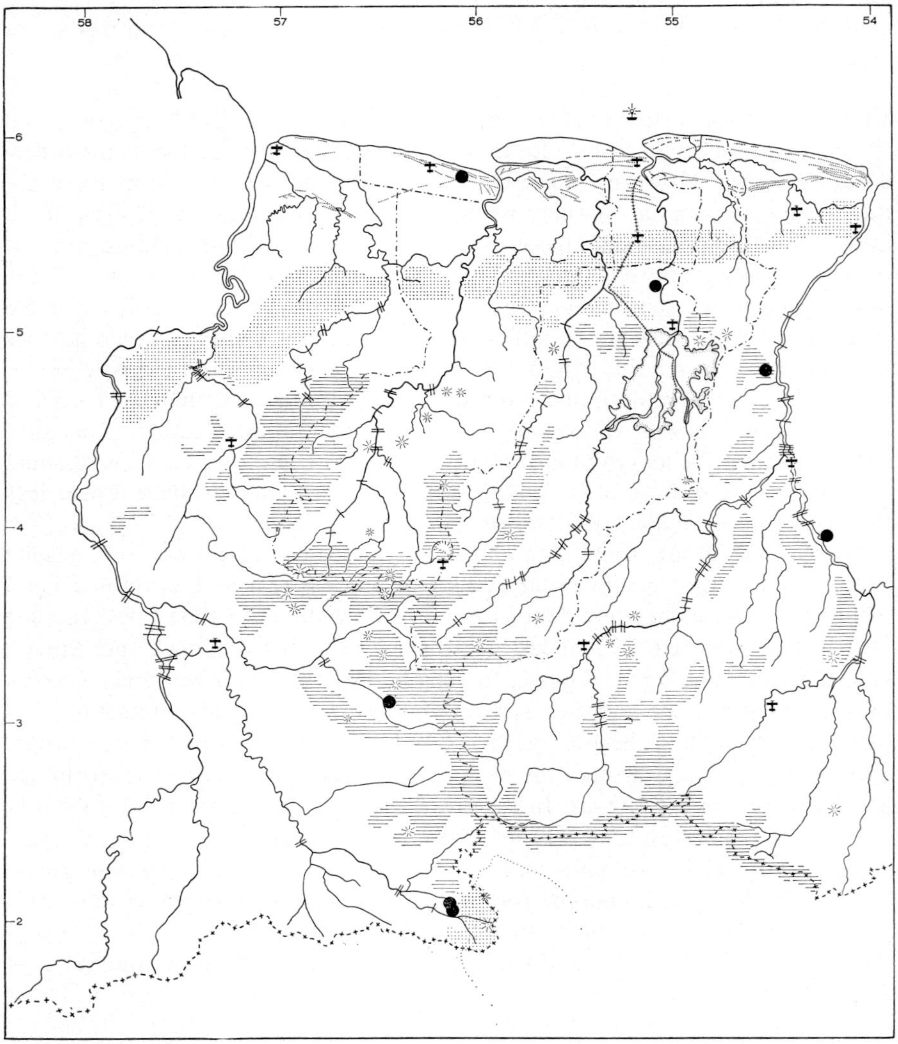

Fig. 44. Distribution of *Arthrosaura reticulata versteegii* Van Lidth de Jeude in Surinam.

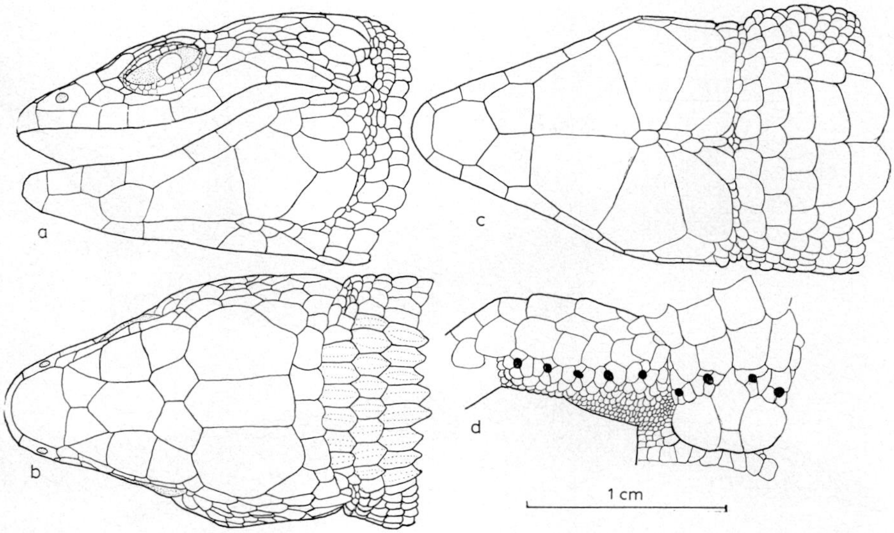

Fig. 45. Arthrosaura reticulata versteegii Van Lidth de Jeude; a, b, c: lateral, dorsal and ventral views of the head, d: anal region of ♂, FMNH 128979, Airstrip Kayser Mountains.

MATERIAL. – Surinam.

Coronie Road: 1 ♂, RMNH 13445, km 0.8, 19–XII–1948, leg. Surinam Expedition 1948–49.

65 km SSE. of Paramaribo, Afobaka Road: 3 juvs. (1 ♀, 2 ♂♂), AMNH 108772–74, 12/13–II–1972, leg. C. W. Myers & J. Daly.

Nassau Mountains: 1 ♀, RMNH 13444, km 9.1, 10–III–1949, leg. Surinam Expedition 1948–49.

Airstrip Kayser Mountains: 1 ♀, 1 ♂, FMNH 128979–80, 1–II–1961, leg. H. A. Beatty.

Airstrip Sipaliwini: 1 ♀, RMNH 13443, February 1961, leg. D. C. Geijskes; 1 ♂, 3 juvs., RMNH 16374, 4 km NW., 10–II–1970, leg. M. S. Hoogmoed & J. J. P. Paats.

Guyana.

Isherton: 1 ♀, AMNH 70623, 5–XI–1937, leg. R. Snedigar.

Kuyuwini landing: 1 ♀, AMNH 57454, leg. R. Snedigar.

French Guiana.

Cottica Mountains: 1 ♂, RMNH 4469, 1903, leg. G. M. Versteeg (Gonini Expedition) (holotype of *Arthrosaura Versteegii* Van Lidth de Jeude).

DIAGNOSIS.—A small lizard with cylindrical body and tail and short legs. Distance between the fore- and hindlimbs 2.0–2.5 times as long as a forelimb. Three supraoculars. Median suture of the second pair of chinshields 2.0–2.5 times as long as that of the first pair. Dorsally and laterally brown, without vertebral stripe, ventral surface yellowish white to orange-yellow.

DESCRIPTION.—Head short and wide, depressed; posterior part slightly swollen in males and consequently wider than the neck, posterior part in females as wide as the neck; neck as wide as the body; snout rounded.

Rostral hexagonal; visible from above; more than twice as wide as deep. A

Plate 27. Arthrosaura reticulata versteegii Van Lidth de Jeude. Figs. 1–3. Dorsal, lateral and ventral aspect of ♂, RMNH 16374 a, Sipaliwini.

large heptagonal frontonasal, separating the nasals. A pair of pentagonal prefrontals, forming a short median suture. Frontal hexagonal, approximately one and a half times as long as wide. Frontoparietals pentagonal, forming a short median suture. Interparietal pentagonal, twice as long as wide, truncate posteriorly. Parietals two; roughly heptagonal, with the lateral margin convex; as long as the interparietal. Three to six occipitals in a transverse row; quadrangular to rectangular (wider than long). Three supraoculars in direct contact with the supraciliaries; second and third subequal; largest; anterior one smallest. Nostril in an undivided nasal. A loreal and a frenoocular; the loreal may or may not be in contact with the supralabials, thus separating the nasal and the frenoocular or not. One small, wedge-shaped preocular. Four or five suboculars; the second

Table 15. Counts and measurements of *Arthrosaura reticulata versteegii* Van Lidth de Jeude. For explanation see table 3.

Reg. no.	A	B	C	D	E	F	G	H	I	J	K	L	M	N	O	P	Q	R	S	T	U	V	W	X
RMNH																								
13443	♀	42	69	8.6	4.9	3.1	9	14	0	31	27	19	10	7	7-7	7-7	4-4	3-3	4-4	3+1	7	4	13-13	21-20
13444	♀	32	50	7.4	4.8	3.4	7	11	0	35	25	18	10	8	7-7	7-7	4-4	3-3	4-4	3+1	6	4	14-14	18-19
13445	♂	54	65(28)	12.5	7.6	6.3	13	21	5/2-5/2	39	26	15	10	8	8-8	7-7	4-4	3-3	4-4	3	6	4	14-13	18-18
16374a	♂	52	90	11.8	6.6	5.0	14	20	6/2-6/2	40	27	19	10	9	8-8	7-7	4-4	3-3	4-5	5+1	6	4	16-16	21-21
b	♀	25	38	6.7	4.4	3.3	·		0	39	27	17	10	8	8-7	7-6	4-5	3-3	4-5	5+1	6	4	15-15	21-21
c	♂	29.5	52	8.1	4.9	3.3			5/2-5/2	38	26	16	10	9	8-8	7-7	5-5	3-3	4-5	5+1	6	4	14-13	18-21
d	♂	27.5	30(27)	7.1	5.1	3.4			6/2-6/2	44	27	16	10	10	7-7	7-7	4-4	3-3	4-5	3	6	4	15-15	19-20
FMNH																								
128979	♂	71	59(21)	16.8	10.9	8.7	18	29	5/2-5/2	40	26	17	10	8	7-7	7-7	5-5	3-3	4-4	3+1	6	4	15-15	20-22
128980	♀	68	81(52)	14.4	9.1	7.0			0	40	26	17	10	8	8-8	8-7	5-6	3-3	5-5	5+1	6	4	16-16	23-22
AMNH																								
57454	♀	25.3	11(11)	6.6	4.3	3.2	7	9	0	43	25	17	10	10	6-7	5-5	4-4	3-3		3+1	6		14-14	19-21
70623	♀	23	32	7.0	4.1	3.2	6	11	0	35	26	16	8	9	8-8	7-7	5-5	3-3		3	6		14-13	18-19
RMNH																								
* 4469	♂	49	86	?	7.8	5.5	?	?	4/2-5/2	36	25	15	10	7	7-7	7-7	5-4	3-3	5-4	3	5	4	?	?

smallest; the fourth largest. Two rows of postoculars; the anterior one short, wedge-shaped, with a variable number of scales; the posterior one wider, with three to four larger scales, decreasing in size from the parietals to the labials. Supraciliaries four to five (rarely six); the anterior one very large, the rest smaller, all elongate; when five or six are present the last one is subtriangular. Temporal region with large, convex scales, increasing in size from antero-ventrally to postero-dorsally; four scales bordering the parietals. Ear-opening small, surrounded by small scales, those anterior of the ear rounded, projecting, thus forming a lobed margin; external auditory meatus short, tympanum clearly visible. Seven or eight supralabials; three to four to the centre of the eye; sixth largest. Lower eyelid with a semitransparent disc of three palpebrals. Pupil round.

Mental trapezoid, with convex anterior margin. A small, heptagonal postmental. Four pairs of chinshields; the anterior two or three in contact; the second pair with a very long median suture, 2.0–2.5 times as long as that of the first pair. Seven, rarely six or eight, infralabials; the anterior five distinctly larger than the posterior two. Anterior three pairs of chinshields in contact with the infralabials, fourth pair separated from them by small scales.

All headscales with a row of small pits along their margin, scales smooth.

In most specimens six (rarely five or seven) rows of gulars between chinshields and collar; the posterior three, or four, rows with a pair of large, smooth, imbricate, central scales. Collar with seven to ten scales; the three or two central ones very large; the scales forming a lobed posterior margin. Scales on the nape hexagonal, with low keels, 1.5 times as long as wide near the head, 2.5 times as long as wide near the forelimbs, imbricate, in transverse rows. Scales on the side of the neck small, round, tubercular, in vertical rows.

Dorsals and laterals identical, hexagonal, keeled, mucronate, more than twice as long as wide, in 25–27 transverse rows. Ventrals large, square to rectangular on the central part of the belly, hexagonal towards the sides, smooth, with rounded posterior margin, slightly imbricate, in ten (rarely eight) longitudinal and in 15–19 transverse rows. Ventrals and laterals very clearly separated by a row of small scales. Total number of scales around midbody 31–44.

In four specimens there are five preanal scales in a semicircle, the laterals and the median one small, enclosing a pentagonal or hexagonal central scale; in four specimens there are three scales enclosing the central one, in four others only three elongate preanals are present.

Preanal pores two in males, absent in females; femoral pores four to six in males, absent in females. Both preanal and femoral pores are in the centre of a group of three to four scales.

Scales on the tail hexagonal, keeled, slightly imbricate, in transverse rows, under the tail wider, smooth. Scales on the regenerated part of the tail rectangular, smooth, juxtaposed, in transverse rows. Tail round in cross-section, the regenerated part tending to be compressed.

Scales on the upper and anterior surfaces of the forelimbs and on the anterior and lower surfaces of the hind limbs large, imbricate, smooth, except on the lower legs and on the antero-dorsal part of the thighs, where they are keeled; on the lower and posterior surfaces of the forelimbs and on the dorsal and posterior surfaces of the hind limbs small, granular. Thirteen to 16 single lamellae under the fourth finger, the proximal ones with a shallow median depression; 18–23, mostly 18–21, lamellae under the fourth toe, divided proximally, single with a shallow median channel distally. Palms and soles with small, flat, granular scales. Fingers and toes subcylindrical.

Colour in life brown above, dorsal surface of tail orange-brown, lateral surface with a row of lighter spots. Chin white, with brown spots on the lower lips. Throat and belly yellowish white to orange-yellow. Ventral surface of tail orange. Iris golden. Anterior part of tongue bluish grey, rest pinkish white.

In preservative the back and dorsal surface of the head are brown, with two longitudinal dark-brown bands, a light dorsolateral band is absent in most specimens. In RMNH 13443 there is a distinct ochre dorsolateral band on the lateral parts of the parietals and on the anterior nuchal scales but it is lost even before it reaches the forelimbs. A black line, two scales wide, from neck to tail bordering on the dorsal colour, tips of the scales white. Ventrally of this line the flanks are grey, scales with white or light-brown tips. A number of white spots on the side of the neck to just posterior of the forelimbs. Upper and lower lips with black and white blotches, the black ones continued on the chinshields. Gulars with or without dark-brown blotches. Belly and underside of tail immaculate or with dark-brown blotches on the chest. Tail brown above, more or less densely spotted with black, with or without a light dorsolateral line or series of light spots; sides black or brown, with or without a longitudinal series of light spots.

Posterior surface of the hind limbs with a number of faint, white spots. Anterior portion of the tongue greyish blue, gradually changing into white posteriorly.

A small lizard, the adult stoutly built, the halfgrown slender. Maximum snout-vent length in females 68 mm, in males 71 mm. Hind limb 1.6 times as long as forelimb; when laid along the body there is overlap of fingers and toes. Distance between forelimbs and hind limbs more than twice the length of a forelimb. The undamaged tail is 1.5–1.6 times as long as the snout-vent length in females, 1.7–1.8 times in males.

HABITAT.—The specimens I collected were in a partly dry creek-bed in rain--forest. The deeper parts of the creek-bed were filled with brown, but clear, water, forming a number of isolated pools. The creek-bed consisted of coarse, white sand, the banks of the creek were fairly steep, not higher than two metres, without vegetation. The deeper parts of the bed showed bare sand, the higher parts near the banks were covered with fallen leaves. *A. reticulata versteegii* Van Lidth de Jeude was present in this layer of fallen leaves, together with *Pseudogonatodes guianensis* Parker. In a more open, sunlit part of the creek-bed *A. reticulata versteegii* Van Lidth de Jeude was absent but there *A. kockii* (Van Lidth de Jeude) was present. From these data it might be concluded that

A. reticulata versteegii Van Lidth de Jeude replaces *A. kockii* (Van Lidth de Jeude) in denser parts of the forest and in wetter situations.

NATURAL HISTORY.—A diurnal species which on one occasion (RMNH 16374) was taken between 10.30 a.m. and 1.00 p.m. No basking behaviour was observed. No pregnant females are known. Five specimens with snout-vent lengths varying from 25–30 mm show traces of an umbilical scar, less distinct when the snout-vent length increases. A specimen with a snout-vent length of 36 mm does not show any traces of an umbilical scar. All these juvenile individuals were collected in the first half of February and we therefore might assume that these individuals hatched from December to early January. Thus, nothing definite about the breeding season can be said. No stomachcontents were examined. One specimen (RMNH 16374a) which was kept alive for two months in a terrarium was seen to chase juvenile *A. kockii* (Van Lidth de Jeude). A large variety of insects and worms was accepted. The specimen was rarely seen and during the greater part of the day was hidden.

RANGE.—Surinam, Guyana, French Guiana and Brazil (Amapá, Amazonas). In Surinam this subspecies is present throughout the country in forested areas from sea level to at least 500 m. From Venezuela the nominate form has been reported to occur at 7100 feet (BURT & BURT, 1931) (fig. 44).

REMARKS.—Hitherto the type locality of this subspecies has been stated to be Surinam. This, however, is not correct. The type specimen was collected by the Gonini Expedition (VAN LIDTH DE JEUDE, 1904) in the Cottica Mountains. From the narrative of this expedition (FRANSSEN HERDERSCHEE, 1905a) it is clear that these mountains are situated on the right bank of the Lawa River, that is, in French Guiana. The confusion probably originates from the presence of the village of Cottica on the left bank of this same river, that is, in Surinam. VERSTEEG, the physician of the expedition who also did the zoological collecting, stayed in the village of Cottica between October 24 and 30, 1903. On October 30 he left to join the rest of the expedition in the Cottica Mountains in French Guiana, stayed with them till November 3 and then again returned to Cottica by way of a *Mauritia* swamp where he stayed overnight till November 4. Thus, the specimen very probably was collected between October 30 and November 4, as the label states expressly Cottica Mountains. The type locality consequently has to be changed into Cottica Mountains, French Guiana.

DA CUNHA (1967) described a new species of *Arthrosaura* from Amapá, *A. amapaense*. The characters he used to distinguish this species from *A. reticulata* (O'Shaughnessy) and *A. versteegii* Van Lidth de Jeude are known to vary extensively in *A. kockii* (Van Lidth de Jeude). The number of preanal scales, which is the most important character used to distinguish it, is very variable in *A. kockii* (Van Lidth de Jeude), in which most specimens have 5 + 1 preanal scales, though specimens with 3 + 1, 4 + 1, 6 + 1 and 5 scales are not too rare. The Surinam specimens of *A. reticulata versteegii* Van Lidth de Jeude exhibit three conditions, namely 3, 3 + 1 and 5 + 1 preanals, the same character being used by DA CUNHA to separate *A. versteegii* Van Lidth de Jeude, *A. amapaense* Da Cunha and *A. reticulata* (O'Shaughnessy), respectively. There are three Surinam localities from which more than one specimen is available. The two specimens from the first locality, Airstrip Kayser Mountains, are a male and a female, exhibiting 3 + 1 and 5 + 1 scales, respectively. From the surroundings of the airstrip Sipaliwini five specimens are available, a male, a female, two juvenile males and one juvenile female. Respectively they exhibit 5 + 1, 3 + 1, 5 + 1, 3, 5 + 1 scales. From these data it can be concluded that the 3 + 1 and 5 + 1 situations are not sexually linked. The 3 scale situation is known from three males and one female, and thus it too is not sexually linked. Since the different situations occur in one locality and constitute the only difference between the specimens, I conclude that this character is not sufficient to distinguish *A. amapaense* Da Cunha. Another character that seems to set *A. amapaense* Da Cunha apart is the number of palpebrals, which is said to be two, in the three other species to be three. In the Surinam, Guyanese and French Guianese material of *A. reticulata versteegii* Van Lidth de Jeude, specimens with two and with three palpebrals occur, one specimen (FMNH 128980) even has two on one, three on the other side. Thus, this character again does not serve to distinguish *A. amapaense* Da Cunha from the other species. In all other characters *A. amapaense* Da Cunha completely agrees with *A. reticulata versteegii* Van Lidth de Jeude. Though I did not see the type of *A. amapaense* Da Cunha, I conclude that *A. amapaense* Da Cunha is not different and is a synonym of *A. reticulata versteegii* Van Lidth de Jeude. I see no reason to doubt BRONGERSMA's (1935) conclusion that *A. reticulata* (O'Shaughnessy) and *A. versteegii* Van Lidth de Jeude are subspecies of the same species, *A. reticulata* (O'Shaughnessy). Consequently I have here adhered to the use of *A. reticulata versteegii* Van Lidth de Jeude for the specimens from Guyana, Surinam, French Guiana and Brazil.

Genus **Bachia** Gray

DIAGNOSIS.—Small worm-like lizards with four rudimentary limbs that bear a varying number of clawed or clawless digits. Body cylindrical, tail long. Ear-opening absent. Frontoparietals absent, prefrontals present or absent. Nostril between the nasal and the first upper labial. Scales on the body arranged in annuli, ventrals forming longitudinal rows. A genus that occurs in northern South America with 18 species, one of which occurs in Surinam.

Bachia cophias (Schneider) (figs. 46, 47; pl. 24; table 16)

Chamaesaura cophias Schneider, 1801: 209.
Chalcides flavescens: Kappler, 1881: 166.
Bachia cophias: Brongersma, 1946: 242; Vanzolini, 1961: 196.
Bachia monodactylus: Dixon, 1973: 40.

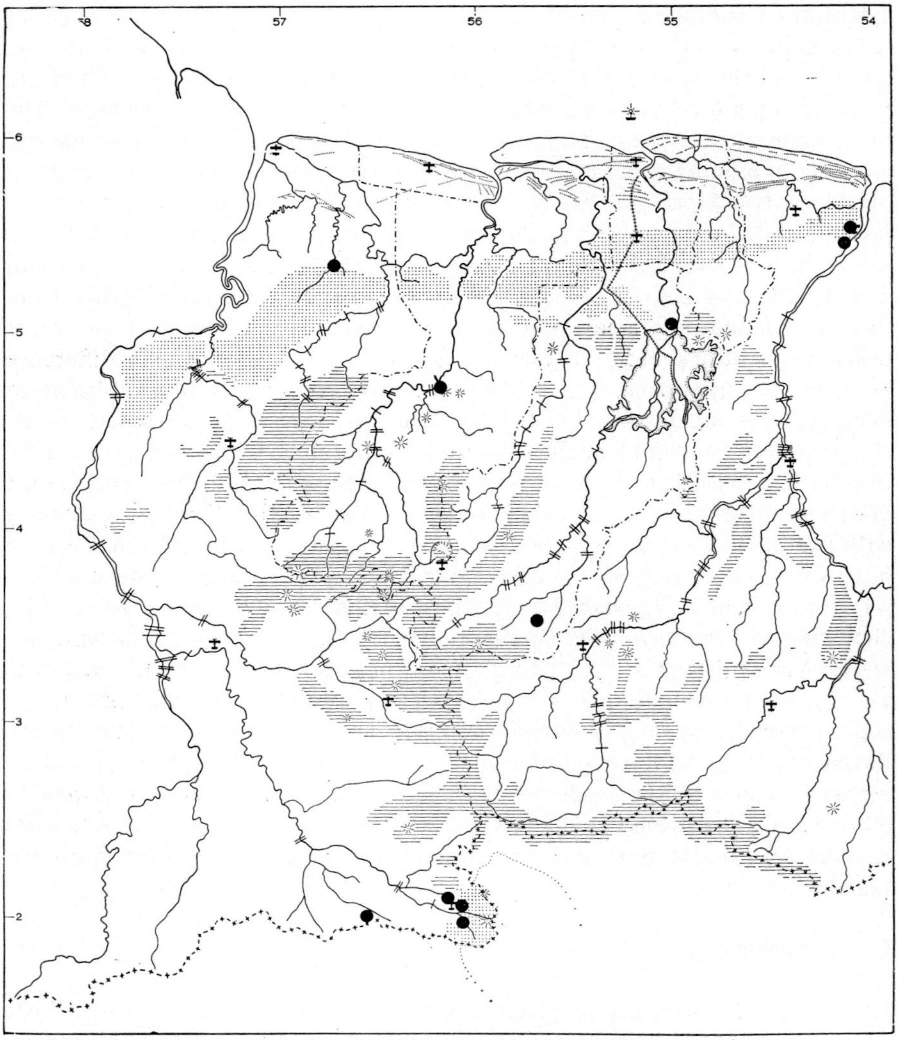

Fig. 46. Distribution of *Bachia cophias* (Schneider) in Surinam.

TYPE-LOCALITY.—None given.

MATERIAL. – Surinam.
Surinam: 1♀, SMNS 1084, 1859, leg. A. Kappler.
Snake Creek, Maratakka River: 1♀, RMNH 13414, May 1965, leg. P. J. M. Maas.
Albina: 1♂, RMNH 13411, 2–XI–1948, leg. Surinam Expedition 1948–49.
Papatam, Marowijne River: 2♂♂, RMNH 15217, 26–VI–1968, leg. M. W. H. de Boer.
Voltz Mountain Reserve: 1♀, RMNH 15216, 5–XI–1968, leg. R. Hoogmoed-Verschoor & M. S. Hoogmoed; 1♀, SM 18, September 1972, leg. P. H. Creutzberg.

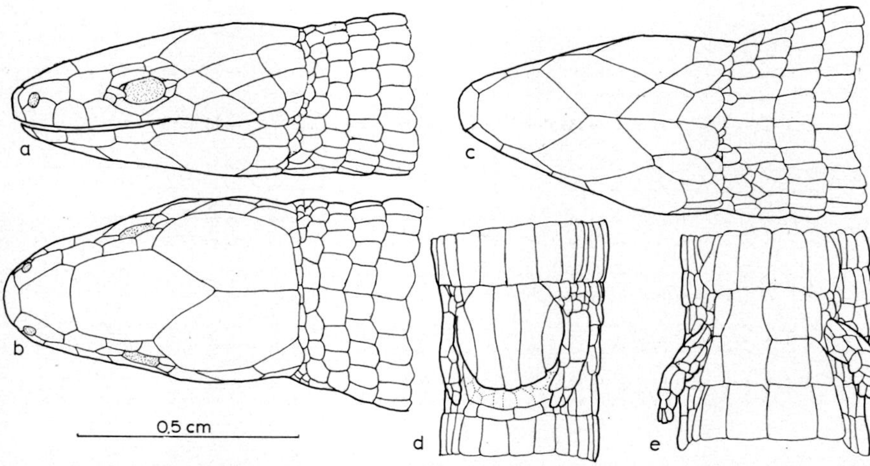

Fig. 47. Bachia cophias (Schneider); a, b, c: lateral, dorsal and ventral views of the head, d: anal region, e: pectoral region of ♀, RMNH 15216, Voltz Mountain Reserve.

Brokopondo: 1♀, RMNH 13413, 15–III–1966, leg. G. F. Mees.

Nassau Mountains: 1♀, 1♂, RMNH 13412, Base Bivouac, 19–II–1949, leg. Surinam Expedition 1948–49.

Ananas Mountain: 1♀, ZMA 11483, leg. J. H. A. T. Tresling (Suriname Expedition 1908).

Airstrip Sipaliwini: 1♀, RMNH 15183, 4 km E., 25–VIII–1968, leg. M. S. Hoogmoed; 1♀, RMNH 16377, 8 km SE., 22–I–1970, 2♀♀, RMNH 16434, 5 km NW., 9–II–1970, all leg. M. S. Hoogmoed & J. J. P. Paats.

Marowijne River: 2♀♀, ZMB 8585, 1♀, ZMB 8586, both leg. A. Kappler.

Surinam/Guyana.

New River: 5♂♂, BM 1939.1.1.76–80, 750 ft, leg. C. A. Hudson.

French Guiana.

French Guiana: 1♂, MP 03–229; 2♀♀, MP 28–109, 110, leg. Guénot.

Cayenne: 1♂, MP 2838, leg. Banon; 2♂♂, MP 2840, leg. Savigny; 1♀, 1♂, BM 53.11.22.16.

Tumuc Humac Mountains: 1♂, MP 99–76, Upper Lunier River, leg. Geay.

South America: 1♀, MP 2839.

DIAGNOSIS.—A small worm-like lizard with four rudimentary legs, a cylindrical body and long tail. Ear-openings absent, few large plates on the head. Brown above, purplish to yellowish brown underneath, dark-ochre dorsolateral stripes.

DESCRIPTION.—Head slightly narrower than the neck from which it is separated by a fold; conical, slightly depressed; 1.5 times as long as wide; snout bluntly pointed. Neck as wide as the body. Tongue lanceolate, without emargination; covered with rounded, scale-like, imbricate papillae in oblique rows; tip bifid, smooth. Teeth conical.

Rostral irregularly hexagonal; once and a half as wide as deep; visible from

Table 16. Counts and measurements of *Bachia cophias* (Schneider). For explanation see table 3.

Reg. no.	A	B	C	D	E	F	G	H	I	J	K	L	M	O	P	Q	R	S	T	U	V
RMNH																					
13411	♂	63	120	7.3	4.4	3.5			0/2-0/2	32	48	36		5-5	5-5	2-2	2-2	1-1	5+1	7	2
13412a	♀	65	84(79)	6.7	4.2	2.9			0	30	51	38		5-5	5-5	2-2	2-2	2-2	5+1	7	2
b	♂	67	51(51)	7.0	4.1	3.2			0/2-0/2	30	48	35		5-5	5-5	2-2	2-2	2-2	5+1	7	2
13413	♀	64	75(68)	6.3	4.0	2.8			0	28	50	37		5-5	5-5	2-2	2-2	?-2	5	7	2
13414	♀	58	31(19)	6.1	3.6	2.8			0	28	50	37		5-5	5-5	2-2	2-2	2-2	5+1	7	2
15183	♀	80	121	7.5	4.4	3.4	1.6	2.7	0	29	51	39		5-5	5-5	2-2	2-2	1-1	5+1	8	2
15216	♀	72	68(56)	6.8	3.9	3.1	2.4	2.6	0	30	51	38		5-5	5-5	2-2	2-2	1-1	5	7	2
15217a	♂	60	18(18)	7.0	4.3	3.2	2.2	3.0	0/2-0/2	32	47	35		5-5	5-5	2-2	2-2	2-2	5	7	2
b	♂	65	61(52)	7.3	4.2	3.2	2.4	3.1	0/2-0/2	32	47	35		5-6	5-5	2-2	2-2	2-2	6	7	2
16377	♀	66	64(57)	6.7	4.3	3.0			0	31	49	37		5-5	5-5	2-2	2-2	1-1	6	7	2
16434a	♀	72	98	6.8	4.1	3.2			0	31	49	39		5-5	5-5	2-2	2-2	1-1	5	7	2
b	♀	67	98						0	31	49	37		5-5	5-5	2-2	2-2	1-1	5	7	2
ZMA																					
11483	♀	72	75(68)						0	28	49	37		5-5	5-5	3-2	2-2	2-2	5	7	2
BM																					
1939.1.76	♂	73	53(39)	7.7	4.8	3.4			0/2-0/2	29	47	35		5-5	5-5	2-2	2-2	2-2	5	7	2
.77	♂	67	60(60)	6.9	4.4	3.3			0/2-0/2	30	48	36		5-5	5-5	2-2	2-2	2-2.	5	7	2
.78	♂	54	41	5.8	4.2	2.9			0/3-0/2	29	47	36		5-5	5-5	2-2	2-2	2-2	5	7	2
.79	♂	59	18(18)	6.6	4.2	3.3			0/2-0/2	30	48	35		5-5	5-5	2-2	2-2	2-2	5	7	2
.80	♂	52	11(11)	6.1	4.0	3.2			0/2-0/2	28	48	34		5-5	5-5	2-2	2-2	2-2	5	7	2
SMNS																					
1084	♀	71	54.5	7.1	4.0	3.2			0	30	50	38		5-5	5-5	2-2	2-3	2-2	5	7	2
ZMB																					
8585a	♀	67	77	6.7	4.2	2.5	1.8	2.3	0	28	50	37		5-5	5-5	2-2	2-2	1-1	5	7	2
b	♀	77	33(16)	7.2	4.5	2.6	1.9	2.3	0	29	49	37		5-5	5-5	2-2	2-2	1-1	5	7	2
8586	♀	78	46(10)	7.5	4.0	2.9	1.8	2.5	0	28	49	38		5-6	5-5	2-2	2-2	1-1	5+2	7	2
MP																					
2838	♂	44	57						0/2-0/2	32	46	34		5-5	5-5	2-2	2-2	1-1	5	7	2
2840a	♂	61	35(?)						0/1-0/1	31	49	36	1/0	5-5	5-5	2-2	2-2	1-1	5	7	2
b	♂	53	44(?)						0/2-0/2	29	46	35		5-5	5-5	2-2	2-2	1-1	5	7	2
03-229	♂	64	71(63)						0/2-0/2	27	46	34		5-5	5-5	2-2	2-2	1-1	5	7	2
28-109	♀	32	32(?)						0	28	48	36		5-5	5-5	2-2	2-2	1-1	5+1	7	2
28-110	♀	55	71						0	32	49	37		5-5	5-5	2-3	2-2	1-1	5+1	8	2
99-76	♂	57	94						0/2-0/2	29	46	34		5-5	5-5	2-4	2-2	1-1	5	6 1/0	2
BM																					
53.11.22.16.a	♂	63	58						0/2-0/2	28	48	35		5-5	5-5	2-2	2-2	2-2	5	7	2
b	♀	72	14(10)						0	30	48	36		5-5	4-4	2-2	2-2	1-1	5+1	7	2
MP																					
2839	♀	69	48(48)						0	31	47	36		5-5	4-5	2-2	2-2	1-1	5+1	7	2

above. A large, irregularly hexagonal frontonasal with concave posterior margin, separating the nasals. Prefrontals absent. A large, pentagonal frontal, longer than wide; narrow anteriorly, widening in the middle and ending in a point posteriorly. Frontoparietals and interparietal absent. A pair of large irregularly hexagonal parietals, forming a long median suture. In SMNS 1084 a small interparietal is present, separating the anterior parts of the parietals. Occipitals absent. Supraoculars two, in contact with the supraciliaries. Nostril on the border between the single nasal and the first supralabial. Loreal longer than deep. One small preocular. One or two suboculars; the only one or the posterior one very long. Postoculars one or absent. Supraciliaries two, the second slightly larger than the first. Temporal region with five large and two small scales. Ear-opening absent. Five supralabials; three to below the centre of the eye; the anterior three scales deeper than long, the posterior two longer than deep. Lower eyelid with an undivided semitransparent disc. Pupil round.

Mental small, trapezoid with convex anterior margin. Postmental very large, polygonal. Two pairs of large chinshields, the anterior pair forming a short median suture; the posterior part of the first pair and the second pair are separated by a few regularly arranged scales; three scales along the posterior margin of the second chinshield, two pairs of scales in the median. Five infralabials; the fourth longest; the posterior two more elongate than the anterior three. Chinshields in contact with the infralabials.

All head scales with a row of small pits along their margin and with scattered pits. All body and tail scales in transverse and longitudinal rows.

Seven, rarely eight, transverse rows of juxtaposed, rectangular gulars between the head and the pectorals. Scales on the nape and on the side of the neck small, wider than long near the head, identical to the dorsals near the forelimbs, smaller than the gulars, juxtaposed. No collar. A row of large, rectangular pectorals.

Dorsals small, rectangular, longer than wide, slightly wider near the median than towards the sides, juxtaposed, smooth, in 47–51 transverse rows. Ventrals larger than the dorsals, nearly quadrangular near the median, rectangular towards the sides, in 34–39 transverse rows between the pectorals and the preanals. In the anterior half of the body a lateral fold without scales is present, posteriorly it is replaced by scales indistinguishable from the adjacent dorsals and laterals. Total number of scales around midbody 28–32. Five elongate preanal scales of which the central one may be divided in an anterior and a posterior scale (in one specimen two posterior scales). ZMA 11484 is aberrant in possessing a very small median preanal (BRONGERSMA, 1946).

Preanal pores absent in females, two on each side in males, femoral pores absent in both sexes.

Scales on the tail identical to the dorsals, scales under the tail slightly wider. Scales on the regenerated part of the tail merely shorter.

Scales on the anterior surface of the limbs smooth, juxtaposed, larger than the scales on the posterior surface. Limbs very much reduced, antero-posteriorly compressed forelimbs with three, hind limbs with one tubercle-like digit, clawless.

Colour in life brown above with dark-brown mottling and dark-ochre dorsolateral stripes. Belly purplish brown anteriorly, yellowish posteriorly, transparent.

In preservative these colours do not really change. The belly is light-brown all over. The dorsolateral stripe starts some distance behind the head on the neck and continues onto the tail; it is bordered dorsally and laterally by narrow dark-brown lines. The mottling of the back is a result of the scales being dark-brown along their margins and light-brown in their central parts. Only the scales in the dorsolateral stripe are entirely ochre and those in the accompanying dark-brown line entirely, or nearly so, dark-brown. Belly light-brown with a very faint mottling of pale yellow. The dark elements expand on the dorsal part of the tail. Ventrally the tail has the same colour and pattern as the belly. Tongue white.

A small lizard, worm-like in appearance with a very slender body and tail and strongly reduced limbs. Maximum snout-vent length in females 80 mm, in males 67 mm. The forelimbs are slightly longer than the hind limbs. Undamaged tail 1.5 times as long as the snout-vent length in one female, 1.9 times in one male.

HABITAT.—Inhabits the forest floor and the upper soil layer in rain-forest, where it may be found under and in dead logs, partly dug into the ground, or among the leaf-litter. DA CUNHA (1958) reports this same type of habitat.

NATURAL HISTORY.—No basking behaviour was observed. Two specimens were trapped at night in a pit, indicating nocturnal activity. One specimen was captured in the early morning in leaflitter. The species uses tunnels that are already present (DA CUNHA, 1958), as also can be concluded from BEEBE's (1945) statement that he found it in *Atta* nests. It probably does not dig tunnels by itself, because if it were capable of doing so, it is not very likely that it would be found in pits without special provisions to prevent its escape into the ground. No data on breeding season or stomachcontents in Surinam are available. BEEBE (1945) mentions a female, collected in June, that contained one egg about to be laid. The stomachs investigated by BEEBE (1925, 1945) contained ants, termites, sow-bugs, grubs, microlepidopteran scales and fine quartz-particles. BEEBE took one specimen from the stomach of a coati-mundi (*Nasua nasua* (Linnaeus)). This same author elaborately discusses the modes of movement used by this lizard.

RANGE.—Surinam, Guyana, French Guiana, Brazil (Pará), Venezuela (Monagas). In Surinam it occurs from sea level to at least 300 m. It is present throughout the country, on both sides of the coastal savannah belt (fig. 46).

REMARKS.—SMNS 1084 is aberrant in possessing a small interparietal scale, separating the anterior parts of the parietals, but in all other characters it agrees with the other Surinam specimens. The keys published by BURT & BURT (1933) and J. A. PETERS & DONOSO-BARROS (1970), in which the absence or presence of an interparietal is used to distinguish between *B. cophias* and *parkeri* on the one hand and, respectively, *boettgeri*, *tridactyla* and *lineata*, and *flavescens*, *bicolor*, *intermedia*, *lineata* and *schlegelii*, on the other hand, thus should be used with reservation.

J. A. PETERS & DONOSO-BARROS (1970) place the name *Chamaesaura cophias* Schneider in the synonymy of *Bachia flavescens* (Bonnaterre) on the grounds that *Chalcides flavescens* Bonnaterre, *Chamaesaura cophias* Schneider and *Chalcides tridactylus* Daudin were based on the figure on plate 32, in Lacépède (1789). This, however, is correct for only two of the descriptions but not for the description of *Chamaesaura cophias* Schneider. From SCHNEIDER's description it is clear that he had before him a specimen from the collection of a certain Mr. LAMPE. In his text he compares it with LACÉPÈDE's description of 'Le Chalcide'. SCHNEIDER's description was fairly good and together with the description and the figures of the same specimen published by GRAVENHORST in 1851 it is clear that the specimen SCHNEIDER described belongs to another species than LACÉPÈDE's 'Le Chalcide'. Thus, the name *cophias* remains available for the species which J. A. PETERS & DONOSO-BARROS (1970) indicate with the name *Bachia monodactylus* (Daudin) and which until recently was known as *Bachia cophias* (Schneider). Thus, the synonymy for this species given by VANZOLINI (1961) remains valid without changes.

As SCHNEIDER's description of *Chamaesaura cophias* was based on two specimens, viz., the specimen from LAMPE's collection and the specimen depicted by LACÉPÈDE (1789) it seems useful to designate a lectotype. Nomenclatorial stability is best served by taking as lectotype the specimen from LAMPE's collection that SCHNEIDER had before him when describing *C. cophias*. In 1851 this same specimen was extensively described and depicted by GRAVENHORST; at that time the specimen was in the Breslau museum. I do not know its present whereabouts.

Genus **Cercosaura** Wagler

DIAGNOSIS.—See diagnosis of species. This genus is composed of only a single species and ranges throughout the greater part of tropical South America. Of the three subspecies the nominate form is present in Surinam.

Cercosaura ocellata ocellata Wagler (figs. 48, 49; pl. 28; table 17)

Cercosaura ocellata Wagler, 1830: 158.
Cercosaura o. ocellata: Ruibal, 1952: 494.

TYPE-LOCALITY.—Asia?, restricted by RUIBAL (1952) to: northeastern South America; Surinam?

MATERIAL. – Surinam.
Surinam: 1♂, ZMB 7222, leg. H. B. Möschler.
Paramaribo: 1♂, RMNH 15230, 3–V–1968, leg. M. S. Hoogmoed.

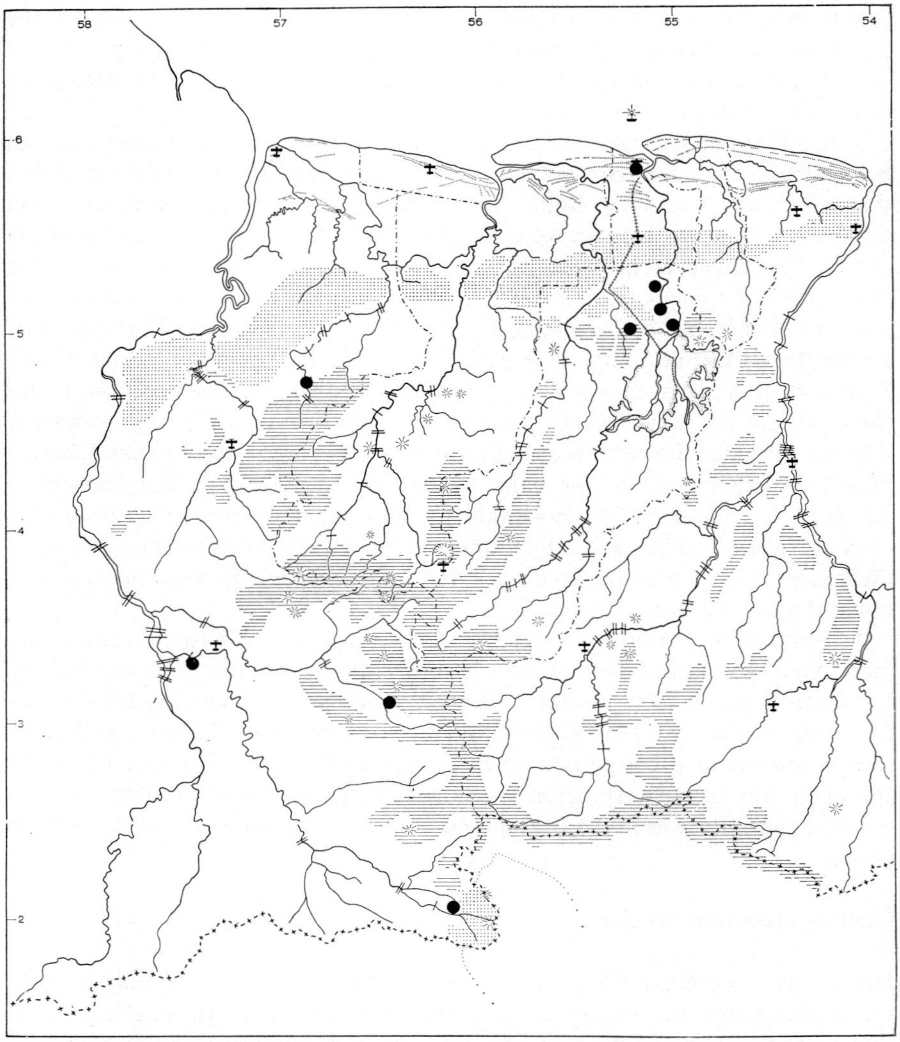

Fig. 48. Distribution of *Cercosaura o. ocellata* Wagler in Surinam.

65 km SSE. of Paramaribo, Afobaka Road: 1♂, AMNH 108787, 12/13–II–1972, leg. C. W. Myers & J. Daly.

Berg en Dal: 1♂, ZMB 8259, leg. H. B. Möschler.

Brown's Mountain: 1♀, RMNH 13862, 19–VIII–1968, 1♂, RMNH 13863, 02–VIII–1968, 1 juv., RMNH 15232, 13–VII–1968, all leg. M. S. Hoogmoed.

Brokopondo: 1♂, RMNH 13384, 19–III–1966, leg. G. F. Mees.

Blanche Marie Cataracts, Nickerie River: 1♂, RMNH 16470, 12–II–1971, leg. D. G. Reeder.

Gonini Encampment, Coeroeni River: 1 piece of tail, RMNH 15231, 26–VI–1968, leg. M. S. Hoogmoed.

Airstrip Kayser Mountains: 1♂, RMNH 13861, 31–VI–1968, leg. M. S. Hoogmoed.

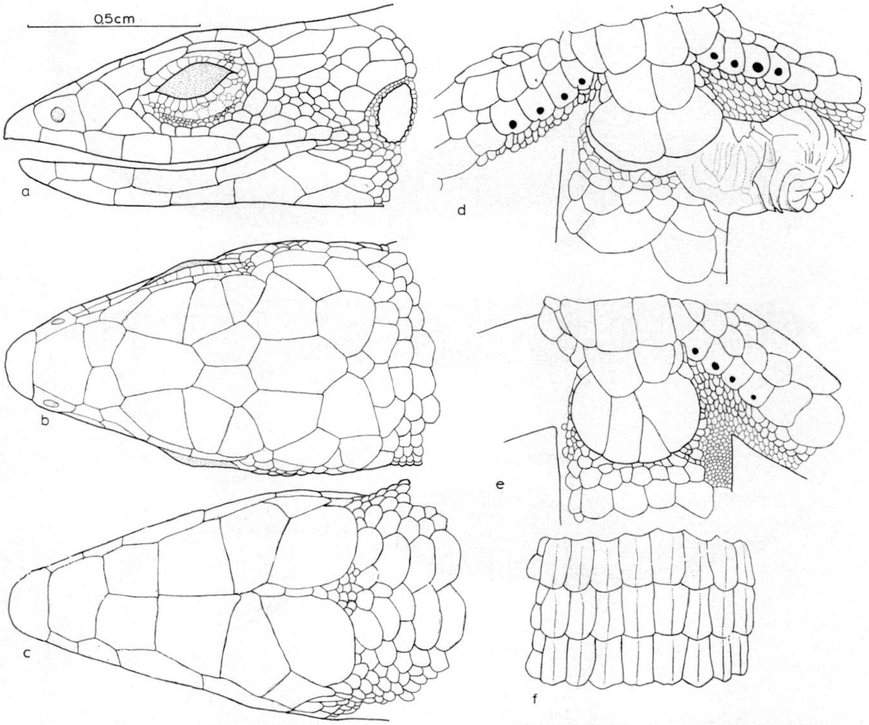

Fig. 49. Cercosaura o. ocellata Wagler; a, b, c: lateral, dorsal and ventral views of the head, d: anal region of ♂, RMNH 15230, Paramaribo; e: anal region of ♀, RMNH 13862, Brown's Mountain; f: dorsals of RMNH 15230.

Airstrip Sipaliwini: 1♂, RMNH 13415, February 1961, leg. D. C. Geijskes.
Guyana.
Kartabo: 1♂, AMNH 21264, 5–V–1919, leg. W. Beebe.
Essequibo River: 1♀, BM 1930.10.10.165, leg. R. N. G. Hingston.
Dunoon, Demerara River: 1♀, UMMZ 46771, 18–VIII–1914, 1♀, UMMZ 46772, 8–VIII–1914, both leg. A. G. Ruthven.
Demerara River: 1♀, UMMZ 53896, 11–XI–1914, leg. E. N. Clarke.
Marudi: 1♂, AMNH 61385, April 1938, leg. R. Snedigar.

DIAGNOSIS.—A small lizard with cylindrical body, very long cylindrical tail (up to 2.5 times as long as the snout-vent length) and well developed legs. Dorsal scales enlarged, rectangular, in 30–41 transverse and in 8–12 longitudinal rows; total number of scales around midbody 25–31. Brown above with three longitudinal black lines. Flanks with a series of white, black-edged ocelli. Ventral parts white, underside of tail red.

DESCRIPTION.—Head pyramidal, depressed; slightly wider than or of equal width as the neck, which is as wide as the body; snout with a rounded tip.

Tongue lanceolate, without emargination; covered with rhomboidal, scale-

Plate 28. Figs. 1–3. *Cercosaura o. ocellata* Wagler; dorsal, lateral and ventral aspect of ♂, RMNH 13861, Airstrip Kayser Mountains. Figs. 4–6. *Tretioscincus agilis* (Ruthven); dorsal, lateral and ventral aspect of ♀, RMNH 13871, Airstrip Kayser Mountains.

Table 17. Counts and measurements of *Cercosaura o. ocellata* (Wagler). For explanation see table 3.

Reg. no.	A	B	C	D	E	F	G	H	I	J	K	L	M	N	O	P	Q	R	S	T	U	V	W	X	Y
RMNH																									
13384	♂	57	123(?)	14.9	8.9	7.4			4/0–4/0	26	30	20	6	6	6–7	6–6	4–4	3–3	4–4	2+2	8	4	16–15	19–20	8
13415	♂	?	161	12.1	7.4	5.4			4/0–4/0	26	?	21	6	7	7–7	6–6	4–4	3–3	3–4	2+2	8	4	17–17	19–20	8
13861	♂	51	130	12.5	7.5	5.9	17	23	4/0–4/0	25	34	20	6	7	7–7	6–6	5–5	3–3	5–6	2+2	8	4	16–16	19–18	8
13862	♀	47	115	11.5	6.8	5.3	16	23	4/0–5/0	26	33	20	6	7	7–7	6–6	4–4	3–3	6–7	4+2	9	4	16–16	18–19	8
13863	♂	46	120	11.7	7.4	5.7	17	24	4/0–4/0	25	33	19	6	5	7–7	6–6	4–5	3–3	8–7	2+2	8	4	16–17	21–23	8
15230	♂	47	102(60)	11.8	6.2	4.9	16	23	4/0–4/0	26	32	20	6	7	7–7	6–6	3–3	3–3	6–4	2+2	8	4	16–15	19–20	8
15232	♀	22	42	6.6	3.4	2.8	7	10	5/0–5/0	31	41	22	6	7	7–7	6–5	5–5	3–3	4–4	4+2	7	4	17–18	19–21	12
16470	♂	57	55(55)	15.4	9.3	7.5			5/0–5/0	31	34	20	6	7	7–7	6–6	3–4	3–3	6–7	2+2	8	4	16–17	20–20	8
ZMB																									
7222	♂	43	117(83)	11.1	6.2	5.1	16	23	5/0–4/0	24	31	19	4	7	7–7	6–6	4–4	3–3	6–5	2+2	8	4	15–16	22–20	8
8259	♂	49	161	11.9	5.9	5.1	18	25	4/0–4/0	24	32	20	4	6	7–7	6–6	4–3	3–3	5–4	2+2	8	4	16–16	19–18	8
BM																									
1930.10.10.65	♀	55	9(9)	13.4	7.6	7.4	19	?	5/0–5/0	27	33	20	6	?	7–7	6–6	3–3	3–3	5–5	4+2	8	4	?	?	8
AMNH																									
21264	♂	48	110	12.0	6.8	5.0	16	22	6/0–5/0	26	30	19	6	?	6–7	5–6	3–3	3–3	?	2+2	8	4	14–14	17–17	8
61385	♂	40	36(36)	11.0	6.1	5.0	14	21	4/0–4/0	29	30	19	6		7–7	5–6	3–3	3–3	?	2+2	8	4	15–16	20–18	8
UMMZ																									
46771	♀	57.5	–	13.4	7.7	6.4	20.5	28	6/0–6/0	25	34	19	6		7–7	6–6	3–4	3–3	5–5	4+2	9		17–17	21–20	8
46772	♀	51	94	12.3	7.3	5.8	19	24	5/0–5/0	27	36	20	6		7–7	6–6	3–3	3–3	5–5	4+2	9		20–19	20–22	8
53896	♀	56	–	13.4	8.9	6.5	20.5	28	3/0–3/0	25	31	21	6		7–7	5–5	4–4	3–3	5–5	4+2	9		16–16	19–20	8

-like, imbricate papillae in oblique rows; tip bifid, smooth. Anterior teeth conical, rest tricuspid, of different sizes, laterally compressed.

Rostral hexagonal with a concave posterior margin; visible from above, more than twice as wide as deep. A large heptagonal frontonasal, separating the nasals. A pair of irregularly hexagonal prefrontals, forming a short median suture, or separated. Frontal hexagonal, wider anteriorly than posteriorly, once and a half as long as wide. A pair of pentagonal frontoparietals, forming a long median suture. Interparietal large, hexagonal. Parietals two, irregularly hexagonal, shorter than the interparietal. Three or, rarely, four occipitals; the central one (or two, one behind the other) very small, the two laterals very large, hexagonal. Supraoculars three; the anterior one largest, the third smallest; in direct contact with the supraciliaries. Nostril in an undivided nasal. A large loreal always in contact with the upper labials, thus separating the nasal and the triangular frenoocular. A small preocular. Five to six suboculars; the last one, or the one before last, largest. Three to four postoculars, the upper one largest. Three to five supraciliaries; the anterior one largest; the second and third subequal, elongate; the fourth or the fourth and the fifth small, quadrangular or granular. Temporal region covered with irregular, flat scales of different sizes, large in the dorsal part, small in the ventral part. Ear-opening large, surrounded by small scales, forming a finely-lobed margin; external auditory meatus short, tympanum well visible. Supralabials seven, rarely six; fourth largest, under the eye; sixth narrow, elongate. Lower eyelid with a semitransparent disc of two to three palpebrals. Pupil round. All dorsal and lateral headscales with scattered, small pits and a row of small pits along their margin.

Mental nearly semicircular. A large, heptagonal postmental. Four pairs of chinshields; the anterior two or three in contact; the posterior pairs separated by one or, rarely, two rows of small scales; suture of the second pair slightly longer than the suture of the first pair. Five to six infralabials; the third and the fourth subequal, elongate, largest. The anterior three pairs of chinshields in contact with the labials, the fourth pair separated from them by a large scale. Scales on the chin with very faint rows of pits along their margins.

Seven to nine, in most specimens eight, pairs of transversely enlarged, smooth imbricate gulars. Five to (mostly) seven collar scales, the median two or three much larger than the rest; the central one smallest, rhomboid; the two accompanying ones transversely enlarged, forming a lobed posterior margin. Scales on the nape in the anterior part wider than long, smooth, juxtaposed; near the forelimbs longer than wide, imbricate, keeled, in transverse and longitudinal rows. Scales on the side of the neck small, granular, in oblique and in vertical rows.

Dorsals large, rectangular, with a convex posterior border and a low keel, longer than wide; in eight (in the juvenile 12) longitudinal and in 30–34 (41 in in the juvenile) transverse rows; the keels forming longitudinal ridges. Laterals smaller than the dorsals, with a sharp demarcation, quadrangular, keeled, juxtaposed, in vertical rows, two rows to one row of dorsals in the posterior part of the flanks; granular, even smaller than the scales on the side of the neck, in the anterior part of the flanks. Ventrals larger than the dorsals, rectangular, wider than long, smooth, imbricate, in six longitudinal and in 19–22 transverse rows. Border between ventrals and laterals distinct, although less well demarcated than between dorsals and laterals. Total number of scales around midbody 25–31. In males there are two, in females four, smooth posterior preanal scales; in both sexes these are preceded by two anterior preanal scales.

Preanal pores absent, femoral pores four to five on each side, both in males and in females; the pores are in the posterior end of a single scale. Hemipenis probably bilobed; with oblique transverse plicae from the base to the top; sulcus spermaticus on the posterior surface.

Scales on the tail rectangular, with convex posterior margin, longer than wide, keeled, imbricate, in longitudinal and transverse rows; under the tail two rows of large, quadrangular, smooth, imbricate scales with rounded posterior margin. Scales on the regenerated part of the tail merely smaller with less distinct keels.

Scales on the upper, anterior and lower surfaces of the forelimbs and on the anterior and lower surfaces of the hind limbs enlarged, more or less hexagonal, smooth, imbricate; on the posterior surface of the forelimbs and on the upper and posterior surfaces of the hind limbs small, granular, juxtaposed. Fifteen to 18 (mostly 16–17) double, tubercular lamellae under the fourth finger, either in pairs or alternating; 18–23 lamellae under the fourth toe, only the three to four distal lamellae single, the rest double, in pairs or alternating. Palms and soles with small, tubercular, granular scales. Fingers and toes subcylindrical.

Colour in life light-brown above, with black longitudinal stripes on the back. A dirty white dorsolateral stripe, flanks light-brown or, in adult males, red in their dorsal part with a series of white, black-edged ocelli and a yellow-brown lateral stripe. Belly, chin and throat white, underside of tail red. Iris golden brown. Tip of tongue blue, rest pink.

In preservative the back is light-brown with three longitudinal black stripes, a vertebral one and two lateral ones bordering on the white dorsolateral stripe. Dorsolateral stripe from the posterior corner of the eye onto the tail white, bordered dorsally and ventrally by black lines, the ventral one narrower than the dorsal one. Dorsal surface of the head brown, with or without a pair of white spots on the interparietal. Flanks grey, with a row of nine to ten small or large ocelli from the side of the neck to the base of the hind limbs; in both cases the white centre occupies one scale; when small, the black rim is narrow, when large the black rim is very wide. A white lateral band from the ear via the base of the forelimb to the groin and continued onto the tail. Ventral parts white, chinshields and gulars immaculate. All ventrals may bear an ill defined grey spot, or the median rows may be immaculate. Infralabials with black spots. Tail above with a median dark-brown stripe and two dorsolateral stripes, a dark-brown and a white lateral stripe. Lower surface of the tail white to pink with two rows of dark-grey spots, one spot on each scale. The juvenile specimen (RMNH 15232) has a dark-brown band on the flanks with white spots in it.

A small lizard, maximum snout-vent length in females 47 mm, in males 57 mm. Hind limb ± 1.5 times as long as forelimb. When laid along the body there is considerable overlap of forelimb and hind limb. The undamaged tail is 2.4 times as long as the snout-vent length in females, 2.5–2.6 times in the males and 1.9 times in the juvenile.

HABITAT.—Inhabits open places along forest margins near creeks where the rain-forest gave way to secondary growth with much grass and weeds on the floor. One specimen (RMNH 13861) was captured 30 cm above the surface of the water among sedges in flooded country.

NATURAL HISTORY.—No pregnant females are available, one recently hatched juvenile (snout-vent length 22 mm) was collected in July. One specimen after collecting regurgitated its stomach contents, consisting of a spider. The species is diurnal and several times was seen basking.

RANGE.—Surinam, Guyana, Brazil (Pará, Amazonas, Amapá). In Surinam this species is present throughout the country, on both sides of the coastal savannah belt, from sea level to at least 200 m (fig. 48).

Genus **Cnemidophorus** Wagler

DIAGNOSIS.—Medium-sized to large lizards with cylindrical bodies, well developed legs with five digits, and long cylindrical tails. Teeth laterally compressed. Tongue arrowhead-shaped, not retractile into a basal sheath. Anterior nasal plates in contact. Dorsal scales small, granular or, on the posterior part of the back, large, imbricate. Supraoculars two to five, parietal region with three to five plates. Tympanum large, clearly visible. Eyelids movable, lower eyelid with a semitransparent disc. Pupil kidney-shaped, with the inversion ventrally. Femoral pores present. Males of some species with a preanal spine. The genus ranges from the United States through Central America south to Argentina, but is absent on the Pacific coast of South America. Of the 36 species in the genus only five occur in South America, with one species in Surinam.

Cnemidophorus lemniscatus lemniscatus (Linnaeus) (figs. 50, 51; pl. 29)

Lacerta coeruleata Merian, 1705a: 23, pl. 23.
'Blauwe Hagadis' Merian, 1705b: 23, pl. 23; Merian, 1719: 23, pl. 23.
'Blaauwe Hagedis' Merian, 1730: 18, pl. 23.
Lacerta lemniscata Linnaeus, 1758: 209; Daudin, 1802a: 175.
?'Pagara of Jamourou' Hartsinck, 1770: 100.
Ameiva murina: Fitzinger, 1826: 51; Gray, 1845: XVII.
Cnemidophorus murinus: Wiegmann, 1834: 27; Schlegel, 1858: 27; Van Lidth de Jeude, 1904: 89; Van Lidth de Jeude, 1917d: 206 (partly); Van Kleef, 1963: 87; Vanzolini & Valencia, 1965: 20.
Cnemidophorus lemniscatus: Duméril & Bibron, 1839: 128; Duméril & Duméril,

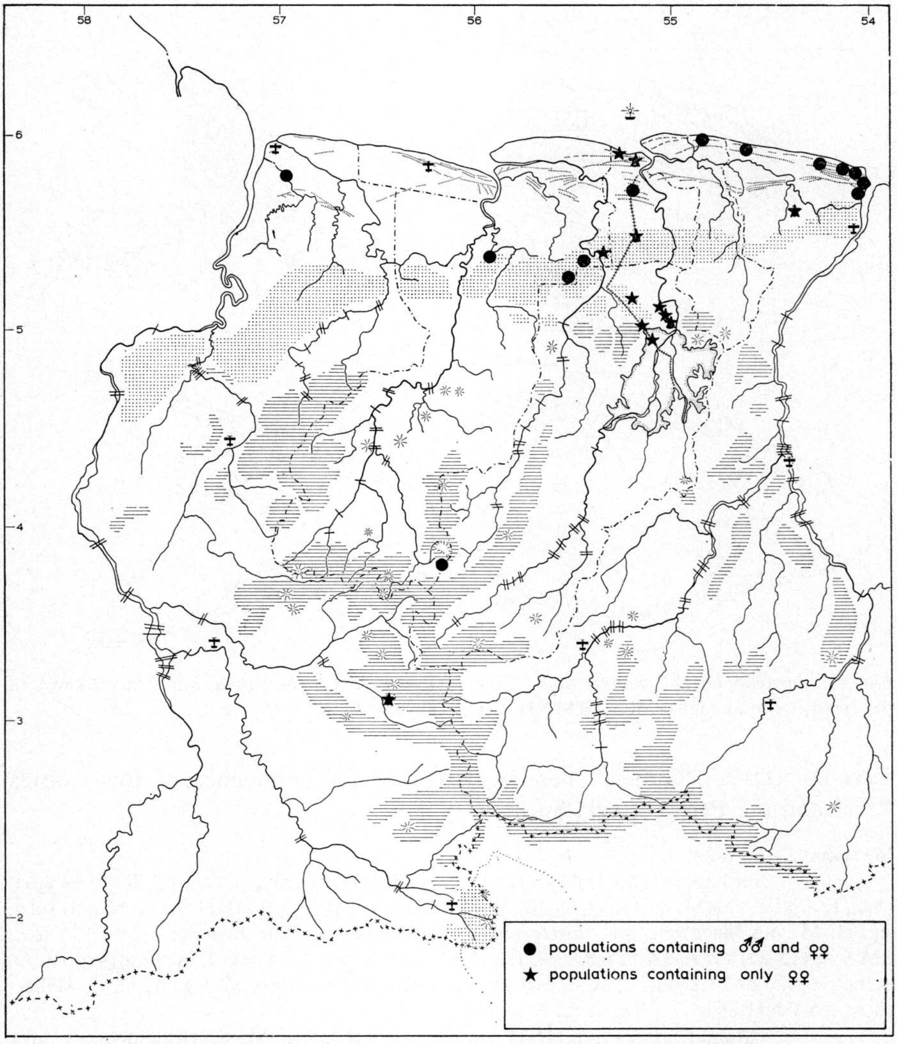

Fig. 50. Distribution of *Cnemidophorus l. lemniscatus* (Linnaeus) in Surinam.

1851: 116; Boettger, 1893: 74; Van Lidth de Jeude, 1917d: 206; Stuhldreher-Nienhuis, 1944: 175; Menne, 1946: 21; Menne, 1952: 10; Schierbeek, 1952: pl. 10; Menne, 1954b: 64.

Ameiva murinus: Gray, 1845: 20.

Cnemidophorus spec. Van Lidth de Jeude, 1898: 24; Rekum, 1962: 50.

'Eidechse' Nebel, 1964: 51.

Cnemidophorus l. lemniscatus: Bergmans, 1965: 59; D'Audretsch, 1968: 51; Donoso-Barros, 1968: 118; Lammerée, 1970: 51; Reed & Borowsky, 1970: 161.

'Thirteen-lined skinks' Walsh & Gannon, 1967: 50.

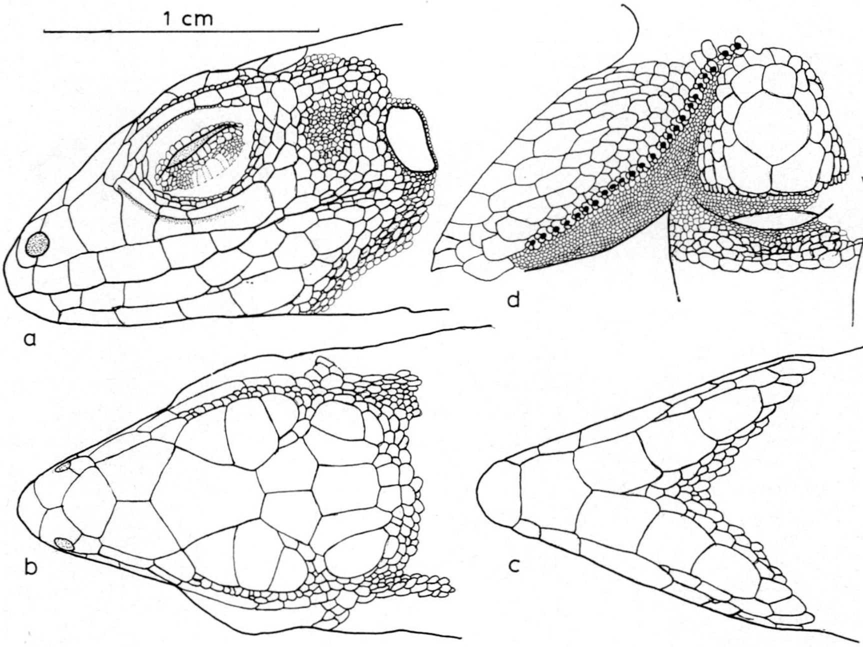

Fig. 51. Cnemidophorus l. lemniscatus (Linnaeus); a, b, c: lateral, dorsal and ventral views of the head, d: anal region of ♀, RMNH 15139, Paramaribo.

TYPE-LOCALITY.—'Guinea', here restricted to the confluence of the Cottica River and the Perica Creek, Surinam.

MATERIAL. – Surinam.

Surinam: 1♀, 2♂♂, RMNH 3400, leg. H. H. Dieperink; 5♀♀, 2♂♂, RMNH 3403, 1844, Frank; 2♀♀, RMNH 15435, leg. G. Witsen Elias; 1♀, 4 juvs., RMNH 16414, March 1969, leg. H. M. van Meeuwen; 4♀♀, ZMA 12336–39, 1887/90, leg. Van Brussel; 1♀, 6♂♂, 1 juv., ZMA 12340–47; 1♀, ZMA 12348, 2♀♀, 1♂, ZMA no number, 1922, leg. J. W. Gonggrijp & G. Stahel; 1 ex., ZMB 26096, 1 ex., ZMB 27135, 7 ex., NMW 18760, all leg. A. O. C. Heller; 18 ex., AMNH 8620–37, leg. J. A. Samuels.

Paramaribo: 2♀♀, RMNH 15139, 3–V–1968, leg. M. S. Hoogmoed; 1 juv., RMNH 15388, 6–VII–1968, leg. W. N. Polder; 1♀, RMNH 15421, November 1952, 1♀, RMNH 15425, August 1938, 1♀, RMNH 15426, 30–III–1950, all leg. D. C. Geijskes; 1♀, RMNH 15427, 12–XII–1962, 1♀, RMNH 15428, 6–I–1963, both leg. A. M. Husson & P. Staffeleu; 2♀♀, RMNH 15434, 19–VII–1938; 1 juv., RMNH 15438, 19–VI–1966, leg. W. Vervoort & C. Cornet; 2♀♀, RMNH 15468, 1965, leg. G. van Vreden; 4♀♀, RMNH 16413, April 1969, leg. H. M. van Meeuwen; 6♀♀, 6 juvs., BM 1946.4.3.60–71, 21–I–1938, 1♀, BM 1946.4.3.72, 25–I–1938, 9♀♀, BM 1946.4.3.73–81, 2–II–1938, 1♀, 2♂♂, 1 juv., BM 1946.4.3.82–85, 3–II–1938, all leg. I. T. Sanderson; 1 ex., ZMB 25984, 1 ex., ZMB 25996, 1 ex., ZMB 26011, 1 ex., ZMB 26082, 12♀♀, 2♂♂, 2 juvs., NMW 18759, 1910, all leg. A. O. C. Heller; 1 ex., CM 44381, 6–V–1966, 6 ex., CM 44385–90, 3–VII–1966, 1 ex., CM 44392, 8–IX–1966, all leg. M. H. de la Fuente.

Nanni Creek: 1♂, RMNH 16456, near dam, 16–III–1971, leg. D. G. Reeder.

Weg naar Zee: 3♀♀, RMNH 15223, 13–X–1968, leg. M. S. Hoogmoed.

Lelydorp: 1♀, RMNH 15433, 19–IX–1957, leg. L. C. M. Wijffels; 9♀♀, 2♂♂, ZMA 10733, 12415–24, 2–V–1963, leg. Van Heusden.

Plate 29. Cnemidophorus l. lemniscatus (Linnaeus). Figs. 1–3. Dorsal, lateral and ventral aspect of ♂, RMNH 15123, Eilanti beach. Figs. 4–6. Dorsal, lateral and ventral aspect of ♀, RMNH 15123, Eilanti beach.

Moengo: 5♀♀, RMNH 15710, June 1969, leg. H. M. van Meeuwen.

Matapica: 1♀, RMNH 15436, 12–I–1966, 1♂, RMNH 15437, 11–I–1966, 1♂, RMNH 17270, 16–IV–1972, all leg. G. F. Mees.

Bigisanti beach: 5♀♀, 3♂♂, RMNH 15037, 28/29–V–1968, leg. M. S. Hoogmoed; 1♀, RMNH 17265, 29–IV–1972, leg. G. F. Mees.

Wia Wia transect: 1♀, 5♂♂, RMNH 15391, Wia Wia, 11–XI–1948, 1♀, 1♂, RMNH 15392, 2nd ridge from sea, 13–XI–1948, 1 juv., RMNH 15393, 4th ridge from sea, 13–XI–1948, all leg. Surinam Expedition 1948–49.

Eilanti beach: 1♀, 6♂♂, RMNH 15122, 0–5 km W., 16–VI–1968, 3♀♀, 5♂♂, RMNH 15123, 11–VI–1968, 3♀♀, 3♂♂, RMNH 15124, 17–VI–1968, all leg. M. S. Hoogmoed.

Galibi: 2♀♀, 10♂♂, RMNH 15418, 4–X–1948, leg. Surinam Expedition 1948–49; 1♀, 5♂♂, RMNH 15420, October 1963, leg. B. Malkin; 1♀, 4♂♂, RMNH 15443, 18–IX–1963, leg. L. D. Brongersma.

Langamankondre: 3♂♂, RMNH 15390, 19–IX–1948, leg. P. H. Creutzberg (Surinam Expedition 1948–49).

Tibiti: 1♂, RMNH 15394, km 2, 6–I–1949, 1♂, RMNH 15395, km 1.4, 6–I–1949, 1♂, RMNH 15396, km 2.8, 8–I–1949, 3♀♀, 4♂♂, RMNH 15397–403, 12–I–1949, 1♀, 1♂, RMNH 15404–05, 2nd transect km 1.4, 12–I–1949, 2♂♂, RMNH 15406–07, 2nd transect km 0.5, 14–I–1949, 1♀, 3♂♂, RMNH 15408–11, 15–I–1949, 1♀, RMNH 15412, 2nd transect km 0.7, 15–I–1949, 2♂♂, RMNH 15413–14, 19–I–1949, 1♀, 1♂, RMNH 15415–16, 3rd transect km 5.9, 18–I–1949, all leg. Surinam Expedition 1948–49.

Encampment Moeroe Creek: 1♂, RMNH 15440, 5 km S., 24–VII–1968, 4♂♂, RMNH 15442, 21 km S., 25–VII–1968, both leg. M. S. Hoogmoed.

Troeli Creek: 1♀, RMNH 15137, 8–V–1968, leg. M. S. Hoogmoed; 1♀, RMNH 15439, 20–VI–1968, leg. H. A. M. de Kruijf.

Zanderij: 1♀, RMNH 16482, 24–IV–1971, leg. P. Staffeleu; 1♀, BM 1946.4.3.52, 7–VIII–1938, 1♀, BM 1946.4.3.53, 9–VIII–1938, 3♀♀, BM 1946.4.3.54–56, 5–X–1938, 1♀, BM 1946.4.3.57, 16–IX–1938, 1♀, BM 1946.4.3.58, 17–IX–1938, all leg. I. T. Sanderson; 5♀♀, FMNH 121075–79, 18–IX–1961, leg. H. A. Beatty; 1 ex., MCZ 66883.

Powakka: 1 ex., CM 44384, 12–V–1966, 1 ex., CM 44391, 13–V–1966, both leg. M. H. de la Fuente.

Railway km 62: 1♀, RMNH 15417, 26/IV–16/V–1949, leg. A. Brouwer (Surinam Expedition 1948–49).

Mama Creek: 1♀, RMNH 15430, 30–IV–1965, leg. G. F. Mees.

Berg en Dal: 2♀♀, RMNH 15389, 14–VII–1968, leg. M. S. Hoogmoed.

Railway km 91.5: 1♀, BM 1946.4.3.59, 20–X–1938, leg. I. T. Sanderson.

Compagnie Creek: 1♀, RMNH 15424, 23–V–1965, leg. G. F. Mees.

Brokobaka: 1♀, ZMA 13222, 31–V–1967, leg. H. Nijssen.

Brokopondo: 1♀, RMNH 15422, 7–IX–1964, 1♀, RMNH 15423, 4–IX–1964, both leg. M. Boeseman; 1♀, RMNH 15429, 31–VII–1965, leg. G. F. Mees; 1♀, ZMA 12691, 11–IX–1966, 2 ex., ZMA 12693–94, 12–IX–1966, 1♀, ZMA 12698, 14–IX–1966, 1♀, ZMA 12701, 25–IX–1966, 1♀, ZMA 12702, 26–IX–1966, 1♀, ZMA 12710, 31–X–1966, 1♀, ZMA 12711, 12–XI–1966, 1♀, ZMA 12968, 23–XI–1966, 1♀, ZMA 13198, 18–II–1967, 3♀♀, ZMA 13201, 3–III–1967, 7♀♀, ZMA 13207, 26/27–III–1967, 1♀, ZMA 13211, 20–IV–1967, 1♀, ZMA 13217, 6–V–1967, 1♀, ZMA no number, stomach contents of ZMA 12696 (*Oxybelis a. aeneus* (Wagler)), 13–IX–1966, all leg. H. Nijssen.

Brownsweg: 6♀♀, RMNH 15125, 14–VII–1968, leg. M. S. Hoogmoed; 1♀, RMNH 17269, 11–II–1972, leg. G. F. Mees.

Afobaka: 1♀, RMNH 15419, November 1963, leg. P. Leentvaar.

Between Kabel and Brownsweg: 1♀, RMNH 15431, 30–III–1965, 3♀♀, RMNH 15432, 15–IV–1965, both leg. G. F. Mees.

Kabel: 1♀, RMNH 15441, 31–VIII–1955, leg. P. Wagenaar Hummelinck.

Airstrip Tafel Mountain: 3♂♂, RMNH 17266, 4–X–1972, leg. G. F. Mees.

Airstrip Kayser Mountains: 1♀, RMNH 15208, 6–VIII–1968, 1♀, RMNH 15209, 31–VII–1968, both leg. M. S. Hoogmoed; 9♀♀, FMNH 128955, 128971–78, 1–II–1961, leg. H. A. Beatty; 2 juvs., PMYU. R. 5909–10, 24–XI–1961, leg. R. Freund.

DIAGNOSIS.—A medium-sized lizard with a cylindrical body, well developed legs and long (up to 2.5 times the snout-vent length) cylindrical tail. Tongue not retractile into a sheath at its base. Anterior nasal plates in contact. Dorsals and laterals small; ventrals larger, quadrangular, smooth, in eight longitudinal rows. Number of scales around midbody 103–128. Brown to grey above with nine or ten cream-coloured and black stripes. In males the flanks and belly are rust red, the head is sky-blue or grass-green, the tail above grass-green, below sky-blue; the flanks with two longitudinal rows of white spots.

DESCRIPTION.—Head pyramidal, slightly depressed; less than twice as long as wide; large, swollen, wider than the neck in males; small, slender, as wide as the neck in females and juveniles; snout pointed, rounded. Canthus rostralis very distinct, angular but rounded. Neck as wide as the body.

Tongue arrowhead-shaped, with a posterior emargination; not retractile into a basal sheath; covered with rhomboidal, imbricate, scale-like papillae in oblique rows; tip bifid, smooth. Anterior teeth conical; lateral teeth tricuspid, laterally compressed.

Rostral pentagonal; visible from above; slightly wider than deep. Nostril in an obliquely divided nasal; the anterior parts forming a short median suture behind the rostral. A transversely oval, hexagonal or rhomboidal frontonasal. A pair of irregularly trapezoid prefrontals, forming a short median suture. Frontal hexagonal, slightly longer than wide, wider anteriorly than posteriorly. A pair of irregularly trapezoid frontoparietals, forming a long median suture. Five, rarely six, small, irregularly shaped parietals in a transverse series. Two or four small occipitals. Back of the head covered with small granules. Supraoculars four or five; when five, the fourth and the fifth together are homologous to the fourth in specimens with four supraoculars; fourth (or fourth and fifth) smallest; second largest; separated from the supraciliaries by one (partly two) row(s) of granules, except the first supraocular which is in contact with the first supraciliary; last supraocular separated from the frontoparietals by a row of small scales. Loreal large, more or less trapezoid, in some specimens longitudinally divided. Preocular absent. A row of four (rarely five) suboculars, forming a suborbital ridge; the anterior subocular more or less in a preocular position, deeper than long; the third longest. An irregular row of four enlarged postoculars. Supraciliaries five to eight; the second one largest; the anterior two elongate, the posterior scales quadrangular. A row of enlarged supratemporals. Enlarged scales in front of the ear-opening; the central region of the temple with small, round to oval granules. Ear-opening large, surrounded by small scales, forming a smooth margin; external auditory meatus short, tympanum clearly visible. Seven (rarely six or eight) large supralabials (forming a lobed upper lip) followed by a variable number of small scales; five supralabials to below the centre of the eye; the third largest, the sixth smallest. Lower eyelid with a semitransparent disc of four to five palpebrals. Pupil kidney-shaped, with the indention ventrally.

Mental trapezoid with convex anterior margin. A heptagonal or pentagonal postmental. Four (rarely five) pairs of chinshields; only the members of the anterior pair in contact, largest; from the second pair on separated from the labials. Six to seven (rarely eight) enlarged anterior infralabials, followed by a variable number of small scales; third infralabial largest.

Gulars small, flat, rounded, juxtaposed, slightly larger in the anterior part of the throat, smaller in the posterior part which starts at the level of the ears. Scales in front of the prepectoral fold enlarged, slightly imbricate, smooth. Scales on the nape and on the side of the neck identical to the dorsal and lateral scales.

Dorsal and lateral scales granular, in indistinct transverse and oblique rows. Ventrals large, rectangular, wider than long, imbricate, smooth, in eight longitudinal and in 29–32 transverse rows. The anterior rows in the median interrupted by a small triangle of smaller scales. Total number of scales around midbody 103–128. Preanal plate with approximately six smooth, irregularly formed, juxtaposed, enlarged scales in two longitudinal rows, and a large number of small scales. Males with a preanal, sharply pointed spine on each side of the preanal plate.

Femoral pores 18–25 on each side, both in females and males. From these pores up to two may be in a preanal position. Pores larger in males than in females; surrounded by three to four scales, of which the one on the median side is largest.

Scales on the tail large, rectangular, obliquely keeled, slightly mucronate; imbricate, in transverse rows. The keels may or may not form longitudinal ridges. Scales under the tail identical but narrower. Scales on the regenerated part of the tail small, irregularly formed, keeled, not in transverse rows. Tail round in cross-section.

Scales on the upper and anterior surfaces of the upper arm, on the anterior and lower surfaces of the forearm, on the anterior and lower surfaces of the thighs and on the lower and posterior surfaces of the lower legs large, smooth, more or less hexagonal, imbricate; on the lower and posterior surfaces of the upper arm, on the posterior and upper surfaces of the forearm, on the posterior and upper surface of the thighs and on the upper and anterior surfaces of the lower legs small, granular, juxtaposed. Sixteen to 20 single, slightly tubercular lamellae under the fourth finger; 30–37 single, proximally distinctly tubercular, distally less tubercular, lamellae under the fourth finger, divided under the articulations. Approximately five triangular, strongly enlarged and tubercular scales under the base of the third toe. Fingers and toes compressed. Palms and soles with small, irregular, juxtaposed, flat scales. Upper surface of hands and feet with large, imbricate flat scales.

In life, the back in females is brown with black and cream-coloured longitudinal stripes, only the anterior part of the dorsolateral stripe is grass-green. Side of the head orange. Chin, throat, lower surface of forelimbs and anterior part of belly white, posterior part of belly, lower surface of hind limbs and tail greenish yellow. Hind limbs black, dark-brown or greyish brown with white spots. Juveniles have this same type of colouration. In life the vertebral region in males is light-brown, on both sides there are two longitudinal black lines surrounded by grass-green. Tail and limbs grass-green, the hind limbs with lighter spots. Flanks rust-red, with three longitudinal series of white spots of which the lower one is on the outer ventrals, groins grass-green. Sides of head grass-green or sky-blue with lighter spots. Chin, throat, chest, lower surface of forelimbs, hind limbs and tail sky-blue. Belly rust-red, connecting the rust-red zones on both flanks. Grass green of the groins just visible on the lateral parts of the posterior part of the belly, which, in the centre, is white.

In preservative the back and flanks of juveniles, young females and young males are longitudinally striped. Individuals obtained from west of the Suriname River have nine light lines, with one exception (RMNH 15137 from Troeli Creek) in which the back is uniform, the flanks with six lines. Along the coast east of the Suriname River individuals have ten light lines, with one exception (RMNH 15418 k from Galibi) which has only nine. The vertebral light line (in the case of nine lines) or the two paravertebral light lines (in the case of ten lines) light-brown, the other light lines white, separated by black bands on the back, on the flanks in juveniles by black bands, in adults by grey bands. The dorsolateral light line starts at the posterior corner of the eye, where it arises from two short lines on upper and lower eyelid, and continues on to the tail. The vertebral (respectively paravertebral) line(s) arise(s) at the back of the head and continue(s) to the base of the tail (where they coalesce). The line between (para) vertebral line and dorsolateral line arises at the posterior margin of the most lateral parietal and continues on to the proximal part of the tail where it coalesces with its fellow from the other side. The upper lateral line starts on the lower eyelid, passes over the ear-opening and continues to the groin; the lower lateral line from the posterior margin of the ear-opening over the forelimb to the groin. In old females and in adult males this striped pattern has undergone reduction. In old females the lower lateral line is vague or has disappeared and in its place a series of small white spots is present. In adult males the two lateral lines have disappeared and two series of large white spots are present. In adult males the back is dark-brown with three (respectively four) light-brown stripes, dorsolateral region bluish with or without two light lines, flanks greyish brown with pale blue spots, head bluish with pale blue spots on the sides. Upper surface of hind limbs and of forearm light-brown via dark--brown to black in females and juveniles, with scattered white spots. Back of the thighs with a horizontal series of white spots, continued on the tail as a white stripe in all specimens. In males the colour of the limbs is light-brown. Ventral parts white, immaculate. Tail above light-brown in females, bluish in males, with white dorsolateral lines on the proximal parts; lower surface of tail bluish in both sexes.

A medium-sized lizard, maximum snout-vent length in females 72 mm, in males 89 mm. Hind limb approximately twice as long as forelimb. When laid along the body there is a considerable overlap of forelimb and hind limb. The undamaged tail is 1.9–2.4 times as long as the snout-vent length in females, 2.2–2.5 times in males.

HABITAT.—Open, sunny places, e.g., beach plains, ridges, savannahs, gardens and forest edges. The habitat on Eilanti and Bigisanti beach is very similar and is formed by the beach plain from the high tide mark to and including the first ridge (HOOGMOED, 1969a). This beach plain is sparsely covered with *Ipomoea pes-caprae* and *Canavalia maritima*, which in some places form clusters from where creepers extend in all directions. The ridge closing off this beach plain towards the landside is covered with parwa trees (*Avicennia*) and shrubs with relatively few weeds among them. The savannahs inhabited by this species are in the coastal savannah belt and near the airstrips Kayser Mountains and Tafel Mountain. The first type of savannah is illustrated by HOOGMOED (1969a). The second type (Kayser Mountains), completely enclosed by rain-forest (HOOGMOED, 1969a), has a dense cover of grasses with low bushes and *Mauritia* palms and some protruding flat rocks; in the rainy season these savannahs are very wet, the rocks being the places where the lizards are found. The species has not been found on the Sipaliwini savannah. In Paramaribo and Moengo *C. l. lemniscatus* (Linnaeus) inhabits gardens and roadsides. When present along forest edges it is confined to roads (Brokopondo, Afobaka, Brownsweg, Berg en Dal) and this might represent a recent dispersion influenced by human activity. Another example of this is the former presence of *C. l. lemniscatus* (Linnaeus) along the (now submerged) railway from Brownsweg to Kabel, a traject which is essentially rain-forest country. In the Amazon valley the species seems to be strictly perianthropic (VANZOLINI, 1970b).

NATURAL HISTORY.—This species is a sun-loving, ground-dwelling lizard, which seeks refuge in burrows when pursued. On Eilanti and Bigisanti beach the lizards run with short quick dashes from one cluster of creepers to another until they finally disappear into a burrow. Elsewhere they have the same type of running, but because of lack of cover may stop in the open. During these short pauses between the subsequent dashes the lizards frantically wave their hands, alternating left and right hand. After a while they run for some distance and the entire performance is repeated again. According to BEEBE (1945) these spurts may be bipedal. Both on beaches and in the gardens the lizards were very numerous. Copulation was observed on Bigisanti beach in late April and was of the normal lizard type, the male biting the female in one of the flanks, throwing one of his hind limbs over her back and turning his cloaca towards hers. Copulation was so intense that when copulating females were collected by means of a noose the males would stay attached, even in mid-air and could be collected together with the female. A pregnant female is known from June, containing two mature eggs. Recently hatched juveniles from April, October and November. BEEBE found that in Kartabo breeding took place throughout the year but seemed to reach peaks during or at the end of the rainy seasons. He also found that courting was especially noticeable in May and September (BEEBE, 1945). My scanty dates agree very well with these findings.

Only in some localities, e.g., Lelydorp, Tibiti, between Encampments Goliath and Moeroe Creek, airstrip Tafel Mountain and in a zone along the coast

(Matapica, Bigisanti, Eilanti, Langamankondre) the populations comprise both males and females. All other localities lack males (see below).

All the stomachs examined were crammed with food and took up a considerable part of the available space in the belly. The contents were very variable, including many species of insects and spiders, which could be expected after a look at BEEBE's (1925, 1945) list of food items taken at Kartabo. LEON *et al.* (1970) studied the diet of three species of lizards in Venezuela, one of which was *C. lemniscatus* (Linnaeus). They found that the majority of its food consisted of Coleoptera, Lepidoptera and Hymenoptera. From their data they conclude that chemoreception is important in locating prey. One specimen (ZMA no number) was regurgitated by an *Oxybelis a. aeneus* (Wagler), after this was collected. BEEBE (1925) also reported it from an *Oxybelis* stomach. He also found that *Cnemidophorus* was eaten by the hawks *Leucopternis* and *Gampsonyx* and by the lizard *Tupinambis*. MEES (personal communication) observed Common Egrets (*Egretta alba* (Linnaeus)) at Bigisanti and Krofajapassie eating *C. l. lemniscatus* (Linnaeus).

RANGE.—From northern Brazil, the Guianas, Venezuela and N.W. Colombia north to Guatemala. In Surinam the species is essentially restricted to the coastal area, to and including the coastal belt of savannahs and only ventures south in two places, viz., the airstrips Kayser Mountains and Tafel Mountains (fig. 50).

REMARKS.—The absence of males in a number of localities is not due to the sampling, but is real. Because of the sexual dimorphism the larger and more colourful males are very obvious and when present must be seen or collected, also by a non-herpetologist. There is no difference in shyness between the sexes. To make quite sure about this matter, I visited the localities, from which only females were known, during my 1968 stay in Surinam and I am convinced males really are absent there. Thus the species in Surinam shows normal bisexual populations and all-female (parthenogenetic) communities.

Very recently VANZOLINI (1970b) reported this same phenomenon for the Amazon valley where most populations are unisexual and in one place there is even an indication that a bisexual population has been replaced by a unisexual one within three years. VANZOLINI supposes that the species has spread into the Amazon valley very recently from the east. Unfortunately no data on the sex ratio in Amapá and French Guiana are available that might indicate whether this spreading of all-female populations from the mouth of the Amazon towards the west also has had an influence in that region. Several remarks can be made concerning the distribution of the all-female populations in Surinam.

1. As in Brazil *C. l. lemniscatus* (Linnaeus) occupies peri-anthropic environments, but not exclusively so, as it is also present on the beach and on inland savannahs. Between at least part of the populations there is little or no geneflow at all. For instance, the bisexual populations along the east coast have contact among themselves but are well separated from those in the rest of Surinam. The same is true for the all-female population on the airstrip Kayser Mountains.

2. The occurrence of an all-female population on the airstrip Kayser Mountains and of a bisexual population on the airstrip Tafel Mountain, seems to suggest development of unisexuality on the spot and not for the recent spreading of a single original strain. The occurrence of *C. l. lemniscatus* (Linnaeus) on the airstrip Kayser Mountains in my opinion is natural and not a result of the human influence there since 1959, when the airstrip was constructed. Reasons for this assumption are that the savannahs present near the airstrip are old and natural, not caused by the recent human influence. The herpetofauna of these savannahs is typical and agrees with that of the coastal savannahs (*Kentropyx s. striatus* (Daudin), *Anolis a. auratus* Daudin, *Leptodactylus* cf. *fuscus* (Schneider)). None of these species is found in the rain-forest, which forms a virtually continuous barrier of about 250 km between the airstrip and the coastal savannah belt. I think it would be too improbable to assume that the total herpetofauna of the coastal savannahs would have migrated to the airstrip Kayser Mountains in nine years, which could have been possible for one species. This leaves the conclusion that the herpetofauna of the airstrip Kayser Mountains (including *C. l. lemniscatus* (Linnaeus)) has been isolated there for a long time and that the unisexuality of the *C. l. lemniscatus* (Linnaeus) population there at least has developed on the spot.

3. The localities with bi- and unisexual populations are spread somewhat randomly throughout Surinam and include some doubtful ones. From Troeli Creek only two females are present and no more data from that locality are available; this might well be a bisexual population. The statement that Lelydorp has a bisexual population, should be regarded with reserve, because the sample on which this was based (ZMA 10733, 12415–24) comes from an animal dealer and there is no guarantee that the animals really originated from that place.

4. Indications of a recent shift from bisexual to unisexual populations have not been found. Several populations (two bisexual, five unisexual) were sampled regularly in the last 30 years, but nowhere was a change noticed. Reliable data are available for Paramaribo (1938, 1950, 1952, 1962, 1963, 1965, 1966, 1968, 1969), Brownsweg (1968, 1972), Brokopondo (1964–1967), Zanderij (1938, 1961), Airstrip Kayser Mountains (1961, 1968), Galibi (1948, 1963) and Langamankondre (1948, 1968).

5. From the aforegoing it will be clear that a lot of interesting research can still be done to solve this problem.

Because of lack of material from French Guiana it is not known how important the difference in number of longitudinal light stripes is. Although the number of ten stripes seems to occur elsewhere (Burt, 1931), I do not know of other populations that exclusively show this number, the normal number being nine. There is another difference between the specimens from the eastern coastal area and the rest of Surinam, namely the total number of scales around midbody in the eastern coastal specimens is slightly higher (117–123) than in specimens from the rest of Surinam (103–117). So, when specimens from French Guiana become available they might give a clue as to whether the eastern coastal Surinam population and other populations further east are subspecifically distinct from the populations west of the Surinam River, or not. If so, the populations west of the Surinam River would have to be renamed.

VERNACULAR NAME.—Wai wai hannoe (= waving hand) because of its peculiar habit of waving its hands. It is also known as 'kamra wientje' or 'kamerawientje' which means chambermaid and indicates that this species may be encountered in or near houses. This last name is also used for *Mabuya m. mabouya* (Lacépède).

Genus **Gymnophthalmus** Merrem

DIAGNOSIS.—Small, slender lizards with cylindrical bodies and tails. Legs short, inner finger absent. Lower eyelid grown over the eye, immovable, transparent. Body covered with large, cycloid, smooth, imbricate scales, 13–17 scales around midbody. Supraoculars one or two, frontoparietals absent, interparietal very large; frontal in contact with internasal or not; prefrontals present. Ear-opening small. Femoral pores present in males. Pupil round. The genus ranges from Central America south to northern Argentina and also is present on some of the Antilles. Six species, of which four are present in South America, one in Surinam.

Gymnophthalmus underwoodi Grant (figs. 52, 53; pl. 30; table 18)

'Kleine skink (vermoedelijk een Mabuia)' Menne, 1954b: 64.
Gymnophthalmus underwoodi Grant, 1958: 228.
Gymnophthalmus lineatus: J. A. Peters & Donoso-Barros, 1970: 137 (partly).

TYPE-LOCALITY.—Barbados.

MATERIAL. – Surinam.

Paramaribo: 1♀, RMNH 13421, June 1965, leg. P. J. M. Maas; 6♀♀, RMNH 15030, 18–V–1968, 1♀, RMNH 15233, 1–VI–1968, 2♀♀, RMNH 15235, 5–V–1968, 7♀♀, RMNH 15245, 3–V–1968, all leg. M. S. Hoogmoed; 1 ex., RMNH 15462, August 1962; 2♀♀, USNM 158973–74, 12–VIII–1961, leg. T. H. Lewis; 2♀♀, CM 44403–04, 23–V–1966, 2♀♀, CM 44405–06, 25–IX–1966, 1♀, CM 44408, 14/15–VII–1966, 1♀, CM 44409, 10–X–1966, all leg. M. H. de la Fuente; 1♀, FMNH 121085, 20–IX–1960, leg. H. A. Beatty.

Wageningen: 1♀, RMNH 16446, 13–III–1971, leg. D. C. Geijskes; 1♀, RMNH 16448, 10–III–1971, leg. P. Staffeleu.

Moengo: 2♀♀, RMNH 15381–82, 12–VI–1969, 1♀, RMNH 15385, 13–VI–1969, 1♀, RMNH 15386, all leg. H. M. van Meeuwen.

Brokopondo: 2♀♀, ZMA 12647–48, 28–IV–1966, 1♀, ZMA 12650, 15–V–1966, 1♀, ZMA 13206, 24–III–1967, all leg. H. Nijssen.

Voltz Mountain Reserve: 1♀, RMNH 15218, Voltz Mountain, 5–XI–1968, 1♀, RMNH 15234, Raleigh Cataracts, 7–XI–1968, both leg. M. S. Hoogmoed; 1♀, AMNH 108784, south slope Voltz Mountain, 22–II–1972, leg. C. W. Myers & J. Daly.

Guyana.

Georgetown: 3♀♀, BM 1929.8.4.6–8, leg. L. D. Cleare; 1♀, MCZ 61030, leg. G. Underwood.

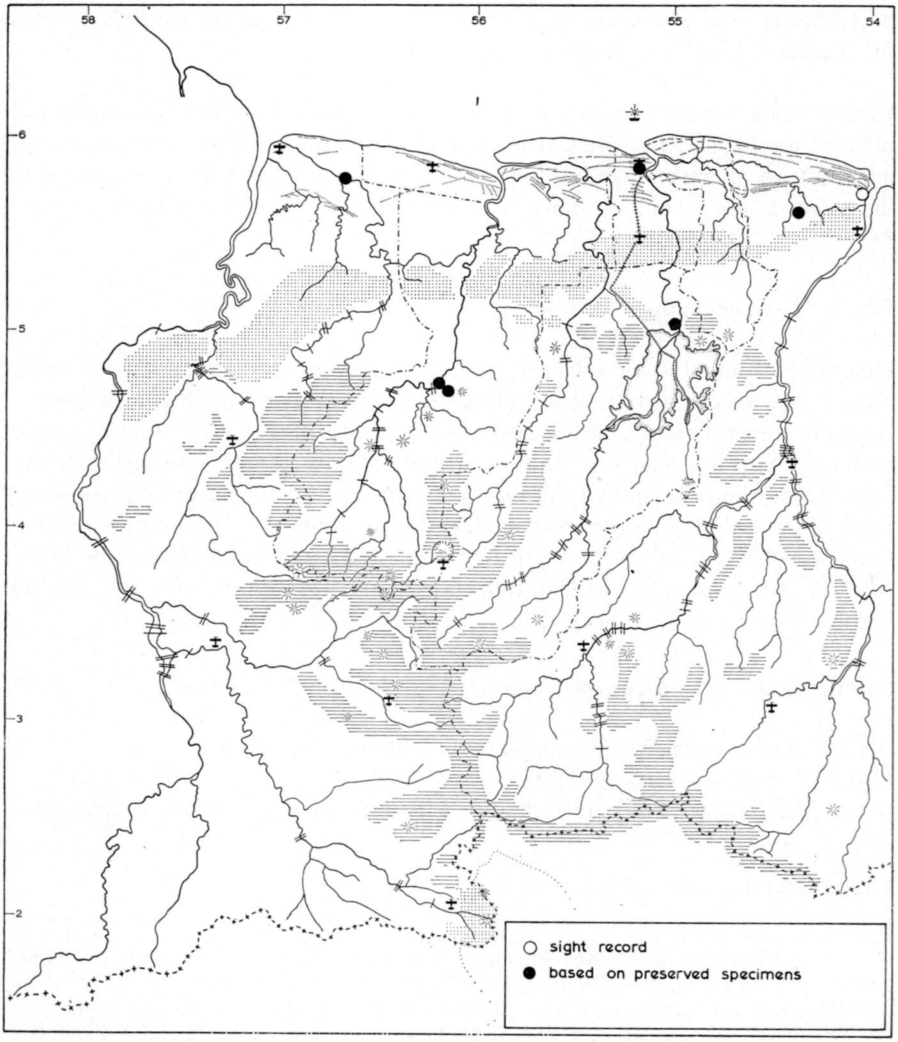

Fig. 52. Distribution of *Gymnophthalmus underwoodi* Grant in Surinam.

Penal Settlement, Mazaruni: 2♀♀, BM 1934.11.1.104–105, leg. G. S. Carter.
Demerara: 1♀, 3 ex., BM 89.9.30.9–12, leg. J. J. Quelch.
Marudi: 3 ex., AMNH 61332–34, 10–IV–1938, leg. R. Snedigar.

DIAGNOSIS.—A small, slender lizard with cylindrical body and tail. Legs short, inner finger absent. Lower eyelid grown over the eye, immovable, transparent. Body covered with large, cycloid, smooth, imbricate scales, 13 around midbody. Three longitudinal rows of dorsals. One large supraocular; frontoparietals absent; a large interparietal. Shining bronze above, slightly lighter, incomplete dorsolateral stripes; flanks black, belly greenish white.

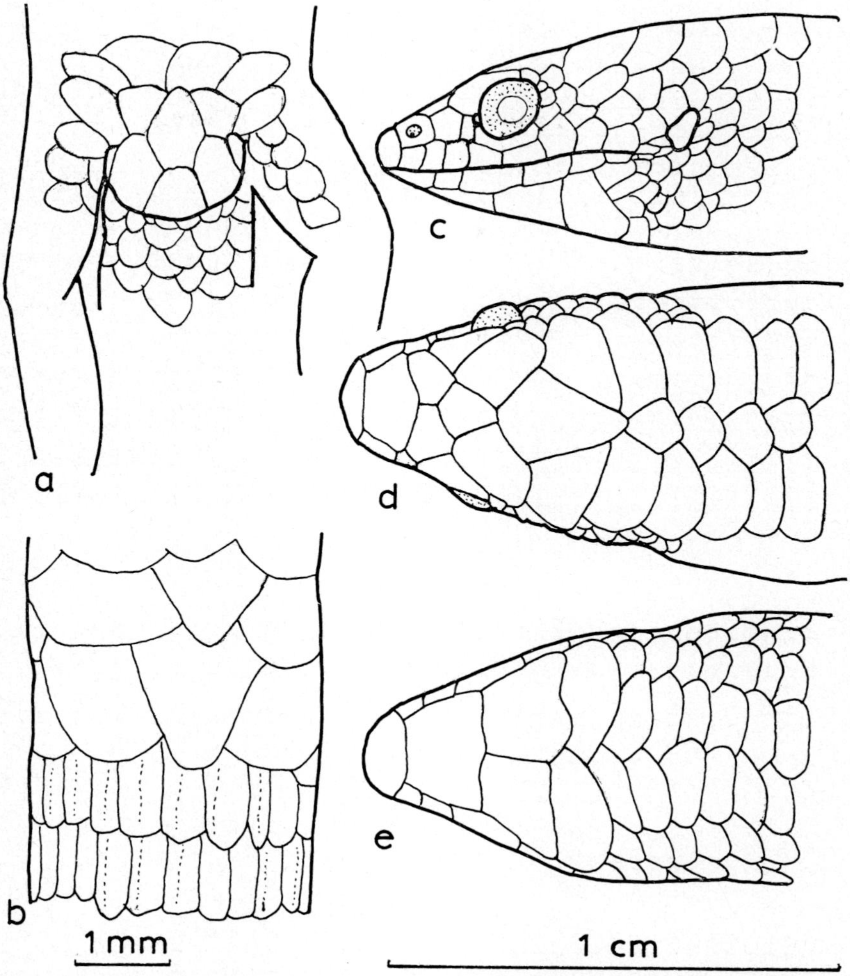

Fig. 53. Gymnophthalmus underwoodi Grant; a: anal region; b: dorsal surface of tail at a former breaking-point; c, d, e: lateral, dorsal and ventral views of the head of ♀, RMNH 15245 c, Paramaribo. a, c, d and e are on the same scale; b is on another scale.

DESCRIPTION.—Head small, pyramidal; 1.5 times as long as wide; slightly depressed; as wide as the neck, which is as wide as the body; snout obtuse.

Tongue lanceolate, without emargination; covered with rhomboidal, scale-like, imbricate papillae in oblique rows; tip bifid, smooth. Teeth conical.

Rostral hexagonal, visible from above, twice as wide as deep. A large, heptagonal frontonasal separating the nasals. A pair of irregularly hexagonal prefrontals, forming a short median suture. Frontal rhomboidal, smaller than a prefrontal. Frontoparietals absent. An enormous, elongate interparietal. Parietals irregularly pentagonal, as wide as long, much smaller than the interparietal. A row of three occipitals; the central one small, rhomboidal; the

Table 18. Counts and measurements of *Gymnophthalmus underwoodi* Grant. For explanation see table 3.

Reg. no.	A	B	C	D	E	F	G	H	I	J	K	L	O	P	Q	R	S	T	U	V	W	X	Y
RMNH																							
13421	♀	34	30(10)	6.7	3.7	2.6			0	13	32	25	8-8	6-6	2-2	1-1	1-1	5+1	11	2	13-13	?-17	3
15030a	♀	42	57(37)	7.9	5.1	3.8			0	13	35	24	8-8	6-5	2-2	1-1	1-1	5+1	10	2	14-14	17-17	3
b	♀	42	78	8.2	5.1	3.7			0	13	34	24	8-8	6-6	2-2	1-1	1-1	5+1	9	2	13-14	17-17	3
c	♀	36	62	6.9	4.3	3.1			0	13	34	23	8-8	6-5	2-2	1-1	1-1	5+1	10	2	13-14	17-17	3
d	♀	40	40(22)	7.8	4.6	3.3			0	13	33	22	8-8	6-6	2-2	1-1	1-1	5+1	9	2	14-14	18-17	3
e	♀	34	55(48)	6.9	4.3	2.9			0	13	35	24	8-8	6-5	2-2	1-1	1-1	5+1	9	2	14-14	18-17	3
f	♀	31	53	6.4	3.8	2.8			0	13	34	23	8-8	6-6	2-2	1-1	1-1	5+1	9	2	14-14	17-17	3
15218	♀	43	74	8.1	5.2	3.6	9	14	0	13	33	22	8-8	4-5	2-2	1-1	1-2	5+1	10	3-2	15-15	19-17	3
15233	♀	39	58(42)	7.4	4.8	3.1	8	13	0	13	34	23	8-8	6-6	2-2	1-1	1-1	5+1	9	2	13-13	17-17	3
15234	♀	29	24(24)	6.0	3.7	3.1	5	10	0	13	33	24	8-8	6-6	2-2	1-1	2-2	5+1	9	2	13-13	16-17	3
15235a	♀	42	40(23)	8.1	5.2	4.0	8	14	0	13	34	23	8-8	6-6	2-2	1-1	1-1	5+1	10	2	13-14	18-17	3
b	♀	39	5(5)	7.7	4.8	3.7	8	13	0	13	33	23	8-8	6-5	2-2	1-1	1-1	5+1	10	2	14-13	17-?	3
15245a	♀	41	61(39)	7.8	4.7	3.9			0	13	34	23	8-8	6-6	2-2	1-1	1-1	5+1	9	2	14-14	17-17	3
b	♀	37	63	7.3	4.4	3.7			0	13	35	24	8-8	6-6	2-2	1-1	1-1	5+1	9	2	14-13	17-17	3
c	♀	38	38(9)	7.4	5.0	3.8			0	13	33	23	8-8	6-6	2-2	1-1	1-1	5+1	10	2	14-13	17-16	3
d	♀	32	55	6.4	4.3	2.8			0	13	33	23	8-8	6-6	2-2	1-1	1-1	5+1	9	2	14-14	17-18	3
e	♀	31	13(11)	6.3	4.3	2.6			0	13	34	22	8-8	6-6	2-2	1-1	1-1	5+1	9	2	14-15	18-17	3
f	♀	31	35(19)	6.2	3.8	2.5			0	13	33	23	8-8	6-6	2-2	1-1	1-1	5+1	9	2	14-14	17-17	3
g	♀	30	33(14)	6.6	4.1	2.5			0	13	33	23	8-8	6-6	2-2	1-1	1-1	5+1	10	2	14-13	16-18	3
16446	♀	26	46	5.8	3.9	2.5			0	13	29	23	8-8	5-6	2-2	1-1	1-1	5+1	9	2	13-13	17-18	3
16448	♀	42	40(15)	7.8	5.5	3.3			0	13	34	25	8-8	6-6	2-2	1-1	1-1	5+1	9	2	13-15	17-17	3
16500	♀	36	47(27)	7.2	4.3	3.1			0	13	34	26	8-8	6-6	2-2	1-1	1-1	5+1	10	2	13-13	16-16	3
ZMA																							
12647	♀	44	46(16)	8.4	5.4	3.8	8	13	0	13	34	25	8-8	6-5	2-2	1-1	1-1	5+1	9		13-14	16-17	3
12648	♀	38	66.5(56)	7.4	4.8	3.4	8.5	12.5	0	13	33	25	8-8	6-6	2-2	1-1	1-1	5+1	8	2	14-15	17-17	3
12650	♀	40	73	7.8	4.9	3.4	6	14	0	13	32	24	8-8	6-5	2-2	1-1	1-1	5+1	8	2	13-13	17-18	3
13206	♀	36	58(43)	7.4	4.4	3.2	8	13	0	13	34	23	8-8	6-6	2-2	1-1	1-1	5+1	11	2	14-14	16-17	3
FMNH																							
121085	♀	42	44(16)	8.0	4.8	3.5			0	13	32	24	8-8	6-7	2-2	1-1	1-1	5+1	10	2	14-13	16-17	3
CM																							
44403		38	27(-)	8.0	4.8	3.4				13	33	24	8-8	6-6	2-2	1-1	1-1	5+1	10	2	?	?	3
44405		36	66	7.5	4.5	3.2				13	33	22	8-8	6-6	2-2	1-1	1-1	5+1	10	2	15-15	17-17	3
44406		38	64	7.6	4.7	3.2				13	33	23	8-8	6-6	2-2	1-1	1-1	5+1	11	2	16-16	17-16	3
44408		29	16(16)	6.2	4.1	2.9				13	33	22	8-8	6-6	2-2	1-1	1-1	5+1	10	2	15-15	18-19	3
44409		35	59	7.1	4.3	3.1					33	25	8-8	6-6	2-2	1-1	1-1	5+1	10	2	14-13	17-18	3
BM																							
89.9.30.9	♀	37	17(7)	7.4	4.5	3.2			0	13	33	22	8-8	6-6	2-2	1-1	1-1	5	10	2	?-13	17-?	3
.10		35	10(10)	7.2	4.8	2.9			0	13	33	20	8-8	6-6	2-2	1-1	1-1	3	11	2	12-?	17-17	3
.11		29	34(15)	6.2	3.8	2.7			0	13	33	21	8-8	6-6	2-2	1-1	1-1	5	10	2	14-13	17-17	3
.12		20	12(12)	4.5	3.0	2.2			0	13	32	22	8-8	6-6	2-2	1-1	1-1	3	9	2	12-12	17-16	3
1929.8.4.6	♀	40	26(5)	7.5	5.3	3.4			0	13	34	?	8-8	6-6	2-2	1-1	1-1	5	9	2	15-15	15-17	3
.7	♀	37	21(21)	7.3	4.8	2.8			0	13	32	26	8-8	6-6	2-2	1-1	1-1	5	10	2	14-12	16-16	3
.8	♀	34	5(5)	7.0	4.7	3.7			0	13	32	22	8-8	6-5	2-2	1-1	1-1	5	9	2	?-13	17-18	3
1934.11.1.104	♀	35	54(32)	7.3	4.5	3.3			0	13	32	24	8-8	6-5	2-2	1-1	1-1	5	11	2	14-13	17-18	3
.105	♀	35	28(22)	7.3	4.6	3.3			0	13	33	26	8-8	6-6	2-1	1-1	1-1	5	10	2	?-14	17-?	3

lateral ones hexagonal, transversely enlarged. One large supraocular in contact with the supraciliaries. Nostril in an undivided nasal. Loreal square. A small frenoocular, deeper than long. One small preocular. One (rarely two) large subocular(s). Two or three (rarely one) postoculars. Two supraciliaries, the anterior one very large, the second minute. Temporal region with large, imbricate scales. Ear-opening small, triangular, one tip pointing ventrally; surrounded dorsally by large scales, the other two sides by small scales, forming a smooth margin; external auditory meatus short, tympanum visible. Supralabials eight; the fifth below the eye; the sixth much deeper than the rest. Lower eyelid transparent, grown over the eye and immovable. Pupil round.

Mental trapezoid, with a convex anterior margin. A large pentagonal postmental. Two pairs of imbricate chinshields, forming median sutures; the anterior pair longer than the second; the second pair wider than the anterior; in contact with the infralabials. Five or six (rarely four) infralabials, the third largest.

Headscales with a row of small pits along their margins and scattered pits; most of them on the anterior dorsal, the lateral and the anterior ventral head scales. All scales of head, body and limbs smooth.

Nine to eleven rows of transversely enlarged, imbricate gulars from the chinshields to the pectorals, no collar. Scales on the nape indistinguishable from the dorsals. Scales on the side of the neck small, rounded, imbricate, in longitudinal and oblique rows.

Dorsals imbricate, in oblique and in three longitudinal rows, one vertebral row of small, rhomboidal scales and two lateral rows of transversely enlarged, hexagonal scales, 29–35 vertebral scales between the interparietal and the posterior surface of the thighs. Laterals imbricate, slightly smaller than the dorsals, hexagonal to rounded, in oblique and in three longitudinal rows. Ventrals imbricate, as large as the transversely enlarged dorsals, hexagonal, in oblique and in four longitudinal rows, 22–26 transverse rows between the gulars and the preanals. Thirteen scales around midbody. Five peripheral and one central preanal scale; the two lateral scales very small.

Males seem to be absent in this species. In females preanal and femoral pores are absent.

Scales on and under the tail smaller than the dorsals, imbricate, rounded, in longitudinal and in oblique rows, smooth in the anterior part, with low, sharp keels forming ten longitudinal ridges in the posterior part. Scales on and under the regenerated part of the tail elongate, hexagonal, keeled, in transverse and in longitudinal rows.

Forelimbs covered with imbricate, rounded scales, larger on the anterior surface than elsewhere, anterior and lower surfaces of the hind limbs with large, rounded, imbricate scales; posterior and upper surfaces of the hind limbs with small, slightly imbricate, granular scales. Four fingers, the inner finger absent; 13–16 single, obtusely keeled lamellae forming a serrated keel under the fourth finger; 16–19 (mostly 17–18) lamellae under the fourth toe. Palms and soles with small, juxtaposed, tubercular, granular scales.

Colour in life shining bronze above, with slightly lighter dorsolateral stripes, flanks black, belly shining greenish white.

In preservative the back is light- to dark-brown. Dorsolateral stripe creamy, from the snout over the eye to the back where it fades out about halfway between the forelimbs and the hind limbs. Flanks black. Scales on the underside of the body with dark-grey centres and white rims; tail lighter, sometimes immaculately white. A dorsolateral series of small, cream-coloured spots on the tail. Tongue entirely grey.

A small, elongate lizard, maximum snout-vent length in females 44 mm. Hind limb 1.5 times as long as forelimb. When laid along the body the toes and fingers are widely separated. The undamaged tail is 1.7–1.8 times as long as the snout-vent length.

HABITAT.—In open, sunny places where it lives between leaf-litter or grass. In Paramaribo, Wageningen, Moengo and Brokopondo it was found in gardens and even in houses. On the Voltz Mountain it lived in an open savannah-type forest, near the Raleigh Cataracts on a large open clearing nearly devoid of vegetation and in Paramaribo it was also taken in a cocoa-grove. In Langamankondre I saw a specimen amongst dense herbs along the river beach.

NATURAL HISTORY.—The species is diurnal and has been seen basking several times. No data on breeding are available. This is another parthenogenetic species.

RANGE.—Surinam, Guyana, Trinidad, Tobago and Barbados. In Surinam the species is probably present throughout the country, but probably because of its small size material is lacking from the deep interior (fig. 52).

REMARKS.—There has been much confusion about the status of this species in the Guianas. PARKER (1935b) reported it as *G. laevicaudus* (Cope) from Guyana, at the same time suggesting that it might be a colour variant of *G. lineatus* (Linnaeus). The next author dealing with this subject, STUART (1939), held the opinion that *G. lineatus* (Linnaeus) and *G. speciosus* (Hallowell) were distinct species and that the Guyanese specimens belonged to this last mentioned species (of which *G. laevicaudus* (Cope) is a synonym). In STUART's view *G. lineatus* (Linnaeus) is confined to 'northeastern South America from western Venezuela to Brazil and the Dutch Leeward Islands' and *G. speciosus* (Hallowell) is widely ranging from 'Central America into northern South America eastward to British Guiana and south to Chile'. GRANT (1958) described a new species, *G. underwoodi*, from Barbados. UNDERWOOD (1962) and THOMAS (1965) demonstrated that *G. underwoodi* Grant is also present on Trinidad. THOMAS (1965) still considered the Guyanese specimens as belonging to *G. speciosus* (Hallowell), but at the same time suggested that the Guyanese specimens might prove to be identical with *G. underwoodi* Grant. PETERS & DONOSO-BARROS (1970) listed *G. lineatus* (Linnaeus) from the Dutch Leeward Islands and Surinam, *G. underwoodi* Grant from Barbados, Tobago and Trinidad, and *G. s. speciosus* (Hallowell) from Central America, northern South America and Margarita Island. After investigating large series from both Surinam and Guyana I have come to the conclusion that these specimens belong to *G. underwoodi* Grant with which they agree in all scale characters and in colour pattern. They are very different from *G. lineatus* (Linnaeus) in having only 13, instead of 15, scales around the body and in colour pattern. They differ from *G. speciosus* (Hallowell) in having a lower number of transverse rows of ventral scales and by the absence of an orange-red tail. From both species it differs in that no males are present. When describing *G. underwoodi*, Grant doubted whether his series really consisted of females only and thought it more likely that males were present but did not have femoral pores. THOMAS (1965) dissected approximately 60 individuals of this species and established beyond doubt that they were all females. He comes to the conclusion that *G. underwoodi* is another lizard with 'unisexual populations'. All Surinam and Guyanese specimens examined are also females, thus males seemingly are also lacking on the mainland. This is another fact in favour of the view that the Guyanese and Surinam specimens belong to *G. underwoodi* Grant. The conclusion is that *G. s. speciosus* (Hallowell) occurs in Central America and adjacent northern South America, probably not further east than Margarita Island, *G. lineatus* (Linnaeus) occurs on the Dutch Leeward Islands and its presence on the adjacent mainland (Venezuela) remains doubtful, *G. underwoodi* Grant occurs on Barbados, Tobago, Trinidad, and in Guyana and Surinam; very probably it occurs also in French Guiana and the eastern part of Venezuela.

Genus **Iphisa** Gray

DIAGNOSIS.—See diagnosis of species. Only a single species, which is present throughout the Amazon basin.

Iphisa elegans Gray (figs. 54, 55; pl. 30; table 19)

Iphisa elegans Gray, 1851, 39; D'Audretsch, 1968a: 51.

TYPE-LOCALITY.—Pará (= Belém do Pará), Brazil.

MATERIAL. – Surinam.

Mooi Wane Creek: 1♂, RMNH 13416, 25–IX–1948, leg. P. H. Creutzberg (Surinam Expedition 1948–49).

Railway km 121: 1♂, RMNH 13420, July 1949, leg. C. Bleys.

Nassau Mountains: 2♂♂, RMNH 13417, Base Bivouac, 14–II–1949, 1♀, RMNH 13418, Base Bivouac, 24–II–1949, 1♀, RMNH 13419, km 0.1, 28–II–1949, all leg. Surinam Expedition 1948–49.

Airstrip Paloemeu: 1♂, RMNH 15240, S., 23–V–1968, 1 ex., RMNH no number, from stomach of *Kentropyx calcaratus* Spix (RMNH 15297b), S., 26–V–1968, both leg. M. S. Hoogmoed.

Guyana.

Kartabo: 1♂, AMNH 21294, 1920, leg. W. Beebe.

Chenapoon River, Upper Potaro: 1♀, 1♂, BM 1905.11.1.12–13, leg. C. Bovallius.

Kuyuwini landing: 1♂, AMNH 70624, 20/30–XI–1937, leg. R. Snedigar.

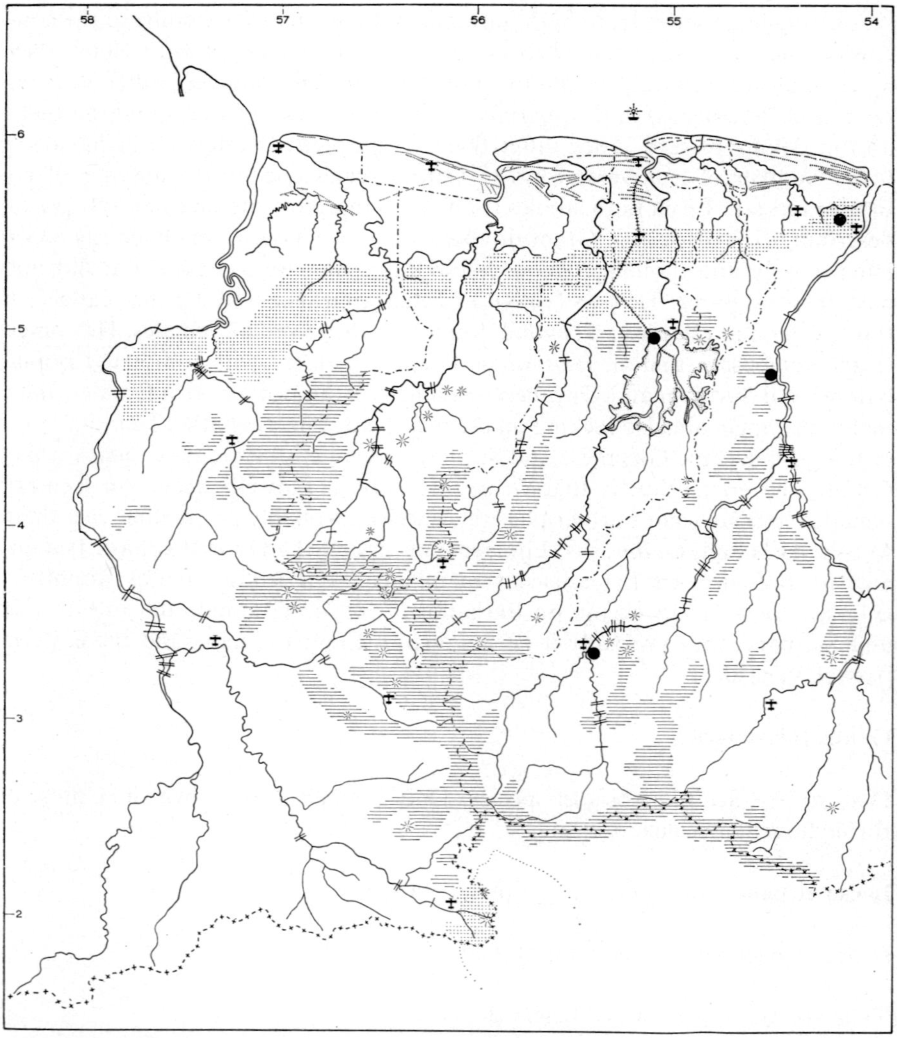

Fig. 54. Distribution of *Iphisa elegans* Gray in Surinam.

French Guiana.

Along trail from Sophie to La Grève: 1♂, MCZ 77500, 10–VIII–1963, leg. T. Monath.

Tumuc Humac Mountains: 1♀, MP 99–74, Upper Lunier River, leg. Geay.

Brazil.

Pará: 1♂, BM 1946.9.1.1 (holotype).

DIAGNOSIS.—A small, slender lizard with cylindrical body and tail. Legs short, inner finger present but small and clawless. Body covered with large, cycloid, smooth, imbricate scales; 12 scales around midbody. Two longitudinal rows of dorsals and two of ventrals. First pair of chinshields very large. Lower eyelid

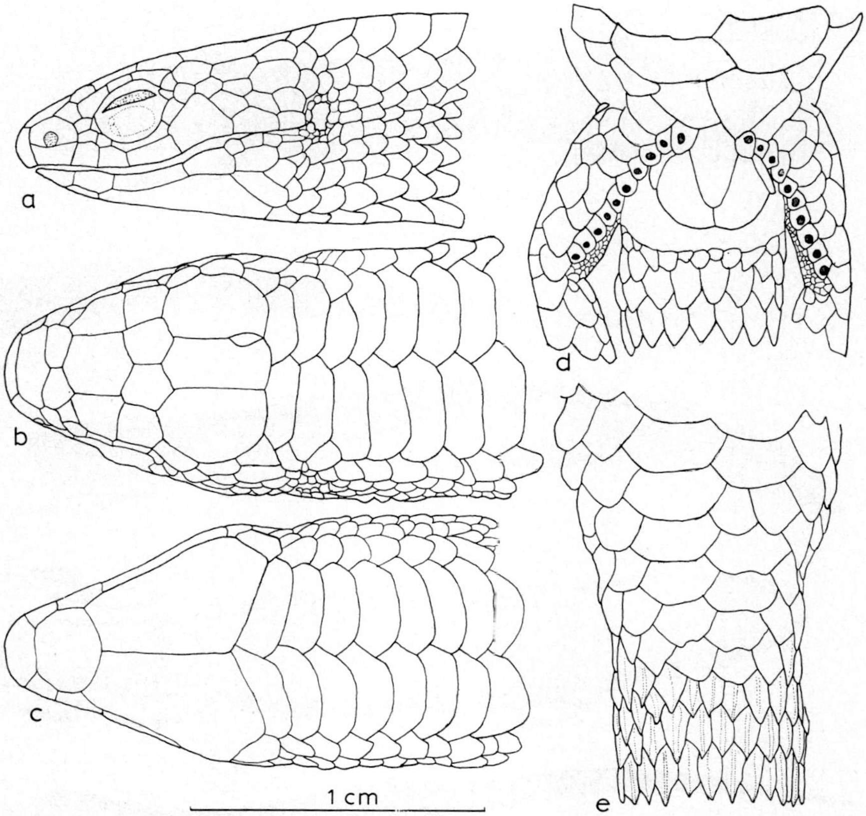

Fig. 55. Iphisa elegans Gray; a, b, c: lateral, dorsal and ventral views of the head, d: anal region, e: dorsal view of the base of the tail of ♂, RMNH 15240, Airstrip Paloemeu.

with a large, undivided transparent disc. Femoral pores present in males and females. Pupil round. Ear-opening small, tympanum recessed. Shiny brown above, pale orange on chin and belly, purplish under the tail.

DESCRIPTION.—Head short, depressed, slightly wider than the neck, which is slightly narrower than the body; snout blunt.

Tongue lanceolate; with imbricate scale-like, rounded, rhomboidal papillae in oblique rows; tip bifid, smooth. Teeth conical anteriorly; bicuspid, laterally compressed, posteriorly.

Rostral more than twice as wide as deep; hexagonal; visible from above. A large heptagonal frontonasal, one and a half times as wide as long, separating the nasals. A pair of irregular hexagonal prefrontals, wider than long. A large, hexagonal frontal, longer than wide. A pair of irregular hexagonal frontoparietals which are as wide as long. A hexagonal interparietal, twice as long as wide. Two polygonal parietals, as long as the interparietal. Occipitals indistinguishable from the nuchals and the dorsals. Three supraoculars; the anterior one

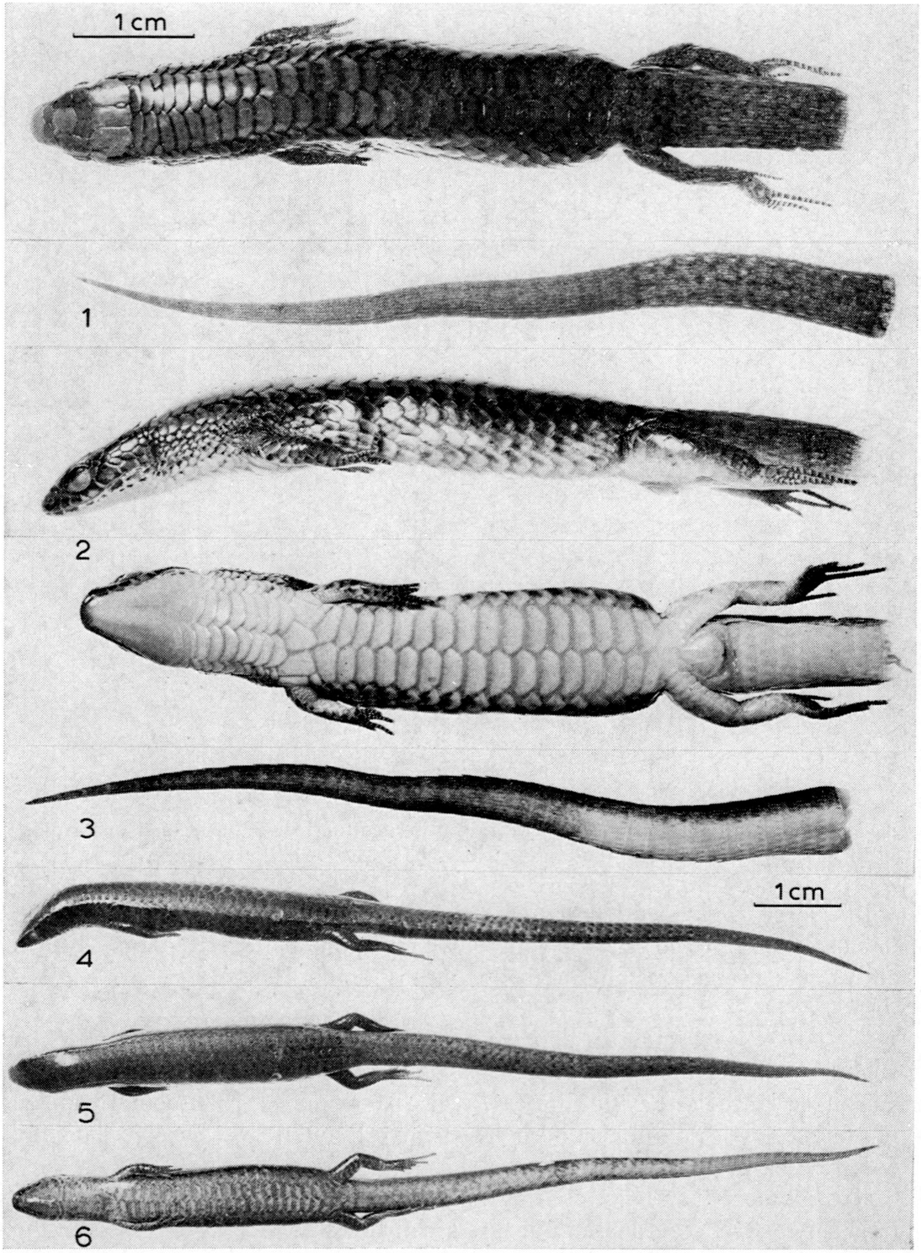

Plate 30. Figs. 1–3. *Iphisa elegans* Gray; dorsal, lateral and ventral aspect of ♂, RMNH 15240, Airstrip Paloemeu. Figs. 4–6. *Gymnophthalmus underwoodi* Grant; lateral, dorsal and ventral aspect of ♀, RMNH 15245 a, Paramaribo.

Table 19. Counts and measurements of *Iphisa elegans* Gray. For explanation see table 3.

Reg. no	A	B	C	D	E	F	G	H	I	J	K	L	M	O	P	Q	R	S	T	U	V	W	X	Y
RMNH																								
13416	♂	52	66(?)	11.5	8.1	5.6			9/1-9/1	12	28	18	2	7-7	6-6	3-3	3-3	1-1	5	8	2	12-12	17-17	2
13417a	♂	34	72	7.8	4.8	3.6			10/1-10/1	12	27	18	2	7-7	6-6	4-4	3-3	2-2	5	7	2	13-13	18-18	2
b	♂	47	105	9.8	6.2	4.6			10/1-10/1	12	28	17	2	7-7	6-7	4-4	3-3	1-1	5	8	2	14-14	18-18	2
13418	♀	46	52(52)	9.6	5.3	4.2			0/1-0/1	12	28	18	2	7-7	6-6	4-4	3-3	2-2	4+1	8	2	12-11	16-17	2
13419	♀	52	46(18)	10.0	6.7	4.7			0	12	28	20	2	7-7	6-6	4-4	3-3	2-2	4+1	8	2	12-14	19-19	2
13420	♂	59	25(25)	12.8	8.8	6.0			9/1-9/1	12	28	18	2	7-7	6-6	4-4	3-3	2-2	4+1	8	2	12-12	17-18	2
15240	♂	55	81(35)	11.5	7.8	5.6	10	18	9/1-9/1	12	28	18	2	7-7	6-6	4-4	3-3	2-2	4+1	8	2	12-12	18-18	2
BM																								
1905.11.1.12	♀	59	121(103)	11.4	7.3	5.3	11	18	0/3-0/3	12	31	20	2	7-7	6-6	2-3	3-3	2-2	5+1	8	2	12-13	19-18	2
.13	♀	57	114	11.3	7.4	5.2	11	?	0/2-0/3	12	31	20	2	7-7	6-6	3-3	3-3	1-2	4+1	8	2	12-12	15-15	2
AMNH																								
21294	♂	54	100.5	11.1	7.2	4.9	10	19	11/1-9/1	12	28	19	2	7-7	6-6	3-3	3-3		4+1	8	2	?	?	2
70624	♂	56	47(15)	10.4	6.5	4.7	9	17	2+5/1-2+4/1	12	31	21	2	7-7	6-6		3-3		4+1	?	2	13-13	19-18	2
MP																								
99-74	♀	50	11(11)				10	17	0	12	27	17	2	6-7	6-6	3-3	3-3	2-1	4+1	8	2	14-13	20-19	2
MCZ																								
77500	♂	46	-						12/0-11/0	14	27	17	2	6-6	5-5	4-4	3-3		5	8	2	15-15	21-20	2
BM																								
* 1946.9.1.1	♂	55	90(90)	11.2	6.7	4.6	10	15	9/1-9/1	12	30	20	2	7-7	6-6	3-3	3-3	4-4	4+1	9	2	13-12	18-19	2

smallest, the second largest. Nostril in an undivided nasal. Two or three loreals; the upper one reaching to the upper surface of the head, deeper than long; the lower one small, longer than deep. One small preocular. Two (rarely one or three) suboculars, the second very long and narrow. One large postocular. Four (rarely three) supraciliaries; the anterior one largest, the fourth smallest; in direct contact with the supraoculars. Temporal region with a few polygonal scales of various sizes. Anterior edge of ear-opening with small scales; denticulate. Ear-opening small; a long external auditory meatus present, tympanum visible. Seven supralabials; fourth under the eye, fifth largest. Lower eyelid with a large undivided transparent disc. Pupil round.

Mental trapezoid with a convex anterior margin. Postmental heptagonal. Two pairs of chinshields, the anterior pair enormous, forming a very long suture; the second pair very small, widely separated; both pairs in contact with the labials. Six (rarely seven) infralabials, the third and the fourth subequal and longest.

All headscales juxtaposed; all scales on neck, throat, body and tail imbricate.

Gulars transversely enlarged, hexagonal, smooth, eight (rarely seven) pairs between the chinshields and the pectorals. No distinct collar. Scales on the nape indistinguishable from the dorsals. Scales on the side of the neck small, ovate.

Dorsals large, transversely enlarged, hexagonal, smooth, in two longitudinal series to the base of the tail, 27–28 pairs between the parietals and the posterior surface of the thighs. Laterals smaller, rhomboidal, smooth, in oblique and in four longitudinal rows. Ventrals slightly larger than the dorsals, transversely enlarged, hexagonal, smooth, in two longitudinal series from the pectorals to the preanals, 17–20 (in most specimens 18) pairs to the preanals (including the pectorals). Twelve scales around midbody. Four preanal scales in a semicircle, enclosing an elongate central one, which in some specimens reaches the cloacal border.

Preanal pores one in males, one or none in females; femoral pores nine to ten on each side in males, absent in females. Preanal and femoral pores in single scales.

Scales on the tail small, hexagonal, mucronate, keeled, imbricate, in transverse and oblique rows; under the tail the keels are low and wide. Scales on the regenerated part of the tail merely smaller. Tail round in cross-section.

Scales on the upper and anterior surfaces of the forelimbs and on the anterior and lower surfaces of the hind limbs large, smooth (except on the lower legs: keeled), imbricate; on the lower and posterior surfaces of the forelimbs and on the posterior and upper surfaces of the hind limbs small, rounded, juxtaposed. Twelve to 15 smooth, single lamellae under the fourth finger; under the fourth toe 16–19 lamellae, smooth, single in the proximal and distal parts, divided in the central part. Fingers and toes subcylindrical, first finger very much reduced in size, clawless. Palms and soles with small tubercular scales.

Colour in life shiny brown above, pale orange on chin and belly, purplish under the tail.

In preservative the back is light- to dark-brown with, anteriorly, a very light-brown dorsolateral stripe bordered dorsally by a narrow black stripe. Dorsals with irregular, black spots extending posteriorly from their anterior margin. Flanks dark-brown, with the centres of the scales darker than the periphery, lighter towards the belly. Infralabials with large greyish brown spots. Lower surface of limbs, ventrals, gulars, chinshields and preanals white to creamy, immaculate. Ventral surface of the tail with a sprinkle of brown towards the tip. Regenerated part of the tail distinctly darker beneath than the original tail. Anterior portion of the tongue dark-grey, with a convex posterior border, posterior portion white.

A small lizard, maximum snout-vent length in females 52 mm, in males 59 mm. The undamaged tail is very long, 2.1–2.2 times as long as the snout-vent length, and slender. Hind limb 1.5–1.8 times as long as forelimb. When laid along the body there is no overlap of forelimb and hind limb.

HABITAT.—Inhabits the tropical rain-forest where it is present among the leaf-litter on the forest floor.

NATURAL HISTORY.—No stomachs were examined. No pregnant females or recently hatched juveniles are known. The only specimen I collected was taken in leaf-litter on the forest floor at noon in a shady part of the forest, where little undergrowth was present. In the stomach of a large diurnal lizard, *Kentropyx calcaratus* Spix, a partly digested specimen of *I. elegans* Gray was found, thus also indicating diurnal activity for this lizard.

RANGE.—The species occurs in Surinam, Guyana, French Guiana, Brazil (Pará), Columbia, Ecuador and Peru; in the last three countries only in the Amazonian lowlands. In Surinam the species is probably present throughout the country, on both sides of the coastal savannah belt, although no data are available for the western part of the country. In Surinam it occurs from sea level to at least 170 m (fig. 54).

Genus **Kentropyx** Spix

DIAGNOSIS.—Lizards of moderate size, with 14–16 longitudinal rows of large, keeled ventrals. Plate of large head-scales narrowing posteriorly, leaving a gap between it and the supratemporals; longitudinally depressed. Anterior nasal plates forming a suture behind the rostral. Tongue long with an emargination; anteriorly covered with imbricate, scalelike papillae, posteriorly of the emargination smooth; anteriorly ending in two smooth, slender tips. Teeth tricuspid, laterally compressed; anterior teeth small, posterior teeth large. Males with two large preanal spurs on each side of the preanal plate. Pupil kidney-shaped. The genus is restricted to South America east of the Andes. Of the nine species three occur in Surinam.

Key to the Surinam species of *Kentropyx*

1. Median dorsal scales much larger than the lateral scales, strongly imbricate, keeled, in 13–27 longitudinal rows 2
   Median dorsal scales only slightly larger than the lateral scales, hexagonal, juxtaposed, keeled, not in longitudinal rows . *Kentropyx calcaratus*, p. 293
2. Enlarged dorsals slightly smaller than the ventrals, the keels forming 13–17 continuous longitudinal ridges. . . . . *Kentropyx striatus*, p. 302
   Enlarged dorsals much smaller than the ventrals, in 22–27 longitudinal rows,

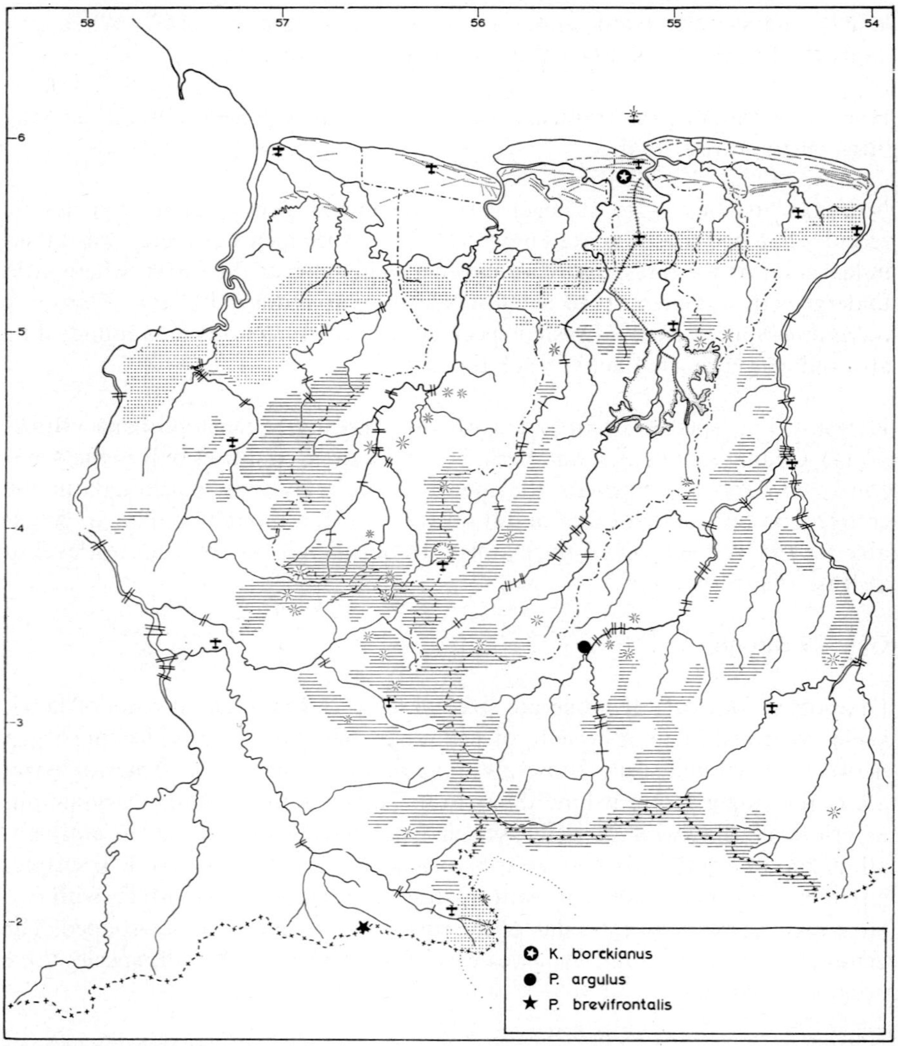

Fig. 56. Distribution of *Kentropyx borckianus* W. Peters, *Prionodactylus argulus* (W. Peters) and *Ptychoglossus brevifrontalis* Boulenger, in Surinam.

keels not continuous and directed towards the median line of the back . . . . *Kentropyx borckianus*, p. 287

Kentropyx borckianus W. Peters (figs. 56, 57; pls. 31, 34; table 20)

'Die Borckische Eidechse' Merrem, 1809: 2.
Lacerta striata: Kuhl, 1820: 122; Merrem, 1820: 65.
Centropyx striatus: Duméril & Bibron, 1839: 151 (partly); Duméril & Duméril, 1851: 117 (partly).
Acanthopyga striata: Gray, 1845: XVII, 23 (partly).
Centropyx intermedius: Cope, 1862: 496.
Centropyx Borckiana W. Peters, 1869: 64.
Kentropyx intermedius: Burt & Burt, 1931: 345; Burt & Burt, 1933: 66; J. A. Peters & Donoso-Barros, 1970: 152.

TYPE-LOCALITY.—Guiana.

MATERIAL. – No locality: 1♀, RMNH 3542.
Guiana: 1♀, ZMB 897.
Surinam.
Surinam: 1♀, RMNH 3375, 1841, Frank; 2♀♀, RMNH 3376, Ancien Cabinet; 2♀♀, RMNH 3381, 1842; 3♀♀, ZMA 12321–23.

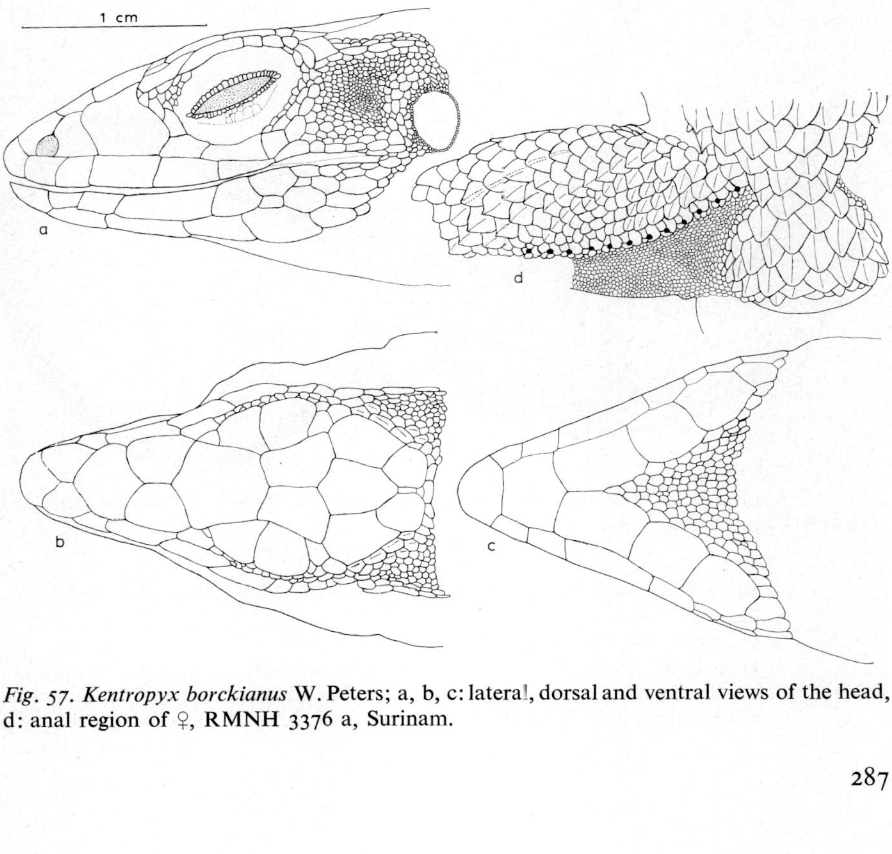

Fig. 57. Kentropyx borckianus W. Peters; a, b, c: lateral, dorsal and ventral views of the head, d: anal region of ♀, RMNH 3376 a, Surinam.

Plate 31. Kentropyx borckianus W. Peters. Figs. 1–3. Dorsal, lateral and ventral aspects of ♀, RMNH 3381 a, Surinam.

Santo Boma, near Paramaribo: 1♀, RMNH 15461, June 1963, leg. G. van Vreden.
Guyana.

Guyana: 1 ex., BM 62.12.15.3.

Demerara: 3♀♀, RMNH 3384, 1849, Frank; 1♀, BM 58.8.28.15; 2♀♀, 4 juvs., BM 89.9.30.3–8, leg. J. Quelch; 1ex. MCZ 31577, leg. J. Quelch.

Georgetown: 2♀♀, UMMZ 80505–06, 1–XI–1936, leg. A. L. Pinkus.

Table 20. Counts and measurements of *Kentropyx borckianus* W. Peters. For explanation see table 3.

Reg. no.	A	B	C	D	E	F	G	H	I	J	L	M	N	O	P	Q	R	S	V	W	X	Y
RMNH																						
3375	♀	96	162(49)	25.3	13.2	12.4	32	59	13/0-12/0	86	32	14	19	10-10	10-9	4-4	4-4	3-3	4	16-17	25-?	26
3376a	♀	96	147(33)	23.6	14.1	11.3			13/0-12/0	89	34	14	17		9-9	5-4	4-4	3-3	4	19-19	?-28	22
b	♀	91	208	22.7	12.7	10.6			12/0-15/0	89	34	14	16		10-9	4-4	4-4	3-3	4	17-18	26-26	27
3381a	♀	100	167(167)	26.3	14.2	12.6	36	67	13/0-13/0	91	33	14	16	11-11	10-10	4-4	4-4	3-3	4	16-16	25-25	25
b	♀	72	137(82)	19.3	11.8	8.9	25	52	12/0-13/0	86	32	14	18	10-11	10-10	4-4	4-4	3-3	4	17-18	26-26	25
15461	♀	100	237	26.2	13.7	11.8			13/0-12/0	91	33	14	15	12-12	10-9	5-6	4-4	6-5	4	20-19	27-27	26
BM																						
58.8.28.15	♀	92	125(55)	22.8	13.8	11.5			12/0-13/0	85	32	14	16			5-5	4-4	2-2		19-19	27-28	27
62.12.15.3	♀	91	183(95)	23.7	13.3	10.7			13/0-12/0	87	33	14	18			5-5	4-4	2-2		?	26-26	26
AMNH																						
37478	♀	75	170	20.0	10.3	9.1			14/0-13/0	88	33	16	16			4-4	4-4			16-17	25-?	?
FMNH																						
26628	♀	85	188(182)	22.2	12.2	10.4			12/0-13/0	90	33	16	17			5-4	4-4	3-3		18-20	29-27	28
26630	♀	77	51(28)	20.0	11.3	9.4			13/0-14/0	94	32	16	17			4-4	4-4	3-3		19-19	27-25	26
26631	♀	96	43(43)	23.8	13.2	12.0			12/0-12/0	92	35	14	17			4-4	4-4	3-3		19-19	29-28	28
26742	♀	80	196	22.0	11.6	9.8			13/0-14/0	87	33	14	19			4-4	4-4	3-3		18-17	26-28	24
MCZ																						
31577	?	?	?	?	?	?			12/0-13/0	?	34	16	16	?	?	?	?	?		19-19	29-28	22
UMMZ																						
80437a	♀	?	?	22.6	11.5	10.1			12/0-13/0		33	16	17			4-4	3-3			19-19	28-26	
b	♀	?	?	21.8	11.9	10.8			12/0-12/0		33	16	16			4-4	4-4			19-19	28-28	
80505	♀	?	?	24.9	13.0	11.5			12/0-13/0		33	16	18			4-4	3-3			19-19	26-?	
80506	♀	86	153	23.3	12.3	10.8			14/0-13/0		35	14	17			5-5	3-3			18-19	26-27	
MP																						
2661	♀	82	66(66)	20.8	12.0	9.8	30	54	14/0-14/0	89	34	14	16	10-10	9-9	3-4	4-4	2-2		19-18	25-26	
ZMB																						
*897	♀	96	221	23.0	13.2	10.5	32	62	12/0-13/0	89	33	16	17	10-11	10-10	5-5	4-4	3-3		19-19	26-28	26

Buxton: 4♀♀, FMNH 26628, 26630–31, 26742.

Wismar: 1♀, AMNH 37478, August 1934, leg. A. S. Pinkus; 2♀♀, UMMZ 80437, 9–III–1936, leg. A. Pinkus.

French Guiana.

Mana: 1♀, MP 2661, leg. Leschenault & Doumerc.

DIAGNOSIS.—A medium-sized lizard with cylindrical body, well developed limbs and long tail. Dorsal scales larger than the lateral scales, keeled, imbricate, smaller than the ventrals, 22–27 in a transverse row at midbody. Ventrals large, keeled, imbricate, in 14 longitudinal and in 32–34 transverse rows. Femoral pores 12–14. In preservative greyish brown above with lighter dorsolateral stripes bordered by indented black bands.

DESCRIPTION.—Head pointed; nearly twice as long as wide; slightly depressed, its depth slightly less than its width; slightly wider than the neck, which is as wide as the body. Central part of the head with a depression, dorsolateral parts slightly raised into shallow ridges.

Rostral pentagonal, visible from above. Nasals divided, anterior parts forming a suture, separating the hexagonal frontonasal from the rostral. A pair of trapezoid prefrontals. A hexagonal frontal, slightly longer than wide. A pair of pentagonal frontoparietals. An elongate pentagonal interparietal; bordered by a pentagonal parietal on each side. Two small pentagonal occipitals. Four supraoculars of which the fourth one is the smallest; the second and the third largest, of nearly equal size. A row of granules between the supraoculars and the supraciliaries. Loreal irregularly rectangular, anterior border concave; in contact with the frontonasal, the prefrontal, the first supraocular, the first supraciliary, the preocular, the anterior subocular, (the second and) the third supralabial(s) and the nasal. Five large supralabials to below the centre of the eye, ten to eleven to the corner of the mouth; posterior four to five scales small. Supraciliaries mostly four, rarely five or six; the first and the second elongate; the third and the fourth small, quadrangular; the second largest. Supraciliaries followed by a series of elongate supratemporals. Space between the supratemporals and the parietals occupied by small polygonal, keeled scales, more than in *K. striatus* (Daudin). A small preocular, nearly always separated from the upper labials by the anterior subocular. Three, rarely five or six, suboculars; the second largest; the third smallest. Two rows of enlarged postoculars. Some scales anterior of the ear-opening enlarged; centre of temple occupied by very small, hexagonal, keeled, irregularly arranged scales, larger near the upper labials. Anterior ear-margin slightly denticulate; external auditory meatus short, tympanum clearly visible. Lower eyelid with several enlarged, semitransparent palpebrals. Pupil kidney-shaped with the indention ventrally.

A pentagonal mental. A heptagonal postmental. Four pairs of chinshields, the first pair in contact in their anterior part; the other pairs separated; the fourth pair very small and barely larger than some of the surrounding scales; the posterior two pairs separated from the infralabials by small scales. Nine to ten infralabials, the anterior five or six much larger than the posterior ones. Scales on chin and throat small, rounded, hexagonal, keeled, with a central patch of slightly enlarged scales on the throat.

Scales on the nape small, keeled, hexagonal, in transverse rows; similar on the side of the neck, smaller towards the throat. Collar strongly denticulate; consisting of 16–19 rhomboidal, keeled, subequal scales. Scales on the back enlarged; larger than the scales on the flanks, smaller than the ventrals; keeled, imbricate, in transverse and oblique rows; 22–27 in a transverse row at midbody; the keels directed obliquely upwards, not forming continuous ridges. Ventrals large, rhomboidal, imbricate, mucronate, keeled, the keels forming continuous ridges; in 14 longitudinal and in 32–34 transverse rows. Scales on the flanks small, rhomboidal, keeled, juxtaposed, smallest near the ventrals. Total number of scales around midbody 86–91. Preanal plate covered with many scales of the same size and form as the ventrals.

Femoral pores 12–14, preanal pores absent.

Scales on the tail large, quadrangular, slightly mucronate, keeled, the keels forming continuous ridges, in transverse and longitudinal rows; scales under the tail identical but narrower. Regenerated part of tail covered with small, irregular, imbricate, keeled scales, the keels forming continuous ridges, longitudinal rows evident, transverse rows confused.

Upper and anterior surfaces of the forelimbs covered with large, rhomboidal, keeled, imbricate scales, the lower surface with small scales. Anterior and lower surfaces of the hind limbs covered with keeled, imbricate scales; from anteriorly to posteriorly the shape of these scales changes from hexagonal to rhomboidal; gradually diminishing in size; more numerous than in *K. striatus* (Daudin); the upper and posterior surfaces with small scales. Sixteen to 19 single, smooth lamellae under the fourth finger; 25–28 single, keeled lamellae under the fourth toe. Outer edge of fourth toe strongly serrated. Fingers subcylindrical, toes laterally compressed. Palms and soles with small, rhomboidal, juxtaposed, tubercular scales.

Beebe (1945) described the colour of a living specimen as follows: 'pinkish--brown above with many broken bands of dark brown which became spots on the tail. A broad band of black extended from the eye down the sides of the body bounded by two pale lines. The head and neck were greenish on the sides, the sides of the body chestnut brown. Below, the head and neck were whitish, the body and tail pinkish-gray'.

In preservative greyish brown above, except the dorsal surface of the head, which has an olive colour. A light dorsolateral line from the posterior corner of the eye, over the ear to the base of the tail, bordered dorsally and ventrally by black bands, starting between ear and forelimb, which may be cut up in transverse bars, connected in the ventral, separated in the dorsal band. The dorsal bands form two rows of transverse bars on the back, towards the tail, which fuse on the tail into one. An indistinct light lateral band and a few light spots near the ventrals may or may not be present. Flanks brown. Ventral parts immaculate, bluish white.

A medium-sized lizard, maximum snout-vent length 100 mm, the undamaged tail 2.3 times as long as the snout-vent length. The forelimbs more than half the length of the hind limbs. Limbs considerably overlapping when laid along the body.

HABITAT.—Nothing is known about the habitat of this species, although the common name BEEBE (1945) cites (Striped Jungle Lizard) for this species might indicate that this is a forest dweller. UNDERWOOD (1962) reports this species from 'beside the creek leading to the carenage' on Barbados and from 'O'Meara Savanna' on Trinidad.

NATURAL HISTORY.—This is another parthogenetic species. No males of this species are known. Only mature females, some of them pregnant (e.g., RMNH 3381a with five and RMNH 15461 with nine mature eggs) are known in museum collections. Some older authors (BOULENGER, 1885b; TROSCHEL, 1848) suggested that the species might represent merely the female of *Kentropyx striatus* (Daudin), although BOULENGER realised that the differences observed 'are not of a kind known to vary according to sexes in other genera of lizards'. Since then females of *K. striatus* (Daudin) have been collected which show a normal type of sexual dimorphism, thus excluding this possibility.

RANGE.—Only known from Surinam, Guyana, French Guiana, Venezuela (Bolivár), Trinidad and Barbados. In Surinam only a single locality is available, so nothing can be said about the distribution in Surinam. In all three Guianas the localities from where the species is known are in the coastal area (fig. 56).

According to WESTERMANN (1953) this species is extinct on Barbados, but UNDERWOOD (1962) says he has seen it 'beside the creek leading to the carenage'. WESTERMANN (1953) mentions the possibility of its introduction on Barbados, but zoogeographically its occurrence on the island is not exceptional. Several other species occurring on Barbados are distributed on mainland South America Trinidad, Tobago and Barbados.

REMARKS.—There has been much confusion about the correct name of this species. In 1861 COPE gave a correct description of a specimen from 'Barbadoes'. He named it *Centropyx intermedius* Gray, but was not sure whether this proceeding was correct, because of the inadequacy of GRAY'S (1831) description of the specimens he used for this name. BOULENGER (1885b) also gave a good description and since that time the species under discussion has been known as *Kentropyx intermedius* (Gray).

W. PETERS (1869), when describing *Centropyx renggerii* also mentioned some other species of the same genus, e.g., *C. Borckiana*, a name which he refers to MERREM. This name is accompanied by a short, but useful description from which the identity of the species can easily be established. In ascribing the name *C. Borckiana* to MERREM (1809) W. PETERS is wrong. MERREM (1809) gives an excellent description and illustration of a new species of lizard which he names 'Die Borckische Eidechse'; however, no binominal Latin name accompanies this description. With this 'Borckische Eidechse' the species under discussion is meant. Consequently the first author to use the species name is W. PETERS, who thus becomes the author of *C. Borckiana*. The species indicated by this name is identical with the species that up till now has been known as *Kentropyx intermedius* (Gray). However, this name, after investigation of the original material, proves to be a synonym of *K. calcaratus* Spix (see below), a possibility already suggested by BURT & BURT (1931). Consequently it is no longer available for the species under discussion. The next name used for this species, being *Centropyx Borckiana* W. Peters, 1869, thus becomes available. W. PETERS based his description on MERREM's description and on a specimen from Guiana (nr. 897) present in the Berlin Museum, which means there are two syntypes. According to W. PETERS the specimen from the collection of BORCKE on which MERREM based his description was lost, so ZMB 897 is now indicated as the lectotype. This specimen is still in existence under this number. It is a pregnant female from Guiana with a snout-vent length of 96 mm, a tail length of 221 mm. The head is 23.0 mm long, 13.2 mm wide and 10.5 mm long. There are 26 enlarged dorsals in a transverse row at midbody. Ventrals in 16 longitudinal and 33 transverse rows. Total number of scales around midbody 89. Subdigital lamellae fourth fingers 19–19; subdigital lamellae fourth toes 26–28.

Kentropyx calcaratus Spix (figs. 58, 59; pls. 32, 34)

Lacerta cauda corpore longiore, squamis omnibus, exceptis dorsalibus longitudinaliter carinatus, aculeatis Gronovius, 1756: 82.
Lacerta vittata Schinz, 1822: 45.
Kentropyx calcaratus Spix, 1825: 21; D'Audretsch, 1968: 51; Reed & Borowsky, 1970: 161.
Teius intermedius Gray, 1831: 31.
Centropyx calcaratus: Duméril & Bibron, 1839: 149; Duméril & Duméril, 1851: 117; Van der Hoeven, 1855: 520; Van Lidth de Jeude, 1904: 89; Van Lidth de Jeude, 1917g: 438.
?'Fraai gekleurde en gestreepte hagedissen' Van Stockum, 1903: 264.
'Wera' Geijskes, 1957: 276.

TYPE-LOCALITY.—Rio Itapicurú, Maranhão, Brazil.

MATERIAL. – Surinam.

Surinam: 2♀♀, 1♂, 1 juv., RMNH 3379, Ancien Cabinet (types of *Teius intermedius* Gray); 1 ex., ZMB 4890, Ehrhardt; 1♀, NMW 18776, 1904, leg. A. O. C. Heller; 1♀, NMW 18777, 1904; 1♀, NMW 18785, 1900; 1♀, NMW 18786, 1909, leg. A. O. C. Heller.

Paramaribo: 1 juv., RMNH 15301, August 1964, leg. S. Ligorie; 1♀, ZMA 12319; 1♀, CM 44409, 8–V–1966, leg. M. H. de la Fuente.

Coronie Road, picket 216: 1♂, RMNH 15371, Encampment, 23–XII–1948, leg. Surinam Expedition 1948–49.

Wia Wia transect: 1♂, RMNH 15346, 4th ridge, 13–XI–1948, 1♀, RMNH 15347, 4th swamp, 13–XI–1948, 1♂, RMNH 15348, km S. 1.85, 16–XI–1948, 1♂, RMNH 15349,

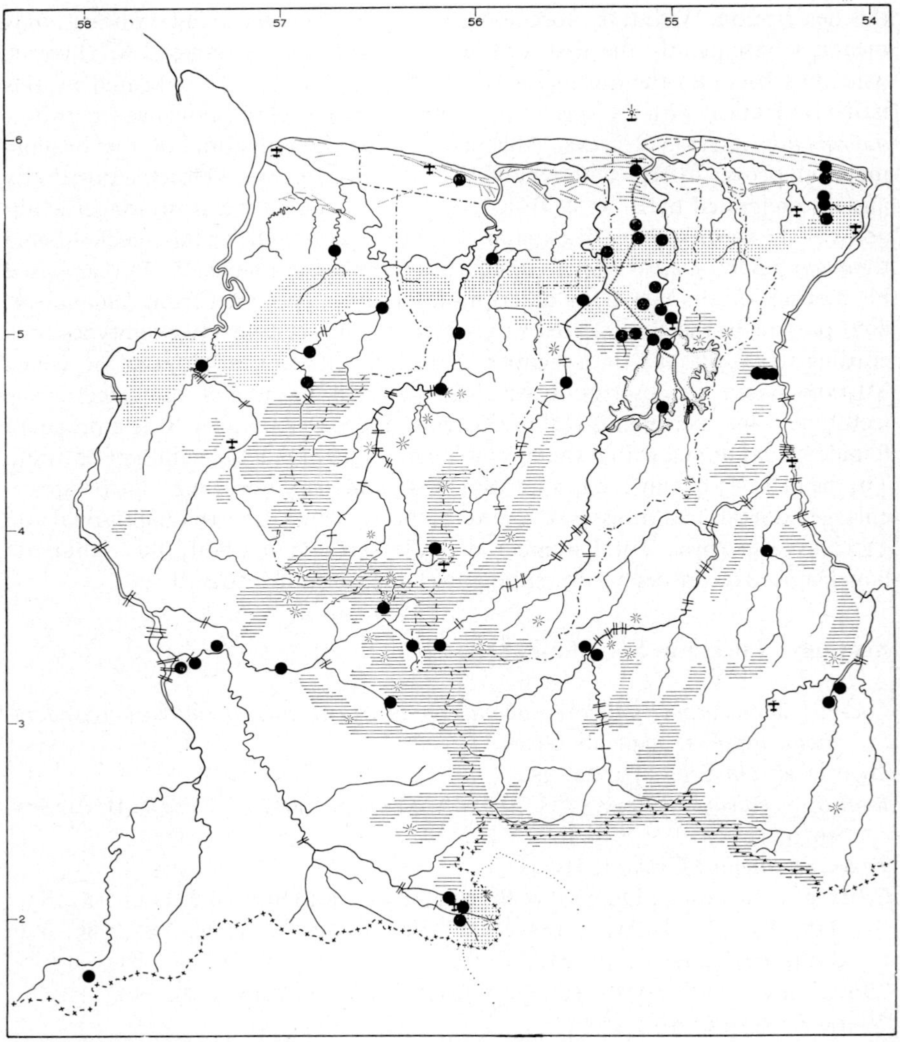

Fig. 58. Distribution of *Kentropyx calcaratus* Spix in Surinam.

km S. 3.75, 17–XI–1948, 1♀, RMNH 15350, 6th ridge, 18–XI–1948, 1♀, 1♂, RMNH 15351, km S. 10.3, 25–XI–1948, all leg. Surinam Expedition 1948–49.

Third Encampment: 1♀, RMNH 15332, km 14.9, 13–X–1948, 2♀♀, RMNH 15333, km 14.2–14.9, 14–X–1948, all leg. P. H. Creutzberg (Surinam Expedition 1948–49); 1♂, RMNH 15335, km 17.1, 15–X–1948, 1♀, RMNH 15338, km 15.1, 18–X–1948, 2♀♀, 1♂, RMNH 15339, km 15.4–15.6, 18–X–1948, 1♂, RMNH 15340, km 14.9, 18–X–1948, 2♂♂, RMNH 15341, km 14.9, 20–X–1948, 1♂, RMNH 15342, km 14.9, 20–X–1948, 1♀, 1♂, RMNH 15343, km 14.9, 22–X–1948, 3♀♀, 1♂, RMNH 15344–45, side transect E., km 15.8, 22–X–1948, all leg. Surinam Expedition 1948–49.

Djai Creek: 3♀♀, RMNH 15321, km 8.1–8.3, 5–X–1948, 1♂, RMNH 15322, E. of encampment, 5–X–1948, 2♂♂, RMNH 15324, 7–X–1948, 1♀, RMNH 15325, km 10.5, 7–X–1948, 2♀♀, RMNH 15326, km 8.6, 8–X–1948, 1♀, RMNH 15327, km 9.1, 9–X–1948,

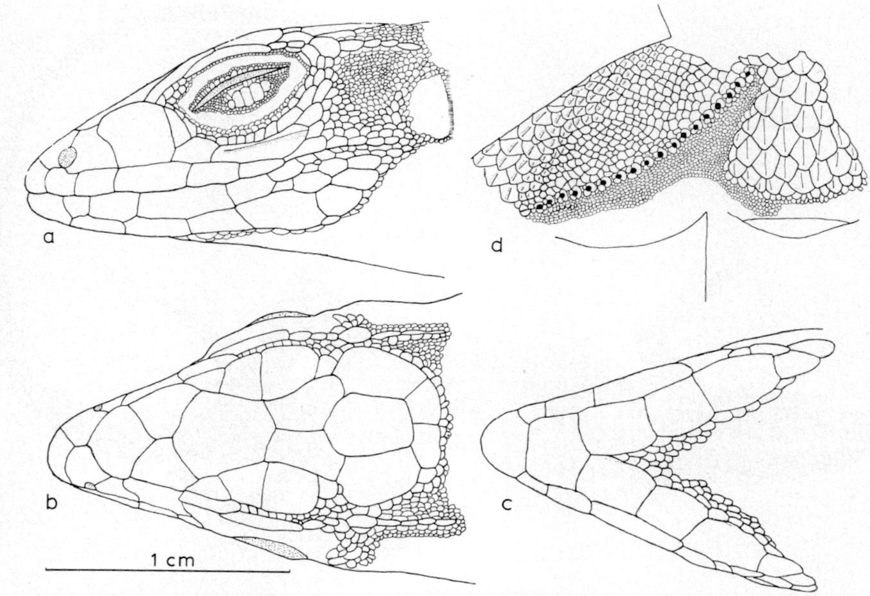

Fig. 59. Kentropyx calcaratus Spix; a, b, c: lateral, dorsal and ventral views of the head, d: anal region of ♀, RMNH 15138, Troeli Creek.

1♂, RMNH 15328, km 8.8, 9–X–1948, all leg. P. H. Creutzberg (Surinam Expedition 1948–49); 2♀♀, 1♂, RMNH 15329, km 8.6, 10–X–1948, leg. Surinam Expedition 1948–49.

Wane Creek N.: 1♂, RMNH 15315, First Encampment, 30–IX–1948, 1♀, RMNH 15316, km 8.4, 2–X–1948, 1♀, RMNH 15317, km 6, 2–X–1948, 1♀, RMNH 15318, km 5.3, 2–X–1948, 1 juv., RMNH 15319, km 5.9, 3–X–1948, 1 juv., RMNH 15320, km 6.2, 4–X–1948, all leg. P. H. Creutzberg (Surinam Expedition 1948–49).

Moengo Tapoe transect: 1♂, RMNH 15308, km 3.1, 22–IX–1948, 1♂, RMNH 15309, km 1.9, 22–IX–1948, 1♂, RMNH 15310, km 2.43, 22–IX–1948, 1♂, RMNH 15311, km 0.8, 21–IX–1948, 1♀, RMNH 15312, km 0.4, 23–IX–1948, 1♂, RMNH 15313, km 0.5, 23–IX–1948, 1♀, 1♂, 1 juv., RMNH 15314, km 0.4, 24–IX–1948, 1♂, RMNH 15323, 1 st ridge-forest, 6–X–1948, 1♀, RMNH 15330, km 12.2, 11–X–1948, 1♂, RMNH 15331, km 14.5, 12–X–1948, all leg. P. H. Creutzberg (Surinam Expedition 1948–49); 1 juv., RMNH 15334, km 15, 14–X–1948, 1♀, 3 juvs., RMNH 15336, km 16.5–16.6, 15–X–1948, 1 juv., RMNH 15337, km 15.4, 15–X–1948, all leg. Surinam Expedition 1948–49.

Republiek: 1 juv., RMNH 15303, 19–X–1963, leg. D. C. Geijskes.

Zanderij: 1♂, BM 1946.4.3.50, 21–VIII–1938, leg. I. T. Sanderson.

Powakka: 1♂, CM 44410, 12–V–1966, 1 ex., CM 44411, 13–V–1966, both leg. M. H. de la Fuente.

Awarra savannah, Maratakka River: 2 juvs., RMNH 16455, 8–III–1971, leg. D. G. Reeder.

Tibiti: 1♀, RMNH 15352, km 0.2, 10–I–1949, 1 juv., RMNH 15353, 3rd transect, km 3, 18–I–1949, both leg. Surinam Expedition 1948–49.

Troeli Creek: 1♀, RMNH 15138, 8–V–1968, leg. M. S. Hoogmoed.

65 km SSE. of Paramaribo, Afobaka Road: 1♀, AMNH 108788, 12/13–II–1972, leg. C. W. Myers & J. Daly.

Loksihati, Saramacca River: 1♀, FMNH 134728, 1962, leg. P. Hershkovitz.

Railway km 91.5: 1♂, BM 1946.4.3.51, 4–XI–1938, leg. I. T. Sanderson.

Plate 32. Kentropyx calcaratus Spix. Figs. 1–3. Dorsal, lateral and ventral aspect of ♀, RMNH 16386, Sipaliwini.

Berg en Dal: 1♂, RMNH 15302, 4–IX–1963, leg. D. C. Geijskes; 1 ex., ZMB 7343, leg. H. B. Möschler.

Compagnie Creek: 1♀, RMNH 15305, 21–III–1966, leg. G. F. Mees.

Stondansi Cataracts, Nickerie River: 1♂, RMNH 16463, 30–I–1971, 1♂, RMNH 16464, 3–II–1971, both leg. D. G. Reeder.

Camp 1, Coppename River: 2 juvs., BM 1946.4.3.48–49, 8–VII–1938, leg. I. T. Sanderson.

Brown's Mountain: 1♂, RMNH 15289, 20–VIII–1968, 1♀, RMNH 15291, 13–VII–1968, both leg. M. S. Hoogmoed; 1 juv., RMNH 15378, 19–IX–1938, leg. D. C. Geijskes; 1♂, RMNH 17267, February 1972, 1 juv., RMNH 17272, 7–XII–1971, both leg. G. F. Mees.

Railway km 121: 1♂, RMNH 15377, 30–XI–1949, leg. C. Bleys.

Aroesoebanja soela, Suriname River: 1 juv., RMNH 15387, 9–XII–1963, leg. S. Ligorie.

Avanavero Cataracts, Kabalebo River: 1♀, 1♂, RMNH 16487, 6–IV–1971, 2♀♀, 1♂, RMNH 16488, 8–IV–1971, all leg. D. G. Reeder; 1♀, RMNH 16499, 11–IV–1971, leg. D. C. Geijskes.

Lombok Cataracts, Nickerie River: 1♂, RMNH 16465, 5–II–1971, leg. D. C. Reeder.

Blanche Marie Cataract, Nickerie River: 1 juv., RMNH 16467, 12–II–1971, leg. D. G. Reeder.

Voltz Mountain Reserve: 1♂, RMNH 15211, Raleigh Cataracts, 4–XI–1968, 1♀, RMNH 15224, 5–XI–1968, both leg. R. Hoogmoed-Verschoor & M. S. Hoogmoed; 1♂, AMNH 108789, Raleigh Cataracts, 19/20–II–1972, leg. C. W. Myers & J. Daly.

Awarradam, Saramacca River: 1 juv., RMNH 15372, 4–IV–1964, leg. M. Boeseman.

Marowijne Creek: 1 juv., ZMA 12662, 43 km S. of Afobaka, 8–VI–1966, leg. H. Nijssen.

Nassau Mountains: 1♂, RMNH 15354, km 0.4, 15–II–1949, 3♀♀, RMNH 15355, km 1.4, 16–II–1949, 1 juv., RMNH 15356, Base Encampment, 20–II–1949, 1♀, RMNH 15357, km 2.2–2.3, 21–II–1949, 1♂, RMNH 15358, km 1.4, 22–II–1949, 1♂, RMNH 15359, km 2.5, 24–II–1949, 1♀, 2♂♂, RMNH 15360, S. of km 7, 3–III–1949, 1♂, RMNH 15361, km 2.4, 7–III–1949, 1♂, RMNH 15362, km 10.5, 11–III–1949, 2♀♀, RMNH 15363, km 10.5, 12–III–1949, 1♀, RMNH 15364, km 10.4, 14–III–1949, 3♂♂, RMNH 15365, 1♂, RMNH 15366, km 10.7, 18–III–1949, 1♀, 1♂, RMNH 15367, km 14.7, 21–III–1949, 1♀, 2♂♂, 1 juv., RMNH 15368, km 14.5, 23–III–1949, 1♂, RMNH 15369, Base Encampment, March 1949, 1♂, RMNH 15370, km 1.2, 17–II–1949, all leg. Surinam Expedition 1948–49.

Tafel Mountain: 1♂, RMNH 15376, km 26, September 1944, leg. L. Schmidt.

Gonini River, confluence of Emma and Wilhelmina Rivers: 1♀, RMNH 15375, 9–X–1903, leg. G. M. Versteeg (Gonini Expedition).

Wilhelmina Mountains: 1♂, 1 juv., RMNH 15298–99, Bivouac 3, km 19, leg. H. P. Pijpers.

Tigri Encampment, New River: 2♀♀, RMNH 15294, E., 28–VI–1968, 1♀, 1♂, RMNH 15295, 29–VI–1968, both leg. M. S. Hoogmoed.

Gonini Encampment, Coeroeni River: 1♀, RMNH 15292, 26–VI–1968, leg. M. S. Hoogmoed.

Airstrip Coeroeni: 2♀♀, RMNH 15293, 15296, 22–VI–1968, leg. M. S. Hoogmoed; 1♀, RMNH 15300, 12–VIII–1959, leg. D. C. Geijskes.

Lucie River: 1♂, RMNH 5634, November 1910, 1♀, 1 juv., RMNH 5693, 1♀, RMNH 15306, 10–X–1910, all leg. J. F. Hulk (Corantijn Expedition).

Upper Gran Rio: 1♀, RMNH 15307, 20–IX–1910, leg. J. F. Hulk (Corantijn Expedition).

Airstrip Kayser Mountains: 1♀, RMNH 15286, NE., 3–VIII–1968, leg. M. S. Hoogmoed; 1♀, FMNH 128956, 1♂, FMNH 128958, 1♂, FMNH 128961, 1♀, 5 juvs., FMNH

128965–70, 1–II–1961, leg. H. A. Beatty; 2 juvs., PMYU R. 5907–08, 25–XI–1961, leg. R. Freund.

Airstrip Paloemeu: 1♀, RMNH 15156, N., 30–V–1968, 2♀♀, 1♂, RMNH 15290, 20–V–1968, 2♀♀, 1♂, RMNH 15297, S., 26–V–1968, all leg. M. S. Hoogmoed.

Janamalé kondre, Litani River: 1 ex., RMNH 15304, 1♂, RMNH 15374, 18–VIII–1939, leg. D. C. Geijskes.

Litani River, near Feti Creek: 1♀, RMNH 15373, 12–VIII–1939, leg. D. C. Geijskes.

Middle Base Camp, Itabu Creek: 1♀, 3 juvs., FMNH 30923–26, September/October 1938, leg. E. R. Blake (Sewell Avery Expedition).

Sipaliwini: 1♂, RMNH 15287, 5 km N. of Base Bivouac, 2–IX–1968, 1♂, RMNH 15288, 4 km E. of airstrip, 29–VIII–1968, both leg. M. S. Hoogmoed; 1♀, 1♂, RMNH 16383, 2–4 km E. of airstrip, 15–I–1970, 1♀, RMNH 16384, near Base Bivouac, 16–I–1970, 1♀, RMNH 16385, surroundings Meyers' farm, 19–I–1970, 1♀, 1♂, RMNH 16386, between airstrip and Meyers' farm, 30–I–1970, 1♂, RMNH 16387, 4 km NW. of airstrip, 10–II–1970, all leg. M. S. Hoogmoed & J. J. P. Paats.

Diagnosis.—A medium-sized lizard with cylindrical body, well developed limbs and long tail. Dorsal and lateral scales small, granular; ventrals large, keeled, imbricate, in 14 longitudinal and in 31–35 transverse rows. Femoral pores 16–20. Brown above with a yellowish middorsal stripe in juveniles and females, flanks with light-blue spots. Belly ochre in males, dull orange in females.

Description.—Head slightly wider than the neck, more so in the males; 1.5–1.8 times as long as wide, distinctly depressed, its depth considerably less than its width. Central part of the head with a shallow depression in females, with a deep depression in males, dorsolateral parts raised into shallow ridges. Snout pointed. Neck as wide as the body.

Rostral pentagonal, visible from above. Nasals divided, anterior parts forming a short suture, separating the hexagonal frontonasal from the rostral. A pair of trapezoid prefrontals. A hexagonal frontal, much longer than wide. A pair of pentagonal frontoparietals. Interparietal elongate, pentagonal, bordered by a pentagonal parietal on each side. Two small pentagonal occipitals. A small interoccipital may be present. Four supraoculars; the fourth very small; the first small; the second and the third large, of nearly equal size. A row of granules between the supraoculars and the supraciliaries. Loreal irregularly rectangular; anterior border concave or straight; in contact with the frontonasal, the prefrontal, the first supraocular, the first supraciliary, the anterior subocular and/or preocular, the second and third supralabials, and the nasal. Five large supralabials to below the centre of the eye, nine to eleven to the corner of the mouth, the posterior three to five scales small. Supraciliaries mostly five, rarely four or six; the first and the second elongate; the third to fifth small, quadrangular; the second largest. Supraciliaries followed by a series of elongate supratemporals, the first of which is large, trapezoid. Space between supratemporals and parietals occupied by small, polygonal, keeled scales, diminishing in size posteriorly. Preocular small and separated from the supralabials by the anterior subocular and the loreal forming a suture, or large and in contact with the supralabials. Three suboculars; the second largest; the third smallest. Two rows of enlarged postoculars. One row of scales anterior of the ear-opening enlarged; temple covered with very small, hexagonal, keeled, irregularly arranged scales, largest near the upper labials, smallest near the supratemporals. Anterior ear-margin not denticulate; external auditory meatus short, tympanum clearly visible. Lower eyelid with several enlarged, semitransparent palpebrals. Pupil kidney--shaped with the indention ventrally.

A pentagonal mental. A heptagonal postmental. Four pairs of chinshields, the first pair in contact in their anterior part; the other pairs separated; the fourth pair smallest, distinctly larger than the surrounding scales; the posterior two pairs separated from the infralabials by small scales. Nine to eleven, rarely twelve, infralabials, the anterior five or six much larger than the posterior ones. Scales on chin and throat small, rounded, more or less hexagonal, keeled, with a central patch of slightly enlarged scales on the throat.

Scales on the nape and on the side of the neck small, keeled, hexagonal, in transverse rows. Collar slightly denticulate, consisting of 14–22 leaf-shaped, keeled scales, the lateral ones smallest.

Scales on the back minute, slightly larger than the scales on the flanks, hexagonal, keeled, juxtaposed, in transverse rows only. Ventrals large, rectangular, imbricate, mucronate, keeled, the keels forming continuous ridges; in 14, rarely 15, longitudinal and in 31–35 transverse rows. Scales on the flanks small, rhomboidal, keeled, juxtaposed, increasing in size from the ventrals to the dorsum. Total number of scales around midbody 124–151. Preanal plate covered with a variable number of rhomboidal, imbricate scales, strongly keeled in females, smooth or with a shallow ridge in males. Males with two preanal spurs, directed dorsally, on each side of the preanal plate.

Femoral pores 16–20, preanal pores absent.

Scales on the tail large, rectangular, imbricate, mucronate, keeled, the keels forming longitudinal ridges, in transverse and longitudinal rows; scales under the tail identical but narrower. Regenerated part of the tail covered with scales of various sizes, rhomboidal, imbricate, keeled, the keels forming longitudinal ridges, in longitudinal rows only.

Upper and anterior surfaces of the forelimbs covered with moderate, rhomboidal, imbricate keeled scales, the keels forming ridges; the lower and posterior surfaces with small scales; anterior and lower surfaces of the hind limbs covered with keeled, imbricate scales, anteriorly largest and hexagonal, posteriorly smallest and rhomboidal; the upper and posterior surfaces with small scales. Fifteen to 20 single, smooth lamellae under the fourth finger, slightly swollen under the articulations; 25–31 single, keeled lamellae under the fourth toe. Outer edge of fourth toe strongly serrated. Fingers subcylindrical, toes laterally compressed. Palms and soles with small, rhomboidal, juxtaposed, slightly tubercular scales.

Colour in life brown above with a green zone on the anterior part of the back, the green being more intense in males than in females; light-blue spots on the flanks; dorsolateral stripe creamy yellow. Throat in males purplish, in females bluish white; posterior part of the belly in males ochre, in females dull orange.

Colour of the iris in males red-brown, in females golden brown. In juveniles a very pronounced middorsal stripe is present from the snout to near the sacral region where it fades out. Its anterior part is reddish gold, posterior to the eyes it is yellowish. The dorsolateral lines anteriorly yellowish green, posteriorly light-brown.

In preservative the back is dark- to light-brown, in females with a more or less distinct remnant of the juvenile middorsal stripe, which in males is absent. On each side of the vertebral column is a series of large, transverse, black bars which fuse on the tail but which may be nearly absent in some specimens. Light dorsolateral lines from the posterior corner of the eye, over the ear to the base of the tail, bordered ventrally by a dentate, dark-brown band. Flanks with one or two rows of grey spots; small and round in females; large, vertically oval in males. Ventral parts dirty white to pale yellow, immaculate.

A medium-sized lizard, maximum snout-vent length in males 103 mm, in females 104 mm. The undamaged tail 2.1–2.3 times as long as the snout-vent length in adults, 2.0–2.1 times in juveniles. Forelimb approximately half as long as hind limb, limbs considerably overlapping when laid along the body.

HABITAT.—The species is a typical forest floor dweller, never venturing from the ground more than one or two metres on horizontal trunks and branches of fallen trees. It prefers open, sunny places and consequently is one of the most frequently observed lizards in the forest, because it is always present along trails, creeks and in clearings in the forest.

NATURAL HISTORY.—The population density in some places is very high and males and females are present together; however, no evidence of pair bonds could be found. Pregnant females with five to seven mature eggs are present in October, March, May and June and recently hatched juveniles from January to April and from July to October, thus indicating breeding occurs throughout the year. Eight stomachs were opened to investigate their contents. Most of them contained spiders, cockroaches and crickets, one stomach contained a caterpillar, apart from the other items, another remains of a snail-shell, whereas a third only contained a small, ground-dwelling lizard, *Iphisa elegans* Gray. BEEBE (1925, 1945) found 'insects of various kinds, especially ants, termites and beetles'. Several of the stomachs contained nematodes which were collected for further investigation. The species is shot with bow and arrow by the Indians who seem to eat it (GEIJSKES, 1957).

RANGE.—Northern South America, east of the Andes, south to Bahia in Brazil: Surinam, Guyana, French Guiana, Brazil (Pará, Amapá, Goias, Amazonas), Venezuela (Amazonas), Colombia, Ecuador and Peru. In Surinam it is present throughout the country, on both sides of the coastal savannah belt, wherever there is forest; it has been found from sea level to at least 500 m (fig. 58).

REMARKS.—In 1825 SPIX described a species of lizard which he called *Kentropyx calcaratus*. Since that time this name has been frequently used. However, J. A. PETERS & DONOSO-BARROS (1970), mentioned a note made by STEJNEGER in a copy of BOULENGER (1885b) which might suggest that *Lacerta vittata* Schinz is a synonym of *Kentropyx calcaratus* Spix. When checking this possibility I came to the conclusion that STEJNEGER was correct in his assumption. The description given by SCHINZ (1822) was short, but leaves no doubt about the identity of the species described. It is identical with the species which three years later was described by SPIX as *Kentropyx calcaratus*. WIED (1825) gives a detailed description of a Brazilian lizard which he names *Lacerta striata* Daudin, but from the description and the accompanying picture it is clear that he is dealing with *Kentropyx calcaratus* Spix (W. PETERS, 1877). In the introductory sentence of the description he says: 'Die Eidechse, welche ich anfangs für eine neue Species hielt, und unter der Benennung der *Lacerta vittata* dem Herrn Dr. SCHINZ mittheilte, scheint mir jetzt nach vielfältiger Vergleichung DAUDIN's *Lacerta striata* zu seyn'. From this it is evident that SCHINZ and WIED were dealing with the same species and that WIED originally was correct in assuming he was dealing with a new species. A consequence of these findings would be that the species until now known as *Kentropyx calcaratus* Spix should be renamed and be known as *Kentropyx vittatus* (Schinz). The type locality of this species is Brazil (SCHINZ, 1822) but with the help of WIED's (1825) data on this species it is possible to restrict it to a smaller area, namely Barra da Vareda, Rio Pardo, province of Bahía, Brazil. Although, since articles 23a and 23b of the International Code for Zoological Nomenclature have been changed (*Bull. Zool. Nomencl.* 29 (4): 177) into article 23a–b and the 50-year rule is no longer valid, the name *vittatus* could be used, I prefer not to do so because *Kentropyx calcaratus* Spix is a well established and widely used name. A proposal to suppress *Lacerta vittata* Schinz will be made to the Commission.

In 1831 GRAY described *Teius intermedius* as 'Back with three pale lines; dorsal scales minute. Surinam, Mus. Leyd.' In 1845 he synonimised it with *Centropyx calcaratus* Spix. Since that time there has been much confusion about this name, mainly because the types of *intermedius* were not studied by subsequent authors. During this study I have been able to trace the specimens GRAY had before him when describing *Teius intermedius*. Gray probably visited the Rijksmuseum van Natuurlijke Historie in the late eighteen twenties. The collections from Surinam at that time were still small, most of the material present dating from the 'Ancien Cabinet' (old cabinet), the collection present at the museum's foundation in 1820. Among the material of *Kentropyx* now present in the RMNH collections are only two reg. nos. from the 'Ancien Cabinet', all other material dating from periods later than 1840. The reg. nos. concerned are 3376 and 3379, and pertain to Surinam specimens. These are the only specimens GRAY can have examined during his visit to Leiden. RMNH 3376 contains two females which in the old catalogue of the RMNH were entered as '*Centropyx striatus*' but belong to the species above described as *Kentropyx borckianus* W. Peters. The other reg. no. (RMNH 3379) contains one adult female, one male, one half grown female and one juvenile specimen belonging to *K. calcaratus* Spix. The specimens from RMNH 3376 do not fit Gray's description at all, because there are only two dorsolateral lines and the scales of the back are distinctly enlarged. The half grown specimen from RMNH 3379 fits Gray's description excellently by having two very distinct dorsolateral and one very distinct middorsal line, the scales on the back are minute and of the same size as the scales on the flanks. I hereby designate this halfgrown female specimen of *K. calcaratus* Spix (RMNH 3379a) as the lectotype of *Teius intermedius* Gray. The other specimens in this registrationnumber are paralectotypes (RMNH 3379b–d).

Vernacular name.—There is none for this species in the Surinam language, apart from the name used for all lizards ('lakkadisja'). The Indians, however, do have a name for this species. According to Geijskes (1957: 276) the Oyanas call it 'wera', according to my own notes 'mira', the difference just being the result of different interpretations of the pronounciation. Trio Indians call this species 'joi' which is the same name as they use for *A. a. ameiva* (Linnaeus).

Kentropyx striatus (Daudin)

Diagnosis.—A moderately large lizard with well developed legs, cylindrical body and long, tapering tail. Back with 12–16 longitudinal rows of enlarged, keeled scales, forming 12–16 longitudinal ridges. Belly with 14 longitudinal rows of enlarged, keeled scales, forming 14 longitudinal ridges. Femoral pores five to nine. Back grass-green, dorsolateral and lateral lines greenish white.

Key to the subspecies of *Kentropyx striatus*

1. Number of scales around midbody 62–75, number of collar scales 12–17, maximum snout-vent length 124 mm, back in life completely green in adults . . . . *K. striatus striatus*, p. 302
   Number of scales around midbody 56–65, number of collar scales 12–15, maximum snout-vent length 92 mm, in life only the nape and the side of the neck are green, rest of back brown . . . . *K. striatus viridicervix*, p. 312

Kentropyx striatus striatus (Daudin) (fig. 60; pl. 33; table 21)

Lacerta striata Daudin, 1802a: 247.
'Lézard strié' Cuvier, 1829: 30; Cuvier, 1834: 19; Cuvier, 1836: 377.
Centropyx striatus: Duméril & Bibron, 1839: 151 (partly); Duméril & Duméril, 1851: 117; Boettger, 1893: 73.
Acanthopyga striata: Gray, 1845: XVII, 23 (partly).
Centropyx decodon Cope, 1862: 495.
Kentropyx striatus: Burt & Burt, 1931: 346; Burt & Burt, 1933: 67; Tamsitt & Valdivieso, 1963: 443; J. A. Peters & Donoso-Barros, 1970: 152.

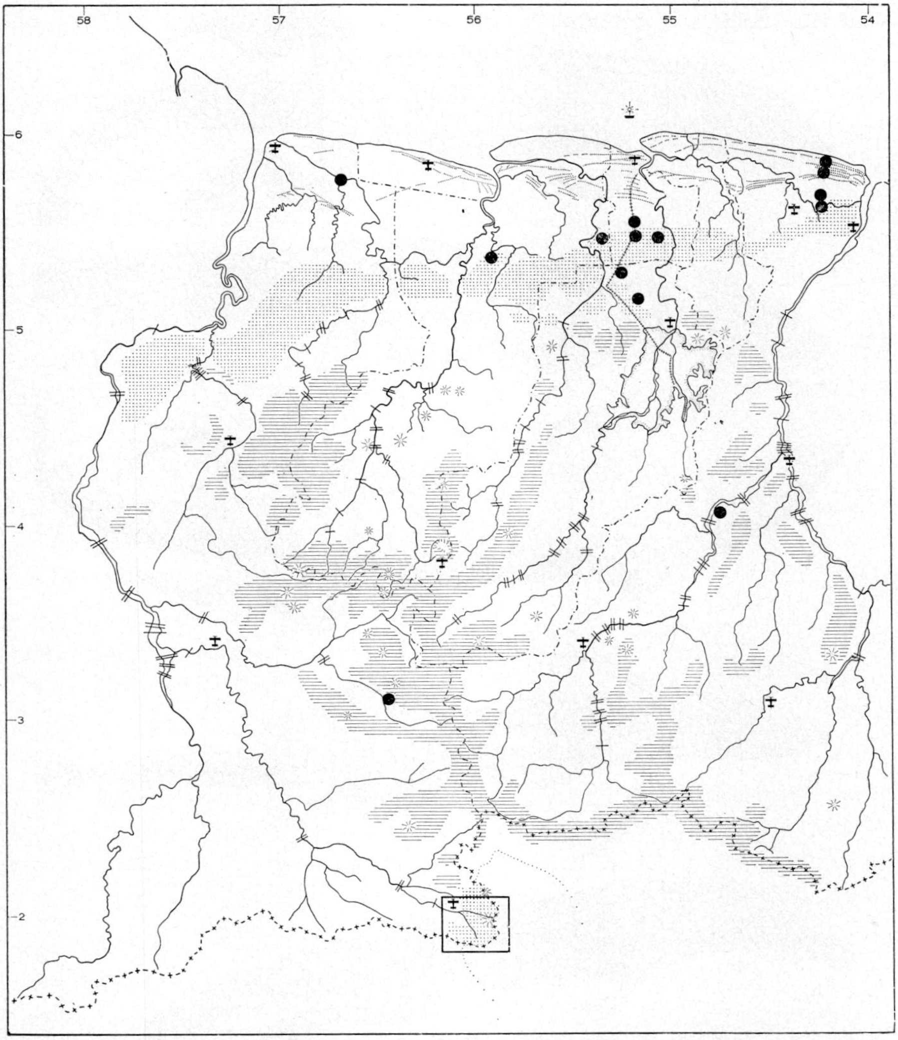

Fig. 60. Distribution of *Kentropyx s. striatus* (Daudin) in Surinam. The framed area is represented on a larger scale in fig. 61.

TYPE-LOCALITY.—Surinam.

MATERIAL. – Surinam.

Surinam: 1♀, AMNH 37478, leg. J. A. Samuels; 1♂, ANS 9557, leg. C. Hering (lectotype of *Centropyx decodon* Cope).

Paramaribo: 1♂, NMW 18761, leg. A. O. C. Heller.

Wageningen: 1♂, RMNH 13454, 9–VI–1963, leg. A. M. Husson & P. Staffeleu.

Wia Wia transect: 1♂, RMNH 13461, 4th swamp, km S. 1.5, 21–XI–1948, 1♀, RMNH 13462, km S. 6.6, 23–XI–1948, both leg. Surinam Expedition 1948–49.

Third Encampment: 1♀, RMNH 13459, km 15.7, 14–X–1948, 1 juv., RMNH 13460, side transect E., km 15.8, 20–X–1948, both leg. Surinam Expedition 1948–49.

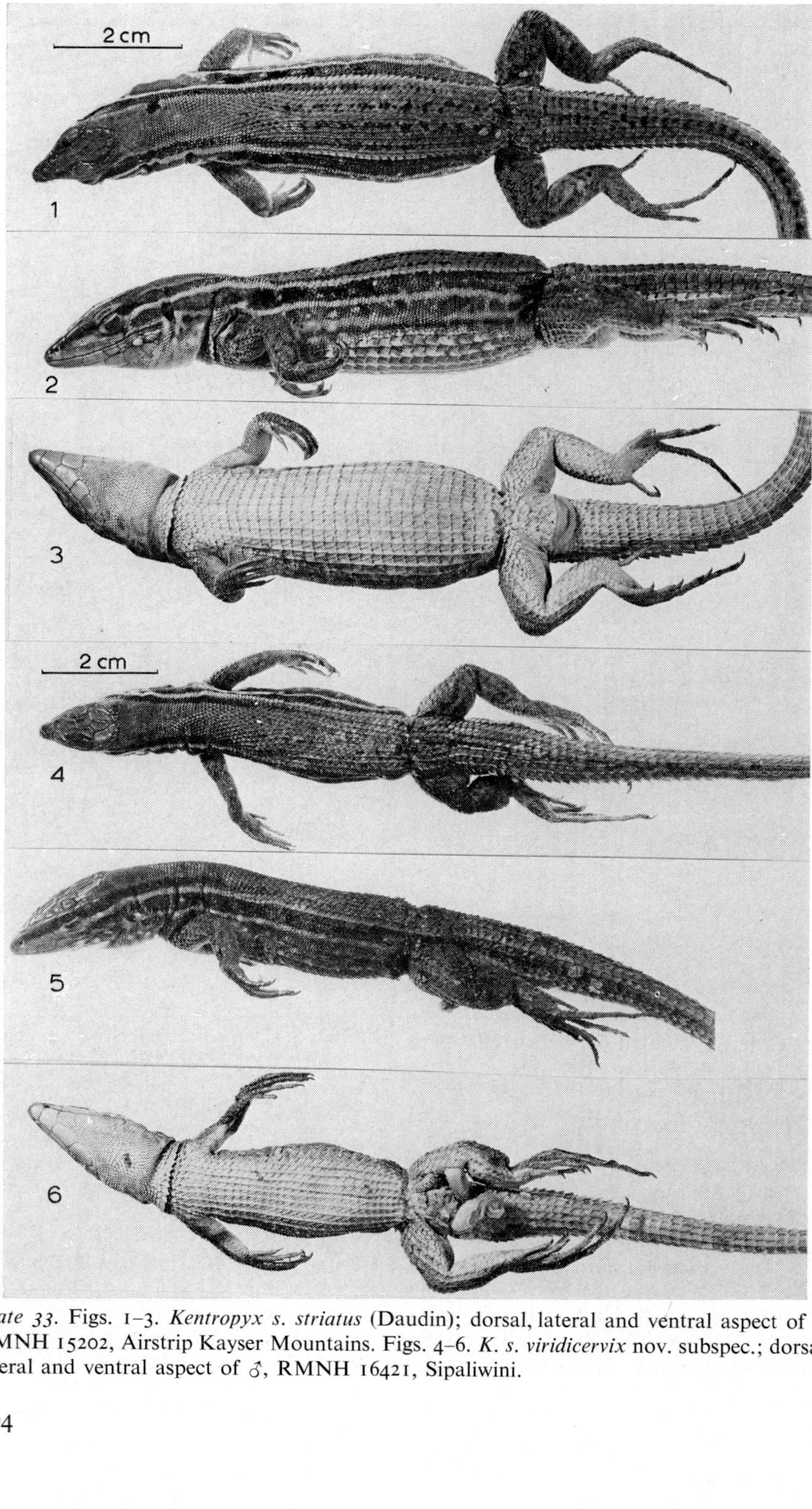

Plate 33. Figs. 1–3. *Kentropyx s. striatus* (Daudin); dorsal, lateral and ventral aspect of ♀, RMNH 15202, Airstrip Kayser Mountains. Figs. 4–6. *K. s. viridicervix* nov. subspec.; dorsal, lateral and ventral aspect of ♂, RMNH 16421, Sipaliwini.

Table 21. Counts and measurements of *Kentropyx s. striatus* (Daudin). For explanation see table 3.

Reg. no.	A	B	C	D	E	F	G	H	I	J	L	M	N	O	P	Q	R	S	V	W	X	Y
RMNH																						
13454	♂	124	173(39)	33.5	19.7	17.8	46	76	6/0-6/0	69	34	14	14	11-10	11-12	4-4	4-5	3-3	4	17-17	26-24	14
13455	♂	?	14(14)	?	?	?	20	49	9/0-9/0	64	32	14	14	?	?	4-4	4-4	?-3	4	?-18	26-26	13
13456	♀	32	72	10.3	6.2	4.9	12	20.5	6/0-6/0	62	31	14	14	10-9	8-7	3-3	4-4	3-3	4	17-16	24-25	13
13457	♀	81	123(30)	20.4	11.0	10.2	17	44	6/0-6/0	64	33	14	14	11-10	9-10	4-4	4-4	3-3	4	16-16	25-26	13
13458	♀	33	11(11)	9.8	5.3	5.0	11	21	8/0-7/0	65	34	14	13	11-11	9-10	4-4	4-4	3-3	4	17-16	26-26	15
13459	♀	76	18(18)	21.3	11.9	?	25	47	7/0-7/0	?	?	14	14	?	?	4-4	4-4	?	4	16-16	?-27	14
13460	♀	32	14	10.4	5.7	4.3	12	21	6/0-7/0	?	34	14	15	11-11	9-11	4-4	4+3-4+2	4-3	4	15-15	23-24	14
13461	♂	92	117(41)	26.4	15.2	13.8	31	60	8/0-7/0	70	34	14	15	10-9	9-10	4-4	4-4	3-3	4	18-?	25-27	15
13462	♀	94	120(120)	23.9	12.7	12.0	30	52	7/0-7/0	68	34	14	16	10-11	11-12	4-4	4-4	3-3	4	15-15	24-24	14
13463	♀	75	199	19.8	10.8	9.7	26	46	6/0-7/0	64	33	14	14	10-10	9-9	4-4	4-4	4-4	4	16-16	24-25	13
13464	♀	73	189	19.0	10.3	9.4	27	45	7/0-7/0	66	33	14	16	10-9	9-8	4-4	4-4	3-3	4	16-17	24-23	13
13465	♂	80	227	22.4	12.0	10.8	28	51	8/0-9/0	73	33	14	16	11-10	9-9	4-4	4-4	3-3	4	18-17	27-27	15
13466	♂	92	251	24.5	13.5	13.9	33	60	7/0-7/0	65	33	14	16	11-10	11-11	4-4	4-4	3-3	4	18-19	26-28	16
15202	♀	82	196	20.2	12.0	10.8	25	45	6/0-7/0	69	34	14	15	10-9	6-6	4-4	4-4	3-3	4	17-16	23-25	16
15243	♀	84	146(71)	21.1	11.7	9.7			7/0-5/0	70	34	14	14	10-10	5-5	4-4	4-4	3-3	4	16-15	26-25	14
15244	♂	103	258	28.8	15.4	14.4			7/0-7/0	73	32	14	16	10-10	5-5	4-4	4-4	4-4	4	17-16	24-24	13
15284	♀	72	115(35)	18.2	9.7	8.7	24	45	7/0-7/0	75	32	14	15	10-10	9-9	4-4	4-4	3-3	4	?-16	24-25	15
15285a	♀	89	133(69)	22.0	12.3	10.3	29	?	7/0-7/0	68	33	14	14	10-11	9-9	4-4	4-4	3-3	4	16-16	24-25	14
b	♂	85	219	24.0	12.5	12.3	26	?	7/0-8/0	72	33	14	16	10-9	8-9	4-4	4-4	3-3	4	16-15	24-?	14
16415a	♂	100	253	28.0	13.4	13.1			7/0-8/0	67	33	14	17	9-10	5-6	4-4	4-4	4-3	4	16-16	23-23	14
b	♀	82	198	21.5	11.3	11.8			6/0-6/0	72	33	14	17	9-9	7-6	4-5	4-4	4-4	4	17-17	25-25	14
BM																						
1946.4.3.45	♀	75	181	19.7	10.5	9.4			7/0-7/0	66	32	14	?	9-9	5-5	4-4	4-4	3-3	4	17-16	25-27	14
.46	♀	47	106	13.2	?	?			9/0-8/0	63	33	14	13	8-8	7-7	4-4	4-4	3-3	4	15-15	24-24	15
NMW																						
18761	♂	85	163	23.3	12.3	10.6	30	56	8/0-7/0		33	14		8-9	7-7	4-4	4-4		5			14
AMNH																						
37478	♀	?	?	25.0	13.3	11.4			8/0-8/0	70	33	16	14	?	7-6	3-3	4-4	?	4	17-16	24-24	16
FMNH																						
121081	♀	74	194	20.3	10.6	9.6			6/0-7/0	65	32	14	12	10-9	6-6	4-4	4-4	3-3	4	16-16	24-25	13
CM																						
44412	♂	88	204	25.3	14.3	12.1			7/0-7/0		34	16	20			4-4	4-4		4	17-17	26-27	16
44413		65	159	19.5	9.9	8.4			8/0-7/0		33	16	14			4-4	4-4		4	17-17	28-26	16
ANS																						
9557	♂	87	79(79)	23.4	11.8	10.8			7/0-7/0	?	32	14	16	10-10	7-7	4-4	4-4	3-3	4	14-16	26-26	14

Second Encampment: 1 juv., RMNH 13458, km 13.9, 13–X–1948, leg. Surinam Expedition 1948–49.

Wane Creek N.: 1♀, RMNH 13457, km 6.4, 30–IX–1948, leg. Surinam Expedition 1948–49.

Republiek: 1♀, RMNH 15284, 19–X–1963, leg. D. C. Geijskes.

Matta: 1 juv., RMNH 13455, 12–I–1963, leg. A. M. Husson & P. Staffeleu.

Zanderij: 1♀, BM 1946.4.3.45, 5–X–1938, leg. I. T. Sanderson; 1♀, FMNH 121081, 18–IX–1960, leg. H. A. Beatty.

Powakka: 1♂, 1 ex., CM 44412–13, 21–IX–1966, leg. M. H. de la Fuente.

Tibiti: 1♀, RMNH 13463, km 2, 8–I–1949, 1♀, RMNH 13464, 2nd transect, km 1.1, 12–I–1949, both leg. Surinam Expedition 1948–49.

Railway km 62: 2♂♂, RMNH 13465–66, 26/IV–16/V–1949, leg. A. Brouwer (Surinam Expedition 1948–49).

Railway km 91.5: 1 juv., BM 1946.4.3.46, 20–X–1938, leg. I. T. Sanderson.

Saniki, Tapanahoni River: 1 juv., RMNH 13456, 10–X–1952, leg. D. C. Geijskes.

Airstrip Kayser Mountains: 1♀, RMNH 15202, 12–VIII–1968, 1♀, 1♂, RMNH 15285, 4–VIII–1968, all leg. M. S. Hoogmoed; 1♀, 1♂, RMNH 15243–44, May 1969, 1♀, 1♂, RMNH 16415, late April 1969, all leg. H. M. van Meeuwen.

DIAGNOSIS.—Number of scales around midbody 62–75, number of collar scales 12–17. Maximum snout-vent length 124 mm. Back in life completely green in adults. Males completely orange below, females partly white, partly orange.

DESCRIPTION.—Head wider than the neck; pointed, nearly twice as long as wide; depressed, its depth slightly less than its width. Central part of the head with a depression, dorsolateral parts slightly raised into shallow ridges. Neck as wide as the body.

Rostral pentagonal, largely visible from above. Nasals divided, forming a suture behind the rostral. A hexagonal frontonasal separated from the rostral by the nasals. A pair of trapezoid prefrontals, which are in contact with the first supraocular. A large, hexagonal frontal, as long as wide or slightly longer. A pair of pentagonal frontoparietals. Interparietal elongate, pentagonal; bordered by a pentagonal parietal on each side; followed by two small pentagonal occipitals. Four supraoculars; the fourth smallest, the third largest. A row of granules between the supraoculars and the supraciliaries. Loreal large, hexagonal, anterior border convex, in contact with the frontonasal, the prefrontal, the first supraocular, the first supraciliary, the nasal, the second and the third supralabials, the preocular and the first subocular or the preocular only. Five supralabials to below the centre of the eye; eight to eleven to the corner of the mouth; the posterior ones small. Supraciliaries mostly four, in a few specimens three; the first and the second elongate; the third and the fourth small, quadrangular; the second largest. Supraciliaries followed by a variable number of enlarged supratemporals. Space between these supratemporals and the parietals occupied by small, polygonal, keeled scales. A preocular of variable size, which is mostly in contact with the supralabials, thus separating the anterior subocular and the loreal; or which may be separated from them by the loreal and anterior subocular forming a suture. Three to four suboculars; the second largest. Two or three rows of postocular scales enlarged. Two or three scale rows anterior of the ear-opening enlarged; centre of the temple occupied by small, hexagonal, keeled scales in horizontal rows. Anterior ear-margin finely lobed. Ear-opening large; external auditory meatus short, tympanum clearly visible. Lower eyelid with several enlarged semitransparent palpebrals. Pupil kidney-shaped with the indention ventrally, sometimes panduriform.

A pentagonal mental. A heptagonal postmental. Four pairs of chinshields, of which only the first pair forms a suture; the posterior two pairs separated from the infralabials by small scales. Nine to eleven infralabials, of which the anterior five to seven are large. Scales on chin and throat small, pointed, hexagonal, keeled, with a central patch of distinctly larger scales on the throat.

Scales on the nape keeled; anteriorly round, posteriorly changing gradually into rhomboids. Scales on the side of the neck granular. Collar strongly denticulate, consisting of 12–17, mostly 14–16, rhomboidal, keeled scales, lateral ones smallest, central ones largest.

Scales on the back large, slightly smaller than the ventrals, rhomboidal, imbricate, keeled, arranged in oblique and in 13–16 longitudinal rows, the keels forming continuous ridges. On both sides this area of large scales may be bordered by a row of smaller keeled scales (marginals), without continuous ridges. Ventrals large, rhomboidal, imbricate, keeled, mucronate, arranged in 31–34 transverse and in 14 (rarely 16) longitudinal rows, the keels forming continuous ridges. Scales on the flanks small, rhomboidal, keeled, juxtaposed, diminishing in size towards the ventrals. Total number of scales around midbody 62–75. Preanal plate covered with a few scales of the same size and form as the ventrals in females, truncate and smooth in males. Males with two, dorsally directed preanal spurs on each side of the preanal plate.

Femoral pores six (rarely five) to nine, no difference in number between males and females. Preanal pores absent. Hemipenis of male an elongate cylinder, smooth near its base, with circling plicae near the slightly expanded top. Sulcus spermaticus starting postero-medial, ending on the posterior part of the top.

Scales on the tail large, quadrangular, imbricate, mucronate, keeled, the keels forming continuous ridges, in longitudinal and transverse rows; scales under the tail identical but narrower. Regenerated part of the tail covered with small, keeled scales, not mucronate, in an irregular pattern.

Upper surface of the forelimbs covered with large, rhomboidal, keeled, imbricate scales; lower surface of the forelimbs with small granules. Anterior and lower surfaces of the hind limbs covered with large, pentagonal, keeled, imbricate scales; upper and posterior surfaces of the hind limbs with small granules. Fifteen to 19 single, smooth lamellae under the fourth finger; 23–28 keeled lamellae under the fourth toe. Outer edge of the fourth toe slightly serrated. Toes compressed, fingers subcylindrical. Palms and soles with small, rhomboidal, juxtaposed, tubercular scales.

In life the back is grass-green in both sexes, although in some females it may be less bright than in the males. The green starts at the nape or at the base of the forelimbs and continues on the anterior part of the tail. Also the upper surfaces of forelimbs and hind limbs are green. Dorsolateral and lateral lines greenish white. Spots on the flanks bluish white. Lateral bands brown. Belly pale orange in the male, white in the female; throat, posterior part of the belly, underside of the hind limbs, preanal plate and underside of the tail orange in both sexes.

In preservative the back is light-brown, with or without two paravertebral rows of dark-brown spots. Tail dorsally with a median row of dark-brown spots. The creamy dorsolateral line starts at the tip of the snout, passes under the eye and over the tympanum and is traceable to the base of the hind limbs. In its anterior part it is bordered dorsally by a light- to dark-brown (in juveniles nearly black) band, narrowing posteriorly and disappearing halfway between the forelimbs and the hind limbs. Ventrally it is bordered by a dark-brown band which starts posterior of the eye and continues on to the tail. In this band, just posterior of the forelimbs, a number of large (in males) or small (in females) white spots, accompanied by some black spots occur. Ventrally this band is bordered by a white line, starting at the corner of the mouth and continuing on to the side of the tail. Between this lateral line and the ventral scales is a light-brown zone with many small white spots. Ventral parts white, immaculate. In juveniles the striped pattern is more distinct than in adults.

A lizard of medium size, maximum snout-vent length in females 84 mm, in males 124 mm. The undamaged tail 2.4–2.7 times as long as the snout-vent length in females, 2.4–3.0 times in males and 2.3–2.8 times in juveniles. Forelimbs just more than half the length of the hind limbs. Limbs considerably overlapping when laid along the body.

HABITAT.—This species is a typical savannah dweller. It occurs in the belt of coastal savannahs and also in small, isolated savannahs surrounded by rain-forest as they are present near airstrip Kayser Mountains (HOOGMOED, 1969a) and near Saniki on the Tapanahoni River. During my visit to the airstrip Kayser Mountains the savannahs there were very wet, in some places even covered with a few centimetres of water. Also two specimens collected by the Surinam Expedition 1948–49 (RMNH 13461, 13462) were taken in very wet habitats, one being described as rush land, the other as wet *Cyperus giganteus* swamp. P. MÜLLER (1971a) observed *K. striatus* (Daudin) on Marajó Island, Brazil, in a savannah habitat that for several months each year is submerged. The lizards there slept in the tops of bushes emerging from the water. BOOS & QUESNEL (1971), in discussing the occurrence of *K. striatus* (Daudin) on Trinidad, say that it is present throughout the cane belt. However, it does not occur in the sugar-cane fields but lives in shrubs and bushes bordering ditches and streams running through the plantations. Moreover, it occurs in two savannah areas: 1. O'Meara Savanna with a poor, waterlogged soil, a scrub vegetation of grasses and small herbs and with only two species of trees (*Byrsonima crassifolia* and *Curatella americana*). 2. Aripo Savanna, a seasonal swamp area with an open grass-sedge community with scattered shrubs and islands of taller vegetation. All Trinidad localities from which *K. striatus* (Daudin) is known agree in being near water and in having a low vegetation. The authors conclude that to *K. striatus* (Daudin) the presence of water is very important and they report having seen specimens take to the water when pursued and in the wet season specimens are present on islands of floating vegetation and hummocks of earth projecting above the water. Boos & Quesnel (1971) also noted that it was arboreal to a certain degree, which might also be concluded from P. Müller's (1971a) data.

Thus, all observations pertaining to the habitat of this form agree on the close link between *K. striatus* (Daudin), water and low vegetation. In Surinam no indications of an arboreal way of life could be found and from my data I conclude that it is terrestrial, which certainly does not exclude the possibility that this form may be arboreal during certain seasons.

Natural history.—A diurnal species I collected on two occasions at about 3.00 p.m. (RMNH 15202, 15285). All other specimens lack time data, but were collected in daytime. Basking in exposed places (rockslates, patches of bare ground) was observed at the airstrip Kayser Mountains. Boos & Quesnel (1971) made similar observations on Trinidad. One specimen (RMNH 15202) when discovered tried to escape attention by pressing itself flat against the ground, between tufts of grass. Pregnant females, containing three to five mature eggs were found in January, October and November. Recently hatched juveniles are as yet only known from October. This might indicate a breeding season approximately coinciding with the long dry period. On Trinidad eggs were laid in captivity in July and August, one batch containing eight or nine eggs. Incubation time was almost exactly three months. Six hatchlings averaged 33 mm snout-vent length and 71 mm taillength (Boos & Quesnel, 1971). The contents of six stomachs were examined. They contained two spiders, two grasshoppers, three beetles, one parasitic wasp, one dragonfly, one cicada, one caterpillar and one cocoon. Moreover they contained some remains that might belong to two flies, one thysanuron and one caterpillar or a fly maggot. Similar observations were made on Trinidad by Boos & Quesnel (1971).

Range.—Surinam, Guyana, Brazil (Amapá, Pará), Venezuela (Amazonas), Trinidad and Colombia east of the Andes. In Surinam this lizard is restricted to the savannah belt parellel to the coast and to some small savannahs in the interior. The localities known for this species in Venezuela and Colombia are over 400 m above sea level (Tamsitt & Valdivieso, 1963). In Surinam it occurs from sea level to at least 200 m (fig. 60).

Remarks.—The Academy of Natural Sciences in Philadelphia possesses a specimen of this form (ANS 9557) originating from Surinam and donated to this museum by Dr. C. Hering. In 1861 Cope described *Centropyx decodon* from Surinam. The description of this species contains only rather general characters except for the number of femoral pores, which are stated to be seven or eight. ANS 9557 agrees in all characters with the description of *C. decodon* except in the number of femoral pores, which is seven on both sides. It is probable that COPE had before him more than one specimen, the series showing a variation in the number of femoral pores of seven to eight. Thus, it seems justified to select ANS 9557 as the lectotype of *Centropyx decodon* Cope, as all further data agree with COPE's description. A short description of the specimen follows: male, snout-vent length 87 mm, the posterior part of the tail is missing, the remaining part is 79 mm. The number of longitudinal rows of dorsals is 14 with a row of smaller marginals on each side, 14 longitudinal rows of ventrals, total number of scales around midbody is 66, but this number may be slightly higher or lower, as the right flank has been cut open and it is difficult to establish how exactly both parts of the cut fit together. There are five supralabials to beneath the centre of the eye, the large subocular is bordering on the mouth directly posterior of the fifth supralabial. Preocular on the left side of the head separated from the supralabials, on the right side in contact with them. Seven femoral pores on each side. The number of lamellae under the fourth finger is 14 and 16 for left and right side, respectively, 26 under each fourth toe. Locality and collector: Surinam, Dr. C. HERING.

There is some noteworthy variation in the shape and position of the preocular. In ten out of 12 specimens from the coastal region, examined for this character, it is in contact with the supralabials, in the other two it is separated from them; the lectotype of *Centropyx decodon* Cope has the preocular in contact with the supralabials on the right, separated from them on the left side. The specimen from Saniki agrees with the majority of the coastal population in having the preocular in contact with the supralabials. The small series from the airstrip Kayser Mountains is remarkable, as only in three specimens the preocular is in contact with the supralabials, while in four specimens it is separated from them.

The specimens from the Sipaliwini savannah, described below as a new subspecies, agree in this respect with the coastal populations. Of 16 specimens examined for this character 15 had the preocular in contact with the supralabials, the holotype of the new subspecies shows this condition on the left side, on the right side the preocular is absent and has probably fused with the first subocular.

TAMSITT & VALDIVIESO (1963) suppose that *K. striatus* (Daudin) has a different habitat preference from that of *K. calcaratus* Spix. This view is based on the fact that the Venezuelan and Colombian localities where *K. striatus* (Daudin) has been found are over 400 meters in elevation and in regions with pronounced wet and dry seasons. According to them *K. calcaratus* Spix has only been reported from regions at lower elevations with heavy rain throughout the year. From my Surinam data it is clear that there is really a difference in habitat, but this is not correlated with the elevation above sea level or the rainfall, at least, not in Surinam. In Surinam there is absolutely no difference in elevation, the species being found sometimes only within metres of each other. The only difference is in the vegetation and the amount of sunshine, *K. calcaratus* Spix being found only in the forest (where, however, it is a heliotherm), *K. striatus* (Daudin) only being found on savannahs. The amount of rainfall for savannahs and adjacent rain-forest is identical (LANJOUW, 1936), as are the pronounced

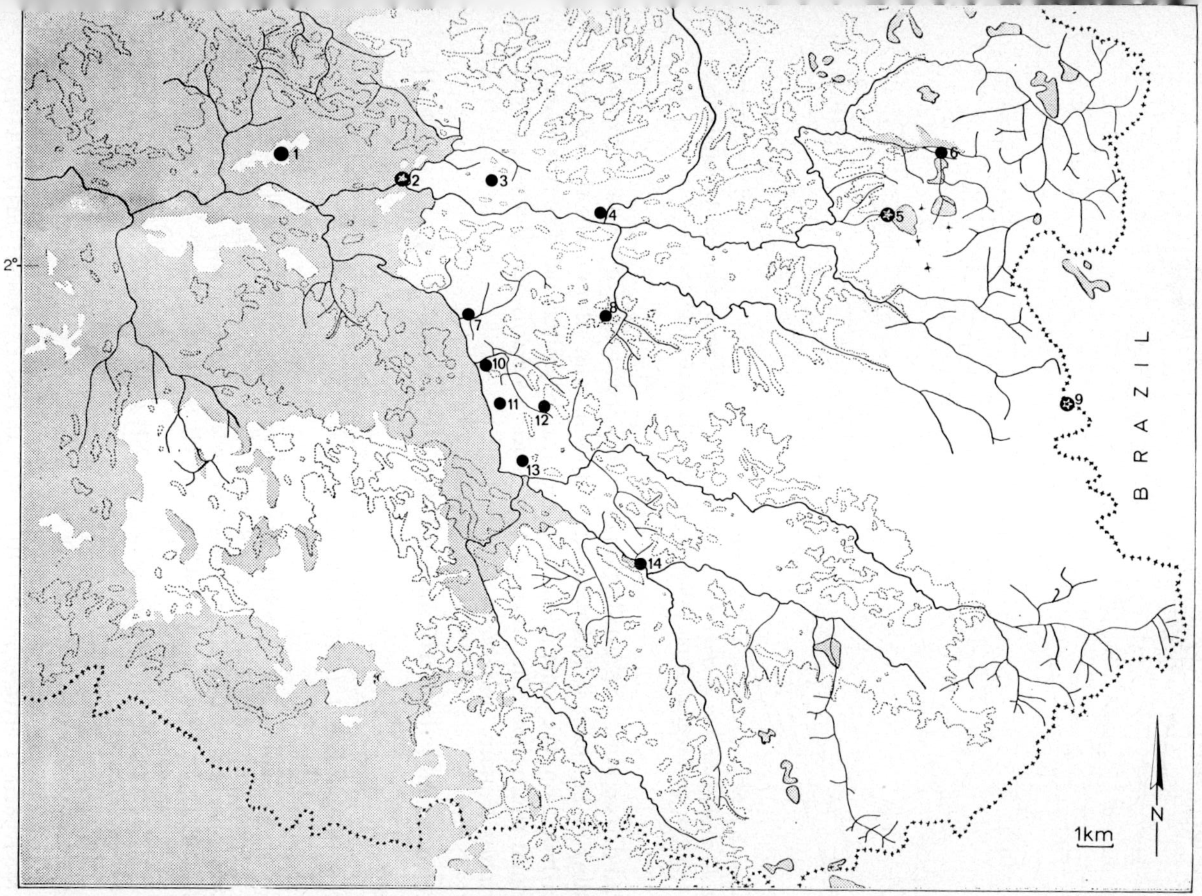

Fig. 61. Detailed map of the Sipaliwini savannah, showing localities where *Kentropyx striatus viridicervix* nov. subspec. was collected (solid circles). For further explanation see fig. 13. 1. airstrip Sipaliwini, 2. Base Bivouac, 3. 6 km E. of airstrip, 4. between Base Bivouac and Vier Gebroeders Bivouac, 5. Vier Gebroeders Bivouac, 6. 1–2 km N. of Vier Gebroeders Mountain, 7. Meyers' farm, 8. 4 km E. of Meyers' farm, 9. airstrip Meyers', 10. 1.5 km S. of Meyers' farm, 11. 3 km S. of Meyers' farm, 12. 4 km SE. of Meyers' farm, 13. between Meyers' farm and South Bivouac, 14. South Bivouac.

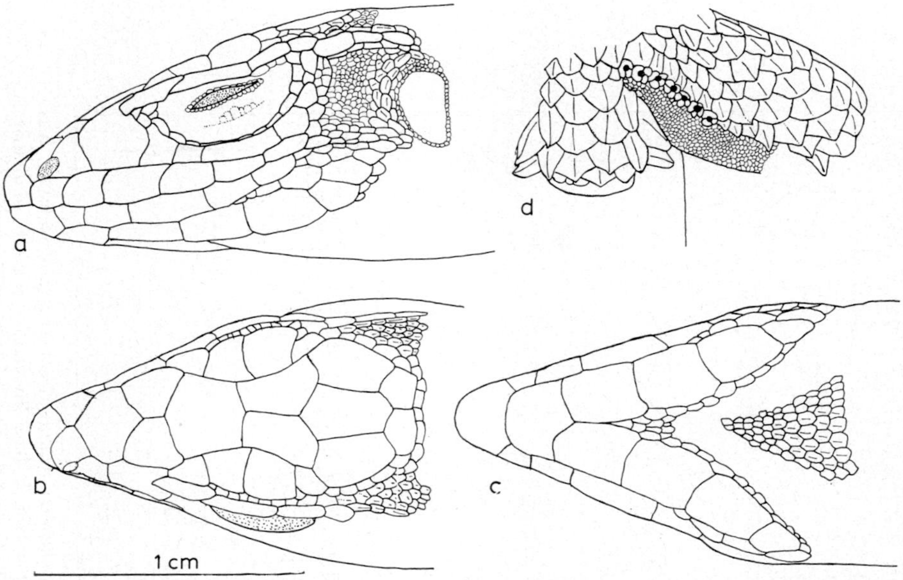

Fig. 62. Kentropyx striatus viridicervix nov. subspec.; a, b, c: lateral, dorsal and ventral views of the head, d: anal region of the holotype, ♂, RMNH 15167, Sipaliwini.

wet and dry seasons. The distribution of both *K. calcaratus* Spix and *K. striatus* (Daudin) in Surinam is independent of the amount of rainfall, the rainfall diminishing from the coast inland in a southeasterly direction. The two lizards are found throughout the country, that is *K. calcaratus* Spix only in forest, *K. striatus* (Daudin) only on savannahs.

Kentropyx striatus viridicervix nov. subspec. (figs. 61, 62; pls. 33, 34; table 22)

MATERIAL.

HOLOTYPE: – 1♂, RMNH 16421, Sipaliwini, 4 km E. of Meyer's farm, Surinam, 4–II–1970 leg. M. S. Hoogmoed & J. J. P. Paats.

PARATYPES.- Surinam.

Sipaliwini: 1 juv., RMNH 15283, February 1961, leg. D. C. Geijskes.

Airstrip Sipaliwini: 1♀, 1♂, RMNH 16419, 13–II–1970, 1♂, RMNH 16423, 23–I–1970, 2♀♀, 2♂♂, RMNH 16430, 11–II–1970, all leg. M. S. Hoogmoed & J. J. P. Paats; 2♂♂, RMNH 15167, 6 km E., 3–IX–1968, 1♀, RMNH 15203, 6 km E., 28–VIII–1968, both leg. M. S. Hoogmoed.

Vier Gebroeders Mountain: 2♂♂, RMNH 15204, 1–2 km N., 20–IX–1968, leg. M. S. Hoogmoed.

Between Base Bivouac and Vier Gebroeders Bivouac: 1♀, RMNH 16425, 6–II–1970, leg. M. S. Hoogmoed & J. J. P. Paats.

Meyers' farm: 1♂, RMNH 16418, 1–II–1970, 1♀, RMNH 16424, 30–I–1970, 2♀♀, RMNH 16428, 31–I–1970, 1♂, RMNH 16429, 5–II–1970, 1♂, RMNH 16431, 28–I–1970, 1♂, RMNH 16417, 4 km SE., 2–II–1970, 1♀, RMNH 16420, 1.5 km S., 20–I–1970, 1♀, 1♂, RMNH 16427, 3 km S., 29–I–1970, all leg. M. S. Hoogmoed & J. J. P. Paats.

Between Meyers' farm and south Bivouac: 1♀, RMNH 16416, 20–I–1970, leg. M. S. Hoogmoed & J. J. P. Paats.

Plate 34. Fig. 1. *Kentropyx borckianus* W. Peters; dorsal aspect of juv., BM 89.9.30.6, Demerara. Fig. 2. *K. calcaratus* Spix; dorsal aspect of juv., RMNH 15301, Paramaribo. Fig. 3. *K. striatus viridicervix* nov. subspec.; dorsal aspect of juv., RMNH 15283, Sipaliwini.

South Bivouac: 1♀, RMNH 16422, 20–I–1970, 1♂, RMNH 16426, 24–I–1970, leg. M. S. Hoogmoed & J. J. P. Paats.

DIAGNOSIS.—Number of scales around midbody 56–65, number of collar scales 12–17. Maximum snout-vent length 92 mm. In life only the nape and the side of the neck are green, the rest of the back is brown. Both males and females below partly white, partly orange.

DESCRIPTION.—In most lepidosis characters this subspecies agrees with *K. s. striatus* (Daudin), so I can refer to the preceding description. Only the differences will be pointed out below. Collar scales 12–15, mostly 13–14. Total number of scales around midbody 56–65, mostly 57–59; enlarged dorsals in

Table 22. Counts and measurements of *Kentropyx striatus viridicervix* nov. subsp. For explanation see table 3.

Reg. no. RMNH	A	B	C	D	E	F	G	H	I	J	L	M	N	O	P	Q	R	S	V	W	X	Y
15167a	♂	58	143	17.0	9.0	7.9	18	37	7/0-7/0	65	33	14	13	11-11	5-5	3-3	4-4	3-3	4	16-17	26-25	14
b	♂	?	141	16.1	9.0	7.1	20	35	8/0-8/0	?	?	14	13	10-10	5-6	3-3	4-4	3-3	4	?-18	27-26	13
15203	♀	45	126	13.5	8.5	6.7	18	32	6/0-7/0	?	33	14	14	?-11	5-5	?-4	?-4	3-3	4	16-16	23-24	14
15204a	♂	65	142(69)	18.2	10.3	8.8	22	42	7/0-7/0	58	33	14	14	10-11	5-5	4-3	4-4	3-3	4	17-19	25-28	15
b	♂	70	208	20.4	11.0	9.1	27	46	8/0-8/0	61	33	14	14	11-11	5-6	4-4	4-4	3-3	4	18-17	26-26	13
15283	♀	44	99	12.7	7.0	5.9	14	27	6/0-7/0	62	34	14	14	10-10	9-9	3-4	4-4	3-3	4	16-18	24-24	14
16416	♀	87	250	23.3	12.0	10.7			8/0-8/0	59	35	14	14	10-10	5-5	3-4	4-4	3-3	4	19-18	26-27	13
16417	♂	85	249	24.8	12.5	12.1			7/0-7/0	60	33	14	13	9-9	5-5	3-3	4-4	3-3	4	17-16	?-24	14
16418	♂	87	200(86)	24.4	12.3	11.1			8/0-8/0	59	33	14	13	9-9	5-5	3-3	4-4	3-3	4	19-18	26-26	13
16419a	♀	66	86(86)	19.3	10.2	8.8			7/0-7/0	56	33	14	14	9-10	6-7	4-4	4-4	3-3	4	17-16	25-24	13
b	♂	81	217	23.1	12.4	11.1			7/0-7/0	65	33	14	14	10-10	6-6	4-4	4-4	3-3	4	17-17	26-23	14
16420	♀	74	136(45)	20.0	10.4	9.2			8/0-7/0	59	33	14	?	?-9	?-6	4-3	4-4	3-3	4	18-18	26-27	13
16421	♂	73	198	19.8	11.3	8.9			? -8/0	58	33	14	14	10-10	5-5	4-3	4-4	3-3	4	18-18	26-24	13
16422	♀	64	179	17.4	9.2	7.3			6/0-7/0	57	33	14	14	10-9	5-5	3-4	4-4	3-3	4	18-17	26-27	12
16423	♂	72	250	20.8	10.6	9.4			8/0-7/0	57	33	14	12	10-10	6-5	4-4	4-4	3-3	4	17-16	28-26	12
16424	♀	78	158(92)	22.0	11.5	10.3			6/0-7/0	63	34	14	13	10-9	5-5	3-3	4-4	3-3	4	16-17	24-25	13
16425	♀	76	201	19.6	11.3	9.6			8/0-8/0	59	34	14	13	10-10	5-5	3-4	4-4	3-3	4	18-18	28-26	13
16426	♂	92	260	25.7	13.7	11.7			8/0-8/0	59	34	14	14	12-11	6-6	4-4	4-4	3-3	4	?-18	27-27	14
16427a	♀	58	136	16.1	8.9	7.2			7/0-7/0	61	34	14	13	10-8	6-6	3-3	4-4	3-3	4	17-17	26-27	13
b	♂	64	120(67)	18.3	9.2	8.1			7/0-7/0	57	32	14	12	11-11	6-6	4-3	4-4	3-3	4	19-17	25-25	13
16428a	♀	80	235	22.0	11.7	10.3			7/0-8/0	63	34	14	14	11-11	6-6	3-3	4-4	3-3	4	17-17	25-28	14
b	♀	65	121(48)	17.5	10.1	8.4			6/0-7/0	58	33	14	13	10-10	6-6	3-3	4-4	3-3	4	16-16	24-25	13
16429	♂	76	175(175)	20.7	10.5	9.4			8/0-8/0	58	34	14	14	10-10	6-6	4-4	4-4	3-3	4	17- ?	29-28	13
16430a	♀	45	27(27)	13.8	7.4	6.1			8/0-8/0	59	34	14	?	12-12	6-6	4-4	4-4	3-4	4	18- ?	28-27	14
b	♂	69	182	19.7	9.7	8.6			8/0-8/0	60	34	14	14	9-?	6-6	4-4	4-5	3-3	4	17-16	25-27	14
c	♂	58	174	?	?	?			8/0-8/0	?	32	14	15	9-9	6-6	4-4	4-4	3-3	4	16-16	25-24	?
d	♀	65	73(73)	17.8	9.4	8.4			7/0-7/0	57	34	14	13	10-10	6-7	4-4	4-4	3-3	4	?	24-?	13
16431	♂	54	145	15.4	8.6	7.4			8/0-7/0	57	33	14	13	?	5-5	4-4	4-4	3-3	4	18-18	28-28	12

12–15, mostly 13–14, longitudinal rows; ventrals always in 14 longitudinal rows. Number of lamellae under the fourth finger 16–19, under the fourth toe 23–29.

In life the nape and the anterior part of the back are dull green, the rest of back and tail brown. The anterior part of the dorsolateral and the lateral lines is pale green, the posterior part creamy. Flanks with pale blue spots. Chin, throat and anterior part of the belly white, posterior part of the belly and underside of the tail orange-brown. Ventral parts immaculate. Iris reddish-brown. Tongue greyish-brown with blue tips.

In preservative this subspecies resembles the nominate form in all colour characters.

Maximum snout-vent length in females 87 mm, in males 92 mm. The undamaged tail is 2.3–2.9 times as long as the snout-vent length in females, 2.5–3.5 times in males and 2.3–2.8 times in juveniles.

HABITAT.—An inhabitant of savannahs enclosed by rain-forest, where it occurs in the more humid parts, near water, such as along creek-banks, in low-lying parts of the savannah where the vegetation is concentrated in tufts, with bare ground between the tufts, and on the lower part of hills where they border swamps of *Mauritia flexuosa*. It occurs rarely on rockslates and only when the rockslates are situated in low-lying parts of the savannah. The lizards have never been found on the higher parts of the hills where *Bulbostylis spadicea* is present.

NATURAL HISTORY.—A diurnal lizard which was taken between 9.00 a.m. and 5.15 p.m. Basking was observed several times. Females were collected in January, February and August, but not one of them was pregnant. Neither are recently hatched juveniles available. Thus, nothing can be said about the breeding season. The contents of eight stomachs were examined. They contained nine spiders, six grasshoppers, two ants and probably one ant larva, two cicadas, two caterpillars, one fly (?), two bugs and one cricket (?). There seems to be a preference for grasshoppers (in six stomachs) and spiders (in five stomachs). In three stomachs grasshoppers and spiders were found together.

RANGE.—So far known only from the Sipaliwini savannah and some small savannahs west of it (near the airstrip) in extreme southern Surinam. Undoubtedly it will also be present on the Brazilian part of this savannah, known as Paru savannah. The general level of the Sipaliwini savannah is about 250 m above sea level (fig. 61).

Genus **Leposoma** Spix *

DIAGNOSIS.—Small lizards with cylindrical bodies and tails, well developed legs with five digits, and rhomboidal, imbricate, keeled, mucronate dorsals, in

* When the manuscript on *Leposoma* was finished, I recieved a copy of a very important paper by UZZELL & BARRY (published in 1971). These authors came to the same conclusions as I did regarding the status of *L. percarinatum* (L. Müller) and *L. guianense* Ruibal.

oblique and transverse series. All dorsal headscales with longitudinal striations. Tongue lanceolate, without emargination, covered with rhomboidal, imbricate, scale-like papillae in oblique rows. Anterior teeth conical; lateral teeth tricuspid, laterally compressed. Ear-opening large, tympanum superficial. Lower eyelid with semitransparent disc. Pupil round. Seven species from Central America to southern Brazil. Two species present in Surniam, of which one is unisexual.

Key to the Surinam species of *Leposoma*

1. Interparietal very large, with diverging lateral margins, prefrontals with a long median suture . . . . *Leposoma guianense*, p. 316
   Interparietal smaller, lateral margins nearly parallel, prefrontals with a short median suture . . . . *Leposoma percarinatum*, p. 323

Leposoma guianense Ruibal (figs. 63, 64; pl. 35; table 23)

Leposoma guianense Ruibal, 1952: 489; Uzzell & Barry, 1971: 17.
'56. Grondhagedissen' Fernandes, 1927: 249.

Type-locality.—Dunoon, Demarara River, Guyana.

Material. – Surinam.

65 km SSE. of Paramaribo, Afobaka Road: 3♀♀, 1♂, AMNH 108775, 108777–79, 12/13–II–1972, leg. C. W. Myers & J. Daly.

Avanavero Cataracts, Kabalebo River: 1♂, RMNH 13399, 20–IX–1965, leg. G. F. Mees; 1♀, RMNH 16478, 10–IV–1971, 1♂, RMNH 16479, 9–IV–1971, both leg. D. G. Reeder.

Camp 1, Coppename River: 1♀, BM 1946.4.3.99, 8–VII–1938, leg. I. T. Sanderson.

Brown's Mountain: 6♀♀, RMNH 13873, 19–VII–1968, leg. M. S. Hoogmoed; 1♂, RMNH 13878, 25–VIII–1968, leg. W. N. Polder; 3♀♀, RMNH 15219–21, 11–XI–1968, 1♀, RMNH 15222, 12–XI–1968, all leg. R. Hoogmoed-Verschoor & M. S. Hoogmoed.

Raleigh Cataracts, Coppename River: 1♂, AMNH 108780, 19/20–II–1972, leg. C. W. Myers & J. Daly.

Voltz Mountain: 1 juv., AMNH 108781, NW. base, 22/23–II–1972, leg. C. W. Myers & J. Daly.

Nassau Mountains: 1♂, RMNH 13400, km 0.6, 17–II–1949, 1♀, RMNH 13401, km 2, 28–II–1949, 1♀, RMNH 13402, km 8.7, 9–III–1949, 1♂, RMNH 13403, km 9.5, 10–III–1949, 1♀, RMNH 13404, km 9.3, 11–III–1949, 1♂, RMNH 13405, km 9.1, 12–III–1949, 1 juv., RMNH 13406, km 15.7, 22–III–1949, 1♀, RMNH 13407, N. of km 11.2, 24–III–1949, all leg. Surinam Expedition 1948–49.

Wilhelmina Mountains: 1♀, ZMA 12235, 1926, leg. Wilhelmina Mountains Expedition.

King Frederik Willem IV Cataracts, Corantijn River: 1♀, FMNH 30938, 15/VIII–23/VIII–1938, leg. E. R. Blake (Sewell Avery Expedition).

E. of Tigri Encampment, New River: 1♀, RMNH 15228, 28–VI–1968, 1♂, RMNH 15229, 2–VII–1968, both leg. M. S. Hoogmoed.

Airstrip Kayser Mountains: 1♂, RMNH 13875, NE., 12–VIII–1968, 1♂, RMNH 13880, E., 8–VIII–1968, 1♂, RMNH 15383, NE., 3–VIII–1968, all leg. M. S. Hoogmoed.

Airstrip Paloemeu: 2♀♀, 3♂♂, RMNH 15007, S., 28–V–1968, 1♀, RMNH 15149, N., 30–V–1968, 2♀♀, 1♂, RMNH 15242, S., 23–V–1968, all leg. M. S. Hoogmoed; 2♀♀,

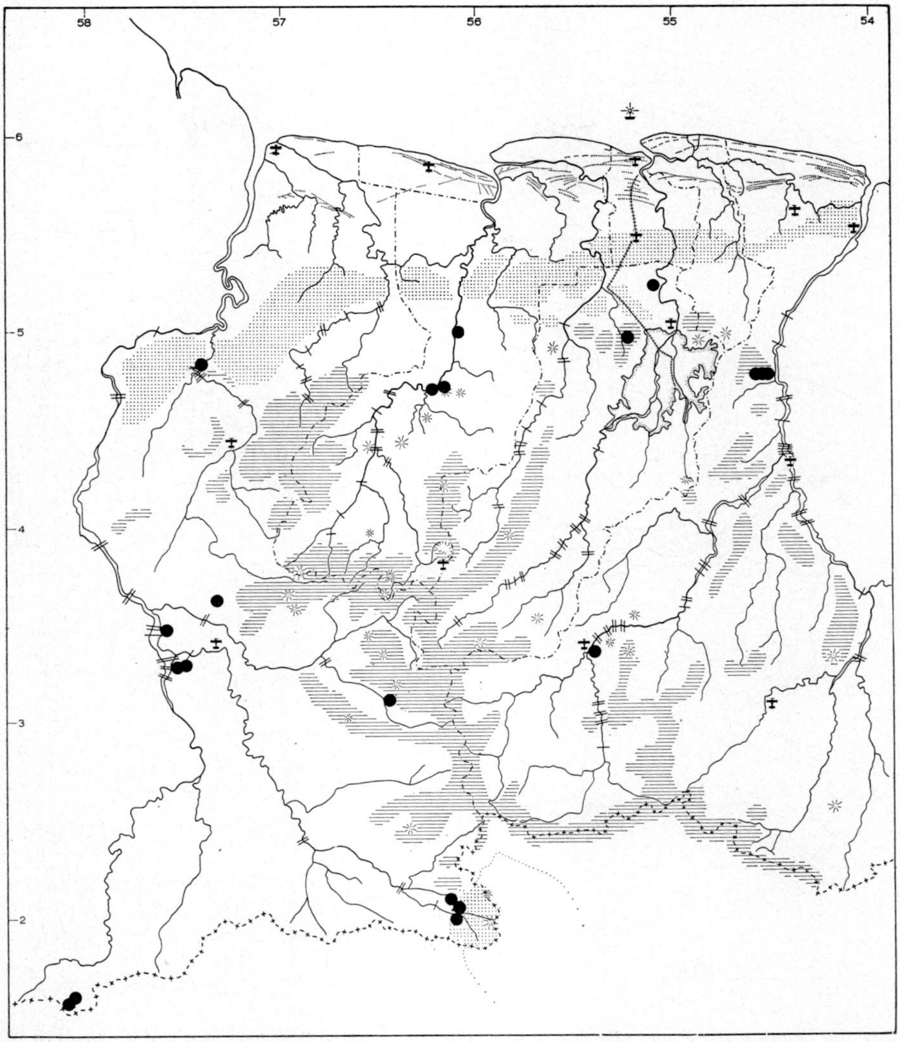

Fig. 63. Distribution of *Leposoma guianense* Ruibal in Surinam.

RMNH 15716, 4–V–1969, leg. H. M. van Meeuwen.

Head of navigation, Itabu Creek: 1 ex., FMNH 30936, 18–IX–1938, leg. E. R. Blake (Sewell Avery Expedition).

Boundary Camp, Itabu Creek: 1♀, 2♂♂, FMNH 30937, 30939–40, 19/IX–19/X–1938, leg. E. R. Blake (Sewell Avery Expedition).

Airstrip Sipaliwini: 1♂, RMNH 13874, 4 km E., 30–VIII–1968, 1♂, RMNH 13879, 4 km E., 28–VIII–1968, 1♀, RMNH 13881, 2 km E., 5–IX–1968, all leg. M. S. Hoogmoed; 1♀, RMNH 16378, 3 km NW., 9–II–1970, 1♀, RMNH 16379, 4 km S., 11–II–1970, 1♀, 3♂♂, RMNH 16380, 3–6 km NW., 12–II–1970, all leg. M. S. Hoogmoed & J. J. P. Paats.

N. of Base Bivouac, Sipaliwini: 1♀, RMNH 13876, 4–IX–1968, 2♀♀, 1 juv., RMNH 13877, 2–IX–1968, all leg. M. S. Hoogmoed.

Sipaliwini: 1♂, RMNH 17279, May–July 1972, leg. G. F. Mees.

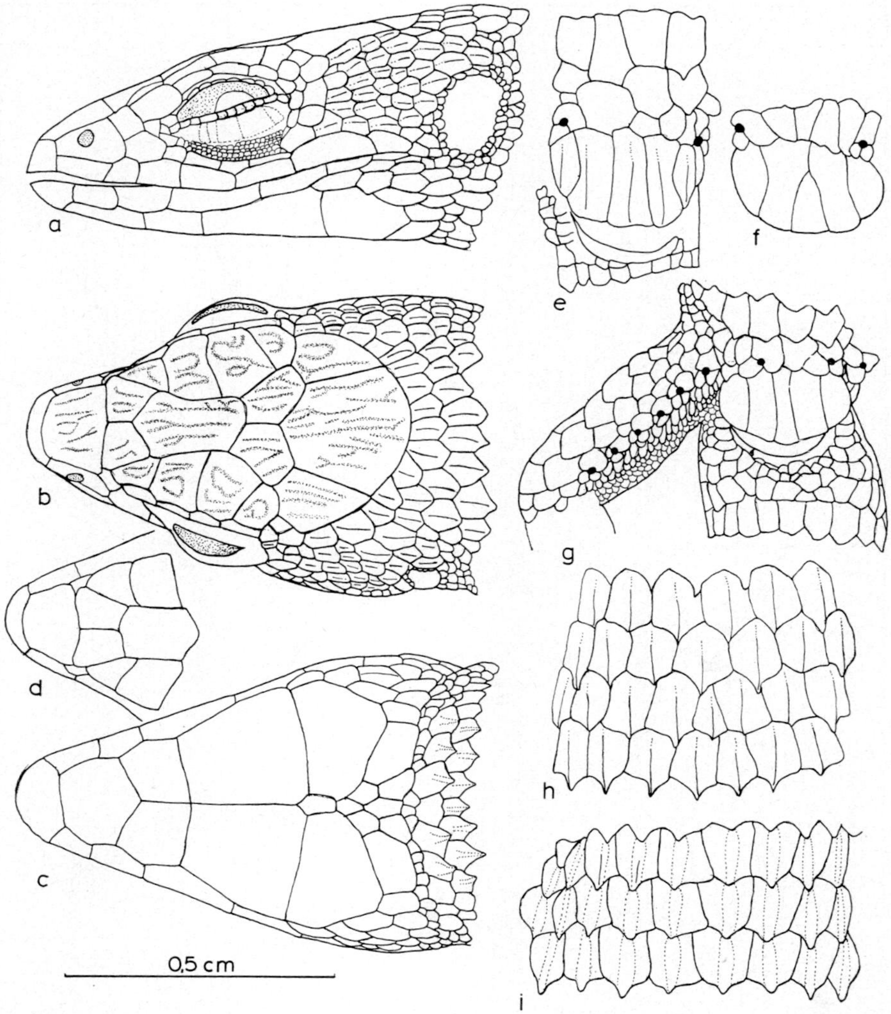

Fig. 64. Leposoma guianense Ruibal; a, b, c: lateral, dorsal and ventral views of the head of ♀, RMNH 15219, Brown's Mountain; d: tip of snout of ♂, RMNH 15242 b, Airstrip Paloemeu; e: anal region of ♀, RMNH 15219; f: anal region of ♀, RMNH 15149, Airstrip Paloemeu; g: anal region of ♂, RMNH 15242 b; h: dorsals of RMNH 15219; i: ventrals of RMNH 15219.

Guyana.
Moraballi Creek, Essequibo River: 1♀, BM 1930.10.10.168, leg. T. A. W. Davis.
French Guiana.
Sophie: 1♂, 1 juv., MCZ 77502–03.
Between Sophie and Dorlin: 1♀, MCZ 77501
Upper Carsevenne: 1♀, MP 98–182, leg. Geay.
Camopi River: 1♀, 3 juvs., MP 02–261–64, leg. Geay.

Plate 35. Figs. 1–3. *Leposoma guianense* Ruibal; dorsal, lateral and ventral aspect of ♂, RMNH 13874, Sipaliwini. Figs. 4–6. *L. percarinatum* (L. Müller); dorsal, lateral and ventral aspect of ♀, RMNH 13882, Sipaliwini.

DIAGNOSIS.—A small lizard with a cylindrical body, well developed, short legs and a long tail. Dorsal scales rhomboidal, in 31–35 transverse rows. Interparietal very large with diverging lateral margins; prefrontals with a long median suture; preanal scales elongate; four pairs of chinshields. Number of scales around midbody 22–26; fourth finger with 10–13, fourth toe with 13–19 subdigital lamellae. Back brown with lighter dorsolateral lines and two series of black spots. Side of tail with a series of white spots.

DESCRIPTION.—Head small, pyramidal, depressed; back of the head slightly narrower than the neck, which is slightly swollen and wider than the anterior part of the body. Snout pointed, rounded.

Rostral hexagonal, visible from above, twice as wide as deep. A large, pentagonal frontonasal separating the nasals. A pair of trapezoid prefrontals; longer than wide; forming a long median suture in most specimens. Frontal elongate, hexagonal; once and a half as long as wide. A pair of pentagonal frontoparietals; forming a short median suture, which is as long as or shorter than that between the prefrontals. Interparietal very large, with four straight and a convex posterior margin, its sides diverging posteriorly. Parietals irregularly pentagonal, much smaller than the interparietal. Occipitals absent. Four supraoculars, in contact with the supraciliaries; fourth smallest; second and third subequal and largest. All dorsal headscales with longitudinal striations. Nostril in an undivided nasal. Loreal deeper than long; separated from the labials by the frenoocular, which is longer than deep. Preocular absent. Four to seven suboculars. Two rows of postocular scales; four to five scales in the posterior row. Supraciliaries four, rarely three, five or six. Temporal region with polygonal, keeled, juxtaposed scales; fewer and larger than in *L. percarinatum* (L. Müller); larger near the parietals. Ear-opening large, surrounded by small, granular scales, forming a finely lobed margin; external auditory meatus absent, tympanum well visible. Supralabials seven, rarely eight; the sixth largest, four to below the centre of the eye. Lower eyelid with a semitransparent disc of three to five palpebrals. Iris round.

Mental nearly semicircular. A large heptagonal postmental. Four (rarely three) pairs of chinshields; the anterior two forming median sutures, that of the second pair once and a half as long as that of the anterior pair; third pair separated by one row of small scales; fourth pair by several larger and smaller scales; second pair largest; fourth pair smallest. Six, rarely five or seven, infralabials; third largest. The third and fourth pairs of chinshields separated from the labials by small scales.

All large anterior dorsal and lateral headscales with a row of small pits along their margin, the posterior dorsal headscales have this row only along their posterior margin; all headscales with scattered pits.

Gulars rhomboidal, mucronate, keeled, imbricate, in oblique and in eight to ten transverse rows. Collar distinct, with eight to twelve scales; the central eight strongly mucronate, forming a serrated posterior margin. Scales on the nape rhomboidal, as long as wide, keeled, mucronate, imbricate, in transverse and in oblique rows.

Dorsals and laterals rhomboidal, keeled, mucronate, imbricate, in oblique, and the dorsals in 31–35 transverse, rows. Ventrals larger than the dorsals, square, mucronate, imbricate, keeled; the keels blunt, wide or narrow; in six longitudinal (keels continuous) and in 20–25 transverse rows. No sharp demarcation between the laterals and the ventrals. Total number of scales around midbody 22–26. Preanal plate with elongate, keeled scales, in few specimens there are five elongate preanals, in most the median is divided in an anterior and a posterior median preanal or the anterior is absent, leaving four elongate preanals with a wedge-shaped median preanal.

Preanal pores zero to two in females, always two in males. Femoral pores absent in females, five or six on each side in males. The pores in the centre of a group of three to four scales.

Scales on the tail rectangular, mucronate, sharply keeled, in transverse and longitudinal rows; scales under the tail narrower; the keels forming 14–17 longitudinal ridges. Scales on the regenerated part of the tail smaller, rectangular, not mucronate, keeled, only slightly imbricate, in transverse and in longitudinal rows. Tail round with 14–17 ridges.

Scales on the upper and anterior surfaces of the forelimbs, on the anterior and lower surfaces of the thighs and on all surfaces of the lower legs large, rhomboidal, keeled and imbricate; on the lower and posterior surfaces of the forelimbs and on the posterior and upper surfaces of the thighs small, granular, juxtaposed. Ten to 13 double lamellae under the fourth finger; 13–19, mostly 14–17, under the fourth toe. Fingers and toes subcylindrical. Palms and soles with small, flat, juxtaposed granules. Fingers connected by traces of a web, most distinct between the second and the third, and between the third and the fourth.

Back brown in life with lighter dorsolateral lines. Belly and underside of tail in females ochre-yellow, chinshields white, sides of throat orange. In males the underside is orange-red, throat, chin and upper lips purplish red. Iris reddish brown.

In preservative the back is orange- to dark-brown with or without scattered black spots. Light-brown dorsolateral lines from the neck onto the tail where they are broken up into two rows of spots. Ventrally they are bordered by a black line or series of black spots. In many specimens they are bordered medially by a series of black spots, which is very much reduced or absent in other specimens. Flanks grey-brown, with or without black spots, a white ocellus above the insertion of the forelimbs, which may or may not be part of a series of ocelli from the ear to halfway along the groin. Lips with diffuse brown spots, which extend onto the lateral parts of the chinshields, rest of the underside immaculate, white. Tail with a dorsolateral series of light-brown spots on each side, joining posteriorly of the proximal quarter of the tail. A lateral series of white spots from the hind limbs to the tip of the tail. Underside of the tail white with brown spots, which frequently form transverse bands, narrow medially, wide laterally enclosing a series of white spots. Hind limbs brown, one or two creamy spots on the posterior surface of the thighs; demarcation between dorsal and ventral colour on the back of the thighs not sharp. Tongue dark-grey, white posteriorly, with a straight transverse border between the two colours.

A small lizard, maximum snout-vent length 39 mm in females, 37 mm in males. Hind limb 1.3–1.7 times as long as forelimb in females, 1.4–1.6 times in males. When the limbs are laid along the body there is overlap of fingers and toes. The undamaged tail is 1.4–1.8 times as long as the snout-vent length in females, 1.5–2.0 times in males.

HABITAT.—Inhabits the rain-forest floor in moist, shady places where it is covered with leaf-litter, mostly near creeks.

NATURAL HISTORY.—A diurnal species, which never was seen basking. The species seems to appreciate high miosture, as several times immediately after heavy showers, specimens were collected, whereas before the shower this species was not to be found. When pursued it tried to escape by hiding among the leaf-litter. Recently hatched juveniles (18–20 mm) were found in February, May, August and September; pregnant females, containing one or two (probably the normal number) mature eggs, in March. These scanty data might indicate that breeding takes place throughout the year, except in the long dry period. Three stomachs were opened and contained: RMNH 13403: two sow-bugs, one cockroach; RMNH 13404: two sow-bugs, one spider, one small homopteron, one beetle-larva; RMNH 13405: legs of grasshopper, one oribatid which probably was eaten accidentally. One stomach also contained a number of nematodes.

RANGE.—Surinam, Guyana, French Guiana and probably northeastern Brazil. In Surinam throughout forested areas south of the coastal savannah belt, to at least 500 m above sea level (fig. 63).

REMARKS.—In describing *L. guianense*, RUIBAL stressed the importance of the elongate median preanal scale and the absence of light dorsolateral stripes in his material. However, among the specimens from Surinam investigated, five preanals are an exception; only eleven out of 55 specimens showed this condition, moreover two have four elongate preanals, one has the middle one divided in three scales, an elongate anterior and two small posterior ones, 23 have the median preanal divided into a long anterior and a short posterior scale and 18 only have a short median preanal. In several series, each from one locality, the three most frequent conditions occur together. Regarding the absence of any pattern in the type specimens this may be either due to preservation or represent a condition of the Dunoon population. In contrast to these pattern-less types, all specimens from Surinam, Guyana and French Guiana studied by me have the distinct pattern as described above. Unfortunately I could not study the types myself, but judging by the data available, the specimens here described undoubtedly belong to the same species and the differences between the types and the Surinam specimens are merely due to individual variation. For a comparison with *L. percarinatum* (L. Müller) see under that species. I cannot give an explanation for the difference in abundance between the two species.

Leposoma percarinatum (L. Müller)

Hylosaurus percarinatus L. Müller, 1923: 146.

TYPE-LOCALITY.—Peixeboi, Estado do Pará, Brazil.

MATERIAL. – Surinam.

Third Encampment: 1♀, RMNH 13409, km 16.8, 16–X–1948, 1♀, RMNH 13410, side transect E., km 15.8, 20–X–1948, both leg. Surinam Expedition 1948–49.

Djai Creek: 1♀, RMNH 13408, km 8.6, 10–X–1948, leg. Surinam Expedition 1948–49.

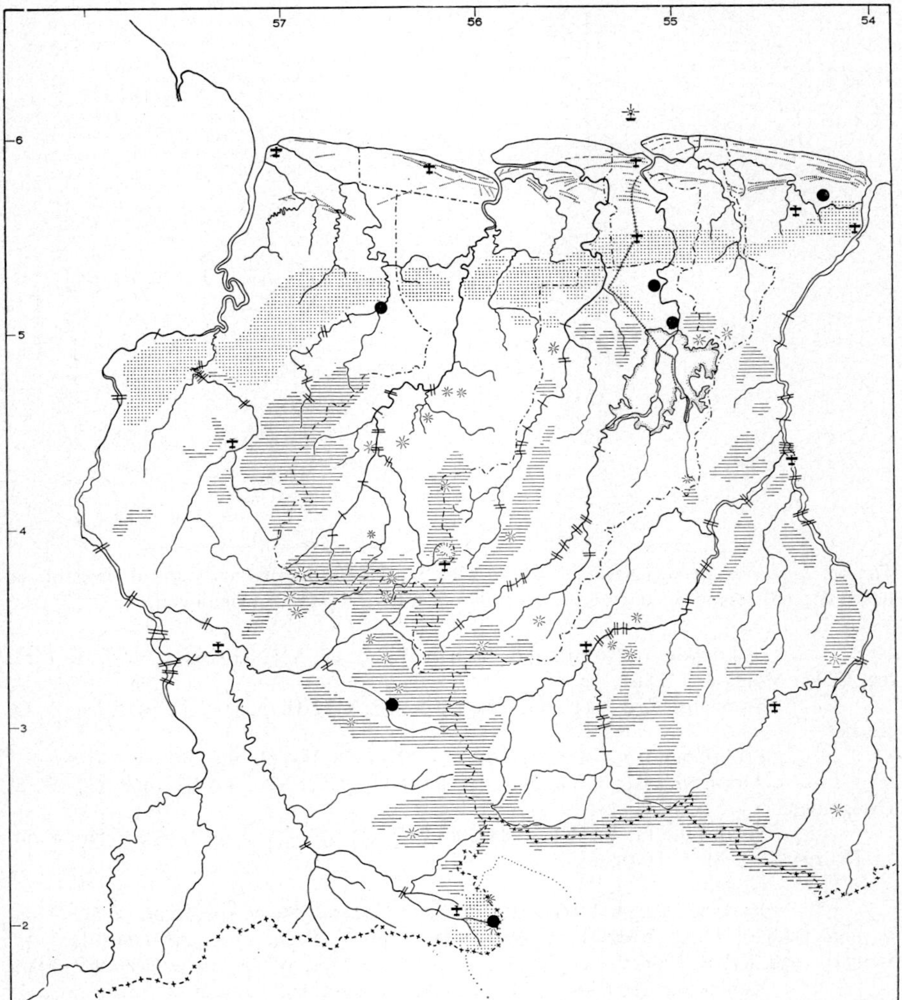

Fig. 65. Distribution of *Leposoma percarinatum* (L. Müller) in Surinam.

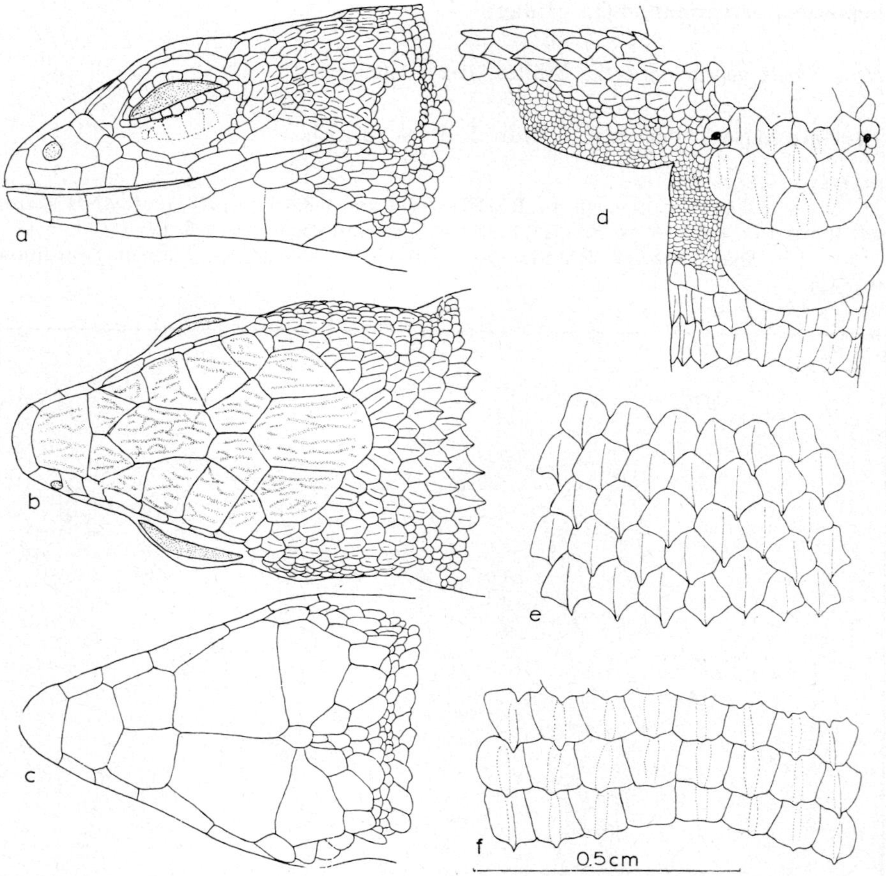

Fig. 66. Leposoma percarinatum (L. Müller); a, b, c: lateral, dorsal and ventral views of the head, d: anal region, e: dorsals, f: ventrals of ♀, RMNH 13882, Sipaliwini.

65 km SSE. of Paramaribo, Afobaka Road: 1♀, AMNH 108776, 12/13-II-1972, leg. C. W. Myers & J. Daly.

Stondansie Cataracts, Nickerie River: 1♀, RMNH 16444, 3-II-1971, leg. D. G. Reeder.

Brokopondo: 1♀, ZMA 12692, 29-VIII-1966, leg. H. Nijssen.

Airstrip Kayser Mountains: 1♀, RMNH 15384, NE., 3-VIII-1968, leg. M. S. Hoogmoed.

Sipaliwini: 1♀, RMNH 13882, 1 km N. of tip of Vier Gebroeders Mountain, 12-IX-1968, leg. M. S. Hoogmoed.

Guyana.

Kartabo: 1♀, AMNH 21263, 1920/21, leg. W. Beebe (paratype of *Leposoma taeniata* Noble); 1 ex., AMNH 21268, 1920/21, leg. W. Beebe; 3♀♀, AMNH 16433-35, January 1930, leg. R. Pinney; 2♀♀, MCZ 45956-57 (paratypes of *Leposoma taeniata* Noble).

Kamakusa: 1♀, 1 juv., AMNH 25082-83, 1922, leg. Lang & La Varra.

Wismar: 1♀, AMNH 58981, December 1934, leg. A. S. Pinkus.

Chenapoon River, Upper Potaro River: 1♀, BM 1905.11.1.6, leg. C. Bovallius.

Kuyuwini Landing: 1♀, AMNH 57455, leg. R. Snedigar.

Table 24. Counts and measurements of *Leposoma percarinatum* (L. Müller). For explanation see table 3.

Reg. no.	A	B	C	D	E	F	G	H	I	J	K	L	M	N	O	P	Q	R	S	T	U	V	W	X
RMNH																								
13408	♀	32	57	7.4	4.3	3.5			0/1-0/1	26	37	23	8	10	6-7	6-6	5-5	4-4	7-6	5+1	9	3	14-15	20-21
13409	♀	32	60	8.1	4.5	3.6			0/1-0/1	26	35	22	?	12	6-7	6-6	6-5	4-4	8-7	5+1	?	3	14-16	?-19
13410	♀	32	31(31)	7.8	4.2	3.8			0/1-0/1	26	38	23	8	10	6-6	6-6	5-5	4-4	5-6	5+1	9	3	14-14	20-20
13882	♀	37	44(23)	8.5	5.3	3.8	9	14	0/1-0/1	25	37	26	8	11	6-7	6-6	5-5	4-4	6-6	5+1	9	3	13-13	19-18
15384	♀	33	38(9)	8.1	5.8	3.5	9	14	0/1-0/1	26	36	24	8	11	6-6	6-6	5-5	4-4	5-5	5+1	9	3	13-13	21-20
16444	♀	37	12(7)	8.5	4.9	3.7	10	16	0/1-0/1	26	37	25	8	11	6-6	6-6	5-5	4-4	6-6	5+1	9	3	13-13	21-20
ZMA																								
12692	♀	33	62				9	14	0/1-0/1	27	36	25	8	12	6-6	6-6	6-5	4-4	7-6	5+1	9	3	14-14	20-20
BM																								
1905.11.1.6	♀	32	35(31)	7.6	4.5	3.4			0	24	39	25	8	8	6-6	6-6	5-4	4-4	6-6	5+1	9	3	12-13	16-17
AMNH																								
16433	♀	34	57	7.6	4.3	3.5	9	14	0	26	36	25	8		6-6	5-5	5-5	4-4		5+1	10	3	13-13	18-18
16434	♀	25	17(17)	6.4	3.7	3.0	6	11	0	26	37	23	8		6-7	5-5	5-5	4-4		5+1	11	3	14-14	20-17
16435	♀	34	61	8.2	4.6	3.7	9	13	0	27	37	24			6-6	6-6	5-5	4-4		5+1	10	3	13-13	20-20
*21263	♀	32	7(7)	7.4	4.3	3.8	9	13	0	26	38	25	8		6-6	6-6	5-5	4-4		5+1	10	3	13-14	16-18
25082	♀	34	55	7.8	4.6	3.7	9	14	0	26	37	26	8		6-?	6-7	4-4	4-4		5+1	9	3	10-11	15-15
57455	♀	22	32	6.0	3.5	2.6	7	10	0	25	38	25			6-6	5-5	5-5	4-4		5+1	10	3	?-12	?-18
58981	♀	29	55	7.1	4.1	3.1	8	13	0	25	37	25			6-6	5-5	5-5	4-4		5+1	10	3	?-12	21-19
MCZ																								
*45956	♀	21	5(5)	6.6	3.6	3.0			0	25	37	25	8	8	6-6	6-6	5-5	4-4	6-6	5+1	9	3	12-14	18-17
*45957	♀	27	15(15)	7.4	3.8	2.5			0	25	36	25	8	8	?	?	5-5	4-4	?	5+1	9	3	12-12	16-?

DIAGNOSIS.—A small lizard with a cylindrical body, well developed short legs and a long tail. Dorsal scales rhomboidal, in 35–38 transverse rows. Interparietal large, with nearly parallel lateral margins; prefrontals with a short median suture; preanal scales rhomboidal; three pairs of chinshields. Number of scales around midbody 25–27; fourth finger with 13–16, fourth toe with 18–21 subdigital lamellae. Back uniformly brown with lighter dorsolateral lines. Side of tail with a white line. This is a unisexual species.

DESCRIPTION.—Head small, pyramidal, depressed; back of the head narrower than the neck, which is slightly swollen and wider than the anterior part of the body. Snout pointed, tip rounded.

Rostral hexagonal, visible from above, twice as wide as deep. A large, pentagonal frontonasal separating the nasals. A pair of irregularly pentagonal prefrontals; wider than long; forming a short median suture (in RMNH 13409 a small scale is present between the prefrontals). Frontal elongate, hexagonal, once and a half as long as wide. A pair of irregularly pentagonal frontoparietals forming a short median suture, which is longer than that between the prefrontals. Interparietal large, with four straight and one convex posterior margin, its sides nearly parallel. Parietals irregularly pentagonal, much smaller than the interparietal. Occipitals absent. Four supraoculars; in contact with the supraciliaries; the anterior one smallest; the second and the third subequal, largest. All dorsal headscales with longitudinal striations. Nostril in a large semidivided nasal. Loreal deeper than long. Frenoocular small; as long as deep; in contact with the nasal. Preocular absent. Five to eight suboculars of variable size. Two rows of postocular scales with a variable number of scales. Supraciliaries five to six; the anterior one largest; the third and the fourth subequal, smallest. Temporal region with small, hexagonal, keeled, juxtaposed scales, larger near the parietals. Ear-opening large, surrounded by small, granular scales, forming a finely lobed margin; external auditory meatus absent, tympanum well visible. Supralabials six, rarely seven; the posterior one largest; four to below the centre of the eye. Lower eyelid with a semitransparent disc of four palpebrals. Pupil round.

Mental nearly semi-circular. A large heptagonal postmental. Three pairs of chinshields; the anterior two forming median sutures, that of the second pair once and a half as long as that of the anterior pair; the third pair separated by a single, small scale; the anterior pair smallest; the second pair largest. Six lower labials; the third largest. The third pair of chinshields separated from the labials by small scales.

All large dorsal and lateral headscales with a row of small pits along their margin, all large headscales with scattered pits.

Gulars rhomboidal, mucronate, keeled, imbricate, in oblique and in nine transverse rows. Collar indistinct, with 10–12 scales; the central eight with pointed tips forming a serrated posterior margin. Scales on the nape like the dorsals, merely smaller; near the parietals not mucronate; in oblique and transverse rows. Scales on the side of the neck small, rounded, tubercular, in vertical and in oblique rows.

Dorsals and laterals rhomboidal, mucronate, keeled, imbricate, in oblique and, the dorsals, in 35–38 transverse rows. Ventrals larger than the dorsals; rhomboidal, mucronate, keeled, imbricate towards the sides; square to rectangular, not mucronate, with low, blunt keels, slightly imbricate towards the median; in eight longitudinal (keels continuous) and in 22–26 transverse rows. No sharp demarcation between the laterals and the ventrals. Total number of scales around midbody 25–27. Preanal plate with six rhomboidal, keeled scales, five in a semicircle (the median one smallest), one enclosed by the other five.

One indistinct preanal pore on each side, in the centre of a group of three scales. Femoral pores absent. All specimens examined are females.

Scales on the tail rectangular, mucronate, sharply keeled, imbricate, in transverse and in longitudinal rows; scales under the tail narrower; keels forming 15–17 longitudinal ridges. Scales on the regenerated part of the tail smaller, rectangular, not mucronate, with indistinct keels, slightly imbricate, in transverse and longitudinal rows. Tail round with 15–17 ridges.

Scales on the upper and anterior surfaces of the forelimbs, on the anterior and lower surfaces of the thighs and on all surfaces of the lower legs large, rhomboidal, keeled, imbricate; on the lower and posterior surfaces of the forelimbs, on the posterior and upper surfaces of the thighs small, granular, juxtaposed. Thirteen to 16 double, rounded lamellae under the fourth finger, 18–21 under the fourth toe. Fingers and toes subcylindrical. Palms and soles with small flat, juxtaposed granules.

Back dark-brown in life with light-brown dorsolateral lines, flanks black. Belly ochre-yellow, chin and throat white. Posterior surface of the thighs black with white spots.

In preservative back and dorsal surface of the head orange-brown to dark-brown. Light-brown dorsolateral lines from the back of the head on to the tail, bordered ventrally by black lines. Side of the neck and the flanks dark-brown to black, with a row of indistinct white spots from the ear to the groin, continued on the tail as a white lateral line. Upper and lower lips white with distinct brown blotches. Ventral parts immaculate, creamy, except for the lateral parts of the chinshields that bear some brown blotches. Hind limbs brown with small, white spots above and on the posterior surface; a sharp demarcation between the brown dorsal and the creamy ventral colouration on the posterior surface of the thighs; no such sharp demarcation on the anterior surface. Dorsolateral line continuous on the anterior part of the tail, broken up into light spots posteriorly. The black line bordering the dorsolateral line ventrally broken up into spots from right behind the hind limbs. The white lateral line bordered ventrally by a brown line. Underside of tail creamy, with brown spots, increasing in number posteriorly. Tongue dark-grey anteriorly, white posteriorly, with a straight transverse border between the two colours.

A small lizard, maximum snout-vent length 37 mm. Hind limbs once and a half as long as the forelimbs. When laid along the body there is overlap of fingers and toes. The undamaged tail is 1.8 times as long as the snout-vent length.

HABITAT.—All specimens were collected on the forest-floor among leaves. The labels on two specimens (RMNH 13409–10) indicate that they were collected in rather wet situations, viz., '*Mauritia*-island in swamp' and 'pina swamp forest'. The specimen collected by me in the Sipaliwini area (RMNH 13882) was among leaf-litter in the border zone between water and land of a creek in a forest-island. The other specimens available are from rain-forest where at the time the ground was dry. PARKER (1935b) reports from hearsay that this species is aquatic. HOEVERS (1967) denies this and says that all specimens were captured in dry situations.

NATURAL HISTORY.—This seems to be a diurnal species which has been taken between 10.45 a.m. and 4.00 p.m. BEEBE (1945), however, in discussing the ecology of this species at Kartabo, British Guiana and Caripito, Venezuela, says this is a chiefly nocturnal lizard, as more than half of the specimens taken in Caripito fell into pits overnight. Contrary to BEEBE's observations HOEVERS (1967) saw specimens basking in open spots in the forest at Laluni, Guyana. Although pits have been dug extensively in Surinam that were also used for zoological collecting, this species was never taken from any of them. In considering this we have to bear in mind that only eight specimens from Surinam are known, and future and more specialised collecting might prove that BEEBE's observations also apply to Surinam representatives of the species. Breeding in Kartabo (BEEBE, 1945) at least lasts from March to June. HOEVERS (1967) collected a pregnant female, containing two mature eggs, at Maduni on July 10. The normal number of eggs per female seems to be two, as two Kartabo females (BEEBE, 1945), also contained two mature eggs each and two separate clutches, each containing two eggs, were found in litter close to a fallen tree. One of these clutches was ready to hatch and a freshly hatched specimen measured 40 mm of which 23 mm tail. The food seems to consist of small ground-dwelling arthropods: craneflies, springtails, homopterons, termites, ants, caterpillars, spiders, roaches and beetles. BEEBE (1945) observed that a specimen was attacked by army ants.

RANGE.—Surinam, Guyana, Brazil (Pará, Amazonas, Amapá), Venezuela (Monagas, Bolívar, Amazonas). In Surinam the species is present throughout the country on both sides of the coastal savannah belt, on places where forest is present. It occurs from sea level to at least 250 m (fig. 65).

REMARKS.—This species and the foregoing *L. guianense* Ruibal are very much alike and in the field can hardly be separated. However, on closer inspection there turn out to be quite a number of differences which, for a clear concept of the species, will be listed below. When collecting in Surinam, I found that the species are sympatric in the surroundings of the airstrip Kayser Mountains where specimens of both species were collected at nearly the same spot on the same day. Also in the Sipaliwini area both species are present, however, *L. guianense* Ruibal only in the rain-forest at the western border of the savannah, whereas the only specimen of *L. percarinatum* (L. Müller) was collected in a forest-island on the savannah. Also the material of both species of *Leposoma* recently collected by MYERS & DALY presents evidence for sympatry of these species. All their *Leposoma* material was continuously numbered, indicating that it was not only collected in the same locality, but on the same spot as well. *L. percarinatum* (L. Müller) seems to be rarer than *L. guianense* Ruibal; so far in Surinam only female specimens of the first species have been collected in the period August–October and in February. RUIBAL (1952) already drew attention to the disproportionate sex ratio among the specimens investigated by him. Among 22 specimens studied there were no males. When describing the Amazonian lizards, DA CUNHA (1961) had before him a series of this species consisting of ten adult and eight juvenile males and eight adult and nine juvenile females from the Territory of Amapá in northern Brazil. However, judging from the counts he gave it seems very likely that in reality he had before him specimens of *L. guianense* Ruibal or, at the best, a mixture of *L. guianense* Ruibal and *L. percarinatum* (L. Müller). The characters mentioned by DA CUNHA that strongly suggest *L. guianense* Ruibal are: six supralabials, two small parietals, a very large interparietal and the presence of males. The one character that suggests *L. percarinatum* (L. Müller) is the number of pairs of chinshields, namely three, but this count may result from not recognizing the fourth pair as such because it is relatively small. All other characters mentioned are applicable to both species. The characters used by RUIBAL (1952) and DA CUNHA (1961) to distinguish between *L. guianense* Ruibal and *L. percarinatum* (L. Müller), namely the presence of five long preanals and the absence of a dorsal pattern in *L. guianense* Ruibal, do not serve to separate the two species, as will be clear from the preceding descriptions and from the following list of differences. Thus, I tend to believe that *L. percarinatum* of DA CUNHA (1961) in reality is *L. guianense* Ruibal. If this is true the presence of males is quite understandable, males for this species being known from its entire range. Should it turn out that DA CUNHA's specimens were *L. percarinatum* (L. Müller) then it would be interesting to know the exact collecting data and have details on the habitat of this series, which might give a clue to the absence of males in the rest of the known range.

Differences between *Leposoma guianense* Ruibal and *L. percarinatum* (L. Müller).

L. guianense Ruibal

1. Interparietal very large, with posteriorly diverging lateral margins.
2. Prefrontals with a long median suture, more than half the length of the prefrontals.
3. Supraciliaries in most specimens four, rarely three, five or six.
4. Nasal undivided.
5. Four pairs of chinshields.

L. percarinatum (L. Müller)

1. Interparietal smaller, lateral margins more or less parallel.
2. Prefrontals with a short median suture, less than half the length of the prefrontals.
3. Supraciliaries five or six.
4. Nasal semidivided.
5. Three pairs of chinshields.

6. Scales of the collar more elongate and strongly mucronate.
7. Upper labials seven, rarely eight.
8. Temporals medium-sized and few.
9. Dorsals in 31–35 transverse rows.
10. Dorsals more elongate.
11. Ventrals distinctly mucronate.
12. Scales around midbody 22–26.
13. Preanals elongate, longer than wide, median divided or not.
14. Subdigital lamellae fourth finger 10–13.
15. Subdigital lamellae fourth toe 13–19.
16. Back with two rows of black spots medially of the dorsolateral lines.
17. Side of tail with a series of white spots.
18. Upper and lower lips, lateral parts of chin-shields white with diffuse brown blotches.

6. Scales of the collar shorter and only slightly mucronate.
7. Upper labials six, rarely seven.
8. Temporals small and numerous.
9. Dorsals in 35–38 transverse rows.
10. Dorsals short and wide.
11. Ventrals near median not mucronate.
12. Scales around midbody 25–27.
13. Preanals rhomboidal, as long as wide, median preanal always small, preceded by a small central one.
14. Subdigital lamellae fourth finger 13–16.
15. Subdigital lamellae fourth toe 18–21.
16. Back uniformly brown between the dorsolateral lines.
17. Side of tail anteriorly with a white line.
18. Upper and lower lips, lateral parts of chin-shields white with solid brown blotches.

Genus **Neusticurus** Duméril & Bibron

DIAGNOSIS.—Medium-sized lizards with a cylindrical body, well developed legs and a compressed tail. Body with rows of enlarged tubercles which form crests on the tail. A pair of frontonasals separating the nasals. Tongue lanceolate, without emargination, covered with rhomboidal, imbricate, scale-like papillae in oblique rows; tip bifid, smooth. Anterior teeth conical; lateral teeth bicuspid, laterally compressed. Ear-opening small or large, tympanum superficial or recessed. Lower eyelid movable, with a semitransparent disc. Pupil round or kidney-shaped. Eight species from Costa Rica to the southern limits of the Amazon basin, in South America only east of the Andes. Two species in Surinam.

Key to the Surinam species of *Neusticurus*.

1. Tympanum deeply recessed; two lateral caudal scales corresponding with two ventral caudal scales; femoral pores on each side 24–31 in males, three to nine in females; lateral ocelli rarely absent . *N. bicarinatus*, p. 330
   Tympanum superficial; three to four lateral caudal scales corresponding with two ventral caudal scales; femoral pores on each side 16–22 in males, two to four in females; lateral ocelli absent . . . . *N. rudis*, p. 337

Neusticurus bicarinatus (Linnaeus) (figs. 67, 68; pls. 36, 38; table 25)

Lacerta bicarinata Linnaeus, 1758: 201.
Neusticurus bicarinatus: Kappler, 1881: 166; Hoogmoed, 1968: 612.
Neusticurus spec. D'Audretsch, 1968: 52 (partly).

TYPE-LOCALITY.—'Indiis', here restricted to the vicinity of Paramaribo, Surinam.

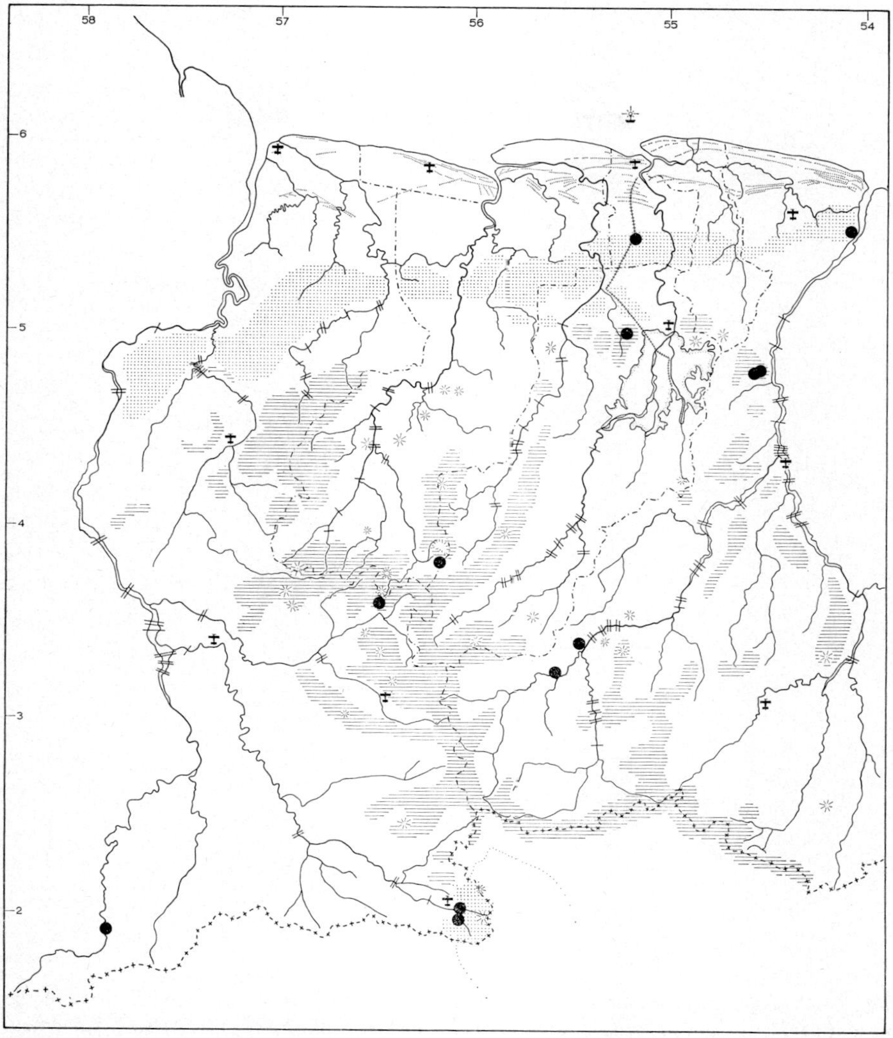

Fig. 67. Distribution of *Neusticurus bicarinatus* (Linnaeus) in Surinam.

MATERIAL. – Surinam.

Surinam: 1♀, RMNH 13388, leg. Surinam Expedition 1948–49; 2♂♂, ZMB 239, leg. A. Kappler; 1♂, SMNS 1002, leg. A. Kappler.

Zanderij: 1♀, 1♂, BM 1946.4.3.91, 98, 21–VIII–1938, 1♀, 2 juvs., BM 1946.4.3.93, 95, 97, 5–X–1938, 1♂, BM 1946.4.3.96, 12–IX–1938, 1♀, 1♂, BM 1946.4.3.92, 94, 16–IX–1938, all leg. I. T. Sanderson.

Albina: 1♂, ZMB 5768, leg. A. Kappler.

Brown's Mountain: 1♂, RMNH 13866, 25–VIII–1968, leg. W. N. Polder; 3♀♀, 2♂♂, RMNH 13868, 20–VIII–1968, 2♀♀, RMNH 13869, 19–VIII–1968, all leg. M. S. Hoogmoed; 1♀, AMNH 108786, Irene Fall, 15–II–1972, leg. C. W. Myers & J. Daly.

Nassau Mountains: 1♀, 1♂, RMNH 13389, km 16.4, 23–III–1949, 2♂♂, RMNH

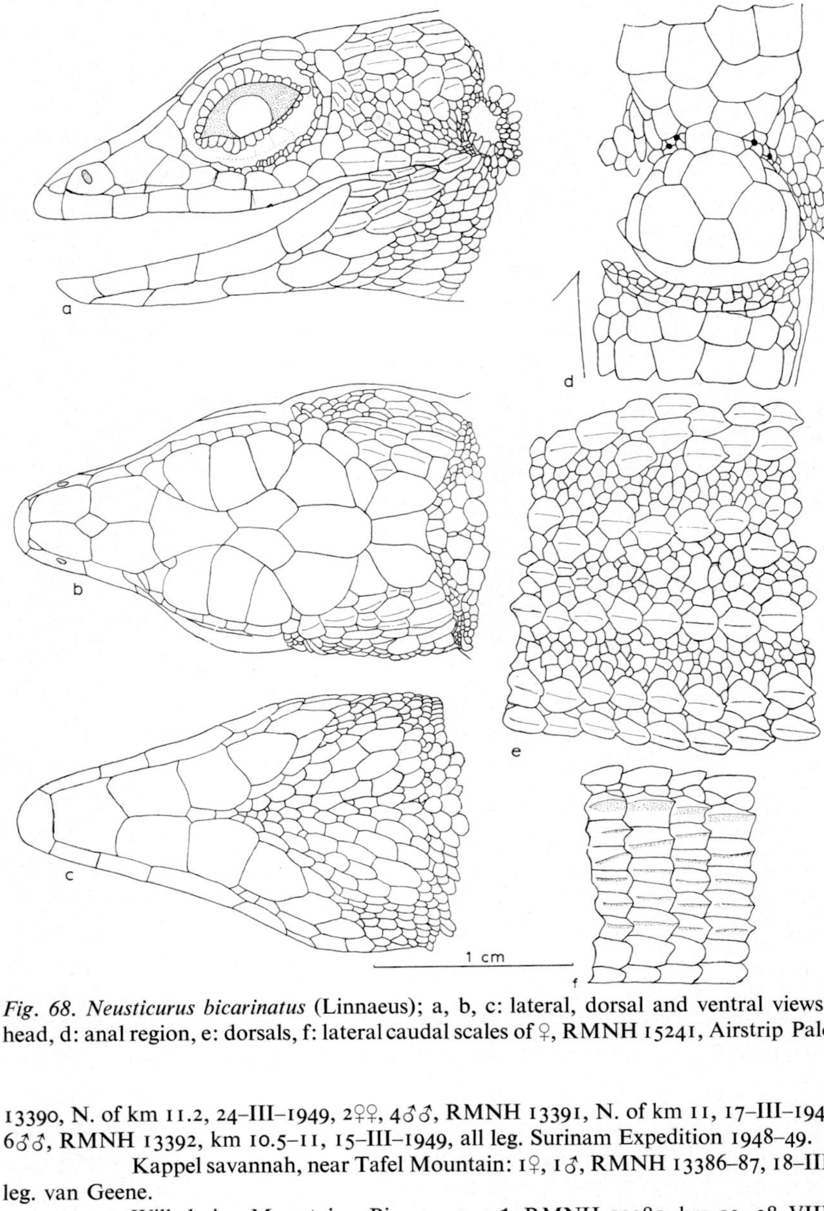

Fig. 68. Neusticurus bicarinatus (Linnaeus); a, b, c: lateral, dorsal and ventral views of the head, d: anal region, e: dorsals, f: lateral caudal scales of ♀, RMNH 15241, Airstrip Paloemeu.

13390, N. of km 11.2, 24–III–1949, 2♀♀, 4♂♂, RMNH 13391, N. of km 11, 17–III–1949, 6♀♀, 6♂♂, RMNH 13392, km 10.5–11, 15–III–1949, all leg. Surinam Expedition 1948–49.

Kappel savannah, near Tafel Mountain: 1♀, 1♂, RMNH 13386–87, 18–III–1958, leg. van Geene.

Wilhelmina Mountains, Bivouac 3: 1♂, RMNH 13385, km 19, 28–VIII–1963, leg. S. Ligorie.

Airstrip Paloemeu: 1♀, RMNH 15241, 22–V–1968, leg. M. S. Hoogmoed.

Between Tepoe and airstrip Paloemeu: 1♀, ZMA no number, 27–II–1970, leg. D. Hillenius.

New River, between camp 7 and 8: 3 juvs., BM 1936.4.2.56–58, 18–XII–1935, leg. J. G. Myers.

Sipaliwini: 1♀, RMNH 13865, 5 km E. of airstrip, 10–IX–1968, leg. M. S. Hoog-

Plate 36. Neusticurus bicarinatus Linnaeus. Figs. 1–3. Dorsal, lateral and ventral aspect of ♀, RMNH 16381, Sipaliwini.

moed; 1♀, RMNH 16381, near Meyers' farm, 19–I–1970, 1♀, RMNH 16382, between airstrip and Meyers' farm, 1–II–1970, both leg. M. S. Hoogmoed & J. J. P. Paats.

Marowijne River: 1♂, ZMB 7606, leg. A. Kappler.

Guyana.

Malalli: 1♀, UMMZ 77818, leg. A. S. Pinkus.

French Guiana.

Guyane: 1♀, MP 02–260, leg. Geay.

Mount Yanioué, Camopi River: 1♀, MP 50–20, leg. Aubert de la Rue.

South America: 1♂, MP 4181.

DIAGNOSIS.—A medium-sized lizard with cylindrical body, well developed legs and a long, compressed tail. Enlarged dorsal scales flat, keeled, slightly imbricate. External auditory meatus long, tympanum deeply recessed. On each side 23–31 femoral pores in males, three to nine in females. Side of tail with two scales per two subcaudals. Reddish to purplish brown above; belly immaculate, white. A distinct white, black-edged ocellus in front of the forelimbs. Over the eye three red or yellow spots, under the eye only one.

DESCRIPTION.—Head large, much wider than the neck in males and in juveniles, as wide as the neck in females; pyramidal, depressed; canthus rostralis sharp; 1.7–1.9 times as long as wide. Snout pointed, in juveniles in front of the eyes with concave lateral margins, in adults these margins are straight. Neck as wide as the body or slightly narrower.

Rostral hexagonal; visible from above; twice as wide as deep. A pair of (exceptionally one large) elongate, rectangular frontonasals separating the nasals. A pair of irregularly hexagonal or trapezoid prefrontals. A rhomboidal median azygous scale between the frontonasals and the prefrontals is present in 18 specimens, absent in 36 specimens. Frontal elongate, hexagonal; wider anteriorly than posteriorly; once and a half as long as wide. A pair of irregularly pentagonal frontoparietals, forming a long median suture. Interparietal polygonal, large; 1.5–2 times as long as wide; posterior border convex; bordered posteriorly by three to seven large scales, which might be called anterior occipitals. Parietals polygonal; slightly longer than wide; smaller than interparietal. Posterior occipitals small. Supraoculars four, the anterior one smallest, divided; the second and the third subequal, largest; in contact with the supraciliaries, but isolated granules, sometimes forming short rows, are present. Nostril in an undivided nasal. Loreal large, longer than deep. A row of small scales between the nasal and the second subocular, separating the loreal from the supralabials. A small preocular. Four to five large suboculars forming a curved subocular ridge. An anterior row of six to seven postoculars, the upper and lower ones large, the rest small; a posterior row of four to five large scales. Six to eight supraciliaries of variable size. Temporal region with flat, keeled, polygonal scales, small in the antero-ventral, large in the postero-dorsal part. Ear-opening relatively small; surrounded by enlarged scales anteriorly and by small scales posteriorly, forming a denticulate margin anteriorly, a smooth one posteriorly. External auditory meatus long, tympanum visible but overhung by part of the ear-margin. Supralabials eight (rarely seven or nine); the anterior five smooth, the posterior three with blunt keels; the fourth largest; the fifth under the centre of the eye; the sixth much higher than the fifth. Lower eyelid with a semi-transparent disc of four to five palpebrals. Pupil round or kidney-shaped with the concave side dorsally.

Mental mushroom-shaped with a large hood and a short, wide stem. A large, pentagonal postmental. Four pairs of chinshields, the anterior pair in contact, the rest separated; the anterior and second pairs subequal in size, largest; the fourth pair smallest, in some specimens hardly larger than part of the surrounding scales. Four (rarely three , five or six) large anterior infralabials, of which the third is the largest, and a variable number of small posterior scales. The third and the fourth pair of chinshields separated from the infralabials by small scales.

Anterior dorsal and lateral headscales may have scattered tubercles, the posterior headscales may show a row of tubercles along their margin or along part of it. In most specimens these tubercles are absent or very indistinct. Few scattered pits, not very distinct, on the ventral headscales.

Gulars relatively large, elongate, flat, juxtaposed, anteriorly near the chinshields; small, rounded, flat, juxtaposed, in transverse rows posteriorly; with the two longitudinal, central rows tending to enlarge and to be imbricate near the collar; the zones delimited sharply. Collar with 10–12 scales, distinct, with a slightly serrated posterior margin. Scales on the nape like those on the back. Scales on the side of the neck large, tubercular; with small granules in between; in vertical and longitudinal rows; only the upper two longitudinal rows are distinct, the lower rows are irregular.

Two kinds of dorsal scales; the majority small, polygonal, flat, imbricate; the rest in more or less regular longitudinal rows, large, more or less hexagonal, flat, keeled, slightly imbricate. Longitudinal scale rows starting halfway the nape, three on each side, a paravertebral one and two dorsolateral ones in contact. Vertebral region and space between the paravertebral and the dorsolateral rows with small scales. Scales on the flanks smaller than the enlarged dorsals, forming vertical rows; polygonal, flat, keeled, slightly imbricate; only few small scales between the large ones. Ventrals large, rectangular to quadrangular with rounded posterior margin; imbricate, smooth, except for the keeled outer rows; in eight (rarely ten) longitudinal and in 23–28 transverse rows. Total number of scales around midbody 37–51. Five (mostly) to seven posterior preanals; the median one very small; forming a semicircle enclosing a large anterior preanal.

Preanal and femoral pores in a continuous row, containing three to nine pores on each side in females, 23–31 in males. Of this total number of pores three or four are in a preanal position. Pores in the centre of a single scale, or the more proximal ones between two or three scales.

Tail compressed, with two denticulate crests, which are continuations of the paravertebral rows of enlarged scales. The upper dorsolateral row continues ons to the anterior part of the tail but soon disappears; the lower dorsolateral row ends on the base of the tail. All scales in the crests sharply keeled, of equal size. Scales between the crests, and one row immediately lateral of them, small, flat, slightly imbricate. Number of scales between the crests decreasing posteriorly until the crests are in contact, finally they fuse. Scales on the side of the tail large, elongate, more or less rectangular, slightly keeled, flat, slightly imbricate, in vertical rows. Two lateral scales per two subcaudal scales. Subcaudals large, smooth, elongate, rectangular with pointed posterior margin, imbricate, the two central rows enlarged. Scales on the regenerated part of the tail quadrangular to elongate, rectangular, slightly imbricate; dorsals and laterals small, keeled; subcaudals larger, smooth. Tail divided in segments, each segment containing one subcaudal.

Scales on the upper and anterior surfaces of the forelimbs, on the anterior part of the upper surface of the thighs and on the upper and anterior surfaces of the lower legs large, rhomboidal, keeled, imbricate, the keels forming continuous ridges; on the anterior and lower surfaces of the hind limbs large, rhomboidal, smooth, imbricate; on the lower and posterior surfaces of the forelimbs and on the posterior surface of the hind limbs small, tubercular, juxtaposed; on the posterior part of the upper surface of the thighs minute, flat, juxtaposed, with scattered, large, rounded, keeled scales in between. Seventeen to 23 (mostly 18–20) single, smooth lamellae under the fourth finger; 24–31 (mostly 26–29) smooth lamellae under the fourth toe; single in the distal and proximal part; in the central part in two rows, of which in the proximal part only the inner one is present. Fingers and toes subcylindrical. Palms and soles with small, slightly tubercular, juxtaposed scales.

In life the back is reddish to purplish brown with indistinct dark-brown to black spots. Belly immaculate, white. Over the eye three, under the eye one light spot; the spots are red in juveniles, pale yellow in adults. The spots on the upper eyelid separated by two black bars. Iris reddish brown.

In preservative back reddish brown with indistinct dark-brown or black spots, tending to form irregular transverse bars. In juveniles a very distinct dorsolateral row of cream-coloured, black-edged ocelli passes from the back of the head onto the tail; in adults these ocelli are less distinct or have disappeared completely. In one old male a number of scales on the nape and the scales of the paravertebral row are pale yellow. A very distinct white ocellus with broad, black edges in front of the forelimb; followed in most specimens by a row of more or less distinct ocelli on the flanks, which is continued on the lower lateral part of the tail. Upper eyelid with three white spots, of which the posterior one may be continued in a light stripe directed posteriorly. A white spot on the suborbitals. Rims of the upper and lower eyelid white. Upper and lower lips white with diffuse brown spots. Ventral parts immaculately white to pale yellow. Proximal part of the underside of the tail white to pale yellow, with or without spots, distal part immaculate, dark-brown to black. Back of thighs with more or less distinct white spots.

A medium-sized lizard, maximum snout-vent length in females 96 mm, in males 109 mm. Hind limb 1.4–1.5 times as long as forelimb. When laid along the body there is a considerable overlap of forelimb and hind limb. The undamaged tail is twice as long as the snout-vent length in juveniles and females and 1.8–2.1 times in males.

The head in old males is strongly swollen, whereas in females it is narrow and slender.

HABITAT.—Banks of creeks and rivers in rain-forest. In the Sipaliwini area this species was collected twice on the savannah bank of a creek forming the border between forest and savannah.

NATURAL HISTORY.—A diurnal species collected between 9.00 a.m. and 5.30 p.m. All specimens were collected in or near creeks or rivers, where they were basking on rocks and branches protruding out of the water. One specimen was collected under a rock near a waterfall. Pregnant females were present in March and May, both specimens containing two large, mature eggs, which is the normal number of eggs per clutch in this genus (UZZELL, 1966). All the juvenile specimens available lack the umbilical scar, thus indicating they have not recently hatched. The 17 stomachs examined contained a large diversity of food items, of which only a few are aquatic, the rest being terrestrial. The stomachs contained one land snail, two diplopods, one larva and seven adults of Hymenoptera (ants), three Diptera (Tipulidae), one butterfly, four caterpillars, three grass-hoppers, one cricket, one waterbug, three beetles, four spiders and the remains of a small, armoured catfish. One stomach contained a soft mass with tough membranes and brown loamy soil, which might represent the remains of an earthworm.

These findings give quite a different picture of the food of this species than that given by BEEBE (1945) who investigated only two stomachs, one of which contained a small poecilid fish and two tadpoles, the other containing fish scales, a water beetle and at least three tadpoles. From BEEBE's data it was concluded that this species fed mainly on aquatic organisms. However, from the larger series investigated by me it is clear that (at least in Surinam) aquatic organisms form only a minor part of the diet.

When pursued these lizards fled into the water, where they swam with undulating movements of body and tail, with the head just above the surface, or dived to the bottom, where they tried to hide under roots, rocks and debris. Thus, this species and the following may be described as being semiaquatic.

RANGE.—Surinam, Guyana, French Guiana, Brazil and Venezuela. In Surinam the species is present throughout the country south of the coastal savannah-belt and reaches at least 300 m altitude (fig. 67).

VERNACULAR NAME.—No specific name for this species is available in the Surinam language. Trio Indians call it 'poededewa'.

Neusticurus rudis Boulenger (figs. 69, 70; pls. 37, 38; table 26)

Neusticurus rudis Boulenger, 1900: 53; Uzzell, 1966: 286; J. A. Peters & Donoso-Barros, 1970: 207.
Neusticurus surinamensis L. Müller, 1924: 295; Burt & Burt, 1933: 70.
Neusticurus dejongi Brongersma, 1927: 543; Burt & Burt, 1933: 69.
Neusticurus spec. D'Audretsch, 1968: 52 (partly).

TYPE-LOCALITY.—Foot of Mount Roraima, 3500 feet, Guyana.

MATERIAL. – Surinam.
Surinam: 1♂, ZMA 10241, 1890, leg. Brussel (holotype of *Neusticurus dejongi* Brongersma).

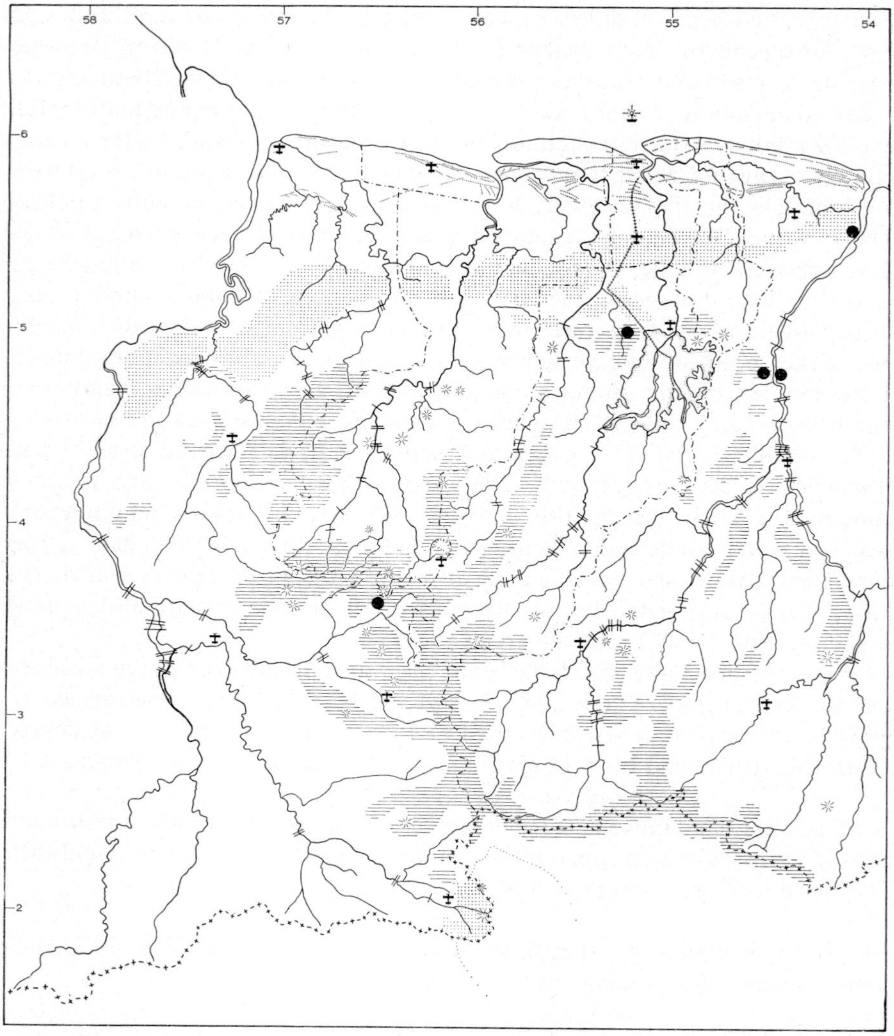

Fig. 69. Distribution of *Neusticurus rudis* Boulenger in Surinam.

Brown's Mountain: 1♂, RMNH 13864, 25–VIII–1968, leg. W. N. Polder; 1♂, RMNH 13867, 20–VIII–1968, 2♂♂, RMNH 13870, 19–VIII–1968, all leg. M. S. Hoogmoed; 1♂, RMNH 15705, Mazaronitop, 11–XI–1968, leg. R. Hoogmoed-Verschoor & M. S. Hoogmoed; 1♀, AMNH 108785, Mazaronitop, 15/16–II–1972, leg. C. W. Myers & J. Daly.

Nassau Mountains: 1♂, RMNH 13393, S. of km 3.6, 8–III–1949, 1♂, RMNH 13394, S. of km 3.3, 7–III–1949, 1♀, RMNH 13397, S. of km 3, 7–III–1949, 1♀, RMNH 13398, km 10.7, 18–III–1949, all leg. Surinam Expedition 1948–49.

Wilhelmina Mountains: 1♀, RMNH 13395, 5–VIII–1963, leg. S. Ligorie.

Guyana.

Mount Roraima: 1♂, BM 1946.8.31.64, 3500 feet, leg. McConnell & Quelch (holotype of *Neusticurus rudis* Boulenger).

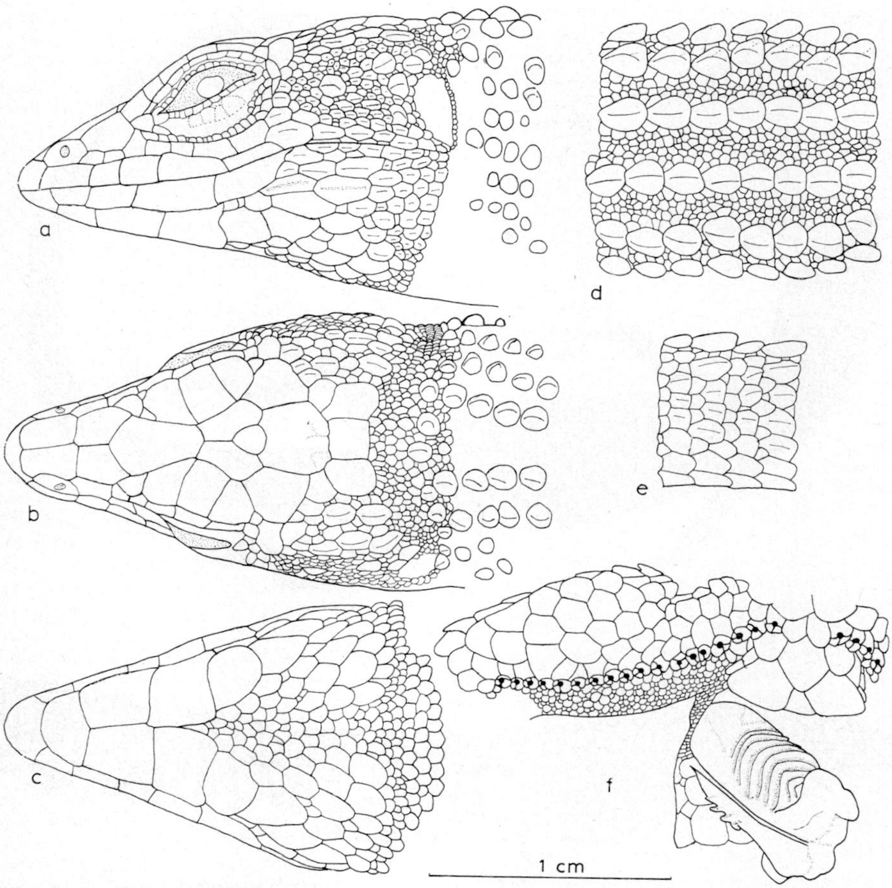

Fig. 70. Neusticurus rudis Boulenger; a, b, c: lateral, dorsal and ventral views of the head, d: dorsals, e: lateral caudal scales, f: anal region of ♂, RMNH 13867, Brown's Mountain.

French Guiana.

Marowijne River: 1♀, RMNH 13396, Opposite Base Encampment, 19–II–1949, leg. Surinam Expedition 1948–49.

Upper Camopi River: 1♂, MP 50–21, leg. Aubert de la Rue.

DIAGNOSIS.—A medium-sized lizard which is smaller than *N. bicarinatus* (Linnaeus), with cylindrical body, well developed legs and a long, compressed tail. Enlarged dorsals trihedral, juxtaposed; forming longitudinal and interrupted transverse rows. External auditory meatus absent, tympanum superficial. Femoral pores on each side 16–22 in males, two to four in females. Side of tail with three to four scales per two subcaudals. Light- or dark-brown above with large, lighter spots on the flanks; belly white with brown spots. No light spots on the eyelids.

Plate 37. Neusticurus rudis Boulenger. Figs. 1–3. Dorsal, lateral and ventral aspect of ♂, RMNH 13867, Brown's Mountain.

DESCRIPTION.—Head large, wider than the neck; pyramidal, depressed; with a sharp canthus rostralis; 1.7–1.9 times as long as wide. Snout pointed, in juveniles in front of the eyes with concave lateral margins, in adults these are straight. Neck narrower than the body.

Rostral more or less rectangular; visible from above; twice as wide as deep. A pair of elongate, rectangular frontonasals separating the nasals. A pair of irregularly pentagonal prefrontals. A rhomboidal median azygous scale between the frontonasals and the prefrontals is present in seven, absent in eight specimens. Frontal elongate, hexagonal; wider anteriorly than posteriorly; once and a half as long as wide. A pair of irregularly pentagonal or hexagonal frontoparietals; either posteriorly in contact, separated by one rhomboidal scale or by two triangular scales. Interparietal large; of irregular polygonal form, more or less resembling a trapezium with a convex posterior margin and posteriorly diverging lateral margins; about as wide as long. Parietals polygonal, smaller than the interparietal; as wide as long. The interparietal is bordered posteriorly by five to seven large scales, which might be called anterior occipitals; the posterior occipitals small. Supraoculars four; first smallest; second and third subequal in size, largest; between the first and the second supraocular a few small granules are present; in contact with the supraciliaries but, isolated granules may be present in between. Nostril in an undivided nasal. Loreal large, longer than deep in most specimens. One or two rows of small scales from the nasal to the anterior or the second subocular; separating the posterior part of the nasal, the loreal and the anterior part of, or the entire, anterior subocular from the supralabials. A small preocular. Four to five large suboculars, forming a curved subocular ridge. A row of four to seven, usually five postoculars; the upper one largest. Supraciliaries five or six, rarely four; anterior one largest. Temporal region with small tubercular polygonal scales and one (usually) or a few larger scales near the centre and in front of the ear. Scales bordering on the parietals larger than the temporals, keeled. Ear-opening large, surrounded by small scales, forming a smooth margin. External auditory meatus absent; tympanum clearly visible. Supralabials eight (rarely nine); the anterior five or six smooth, the posterior two or three with blunt keels; fourth largest, fifth under the centre of the eye. Lower eyelid with a semitransparent disc of three to five palpebrals. Pupil round.

Mental trapezoid or semicircular. A large pentagonal postmental. Four pairs of chinshields, the anterior pair forming a median suture; the second pair of chinshields either in contact, partly in contact or separated by one row of small scales; the posterior two pairs widely separated; the anterior pair largest, the others diminishing in size posteriorly. Four large anterior infralabials, fourth largest, and a variable number of small posterior scales. The third and the fourth pair of chinshields separated from the infralabials by small scales.

Anterior dorsal and lateral headscales with many scattered pits, the posterior headscales with a row of pits along their margin or along part of it. Few scattered pits, less distinct, on the ventral headscales. In some specimens the pits are more distinct than in others.

Gulars anteriorly, near the chinshields, relatively large, elongate, flat, juxtaposed; posteriorly small, round, tubercular, juxtaposed, in transverse rows; imbricate near the collar; the two groups delimited sharply. Collar indistinct, with nearly smooth posterior margin, containing nine to ten, rarely eight, scales. Scales on the nape like those on the back. On the side of the neck small, tubercular scales with four longitudinal rows of larger scales.

Two kinds of dorsal and lateral scales, the majority small, polygonal, flat, juxtaposed, the rest in more or less regular longitudinal rows, large, oval, keeled, trihedral, juxtaposed. The large dorsals also forming interrupted transverse rows. On the nape four regular, longitudinal rows from the back of the head to above the forelimbs, and from there continued on the back and onto the tail, two on each side, a paravertebral and a dorsolateral one. Vertebral region with small scales, the interspace between the paravertebral and dorsolateral rows of enlarged scales may be free of large scales or may contain several. Enlarged scales on the flanks smaller than those on the back, only the upper ones forming a longitudinal row, which accompanies the dorsolateral row. Ventrals large, rectangular to quadrangular, with rounded posterior margin; slightly imbricate, smooth, except for the keeled outer rows; in eight longitudinal and in 21–25 transverse rows. Total number of scales around midbody 41–53. Three to six posterior preanals, the median one or two smallest, forming a semicircle enclosing one or two anterior preanals.

Preanal and femoral pores in a continuous row, containing two to four pores on each side in females, 16–22 in males. Of this total number of pores two are in a preanal position. Pores in the centre of a group of two to four scales. Everted hemipenis bilobed, the sides with curved lamellae, the convexity directed towards the base; no spines, sulcus spermaticus on the posterior surface, bifurcate near the top.

Most scales on the tail small, polygonal, flat. The paravertebral and dorsolateral rows of large dorsals continued on the tail, the paravertebral one on the top, the dorsolateral one more on the side. Scales in the paravertebral row keeled, forming a crest on the top of the tail; occuring in pairs, the anterior scale smallest. Scales on the sides of the tail small, polygonal, flat, mixed with larger keeled ones. Three to four lateral scales per two subcaudal scales. Subcaudals large, rectangular, smooth, imbricate, with rounded posterior margin, the two central rows enlarged. Scales on the regenerated part of the tail all small, rectangular; dorsals and laterals keeled; ventrals smooth. Tail compressed with two denticulate crests; divided in segments, each segment containing one subcaudal.

Scales on the upper and anterior surfaces of the forelimbs, on the anterior surface of the thighs and on the posterior surface of the lower legs enlarged, rhomboidal, keeled; on the lower surface of the thighs and lower legs smooth; on the lower and posterior surfaces of the forelimbs small, tubercular; on the upper surface of the hind limbs, on the anterior surface of the lower legs and on the posterior surface of the thighs small with scattered large, tubercular scales. Fourteen to 18 single, smooth lamellae under the fourth finger; 23–26

Plate 38. Figs. 1–2. *Neusticurus rudis* Boulenger; lateral and dorsal aspect of juv., RMNH 13870 a, Brown's Mountain. Figs. 3–4. *Neusticurus bicarinatus* (Linnaeus); lateral and dorsal aspect of juv., RMNH 13869 b, Brown's Mountain.

Table 26. Counts and measurements of *Neusticurus rudis* Boulenger. For explanation see table 3.

Reg. no.	A	B	C	D	E	F	G	H	I	J	L	M	N	O	P	Q	R	S	T	V	W	X	Z
RMNH																							
13393	♂	73	70(13)	20	10.6	8.8	23	38	17-18	46	23	8	9	8-8	4-4	5-6	4-4	5-5	3+2	4	18-18	24-25	3-4
13394	♂	61	107	16.4	9	7.2			16-17	42	23	8	10	8-8	4-4	5-6	4-4	5-5	3+1	4	16-16	25-24	4
13395	♀	64	108	16.9	9.5	7.5			2-2	45	23	8	9	8-8	4-4	5-6	4-4	5-5	5+1	4	16-17	23-24	4
13396	♀	53	68(35)	14.2	7.8	6.2			4-4	42	22	8	9	8-8	4-4	6-6	4-4	5-5	3+1	4	18-17	24-24	4
13397	♀	54	74(36)	15.2	8.5	6.7	18	28	3-3	47	23	8	10	8-8	4-4	6-6	4-4	5-5	5+1	4	17-16	24-24	3-4
13398	♀	44	63	11.8	6.4	5.4			3-4	?	23	8	10	9-9	4-4	6-6	4-4	5-5	5	4	14-16	22-22	4
13864	♂	46	71(64)	12.9	7.7	6.0	15	24	19-19	41	25	8	9	8-8	4-4	5-5	4-4	4-4	5+1	4	16-16	25-25	3-4
13867	♂	71	114(111)	18.4	10.8	9.0	22	39	19-18	47	23	8	10	8-8	4-4	5-5	4-4	5-4	3+1	4	16-16	25-24	4
13870a	♂	43	65	12.3	6.9	5.4	13	22	21-22	53	24	8	10	8-8	4-4	5-4	4-4	5-5	3+2	4	18-18	26-25	4
b	♂	50	71(50)	14.7	8.0	6.3	15	26	19-18	49	24	8	8	8-8	4-4	6-6	4-4	5-5	5+1	4	17-16	23-24	2-3
15705	♂	78	107(44)	21.4	10.9	9.7			21-20	49	24	8	9	8-8	4-4	?	4-4	?	6+2	4	15-16	23-23	3-4
ZMA																							
*10241	♂	49	86						17-17		21	8		8-8	4-4				5	5	17-15	23-23	3-4
BM																							
*1946.8.31.64	♂	59	122	14.8	8.4	6.7			20-20	48	28	8		8-8	4-4	5-6	4-4	5-4	6	3	20-20	30-29	4
MP																							
50-21	♂	87	110(21)						18-18	41	22	8		8-8	4-4				3		16-16	24-23	2

lamellae under the fourth toe; single in the distal and proximal part; in the central part in two rows, of which in the proximal part only the inner row is present; keeled in the proximal part, smooth distally. Fingers subcylindrical, toes compressed. Palms and soles with small, tubercular, juxtaposed scales.

In life light- to dark-brown above with lighter lateral spots. Ventral parts white with brown spots; underside of the posterior end of the tail dark-brown. Iris orange-brown.

In preservative light- to dark-brown, mottled with black. A series of large, square, white to light-brown spots, from the back of the head via the flanks onto the side of the tail; especially obvious in juveniles; in adults only the spots on the tail are conspicuous, on the body they are faint. Lips in adults uniformly dark-brown, in juveniles and halfgrowns with some white bars. Ventral parts white to dark-ochre with brown or grey spots; under the base of the tail light with brown spots, the rest of the tail very dark-brown with a few faint white spots. Posterior surface of hind limbs with round white spots. Tongue bluish grey, only a small cream-coloured patch posteriorly.

A medium-sized lizard, maximum snout-vent length in females 64 mm, in males 78 mm. Hind limb 1.5–1.7 times as long as forelimb. When laid along the body there is a considerable overlap of forelimb and hind limb. The undamaged tail is 1.4–1.5 times as long as the snout-vent length in juveniles, 1.5–1.7 times in adults.

HABITAT.—Banks of creeks and rivers in rain-forest.

NATURAL HISTORY.—A diurnal lizard, collected between 9.00 a.m. and 5.30 p.m. Most specimens were collected in or near creeks with clear, running water, where they were basking on rocks and branches protruding out of the water. On two occasions they were found under logs on creek banks or in dry creek beds, during the dry season, which might indicate aestivation. No pregnant females or recently hatched juveniles are available. Two stomachs investigated contained a winged ant, a caterpillar, remains of a spider and the distal part of the first pereopod of a freshwater crab (*Pseudothelphusa* sp.). From this it may be concluded that food is captured both on land and in water. When pursued the animals dive into the water and swim to the bottom where they try to hide under logs or in holes. Swimming is by means of an undulating movement of body and tail, the legs are laid alongside the body and tail. RMNH 13395 was taken from the stomach of a snake, but unfortunately this snake did not reach the Rijksmuseum van Natuurlijke Historie and consequently it could not be identified.

RANGE.—Surinam, Guyana, French Guiana and Venezuela (Bolivár). In Surinam the species occurs in more or less hilly country covered with rain-forest to at least 300 m above sea level. An exception is formed by the type-locality of *N. surinamensis* (L. Müller), Albina, which is situated in the lowlands, but the type may have been collected farther inland, transported to Albina and from there sent to Germany (probably by KAPPLER) (fig. 69).

REMARKS.—BEEBE (1945) discusses part of the life history of a *Neusticurus* species which is called *N. rudis* Boulenger by him, but in my opinion is *N. bicarinatus* (Linnaeus), for the measurements he gives for these specimens are in the range of *N. bicarinatus* (Linnaeus), but are too high for true *N. rudis* Boulenger, and because of his colour description of living specimens. The last sentence concerning this subject is: 'Three supraorbital and a single suborbital spot bright yellow', a character applicable only to *N. bicarinatus* (Linnaeus) and certainly not to *N. rudis* Boulenger.

Genus **Prionodactylus** O'Shaughnessy

DIAGNOSIS.—Small, slender lizards with cylindrical bodies, well developed legs with five digits and long tails. Tongue lanceolate, covered with rhomboidal, scale-like, imbricate papillae. Teeth conical anteriorly, bicuspid, laterally compressed posteriorly. Nasals separated by a divided or an undivided frontonasal; prefrontals, occipital and postparietals present. Lower eyelid movable, with or without a divided transparent disc. Ear-opening moderately large, tympanum superficial. Pupil round. Scales on the neck granular; dorsals enlarged, hexagonal, imbricate, keeled, in transverse, but not in longitudinal rows. Femoral pores present. Four species from Panama south to Bolivia, on both sides of the Andes, east to the Guianas. Only one species in Surinam.

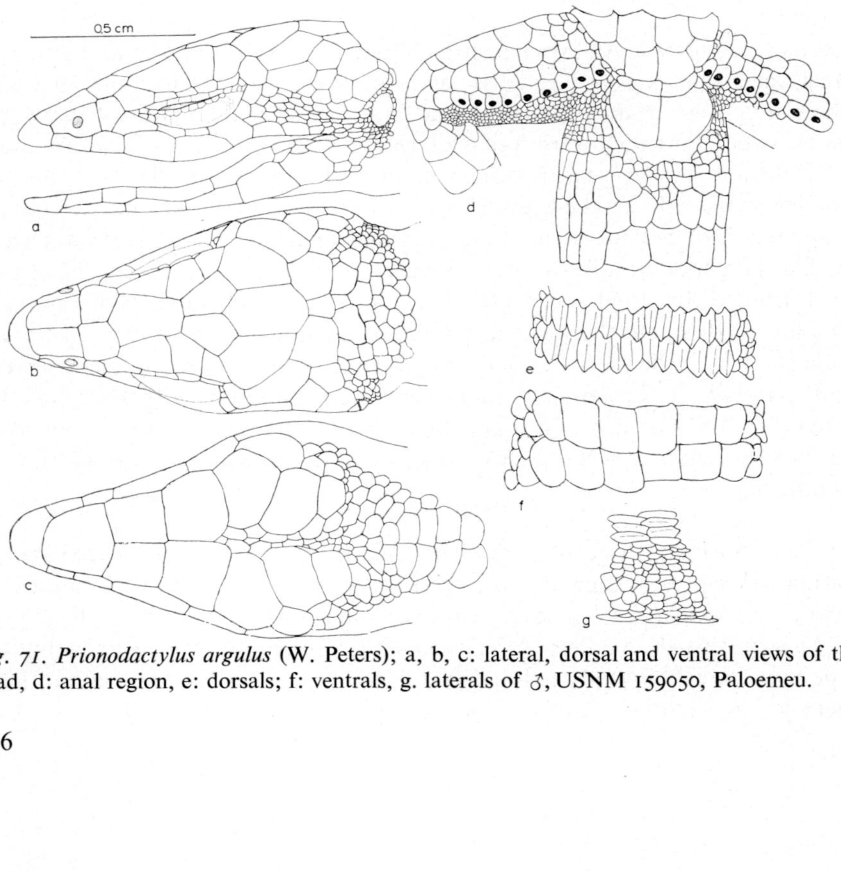

Fig. 71. Prionodactylus argulus (W. Peters); a, b, c: lateral, dorsal and ventral views of the head, d: anal region, e: dorsals; f: ventrals, g. laterals of ♂, USNM 159050, Paloemeu.

Prionodactylus argulus (W. Peters) (figs. 56, 71; pl. 39; table 27)

Cercosaura argulus W. Peters, 1862: 184.

TYPE-LOCALITY.—Santa Fé de Bogotá, Colombia.

MATERIAL. – Surinam.
Paloemeu: 1♂, USNM 159050, 20–VIII–1964, leg. T. H. Lewis.
Guyana.
Bartica district: 1♂, AMNH 21265, 1916, leg. W. Beebe.
Marudi: 1♀, AMNH 61386.
Shudikar-wau: 1♀, AMNH 61434, April 1938, leg. R. Snedigar.
French Guiana.
Tumuc Humac Mountains, Upper Lunier River: 1♀, MP 99–73, leg. Geay.
Ecuador.
Canelos: 1♀, 2♂♂, BM 1946.8.31.18–20, leg. Buckley (types of *Prionodactylus oshaughnessyi* Boulenger).

DIAGNOSIS.—A small, slender lizard with cylindrical body, well developed legs and long tail. Dorsals enlarged, in 40 transverse rows, 13 dorsals in a row at midbody level. Postparietals present, parietals trapezoid. Brown above with an indistinct dorsolateral line. Flanks with a row of ocelli of which the anterior two or three are large, the others small. Upper lips white, sharply delimited against the rest of the dorsal head colour.

DESCRIPTION.—Head long with pointed snout, slightly depressed; posterior part rounded, slightly wider than the neck, which in turn is narrower than the body.

Tongue lanceolate, narrow anteriorly, covered with rhomboidal, scale-like, imbricate papillae; tip bifid, smooth. Teeth conical anteriorly; bicuspid, laterally compressed, posteriorly.

Rostral hexagonal, visible from above, more than twice as wide as deep. Two longitudinally rectangular frontonasals. A pair of irregularly hexagonal prefrontals. Frontal hexagonal, longer than wide. A pair of pentagonal frontoparietals. A heptagonal, elongate interparietal. Two trapezoid parietals with convex lateral margins; shorter than the interparietal. One small, trapezoid central and two large, irregularly hexagonal, lateral occipitals (postparietals). Three supraoculars; the anterior one largest; the posterior one smallest; in direct contact with the supraciliaries. Nostril in an undivided nasal. One large loreal, irregularly pentagonal, deeper than long. Preocular absent. Four suboculars; the anterior and the posterior subequal, largest; the third smallest. Three postoculars, the upper one much larger than the other two. Four or five supraciliaries; the anterior one very large; the others long and narrow, subequal. Temporal region divided in an antero-ventral part with small, more or less hexagonal scales and a postero-dorsal part with eight large, polygonal scales (one of which borders on the parietal). Ear-opening moderately large, surrounded by small scales, forming a smooth margin. External auditory meatus short;

tympanum well visible. Nine supralabials; the third and the fourth subequal, largest; the fourth under the eye. Lower eyelid with a semitransparent disc of three palpebrals. Pupil round.

Mental crescent-shaped, with blunt ends. Postmental large, pentagonal. Five pairs of chinshields, the members of the anterior two pairs forming a median suture; the members of the posterior three pairs separated by several rows of small scales; the second and third pairs subequal, largest; the fifth pair smallest. Eight infralabials; the third and the fourth very long, subequal, largest; the posterior four scales small. The posterior three pairs of chinshields separated from the infralabials by small to large scales.

Seven pairs of transversely enlarged gulars, increasing in size towards the collar. Collar consisting of six scales, the two central ones very large; its posterior margin with round lobes. Scales on the nape enlarged, hexagonal; wider than long anteriorly, longer than wide near the forelimbs; the anterior scales smooth, the posterior ones with keels; in transverse rows. Scales on the side of the neck small, rounded, in vertical rows.

Dorsals larger than the laterals, smaller than the ventrals, hexagonal, keeled, imbricate; 13 transversely at midbody; in transverse rows only, 40 rows from the occipitals to the posterior surface of the thighs. Laterals small, granular, juxtaposed on the anterior part of the flanks; larger, rhomboidal, imbricate, in vertical rows on the posterior part of the flanks, two rows corresponding to one row of dorsals. Ventrals large, square, slightly imbricate, smooth, in 20 transverse and in six longitudinal rows. Total number of scales around midbody 34. Two very large preanal scales, preceded by four small scales.

Preanal pores absent; femoral pores nine on each side, in the posterior part of single scales.

Scales on the tail hexagonal, keeled, imbricate, in transverse rows; under the tail square, smooth, slightly imbricate, also in transverse rows. Scales on the regenerated part of the tail rectangular, smooth, slightly imbricate, in transverse rows, smaller than those on the original part of the tail. Tail round in cross--section.

Scales on the upper and posterior surfaces of the forelimbs, on the anterior and ventral surfaces of the hind limbs large, smooth, imbricate; on the ventral and anterior surfaces of the forelimbs, on the posterior and dorsal surfaces of the hind limbs small, granular, juxtaposed. Seventeen to 18 divided, smooth lamellae under the fourth finger; 20–21 single, keeled lamellae under the fourth toe. Palms and soles with small granules. Fingers subcylindrical; toes compressed; all fingers well developed and provided with claws.

Colour in life unknown.

In preservative the back is light-brown; the dorsal and lateral surface of the head grey-brown, forming a sharp border with the cream-coloured upper lips. A whitish dorsolateral line with a narrow dark-brown line medially, continued on the tail. The flanks have the same colour as the back; with two (left side) or three (right side) large, round, black spots, with in the centre a single white scale; followed at some distance by four smaller, separate, identical spots,

Plate 39. Figs. 1–3. *Prionodactylus argulus* (W. Peters); dorsal, lateral and ventral aspect of ♂, USNM 159050, Paloemeu. Figs. 4–6. *Ptychoglossus brevifrontalis* Boulenger; dorsal, lateral and ventral aspect of ♂, BM 1939.1.1.75, New River.

Table 27. Counts and measurements of *Prionodactylus argulus* (W. Peters). For explanation see table 3.

Reg. no.	A	B	C	D	E	F	G	H	I	J	K	L	M	N	O	P	Q	R	S	T	U	V	W	X	Y
USNM																									
159050	♂	42	49(6)	11.3	6.1	4.9	12	17	9/0-9/0	34	40	20	6	7	9-9	8-8	5-4	3-3	3-3	2+4	7	5	17-18	21-20	13
AMNH																									
21265	♂	38	?	10.2	5.7	4.3	11	15	8/0-8/0	33	37	20	6	4	8-8	6-6	5-5	3-3		2+2	8		?	?	13
61386	♀	40	?	10.6	6.1	4.3	?	?	2/0-3/0	34	36	19	6	4	8-8	6-7	5-4	3-3		4+2	8		17-17	20-22	13
61434	♀	40.5	91	10.3	5.8	4.3	12	15	2/0-3/0	32	36	18	6	5	8-8	6-6	4-4	3-3		4+2	7		18-18	22-19	13
MP																									
99-73	♀	43	100				13	19	0	37	39	20	6	9	8-8	7-7	3-4	3-3	5-5	4+2	7		20-19	23-23	12
BM																									
*1946.8.31.18	♂	46	59(31)	12.1	6.6	5.8	15	21	9/0-9/0	35	40	17	6	8	8-8	6-7	4-4	3-3	4-4	2+2	7		16-16	21-19	13
* .19	♂	48	98	11.6	5.2	4.5	16	19	6/0-7/0	34	40	18^{1}/1	6	7	8-8	7-7	4-4	3-3	?	4+2	7		17-17	20-?	12
* .20	♀	40	40(16)	10.1	5.2	4.0	12	18	0	33	41	19	6	7	8-8	7-7	4-4	3-3	3-4	4+2	7		16-15	21-20	12

Table 28. Counts and measurements of *Ptychoglossus brevifrontalis* (Boulenger). For explanation see table 3.

Reg. no.	A	B	C	D	E	F	G	H	I	J	K	L	M	N	O	P	Q	R	S	T	U	V	W	X	Y
BM																									
1939.1.1.75	♂	53	30(30)	11.1	7.2	5.1	11	18	11/2-12/2	29	30	18	8	7	6-6	6-6	4-4	4-4	3-3	4	6	3	11-11	18-19	22
*1946.8.31.63	♂	64	17(17)	13.6	9.4	5.6	13	23	12/2-12/2	36	31	18	8	7	7-7	6-6	4-4	4-4	4-3	4	6		13-13	18-18	24

diminishing in size posteriorly, the most posterior one in the groin. Colour of the flanks separated from that of the belly by a white lateral band, which starts at the upper lip, passes over the forelimb and continues to the groin; ventrally it is bordered by a band composed of small grey dots; dorsally there is a black line from the posterior corner of the eye to the first ocellus; posterior of the ear it is wider than anterior of it. Base of the forelimb with a black spot. Ventral surface cream-coloured; all ventrals, preanals, lower labials and posterior gulars with central clusters of grey dots; chinshields with the clusters near their lateral margin. Anterior part of tongue dark-grey, with a concave posterior margin, posterior part white.

A small, slender lizard, with a snout-vent length of 42 mm. Hind limb 1.4 times as long as forelimb, when laid along the body there is overlap of fingers and toes.

HABITAT.—Unknown, probably rain-forest.

NATURAL HISTORY.—Nothing is known about this subject.

RANGE.—Surinam, Guyana, French Guiana, Brazil (Amapá) and in Ecuador east of the Andes. In Surinam known only from one locality in the interior (170 m) (fig. 56).

REMARKS.—The single specimen available from Surinam differs in some characters from specimens from other localities (Guyana, French Guiana, Ecuador). For instance, all other specimens studied had eight upper labials, six or seven lower labials, three large central collar scales. Most specimens do have four large preanals, only three out of eight specimens have two. Until recently this species was known as *Prionodactylus oshaughnessyi* Boulenger, but UZZELL (1970) referred this name to the synonymy of *P. argulus* W. Peters.

Genus **Ptychoglossus** Boulenger

DIAGNOSIS.—Small lizards with cylindrical bodies, short, well developed legs and long tails. Tongue lanceolate, covered with two longitudinal rows of obliquely transverse plicae. Anterior teeth conical; the rest bicuspid or tricuspid, laterally compressed. Nasals separated by a frontonasal; prefrontals and frontoparietals present; a transverse row of small occipital scales; dorsals hexagonal, smooth, imbricate; ventrals rectangular, smooth, juxtaposed; gulars subequal, in transverse rows. Femoral pores present. Lower eyelid movable, with a divided semitransparent disc. Ear-opening large, tympanum recessed. Pupil round. Seven species from Costa Rica to southern Ecuador, east to the Guianas, not present on the Pacific slopes of the Andes. One species present in Surinam.

Ptychoglossus brevifrontalis Boulenger (figs. 56, 72; pl. 39; table 28)

Ptychoglossus brevifrontalis Boulenger, 1912: 421.

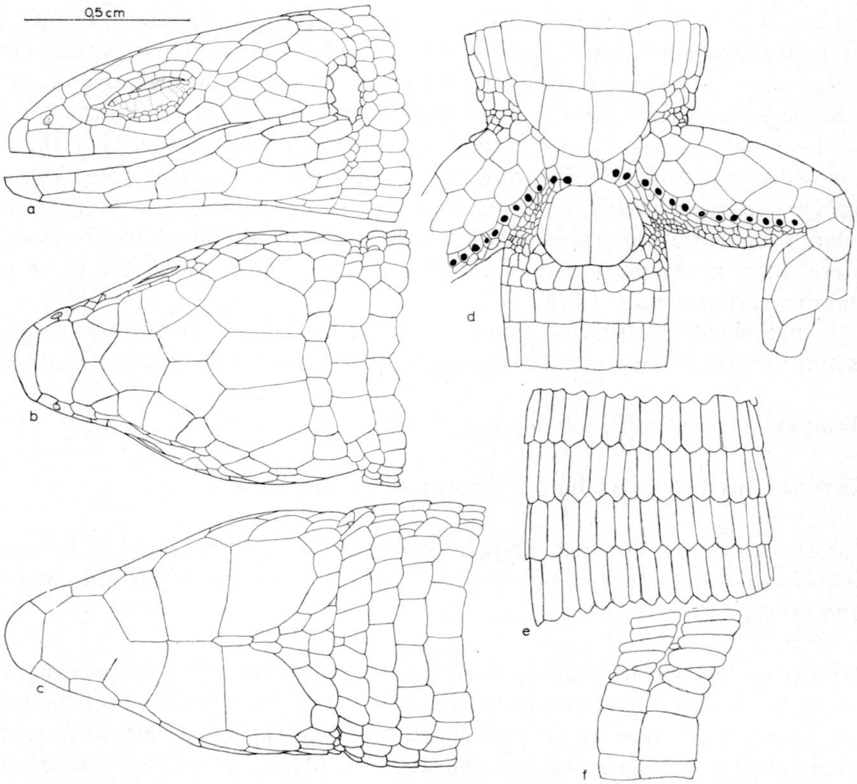

Fig. 72. Ptychoglossus brevifrontalis Boulenger; a, b, c: lateral, dorsal and ventral views of the head, d: anal region, e: dorsals, f: border-zone between laterals and ventrals of ♂, BM 1939.1.1.75, New River.

TYPE-LOCALITY.—El Topo, Río Pastaza, Ecuador.

MATERIAL. – Surinam/Guyana.
New River, 750 feet: 1♂, BM 1939.1.1.75, leg. C. A. Hudson.
Ecuador.
El Topo, Río Pastaza, 4200 feet: 1♂, BM 1946.8.31.63, leg. M. G. Palmer (holotype of *P. brevifrontalis* Boulenger).

DIAGNOSIS.—A small lizard with a cylindrical body; short, well developed legs and a (probably) long tail. Dorsals and laterals enlarged, rectangular; in 30 transverse rows. Ventrals rectangular, juxtaposed. Three pairs of chinshields; a pair of prefrontals; four supraoculars. Brown above with some dark-brown spots. Ventral parts immaculate, cream-coloured. Side of tail with a row of cream-coloured spots.

DESCRIPTION.—Head pyramidal, depressed; as wide as the neck which is as wide as the body. Snout rounded.

Tongue lanceolate; without emargination; covered with two longitudinal rows of oblique plicae, the pairs converging anteriorly; tip bifid, smooth. Anterior teeth conical; the rest bicuspid, laterally compressed.

Rostral hexagonal, visible from above, slightly more than twice as wide as deep. A large pentagonal frontonasal, wider than deep. A pair of irregularly pentagonal prefrontals, forming a short median suture. Frontal hexagonal, slightly longer than wide. A pair of irregularly pentagonal frontoparietals, forming a long median suture. Interparietal elongate, pentagonal with parallel lateral margins. Parietals irregularly hexagonal, slightly longer and larger than the interparietal, truncate posteriorly. A transverse row of five rectangular occipital scales, which are smaller than the nuchals. Four supraoculars; the anterior one smallest; the third one largest; in direct contact with the supraciliaries. Nostril in a divided nasal, the anterior part distinctly smaller than the posterior part. Loreal small, rectangular, deeper than long; narrowly separated (or not) from the supralabials by the posterior nasal and the anterior subocular forming a short suture. Preocular absent. Four suboculars, the third very large. Two postoculars. Four supraciliaries, the anterior one largest. Temporal region with large, flat, smooth scales of different sizes and shapes. Ear-opening large; surrounded by small scales, the anterior margin denticulate, posterior margin smooth. External auditory meatus long, tympanum well visible. Six supralabials; the third largest; the fourth smallest; three labials to below the centre of the eye. Lower eyelid with a semitransparant disc of three palpebrals. Pupil round.

Mental trapezoid with a convex anterior margin. A large heptagonal postmental. Three pairs of chinshields; the third pair largest; the first pair smallest; the anterior two pairs in contact; the members of the third pair separated by a wedge-shaped group of small scales. Six infralabials, the posterior three distinctly narrower than the anterior three. Third pair of chinshields separated from the infralabials by a few large scales.

A row of small pits along the margins of all dorsal and lateral headscales, on ventral scales restricted to part of their margins. Scales on the anterior part of the head also with a scattering of pits on the rest of their surface.

Five rows of transversely rectangular, smooth, slightly imbricate gulars, and a row of smaller scales between the chinshields and the collar. Collar with seven scales, the central three largest. Scales on the nape rectangular, wider than long near the head; hexagonal, longer than wide near the forelimbs; smooth, slightly imbricate; in transverse rows. Scales on the side of the neck rectangular, with rounded posterior margins; narrower near the gulars than near the scales on the nape; smooth, juxtaposed; in vertical rows.

Dorsals hexagonal, about three times as long as wide; smooth, imbricate; in 30 transverse rows between the parietals and the posterior surface of the thighs. Laterals rectangular, about twice as long as deep; smooth, imbricate; in transverse rows. Ventrals large, rectangular, smooth, juxtaposed, in eight longitudinal and in 19 transverse rows. Ventrals and laterals of the same annulus separated by two small, wedge-shaped scales. Total number of scales around midbody 29. Four elongate preanals, the two centrals very large, the two laterals minute.

Preanal pores two on each side, separated from the 11–12 femoral pores by a small gap. Pores in the latero-posterior corner of a scale which may be accompanied by a smaller one.

Scales on the tail rectangular; slightly keeled on the anterior part, smooth on the rest; juxtaposed, in transverse and in longitudinal rows; under the tail quadrangular, smooth. Tail round in cross-section.

Scales on the limbs smooth, on the upper and anterior surfaces of the forelimbs and on the anterior and lower surfaces of the hind limbs large, rhomboidal, imbricate; on the ventral and posterior surfaces of the forelimbs small, rhomboidal, juxtaposed; on the posterior and upper surfaces of the hind limbs small, granular, juxtaposed. Eleven single lamellae under the fourth finger; 18–19 lamellae under the fourth toe, divided proximally, single distally. Palms and soles with small, flat, granular scales. Fingers and toes subcylindrical.

In preservative the back is brown with a few dark-brown spots which are more numerous on the flanks. Dorsal surface of the head brown with a dark-brown mottling, side of the head with dark-brown spots; upper and lower lips with dark-brown and cream-coloured spots. Chin, throat and belly cream-coloured, immaculate. Underside of the tail cream-coloured, with irregular dark-brown spots invading the light part from the sides. Side of the tail with a row of cream-coloured spots.

A small lizard, snout-vent length 53 mm. Hind limb 1.6 times as long as forelimb. When laid along the body fingers and toes do not overlap or touch; distance between forelimbs and hind limbs 2.7 times the length of a forelimb. No complete tails are available.

HABITAT AND NATURAL HISTORY.—Nothing is known about these subjects.

RANGE.—Surinam and the Amazonian slopes of Ecuador. In Surinam occurring to at least 750 feet above sea level, in Ecuador to at least 4200 feet (BOULENGER, 1912) (fig. 56).

REMARKS.—Although the species has not been found between Ecuador (type-locality) and Surinam, the identification of the Surinam individual is beyond doubt. It agrees very well with the type specimen (BM 1946.8.31.63) in most characters. However, there is a difference in pattern which is not due to a difference in sex, for both specimens are males. The holotype has very distinct light-brown dorsolateral stripes which start on the snout and continue to just posterior of the forelimbs, on the rest of the body they are absent but are present again on the anterior part of the tail. Moreover, the scales on the flanks of the holotype are nearly completely dark-brown with only small light-brown spots left, giving rise to a dark-brown flank with light-brown spots. The only important differences in scale characters are that in the holotype the four preanal scales are of nearly equal size and that one of the femoral pores forms a group with the two preanal pores of the same side and which is separated from the rest of the femoral pores by a gap. Should these differences prove to be constant when series are investigated, recognition of a Guyanese subspecies may be justified.

At first glance there is a striking resemblance between the Surinam specimen of *Ptychoglossus brevifrontalis* Boulenger and half-grown specimens of *Arthrosaura reticulata versteegii* Van Lidth de Jeude, which, however, soon disappears on closer investigation.

The exact collecting locality of the specimen described above is not known, but whether it was actually collected in Surinam or in Guyana does not matter very much, because it will presumably also be present in other parts of the Acarai Mountains west and east of the New River.

Hitherto the species has been reported only from eastern Ecuador. The specimen from the border between Surinam and Guyana thus extends the range of the species several thousands of kilometers to the east. Whether this species inhabits the entire Amazon basin or is restricted to mountain ranges (high or low) along its northern border remains to be solved. A relative, *P. kugleri* Roux, is known from mountain ranges in northern Venezuela (Donoso-Barros, 1968) which might indicate that the genus is restricted to mountainous areas. However, some other species which were known only from eastern Ecuador and now are also known from the Guianas proved to be present in lowland as well as in mountains. This may very well be true for this species as well and than it might be expected to be found in the entire Amazon basin. That it has escaped notice in that large area until now is not very surprising, because it probably has a secretive way of life on the forest floor.

Genus **Tretioscincus** Cope

Diagnosis.—Small lizards with cylindrical bodies and tails, with short but well developed legs. The inner finger is rudimentary and clawless. Tongue lanceolate, covered with imbricate, scale-like rhomboidal papillae in oblique rows. Teeth conical anteriorly; bicuspid, laterally compressed posteriorly. Nasals separated by a single frontonasal, prefrontals widely separated, two supraoculars, parietals transversely enlarged. Lower eyelid movable, with an undivided transparent disc. Ear-opening small, tympanum recessed. Pupil very peculiar, round with a dorsal and a ventral flap projecting inwards. Femoral pores present in males. Two species in northern South America, from Colombia to Amazonian Brazil. One species occurs in Surinam.

Tretioscincus agilis (Ruthven) (figs. 73, 74; pl. 28; table 29)

Calliscincopus agilis Ruthven, 1916: 2; D'Audretsch, 1968: 51; Hoogmoed, 1968: 611.
?'62. Grondhagedis' Fernandes, 1927: 249.
'Klein hagedisje' Van Lynden, 1939: 844.

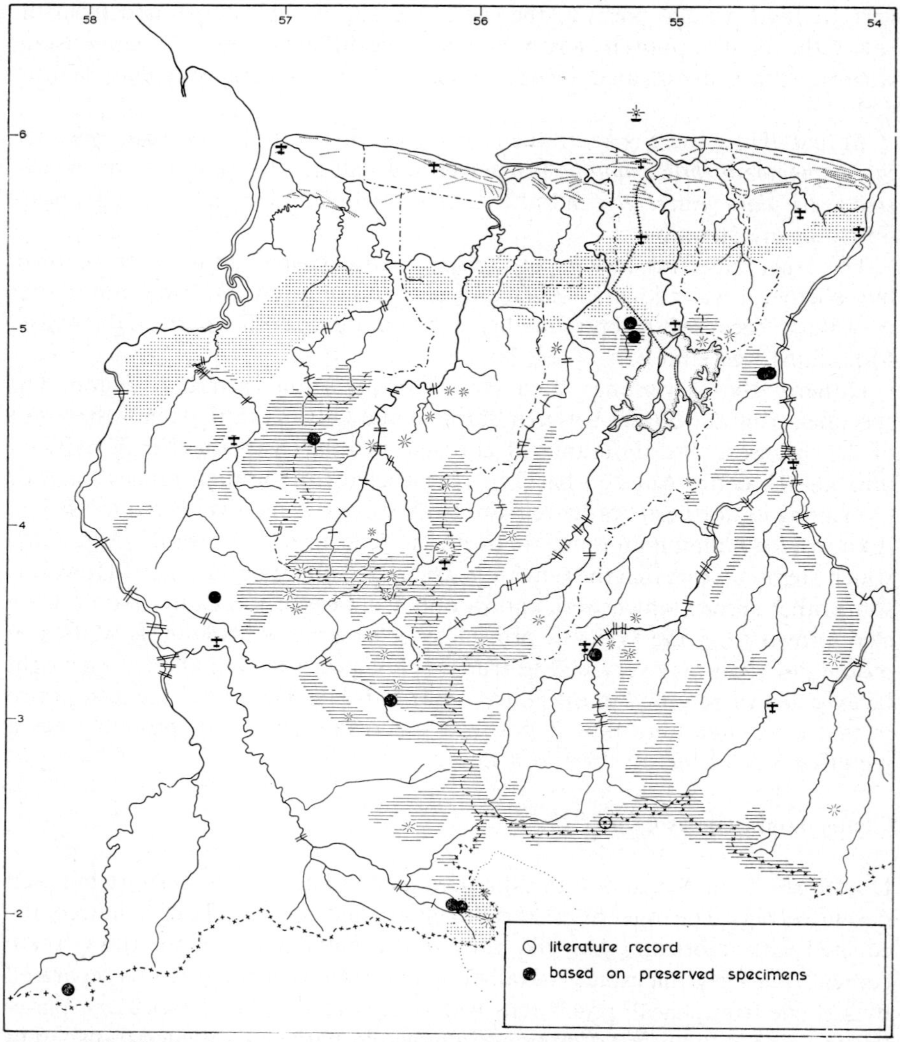

Fig. 73. Distribution of *Tretioscincus agilis* (Ruthven) in Surinam.

TYPE-LOCALITY.—Dunoon, Demerara River, Guyana.

MATERIAL. – Surinam.

Bakhuis Mountains: 1♂, RMNH 13447, bivouac IV, Adampadda Creek, 5–VII–1963, leg. R. van Aerde.

Brown's Mountain: 1♀, RMNH 13872, 20–VIII–1968, leg. M. S. Hoogmoed.

Nassau Mountains: 1♂, RMNH 13449, km 8.9, 9–III–1949, 1♂, RMNH 13450, km 10.4, 11–III–1949, 2♂♂, RMNH 13451, km 9.3 and 9.4, 11–III–1949, 1♂, RMNH 13452, km 9.7, 12–III–1949, 1♀, 1♂, RMNH 13453, km 10.2, 14–III–1949, all leg. Surinam Expedition 1948–49.

Wilhelmina Mountains: 1♀, 1♂, ZMA 12313–14, 1926, leg. Wilhelmina Mountains Expedition.

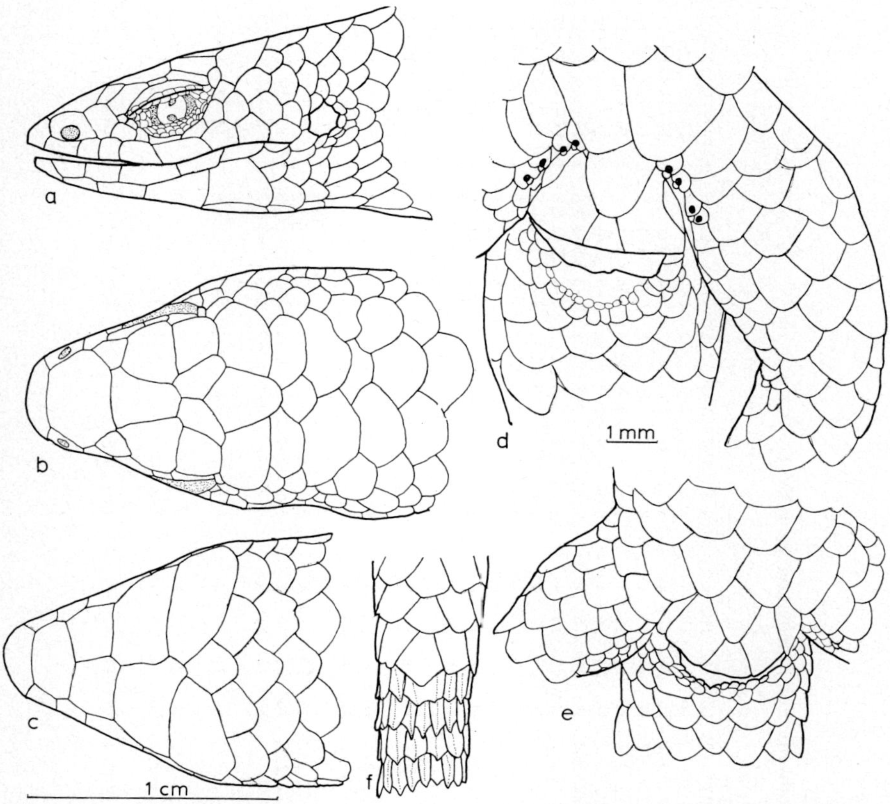

Fig. 74. Tretioscincus agilis (Ruthven); a, b, c: lateral, dorsal and ventral views of the head of ♀, RMNH 13871, Airstrip Kayser Mountains; d: anal region of ♂, RMNH 15239, Airstrip Paloemeu; e: anal region of RMNH 13871; f: dorsal view of former breaking-point in the tail of RMNH 13871.

Airstrip Kayser Mountains: 1♀, RMNH 13871, E., 8–VIII–1968, leg. M. S. Hoogmoed.

Airstrip Paloemeu: 2♂♂, RMNH 15146, S., 28–V–1968, 1♂, RMNH 15239, S., 23–V–1968, both leg. M. S. Hoogmoed; 1♀, USNM 159051, August 1964, leg. T. H. Lewis.

Itabu Creek Head: 1♀, 2♂♂, FMNH 30932–34, Boundary Camp, 1938, leg. E. R. Blake (Sewell Avery Expedition).

Airstrip Sipaliwini: 1♀, RMNH 13448, 12–VI–1963, leg. P. H. van Doesburg, 1♂, RMNH 16376, near Base Bivouac, 16–I–1970, leg. M. S. Hoogmoed & J. J. P. Paats.

Guyana.

Dunoon, Demerara River: 2♂♂, UMMZ 47798–99, 24–VIII–1914, leg. A. G. Ruthven (holo- and paratype of *Calliscincopus agilis* Ruthven).

French Guiana.

Tumuc Humac Mountains: 1♂, MP 99–75, Upper Lunier River, leg. Geay.

Diagnosis.—A small lizard with a cylindrical body and tail and short, but well developed legs. All scales large, smooth and shining, imbricate, 16 around the

Table 29. Counts and measurements of *Tretioscincus agilis* (Ruthven). For explanation see table 3.

Reg. no.	A	B	C	D	E	F	G	H	I	J	K	L	M	O	P	Q	R	S	T	U	V	W	X
RMNH																							
13447	♂	39	46(33)	9.4	5.9	4.3	13	19	3/2-3/2	16	25	16	4	8-8	7-7	4-3	2-2	4-4	5+1	8	3	15-15	17-17
13448	♀	?	54	?	?	?	?	?	0	16	24	18	4	8-8	7-7	4-3	2-2	4-4	5+1	?	3	14-16	15-17
13449	♂	37	53	8.6	5.8	3.9	11	18	3/2-3/2	16	24	15	4	8-8	7-7	3-3	2-2	4-4	5+1	8	3	14-14	17-17
13450	♂	54	51(9)	12.1	8.2	7.0	17	30	2/2-2/2	16	25	15	4	8-8	7-7	3-3	2-2	4-4	5+1	8	3	14-12	18-17
13451a	♂	44	65(30)	11.4	8.0	?	18	27	2/2-2/2	16	24	16	4	8-9	7-7	3-3	2-2	4-4	5+1	7	3	14-14	17-17
b	♂	47	71	10.8	7.2	5.3	16	22	2/2-2/2	16	24	14	4	8-8	7-7	3-3	2-2	4-4	5+1	8	3	14-14	17-17
13452	♂	59	87(63)	13.3	8.3	6.5	20	31	3/2-3/2	16	26	16	4	8-8	7-7	3-3	2-2	4-3	5+1	8	3	14-14	17-17
13453a	♀	50	8(8)	11.4	7.3	5.8	18	23	0	16	26	19	4	8-8	7-8	3-3	2-2	4-4	5+1	8	3	14-14	17-17
b	♂	55	77(54)	13.8	9.7	7.6	20	27	2/2-2/2	16	25	15	4	8-8	7-7	3-3	2-2	3-4	5+1	8	3	14-13	15-16
13871	♀	62	66(28)	12.7	9.2	6.6	16	25	0	16	25	18	4	8-8	7-7	3-3	2-2	4-5	5+1	8	3	13-15	19-18
13872	♀	59	74(64)	12.4	9.1	6.6	17	27	0	16	27	18	4	8-8	7-7	3-3	2-2	4-4	5+1	8	3	15-15	18-18
15146a	♂	51	30(14)	11.7	8.3	5.9	15	25	2/2-2/2	16	26	16	4	8-8	7-7	3-3	2-2	4-4	5+1	8	3	14-?	17-17
b	♂	55	85	11.3	8.3	6.0	18	27	2/2-2/2	16	26	16	4	8-8	7-?	3-3	2-2	4-4	5+1	8	3	15-14	19-18
15239	♂	38	55	8.9	6.1	4.6	12	17	2/2-2/2	16	26	17	4	8-8	7-7	3-3	2-2	4-4	5+1	8	3	14-14	19-19
16376	♂	54	57(40)	?	?	?	17	27	3/2-3/2	16	24	15	4	?	?	?	?-2	?	5+1	8	3	14-14	18-18
ZMA																							
12313	♂	45	50(26)				14	23	2/2-2/2	16	26	17	4	8-8	7-8	3-3	2-2	4-3	5+1	8	3	15-13	17-17
12314	♀	57	36(27)				19	26	0	16	26	18	4	8-8	7-7	3-3	2-2	4-4	5+1	8	3	14-15	17-17
FMNH																							
30932	♂	57	82	13	9.8	7.1	18	29	2/2-2/2	16	25	16	4	8-8	8-8	4-3	2-2	4-3	5+1	8	3	14-13	17-17
30933	♂	29	5(5)	7.8	4.6	3.9	8	12,5	2/2-2/2	16	25	16	4	8-8	7-7	3-3	2-2	2-3	5+1	8	3	15-15	17-17
30934	♀	51	61(36)	10.9	7.6	5.6	15	24	0	16	26	18	4	8-8	7-7	4-4	2-2	3-3	5+1	8	3	14-14	17-17
USNM																							
159051	♀	40	30(30)	9.2	6.4	4.3	12	18	0	16	?	18	4	8-8	7-7	3-3	2-2	4-4	5+1	8	3	14-14	17-18
UMMZ																							
*'47798	♂	55	88	12.2	8.2	6.4	19	28	2/2-2/2	16	26	16	4	9-8	7-7	3-3	2-2	3-3	5+1	8	3	14-14	19-18
*'47799	♂	44	64	10.3	6.8	4.6	14	22	2/2-2/2	16	24	15	4	8-8	7-7	3-3	2-2	3-3	5+1	9	3	15-14	17-16
MP																							
99-75	♂	34	43(39)	8.2	5.3	4.4	11	16	3/2-3/2	16	24	15	4	8-8	7-6	3-3	2-2	4-5	5+1	8	3	14-14	18-19

middle of the body. Inner finger short, clawless. Purplish brown above with yellowish dorsolateral stripes, tail blue. Lateral parts black.

DESCRIPTION.—Head and neck as wide as the body; snout blunt; head short, distinctly depressed.

Tongue lanceolate, without emargination; covered with imbricate, scale-like, rhomboidal papillae, in oblique rows; tip bifid, smooth. Teeth conical anteriorly; laterally compressed, bicuspid posteriorly.

Rostral more than twice as wide as deep, hexagonal, visible from above. A large, heptagonal frontonasal, wider than long. Prefrontals small, trapezoid, widely separated, in most specimens (14) separated from the frontal, in four forming a short suture with the frontal. Frontal elongate, pentagonal, longer than wide. Two pentagonal frontoparietals. A rhomboidal interparietal. Two transversely enlarged parietals. Three occipitals; the central one small, rhomboidal; the two lateral ones transversely enlarged, identical in shape to the dorsals, but smaller. Supraoculars two; the first largest, the second smallest; directly in contact with the supraciliaries. Nostril in an undivided nasal. Loreal subquadrangular. A small, rectangular frenoocular between loreal and anterior subocular. Three or four suboculars of variable size. One or several small preoculars. Two or three large postoculars. Mostly three, sometimes four supraciliaries; the first largest, the second smallest. Temporal region with a few, large, hexagonal, smooth, imbricate scales. Eight, rarely nine supralabials, the fifth largest and below the centre of the eye. Ear-opening small, with a smooth margin, composed anteriorly of large scales, posteriorly of small scales; external auditory meatus moderately long, tympanum hardly visible. Lower eyelid with a large undivided transparent disc. Pupil round with a ventral and a dorsal flap projecting inwards.

Mental trapezoid with a convex anterior margin. Postmental heptagonal. Three pairs of imbricate chinshields; the first pair smallest, the third pair largest; the members of each pair in contact, or the posterior pair separated. Seven, rarely eight, infralabials, the third largest. All chinshields in contact with the infralabials.

Gulars large, smooth, imbricate, in longitudinal rows, mostly eight between the posterior pair of chinshields and the pectoral scales. Collar absent. Scales on the nape identical to the dorsals, scales on the side of the neck smaller, rounded, imbricate, smooth.

Dorsals and laterals rounded rhomboidal, smooth, imbricate, in longitudinal and oblique series, 23–26 dorsals from the small central occipital to the posterior surface of the thighs. Ventrals hardly larger than the dorsals, in longitudinal and oblique series, 14–19 between the last row of gulars and the preanals. The first row of ventrals (sometimes called pectorals) is the row of scales between the bases of the forelimbs. Sixteen scales around midbody. Preanals forming a semicircle of five posterior scales bordering the cloaca and one central, anterior scale.

Preanal pores absent, femoral pores in a single scale, four or five on each side in males, absent in females. Hemipenis with a narrow zone of small spines near the base, medial from the sulcus spermaticus; distally of these spines it is covered with oblique plicae, sloping proximally from the sulcus spermaticus, changing into longitudinal plicae on the median side of the inner lobe of the divided tip; sulcus spermaticus bifid.

Scales on the tail rhomboidal, imbricate, smooth, in oblique and longitudinal series; scales under the tail somewhat more pointed. Scales on the regenerated part of the tail small, irregularly hexagonal, elongate, in transverse and in longitudinal series, keeled, the keels more distinct on the scales near the tip of the tail. Tail round in cross-section.

Scales on the forelimbs and hind limbs large, rhomboidal, rounded, smooth, imbricate, only the scales on the posterior surface of the upper arms and thighs smaller. Twelve to 16 (mostly 14) tuberculate, single lamellae under the fourth finger; 15–19 (mostly 17) keeled, single lamellae under the fourth toe. Scales on the palms and the soles tuberculate. Fingers and toes compressed, first finger rudimentary, clawless.

The smooth scales give the animal a very glossy appearance. Colour in life very beautiful. The back is purplish brown to the base of the tail. A light dorsolateral stripe from the snout to the tail where it fuses with its partner, ochre--yellow in its anterior part, greenish white on the body and bright blue on the tail, resulting in the tip of the tail being entirely blue. Sides of head and body with a black band, bordered ventrally from tip of snout to base of forelimb by an ochre-yellow stripe. Ventral parts greenish blue. For a coloured plate see HOOGMOED (1968).

In preservative the colour of the back varies from chestnut-brown to olive--brown. The dorsolateral stripe is greenish to bluish white in its anterior, pale blue in its posterior portion. Medially it is bordered by a narrow black line. Upper lips bluish white. No sharp border between the black lateral band and the ventral parts, the black part on each scale decreasing in size towards the median. Throat and chin immaculate, white to blue, a row of scales with black centres from the corner of the mouth to the base of the forelimb. Ventral parts white to blue, the ventrals with darker centres. Tail pale blue. Tongue dark-grey with a light median stripe.

A small lizard, maximum snout-vent length in females 62 mm, in males 59 mm. The undamaged tail is 1.4–1.5 times as long as the snout-vent length. Hind limb 1.4–1.8 times as long as the forelimb. When laid along the body there is overlap of hind limb and forelimb, the hind limb just reaching to the axilla.

HABITAT.—An inhabitant of the rain-forest where it lives on the lower part of trees or on the ground near them. A very different habitat is found in Oriximiná (Brazil) where the species was found in the town itself in 'leafpalm walls of huts, empty lots, backyards, rubbish piles, etc.' (VANZOLINI & REBOUÇAS-SPIEKER, 1969). In Guyana the species was reported from rain-forest by BEEBE (1945) and from ridge-forest, which is considerably more open than rain-forest, by RUTHVEN (1916).

NATURAL HISTORY.—A diurnal lizard, which was collected between 11.00 a.m. and 3.00 p.m. Basking in open places was observed for all specimens collected by me. This species is semi-arboreal, living on the trunks of living trees near the base, and on felled trees. It may also be encountered on the forest-floor, always near the base of a tree. One specimen (RMNH 13448) was taken from the interior of a rotten log. According to RUTHVEN (1916) this species is a climber but according to BEEBE (1945) it is terrestrial. From the data of VANZOLINI & REBOUÇAS-SPIEKER (1969) it can be concluded that this species is an occasional climber. The type specimen of *Tretioscincus romani* Andersson (1918) (a synonym of *agilis*) was taken from a rotten stump. When pursued these lizards try to escape by circling the tree, always trying to keep this between them and the observer. No pregnant females, nor recently hatched juveniles were available to me. VANZOLINI & REBOUÇAS-SPIEKER (1969) report five females with mature and six with immature eggs from January. Two stomachs examined by me contained spiders (e.g., Clubionidae), crickets and bugs. VANZOLINI & REBOUÇAS-SPIEKER (1969) examined 25 stomachs and list the results in tables. From their data they conclude that *T. agilis* (Ruthven) searches for its prey under objects, a conclusion that also holds true for the two Surinam specimens examined by me.

RANGE.—The species is known from Guyana, Surinam, French Guiana and Brazil (Pará, Amazonas). In Surinam it is present in the interior south of the coastal savannah belt. It reaches to at least 500 m above sea level. The species was reported from Ecuador by FITCH (1968), but probably this record is based on *T. bifasciatus* (Duméril) (fig. 73).

REMARKS.—VANZOLINI & REBOUÇAS-SPIEKER (1969) described a large sample of 212 specimens collected in Oriximiná in January 1967 in the course of three days. Surprisingly this species had not been collected in that town during a previous stay in the dry season (September and October). On the basis of their material they conclude that *Calliscincopus* cannot be separated from *Tretioscincus* and consequently the species under discussion has to be named *Tretioscincus agilis* (Ruthven). The population described by these authors differs from all other known specimens (including the Surinam material) in several characters: the snout-vent length is smaller (maximum snout-vent length for both sexes 48 mm), the number of dorsals between the occipital and the posterior margin of the thighs is higher (29–34), the number of ventrals between the pectorals and the 'anterior median preanal plate', is higher (17–22), the tail is relatively longer and the head is relatively shorter. VANZOLINI & REBOUÇAS-SPIEKER do not name this population as a separate subspecies, but in my opinion this population is well enough differentiated to deserve subspecific rank.

There is no indication that populationoutbursts, comparable to that in Oriximiná in January occur in Surinam. During the periods when I collected in Surinam (January–February, April–November) the species was scarce in all localities visited. There is one datum that indicates a slightly larger abundance and that is that the Surinam Expedition 1948–49 collected seven specimens between

March 9–14 on a transect of just 1.5 km. All other Surinam material indicates that the species is rare. The same is true for Guyana where BEEBE (1945) saw only five or six specimens in Kartabo in all the years he spent there, and RUTHVEN (1916) observed only six specimens during the summer of 1914 in Dunoon.

Genus **Tupinambis** Daudin

DIAGNOSIS.—Very large lizards with cylindrical bodies and tails, and well developed legs with five digits. Tongue lanceolate, retractile into a basal sheath, covered with rounded, imbricate, scale-like papillae in oblique rows. Anterior teeth conical, posterior teeth tricuspid, laterally compressed in young specimens, bluntly conical in adults. Anterior parts of the nasals in contact, one or two loreal plates. At midbody there are 30–50 ventrals across the belly. Flanks with wedge-shaped groups of scales arising on the back, alternating with similar groups originating from the belly. Lower eyelid movable, with a divided semi-transparent disc. Ear-opening large, tympanum superficial. Pupil kidney-shaped. Femoral pores present in males and females. Four species in South America east of the Andes, south to northern Argentina. Only a single species in Surinam.

Tupinambis nigropunctatus Spix (figs. 75, 76; pl. 40)

'Sauvegard' Merian, 1705a: 4, pl. 4; Merian, 1705b: 4, pl. 4; Merian, 1719: 4, 70, pl. 4, 70; Merian, 1730: 4, 49, pl. 4, 70.
Lacerta, Tejuguacu, *Americana, maxima*, Sauvegarde *dicta, marmorei coloris, amphibia* Seba, 1734: 154, pl. 99 fig. 1.
Lacerta cauda tereti, corpore duplo longiore, pedibus pentadactylis, crista nulla, hypochondriis plicatis: Sundius, 1749: 501.
Lacerta caudâ tereti, corpore parum longiore, squamis laevissimus, abdominalibus subquadratis, minimis Gronovius, 1756: 81.
'Sabegaal' Pistorius, 1763: 86.
'Surinaamsche Sauvegarde' Houttuyn, 1764: 99.
'Le Sauve-garde' Fermin, 1765: 31; Fermin, 1769: 205; Fermin, 1770: 181.
?'Salamander' Hartsinck, 1770: 99.
?'Tamapara' Hartsinck, 1770: 99.
'Sauvegarde' Merian, 1771: 4, 70, pl. 4, 70.
'Le Tupinambis' Lacépède, 1788a: 251 (partly); Lacépède, 1788b: 305 (partly); Lacépède, 1799: 100 (partly); Desmarest, 1833: 397 (partly); Lacépède, 1847: 175 (partly).
'Sapagala' Stedman, 1796a: 308; Stedman, 1796b: 167; Stedman, 1799b: 68; Stedman, 1800a: 113.
'Die Warn-Eidechse' Bechstein, 1800a: 460 (partly).
Lacerta monitor: Latreille, 1802a: 220.
Tupinambis monitor: Daudin, 1802a: 20.
Lacerta Teguixin: Shaw, 1802: 235.

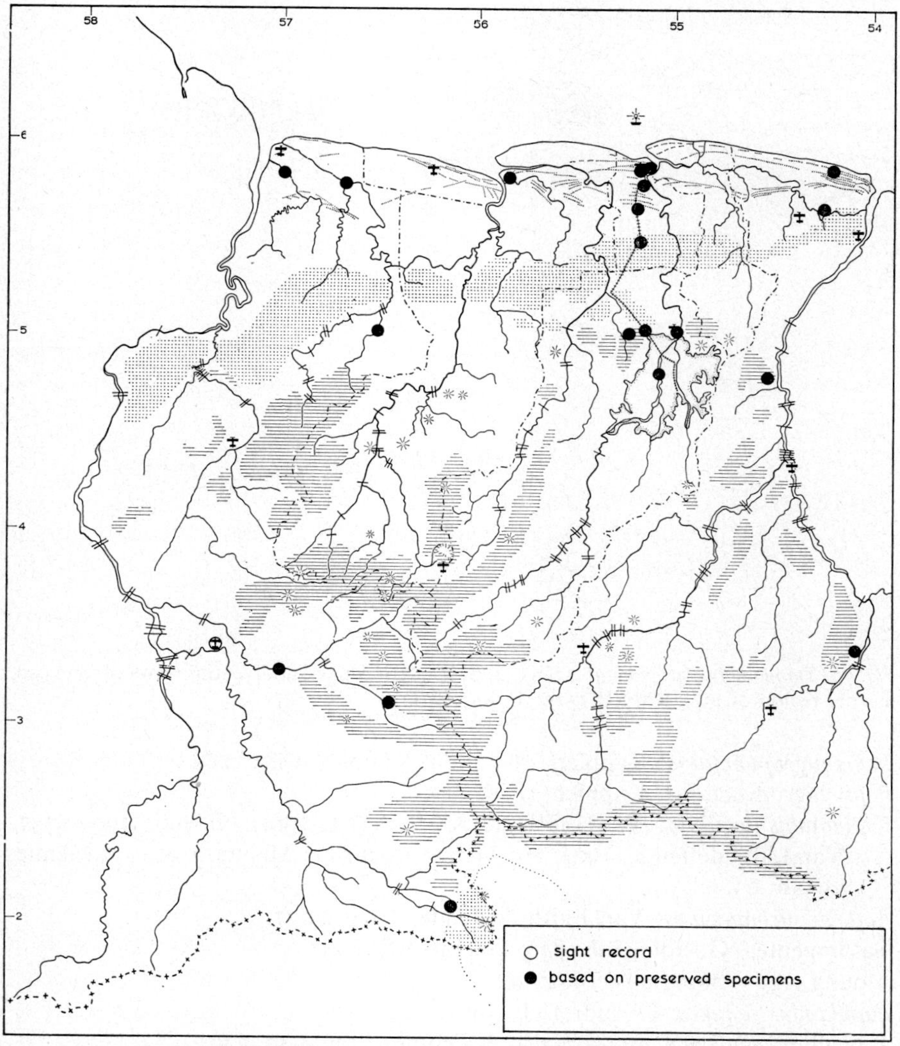

Fig. 75. Distribution of *Tupinambis nigropunctatus* Spix in Surinam.

Tupinambis nigropunctatus Spix, 1825: 18; Van Lidth de Jeude, 1917h: 688; Baylis, 1947a: 394, 395; Baylis, 1947b: 406, 407; Boschma, 1947: 40; Hoogmoed, 1967: 82.
Lacerta teguixin: Cloquet, 1827: 526.
'Great American Safe-guard' Griffith & Pidgeon, 1830: 204.
'Sapagara' Teenstra, 1835: 438.
Salvator Merianae Duméril & Bibron, 1839: 85 (partly).
'Sabbacarra' Kappler, 1854a: 94.
Tejus monitor: Schlegel, 1858: 27.
Tajus monitor: Kappler, 1881: 140.

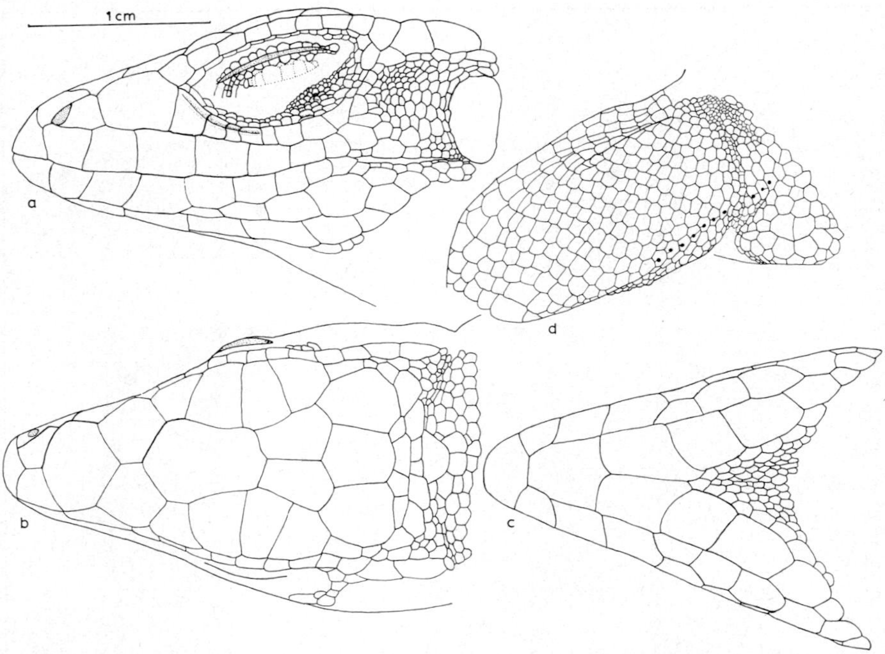

Fig. 76. Tupinambis nigropunctatus Spix; a, b, c: lateral, dorsal and ventral views of the head, d: anal region of juv., RMNH 15256 b, Surinam.

Tejus nigropunctatus: Kappler, 1881: 166; Kappler, 1887: 125.
Pejus nigropunctatus: Kappler, 1885: 817.
Tupinambis teguixin: Van Lidth de Jeude, 1898: 23(a); Goeldi, 1902: 537; Van Lidth de Jeude, 1904: 89; Menne, 1946: 21; Menne, 1952: 10; Menne, 1954b: 63.
Ameiva surinamensis: Van Lidth de Jeude, 1898: 23(a).
'Salompenter' Goeldi, 1898: 648; Goeldi, 1902: 538.
'Young tegu' Anonymus, 1962: 39.
Tupinambis tequixin D'Audretsch, 1965c: 23.
Tapinambis nigropunctatus: Walsh & Gannon, 1967: 50, 246.
'Tegu lizard' Walsh & Gannon, 1967: 65, 66, 246.
Tupinambis spec. Haverschmidt, 1968: 59.

TYPE-LOCALITY.—Brazil.

MATERIAL. – Surinam.

Surinam: 2 juvs., RMNH 3390, 1826, leg. H. H. Dieperink; 1 ex., RMNH 3391; 1 ex., RMNH 3408, leg. H. H. Dieperink; 1 ex., RMNH 5378; 1 ex., RMNH 7529, 4–VI–1910, leg. D. Bolten; 1 ex., RMNH 7539, 1910, leg. D. Bolten; 1♂, 2 juvs., 3 eggs, RMNH 15256, 1864, leg. F. F. Steenberghe; 1♀, 1♂, RMNH 15459, leg. J. T. Noordijk; 1♀, RMNH 15460, leg. G. Versteeg (Gonini Expedition); 2 juvs., ZMA no number, leg. F. Gonggrijp & G. Stahel; 4 juvs., SMNS 142, 1845, leg. A. Kappler; 1 juv., NMW 13114; 1♀, ANS 9282, leg. C. Hering.

Paramaribo: 1 juv., RMNH 7526, 1911, leg. Jhr. W. C. van Heurn; 2 ex.,

Plate 40. Tupinambis nigropunctatus Spix. Figs. 1–3. Dorsal, lateral and ventral aspect of ♀, RMNH 15444, Brown's Mountain.

RMNH 7528, 5–III–1939, leg. H. W. C. Cossee; 1 ex., RMNH 7538, 1911, 2 ex., RMNH 7540, July 1911, both leg. Jhr. W. C. van Heurn; 1♂, RMNH 15445, 20 km S., 6–VI–1968, 1♂, RMNH 15446, 4 km S., 1–XI–1968, both leg. M. S. Hoogmoed; 1 juv., RMNH 15463, September 1962, 1♀, RMNH 15464, June 1966, 1♂, RMNH 15465, October 1966, all leg. G. van Vreden; 1 ex., BM 1946.4.3.39, 2–III–1938, leg. I. T. Sanderson; 1 ex., ZMB 25978, 1♀, NMW 13097, 1 juv., NMW 13113, 1910, all leg. A. O. C. Heller; 1 ex., FMNH 121083, 18–IX–1960, leg. H. A. Beatty; 1 ex., CM 44414, 6–V–1966, 1 ex., CM 44415, 3–VIII–1966, 1 ex., CM 44416, ± 10 km W., 31–VII–1966, all leg. M. H. de la Fuente.

Wia Wia transect: 1♂, RMNH 15454, 4th swamp from sea, 16–XI–1948, 1 ex., RMNH 15456, November 1948, both leg. Surinam Expedition 1948–49.

Nanni Creek: 1♀, 1♂, RMNH 16473, 18–III–1971, 1 juv., RMNH 16489, 16–III–1971, all leg. D. G. Reeder.

Wageningen: 1♀, RMNH 15449, 9–VI–1963, 1♂, RMNH 15450, 12–VI–1963, both leg. A. M. Husson & P. Staffeleu.

Boskamp: 1♂, RMNH 15448, 15–XI–1968, leg. M. S. Hoogmoed.

Djai Creek: 1♂, RMNH 15453, km 8.6, 6–X–1948, leg. P. H. Creutzberg (Surinam Expedition 1948–49).

Zanderij: 1♀, 1♂, BM 1946.4.3.40–41, 19–IX–1938, leg. I. T. Sanderson.

Transect Fallawatra-Coppename River: 1 ex., RMNH 7527, 1901, leg. H. van Capelle.

Brown's Mountain: 1♀, RMNH 15444, 13–VII–1963, leg. M. S. Hoogmoed.

Railway km 121: 1 ex., RMNH 15457, 30–XI–1949, leg. C. Bleys.

Afobaka: 1♂, ZMA 12783, 1–XI–1966, 1♂, ZMA 12784, 2 km N., 3–XI–1966, both leg. H. Nijssen.

Gania kondre, Surinam River: 1♀, RMNH 15451, 22–I–1963, leg. A. M. Husson & P. Staffeleu.

Between Gansee and Lombé: 1♀, RMNH 15452, 24–I–1963, leg. A. M. Husson & P. Staffeleu.

Nassau Mountains: 1 juv., RMNH 15455, km 3.1, 28–II–1949, leg. Surinam Expedition 1948–49.

Lucie River: 1 ex., RMNH 5658, November 1910, leg. J. F. Hulk (Corantijn Expedition).

Airstrip Kayser Mountains: 1 juv., RMNH 15446, 9–VIII–1968, leg. M. S. Hoogmoed.

Anapaike kondre: 1♂, RMNH 15458, September 1963, leg. B. Malkin.

Airstrip Sipaliwini: 1♂, RMNH 16388, 12–II–1970, leg. M. S. Hoogmoed & J. J. P. Paats.

DIAGNOSIS.—A very large lizard with cylindrical body, well developed legs and a long tail. Dorsal and lateral scales small; ventrals small, larger than the dorsals, in 22–27 longitudinal rows. One large loreal. Frontonasal not separating the nasals. Back black with white and brown spots in adults, olive-brown with black transverse bands in juveniles. Large headscales with black centre and brown margins. Belly without black transverse bands.

DESCRIPTION.—Head pyramidal, slightly depressed; less than twice as long as wide; as wide as the neck in juveniles and females; wider than the neck in males; canthus rostralis round, distinct; snout pointed, rounded. In adult males the posterior part of the head and the anterior part of the neck are strongly swollen, giving the head an enormous size. Neck as wide as the body.

Tongue lanceolate, without posterior emargination; covered with rounded, imbricate, scale-like papillae in oblique rows; tip bifid with a distinct longitudinal sulcus in each tip, smooth. Ventral surface of the tongue with two oblique rows of imbricate, rounded plates, decreasing in size posteriorly; sides with small oblique plicae. Anterior teeth conical; posterior teeth tricuspid, laterally compressed, of various sizes; in old males all teeth are bluntly conical.

Rostral pentagonal to nearly semicircular, visible from above, nearly twice as wide as deep. Nostril in an obliquely divided nasal, the anterior parts of both sides forming a short to long median suture behind the rostral. A large, regularly hexagonal frontonasal. A pair of slightly irregular prefrontals, forming a short median suture. Frontal irregularly octagonal, elongate; slightly longer than wide; anteriorly wider than posteriorly. A pair of irregularly pentagonal frontoparietals forming a long median suture. A pentagonal interparietal and two slightly larger parietals of variable shape; followed posteriorly by a variable number of transversely enlarged scales. Four supraoculars, in direct contact with the supraciliaries or partly separated from them by a few isolated granules; followed by a few small scales bordering the parietals laterally. Second supraocular largest; third smallest. One large, nearly rectangular loreal, slightly longer than deep. Preocular absent. Six to seven (rarely four, five or eight) suboculars, forming a suborbital ridge; the anterior one largest. Two rows of four or five enlarged postocular scales. Supraciliaries seven to nine, rarely six; the first largest, the rest subequal and quadrangular. Three or four large supratemporals. Temporal scales small, flat, hexagonal; smallest in the centre, larger towards the supratemporals and supralabials. Ear-opening large; surrounded by small flat scales, forming a smooth margin; external auditory meatus absent, tympanum well visible. Nine to ten (rarely eight) large supralabials, followed posteriorly by a few small scales; five labials to below the centre of the eye; third labial the largest. Lower eyelid with a semitransparent disc of five enlarged or numerous, small palpebrals. Pupil kidney-shaped with the indentation ventrally.

Mental trapezoid with a convex anterior border. A heptagonal postmental. Four to six pairs of chinshields; the anterior pair largest, forming a long median suture; the posterior pair smallest; relatively many specimens have different numbers of chinshields on each side; only the first and the second pair in contact with the lower labials. Six to nine (mostly seven) infralabials; followed posteriorly by a variable number of small scales; the third largest.

Gulars flat, juxtaposed, angular; small in the anterior part, larger posteriorly. A gular and a prepectoral transverse fold. A row of scales just posterior of the gular fold very small, scales between the prepectoral fold and the ventrals very small. Scales on the nape more or less hexagonal, juxtaposed, flat; wider than long near the head, as wide as long near the forelimbs; in transverse rows. Scales on the side of the neck more or less hexagonal, juxtaposed, flat, longer than deep, in vertical rows.

Dorsal scales small, quadrangular to rectangular with rounded edges, juxtaposed, flat, in transverse rows. Scales on the flanks smaller, in downwardly wedge-shaped groups which alternate with identical groups arising on the belly, the scales near the tips smallest, minute. Ventrals small, larger than the dorsals; the scales in the four or five anterior transverse rows hexagonal, the rest rectangular, longer than wide; slightly imbricate, in 32–36 transverse rows and in 22–27 longitudinal rows. Total number of scales around midbody 103–112. Preanal scales in three transverse rows of enlarged scales, surrounded by small and minute scales.

Preanal pores three to five (mostly four) on each side, separated from the femoral pores on the same side by a wide gap. Femoral pores five to nine. Pores in a scale which is divided into one large and two small parts, slightly larger in adult males than in females. Everted hemipenis bifurcate, smooth near its base, with fine horizontal plicae towards the bifurcation on the anterior part, and on the sides of both tips. Space between the tips smooth. Sulcus spermaticus on the posterior surface, bifurcating in the saddle, a branch to each tip, ending in a flat, projecting structure. The plicae are oblique and converging towards the tip of the hemipenis in three lines, one on the anterior and one on each of the lateral surfaces.

Scales on the tail rectangular, juxtaposed; proximally with an indistinct keel, distinctly keeled distally; in transverse rows. Tail with verticils, transverse rows arranged in groups of three, the anterior row shortest, the posterior row longest. On the side of the tail the anterior two rows fuse into one and the posterior row widens. Thus, two ventral scale rows equal three dorsal scale rows. Scales under the tail identical to those on the dorsal surface, but larger. Scales on the regenerated part of the tail rectangular to polygonal, keeled, juxtaposed, in irregular transverse rows. Tail round proximally, compressed distally.

Scales on the anterior surface of the forelimbs, on the anterior surface of the thighs and on the lower surfaces of thighs and lower legs enlarged, hexagonal, flat, juxtaposed; scales on the lower, posterior and upper surfaces of the forelimbs, on the posterior and upper surfaces of the thighs and on the anterior, upper and posterior surfaces of the lower legs small, rounded, juxtaposed, granular. Fourteen to 18, flat to slightly tubercular lamellae under the fourth finger, proximally in transverse series of small scales, distally in pairs. Thirty one to 38 rows of flat lamellae under the fourth toe; the row of lamellae on the inner side large, the rest small; proximal scales of the inner row triangular, elevated above the surrounding scales, thus forming a serrated edge. Two proximal scales of the inner row of scales under the third toe enormously enlarged. Fingers slightly compressed, toes cylindrical. Palms and soles with small, rhomboidal, narrowly separated, flat scales. A serrated row of scales on the inner side of the sole which is continued under the first toe.

Colour in life and in preservative identical. There is a change of pattern during ontogeny, the juveniles being distinctly transversely banded, the adults lacking this distinct pattern and having a uniformly black back with lighter spots in which the transverse bands can hardly be recognised, except on the tail where they are still distinct. Juveniles with ten to twelve transverse black bands from the nape to the base of the tail, from the back via the flanks extending onto the lateral parts of the belly where they end in sharp points. The colour between the bands is olive-brown. A dorsolateral and two indistinct paravertebral series of isolated white spots. Large headscales with black centre and brown margins. Upper part of temple with a wide black band, continued over the ear on the side of the neck where it turns downwards; bordered ventrally by a brown line. Tail with alternating black and brown bands; the anterior black bands interrupted ventrally, the posterior bands continuous; narrower below than above; the brown bands turn white under the tail. Hind limbs anteriorly with black and white vertical bands; dorsally and posteriorly dark-brown with white spots; lower posterior surface immaculate, white. Ventral parts white, chin and throat with large black spots, belly with transverse black lines (see above) on the lateral parts and large black spots in the centre, no continuous black bands.

In halfgrown specimens the black bands have expanded at the expense of the brown ones, which are heavily invaded by black scales. The dorsolateral line is much more distinct and now consists of the white dots and connecting stripes of brown. The black temporal band is (or rarely is not) continued on the side of the neck and the anterior part of the flanks. The anterior bands on the tail are obliterated, the posterior ones are still distinct.

In adults the pattern has undergone still more modifications, the transverse bands have nearly disappeared and instead, the back is black with white and brown spots. The dorsolateral line is evident, the paravertebral line may or may not be present, but between them numerous scattered light spots are present. The anterior bands on the tail have nearly disappeared, the posterior ones are still distinct, but in the white bands black spots are present. The tongue is cream-coloured. Tympanum chestnut-brown.

The largest Surinam teiid lizard, which does not reach the total length of *Iguana*, but in body bulk is much larger. Maximum snout-vent length 209 mm in females, 333 mm in males. The undamaged tail in females is 1.8–2.0 times as long as the snout-vent length, 1.9–2.2 times in males and 1.7–2.0 times in juveniles. Hind limb 1.7–1.9 times as long as forelimb in females, 1.6–2.0 times in males and 1.8 times in juveniles. When laid along the body there is a considerable overlap of forelimb and hind limb.

HABITAT.—The species is present in a wide variety of habitats and can be found in gardens in Paramaribo, on the sea beach, on shifting grounds, in the forest along trails, along forest edges, in patches of secondary growth and on savannahs. However, some bushes in which the species can take refuge must be present.

NATURAL HISTORY.—The species can be observed basking throughout the day, always taking the precaution that its burrow, which is generally situated under a bush and in which it takes refuge when pursued, is near by. Pregnant females with correct collecting dates are not available. Recently hatched juveniles with a snout-vent length of 87–100 mm are known from February and August, which might indicate that breeding occurs throughout the year. BEEBE (1945), GOELDI (1898), HAGMANN (1906), KAPPLER (1845a) and REESE (1922) reported

Tupinambis eggs being found in termite nests. BEEBE found eggs in April, June and August, HAGMANN in March and October, REESE in June.

One stomach examined contained a large spider, a centipede and a cockroach. BEEBE (1925) reports the stomach contents for this species as crabs, spiders, long-horned grasshoppers, beetles, wasps, ants, *Cnemidophorus* and a spiny rat (*Proechimys*). In 1945 he adds to this list berries, fruits, centipedes, roaches, termites, a *Leptodactylus* frog, *Ameiva*, a small snake and a jungle mouse. HAGMANN (1909) reports as food large insects and probably frogs and mice.

HAVERSCHMIDT (1968) reports small specimens of the species among the lizards found in the stomachs of grey hawks (*Buteo nitidus* (Latham)).

RANGE.—Northern South America, south to central Brazil. In Surinam throughout the country, from sea level to at least 200 m (fig. 75).

VERNACULAR NAME.—In the Surinam language this lizard is called 'sappakara'.

REMARKS.—Recent faunal lists (J. A. PETERS & DONOSO-BARROS, 1970) still list *Tupinambis teguixin* (Linnaeus) as occurring in the Guianas. This is incorrect and as far as I could ascertain, all data in literature concerning the presence of *T. teguixin* (Linnaeus) in the Guianas are based on a mistake made by DUMÉRIL & BIBRON. In their 1839 publication they distinguish between *Salvator Merianae* Duméril & Bibron and *Salvator nigropunctatus* (Spix), the first with two, the second with one loreal plate. *Salvator Merianae* Duméril & Bibron is commonly considered a synonym of *Tupinambis teguixin* (Linnaeus), but this is not completely correct. The description of *S. Merianae* Duméril & Bibron agrees with that of *T. teguixin* (Linnaeus), not that of one of the synonyms. They cite as synonym 'sauvegarde' of MERIAN (pl. 70). In the text they say under 'observations' that 'Cette espèce est très-bien représentée sous les deux états de jeune et d'adulte, dans l'ouvrage de mademoiselle SYBILLE DE MÉRIAN'. Here undoubtedly they refer to both plates in MERIAN's work (1719) representing the 'Sauvegarde' of MERIAN, namely pl. 4 with a juvenile and pl. 70 with an adult specimen. In my opinion it is clear that the specimens depicted there really belong to *T. nigropunctatus* Spix for the following reasons:

1. They show one single, large loreal, although the first subocular has been depicted somewhat exaggerated and might be interpreted at first glance as a second loreal. However, in none of the specimens of *T. teguixin* (Linnaeus) examined by me is part of the second loreal situated beneath the eye. In *T. nigropunctatus* Spix, on the contrary, the first subocular may be enlarged and reach somewhat anteriorly of the eye, contacting the first supraciliary.
2. The pattern on the head of the specimens depicted by MERIAN consists of large black dots in the centre of the scales covering the dorsal surface of the head, a characteristic of *T. nigropunctatus* Spix. In *T. teguixin* (Linnaeus) the centres of the scales are light, the sutures between them are edged by black.

Under 'patrie' of *S. Merianae* Duméril & Bibron, the authors say that the species occurs in the entire South America and on several of the Antilles. As proof they list specimens in the Paris museum of this species originating from Brazil (DELALANDE & A. DE SAINT-HILAIRE), Montevideo (D'ORBIGNY) and Cayenne (LESCHENAULT & DOUMERC). Of *S. nigropunctatus* (Spix) the authors say 'cette espèce habite les mèmes contrées que la précédente', but do not list the specimens.

DUMÉRIL & DUMÉRIL some years later (1851) in the 'Catalogue méthodique de la collection des Reptiles' of the Muséum d'Histoire Naturelle de Paris, list all specimens of each species available at the time. Under *S. Merianae* Duméril & Bibron they mention specimens from Brasil received from DELALANDE, A. SAINT-HILAIRE and POYER and from Montevideo received from D'ORBIGNY. Under *S. nigropunctatus* (Spix) they only say: 'Amér. merid., et en particulier, Cayenne et La Mana: *Leschenault* et *Doumerc*. Ages divers. V.V.'. From these data it is evident that in 1839 a mistake was made and all material of the genus *Salvator* was listed under *S. Merianae* Duméril & Bibron, instead of being divided among *S. Merianae* Duméril & Bibron and *S. nigropunctatus* (Spix) as was done in 1851. DUMÉRIL & DUMÉRIL's 1851 data, that *S. Merianae* Duméril & Bibron occurs in southern Brazil and Uruguay and *S. nigropunctatus* (Spix) in French Guiana, and the fact that MERIAN's 'Sauvegarde' should be cited under *Tupinambis nigropunctatus* Spix, together with the present knowledge that only *T. nigropunctatus* Spix occurs in the Guianas, all point to one conclusion: that *T. nigropunctatus* Spix is an Amazonian species reaching Trinidad and Tobago and that *T. teguixin* (Linnaeus) is a southern form inhabiting southern Brazil, Uruguay and northern Argentina, which does not occur north of the Amazon and probably far less north. MERTENS (1969, 1972) considers *nigropunctatus* as a subspecies of *teguixin*. However, he does not explain his use of trinominals and therefore I prefer to consider both as separate species until more details on geographic distribution and interbreeding are available.

Suborder AMPHISBAENIA

Usually this group is considered as belonging to the Sauria, but recently CARROLL (1969), GANS (1960, 1969) and GUIBÉ (1970) separated them from that group and treated them as a distinct suborder in the order Squamata. For convenience's sake I treat them with the lizards.

Family AMPHISBAENIDAE

In Surinam only three members of the family, belonging to the genus *Amphisbaena*, are known. In general habitus they are very much alike, being legless, having reduced eyes covered by the oculars and the scales arranged in annuli. From Surinam lizards they can easily be distinguished by the total absence of legs and the reduced, subcutaneous eyes. From Surinam snakes they can be distinguished by the small, juxtaposed, rectangular scales, forming transverse series, by the presence of the lateral sulcus and by the preanal plate which comprises six scales at least.

The two largest representatives are often found in the nests of leaf-cutter ants, the smallest representative has been found mostly in rotten logs and once in the uppermost soil layer.

Genus **Amphisbaena** Linnaeus

DIAGNOSIS.—Small to large reptiles without legs, reduced eyes, a lateral body sulcus, the smooth rectangular scales arranged in annuli. Head covered with large plates, symmetrically arranged with regard to a middorsal depression; one scale covering the reduced eyes; ear-opening absent. The tail may or may not show an autotomy constriction. Eight teeth in each jaw-half and an unpaired anterior median tooth in the upper jaw. Forty-five species in South America, Panama, the Antilles and may be one species in southern Africa. Of these, 35 species in Panama and in South America on both sides of the Andes, south to central Argentina. Only three of these species occur in Surinam.

VERNACULAR NAME.—All representatives of this genus are called 'to-hedde-sneki' (= two headed snake) in Surinam.

Key to the Surinam species of *Amphisbaena*

1. Scales around midbody 26–30, preanal pores four, supralabials two . . . . *A. vanzolinii*, p. 384
   Scales around midbody at least 42, preanal pores seven to twelve, supralabials three or four 2
2. Scales around midbody 70–86, tail annuli 15+1, supralabials four, caudal autotomy annulus absent . . . . *A. alba*, p. 372
   Scales around midbody 42–55, tail annuli 24+1 – 28+1, supralabials three, caudal autotomy annulus present . . . . . . *A. f. fuliginosa*, p. 378

Amphisbaena alba Linnaeus (figs. 77, 78; pl. 41)

Amphisbaena alba Linnaeus, 1758: 229; Daudin, 1803: 401; Eichelberg, 1847: 496; Van der Hoeven, 1855: 504; Kappler, 1881: 166; Kappler, 1885: 817; Kappler, 1887: 127; Matschie, 1897: 144 (partly); Van Lidth de Jeude, 1898: 24; Penard & Penard, 1906: no. 83; Van Lidth de Jeude, 1917b: 43; Brongersma, 1932b: 133; Stahel & Geijskes, 1939: 46; Stahel & Geijskes, 1940: 17; Gans, 1962: 17; Gans, 1963: 2; D'Audretsch, 1965b: 15.
'Zweiköpfige Schlange' Von Sack, 1821a: 120.
'Tweehoofdige slang' Von Sack, 1821b: 215.
'White Amphisboena' Griffith & Pidgeon, 1830: 325.

TYPE-LOCALITY.—'America', here restricted to the confluence of the Cottica River and the Perica Creek, Surinam.

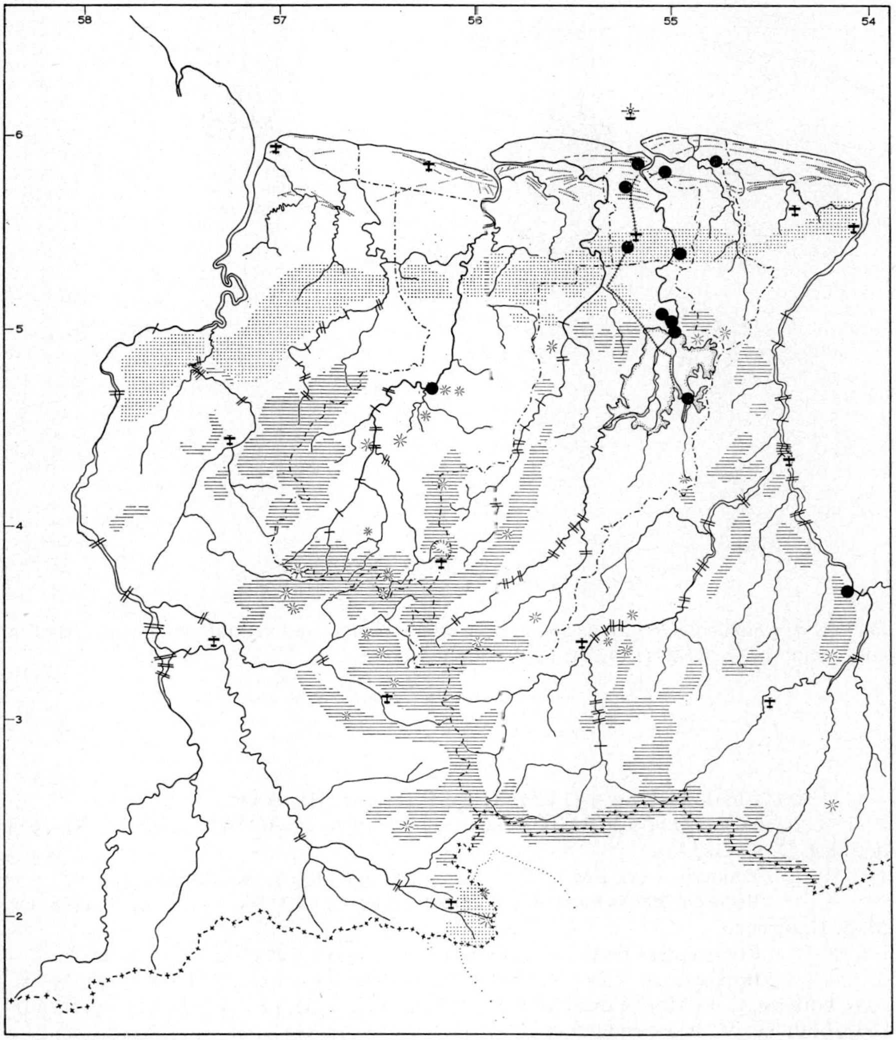

Fig. 77. Distribution of *Amphisbaena alba* Linnaeus in Surinam.

MATERIAL. – Surinam.

Surinam: 2 ex., RMNH 3551, 1866, Cabinet Van Lidth de Jeude; 2 ex., RMNH 3557, Ancien Cabinet; 1 ex., RMNH 7558, November 1905, leg. P. Buitendijk; 1 ex., RMNH 7559, 1904, leg. G. M. Versteeg (Tapanahoni Expedition); 2 ex., RMNH 10652, February 1888, leg. J. H. Spitzly; 1 ex., ZMA 10175, leg. E. Schulz; 1 ex., ZMA 10817; 1 ex., ZMA 10819; 1 ex., ZMA 10820, leg. J. F. Pool; 1 ex., ZMA 11270, March 1891; 1 ex., ZMA 11793; 1 ex., ZMA 12353; 3 ex., ZMA 12356–58, 1887–90, leg. Van Brussel; 1 ex., ZMA 12359; 1 ex., ZMA 12360, leg. D. Bolten; 1 ex., ZMA 12361; 1 ex., ZMA 12431.

Paramaribo: 1 ex., RMNH 7560, leg. Summels; 3 ex., RMNH 10651, 3–XI–1938; 1 ex., RMNH 10654, leg. Jhr. W. C. van Heurn; 2 ex., RMNH 10656–57, August 1940, leg. D. C. Geijskes; 1 ♂, RMNH 15259, 6 km E., south of plantation Marienburg, 29–X–1968, leg. M. S. Hoogmoed; 2 ex., ZMA 12354–55; 1 ex., ZMB 26281, leg. A. O. C. Heller.

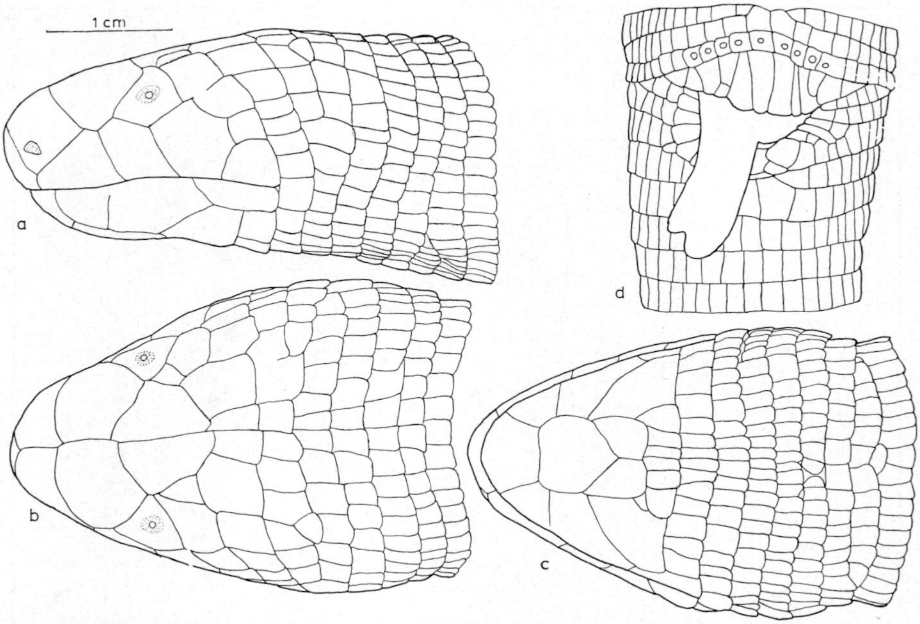

Fig. 78. Amphisbaena alba Linnaeus; a, b, c: lateral, dorsal and ventral views of the head, d: anal region of ♂, RMNH 15259, E. of Paramaribo.

Charlottenburg: 1♂, RMNH 3558, 1861, leg. H. Holm.

Lelydorp: 1 ex., RMNH 10653, 27–IV–1938, 1♂, RMNH 10658, 17–XI–1938, both leg. D. C. Geijskes.

Zanderij: 1 ex., ZMA 12362, S., 29–IV–1956, leg. J. van der Kamp.

Between Blakkawatra and Jodensavanne: 1♂, RMNH 15272, 14–V–1968, leg. M. S. Hoogmoed.

Compagnie Creek: 1 ex., RMNH 15273, 13–IV–1965, leg. G. F. Mees.

Brokopondo: 1 ex., RMNH 15274, 9–XII–1965, 1 ex., RMNH 15275, 23–III–1965, both leg. G. F. Mees; 1 ex., RMNH 15276, 14–VIII–1964, 1 ex., RMNH 15279, 20–VII–1964, both leg. M. Boeseman.

Afobaka: 1 ex., RMNH 15278, 14–IV–1964, leg. M. Boeseman.

Koffiekamp: 1 ex., RMNH 15277, 1–IV–1964, leg. M. Boeseman.

Raleigh Cataracts, Coppename River: 1♀, RMNH 17283, early 1971, leg. Pomiran.

Maripa Hill, Upper Sara Creek: 1 ex., RMNH 15280, 20–IX–1959, leg. J. van der Drift.

Benzdorp: 1 ex., RMNH 10655, 1940?, leg. A. M. Hermans.

Marowijne River: 1 ex., ZMB 9392, leg. A Kappler.

Diagnosis.—A large amphisbaenid with a short tail. Number of body annuli 219–243, tail annuli 13+1 – 20+1; 70–86 scales around midbody; four supralabials; ocular large; preanal pores 9–12, no caudal autotomy constriction. Light- to dark-brown above, dirty white below.

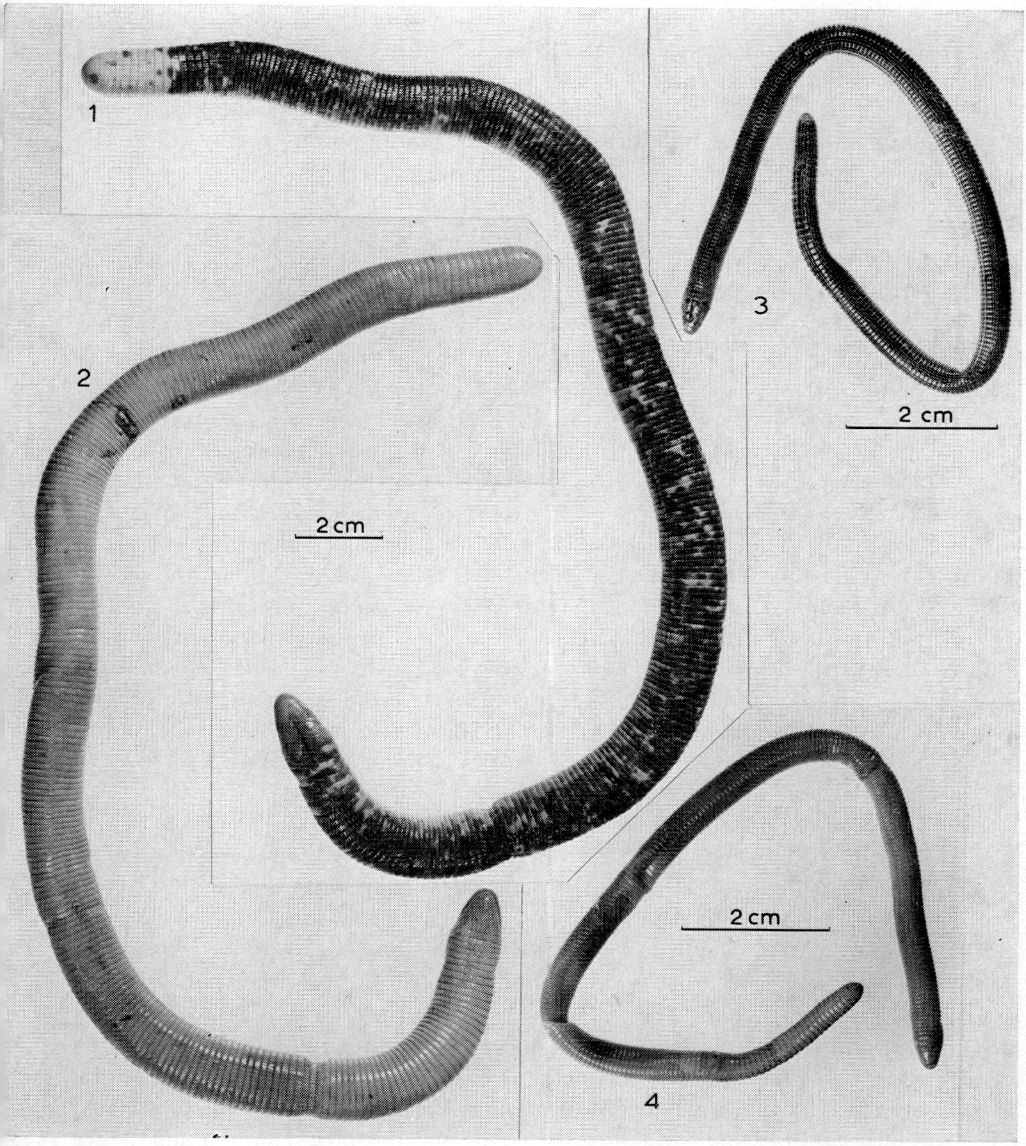

Plate 41. Figs. 1–2. *Amphisbaena alba* Linnaeus; dorsal and ventral aspect of RMNH 15277, Koffiekamp. Figs. 3–4. *A. vanzolinii* Gans; dorsal and ventral aspect of ♀, RMNH 15186, Sipaliwini.

Description.—Head wider than the body, distinctly set apart, depressed; posterior part swollen, with a deep median dorsal sulcus. Snout rounded, blunt; projecting over the mouth.

Tongue lanceolate, covered with oblique rows of rhomboidal, scale-like papillae; tip bifid, ending in two long, sharp, smooth points. Teeth conical, the 'canines' larger than the rest; decreasing in size posteriorly.

Rostral triangular or pentagonal, large, deeper than wide, well visible from above. Four supralabials; the first more or less triangular; the second and third elongate, pentagonal; the fourth small, quadrangular. Ocular large, more or less rhomboidal. Subocular absent. One, two or three postoculars. A pair of large, roughly rectangular nasals bordering on the rostral. A pair of rectangular prefrontals, equal in size to the nasals. A pair of trapezoid frontals, smaller than the prefrontals or the nasals. Nasal suture longer than the prefrontal suture, shorter than the frontal suture. Posterior part of the head covered with small, irregularly shaped to quadrangular scales, forming transverse rows.

One large mental, as wide as long, trapezoid in shape with slightly concave sides, or rectangular with the frontal surface produced in lateral direction; followed by a pentagonal or heptagonal postmental which is as long as wide or longer than wide. Three infralabials, the first moderate, triangular; the second large, pentagonal; the third small, elongate. Postmental frequently in contact with the large malars; followed by one or two rows of postgenials, a first row of two large scales and a second row of four to six, mostly five, scales, of which the central one may be elongated and in contact with the postmental, the second row also may be absent. A row of rectangular postmalars of which the central scales are narrower than the lateral scales. Throat with transverse rows of rectangular scales, identical to the ventrals.

Scales on the nape and on the side of the neck identical to the dorsals.

Dorsals small, rectangular with rounded edges, longer than wide, 31–39 in an annulus at midbody level. Ventrals small, rectangular with right angles; longer than wide, though near the median they are nearly as wide as long; 39–48 in an annulus at midbody level. Always more ventrals than dorsals in the same annulus. Total number of scales around midbody 70–86. Dorsals and ventrals separated by a lateral sulcus. The scales bordering on the lateral sulci have rounded lateral margins, the space between dorsal and ventral part of an annulus is filled by two small scales. A distinct dorsal sulcus, without differentiated scales, is present. One intercalated dorsal annulus between the first and the second annulus; 219–243 body annuli; two to four, mostly three, lateral annuli in the cloacal region.

Preanal pores 9–12, preanal segments 10–13, postanal segments 8–16, mostly 11–14. Hemipenis cylindrical smooth, tip bifid; sulcus spermaticus shallow, on the posterior surface, forming a straight line to between the two horns; there bifurcating, one branch on the inner surface of each horn.

Caudal annuli 13–20, followed by an undifferentiated knob; no autotomy constriction.

Colour in life and in preservative light- to dark-brown above, fading to a dirty white on the ventral parts (GANS, 1969). Juvenile specimens show white spots in the brown dorsal area, which, in adults, is uniform. Tip of the tail entirely white.

A large amphisbaenian with a thick cylindrical body and a short, cylindrical, blunt tail. Maximum snout-vent length 609 mm. The tail is about 1/10 of the snout-vent length.

HABITAT.—A burrowing species which occurs both in cultivated areas, in the coastal savannah-belt and in undisturbed rain-forest. There are several reports on the association between these reptiles and ants and some of their popular names express it ('King of ants'). GANS (1962) doubts whether this species and the related *A. fuliginosa* Linnaeus really are associated with ants, and supposed that all stories on this subject have been copied from an ancient writer. In a recent paper (GANS, 1969) he seems to have abandoned this opinion as he states that they (the amphisbaenians) are commonly found in the nests of leaf-cutter ants. Among the Surinam material there are specimens (RMNH 10653, 10656, 10657 and 10658) which were collected in the nests of *Atta cephalotes* Linnaeus. The amphisbaenians do not occur in the actual living-nest of the ants, but in refuse pits (STAHEL & GEIJSKES, 1939; 1940) where the ants dispose of the leaves that have been used in their moul-culture. In these pits the larvae of a beetle (*Coelosis biloba* Linnaeus) occur which form an important part of the diet of *A. alba* Linnaeus (STAHEL & GEIJSKES, 1939). Thus, there does not seem to exist a direct relationship with the ants, but an indirect one is definitely present. From the papers by STAHEL & GEIJSKES (1939, 1940) it can be concluded that the relationship between *A. alba* Linnaeus and the ants is restricted to *Atta cephalotes* Linnaeus and that this amphisbaenian is not associated with *Atta sexdens* Linnaeus, the other species of leaf-cutter ant in Surinam. A reason for this might be the different shape of the refuse-pits, which, in *Atta cephalotes* Linnaeus are wide and large and in *Atta sexdens* Linnaeus are narrow and small (STAHEL & GEIJSKES, 1939: pl. 9, 18).

NATURAL HISTORY.—Most of the specimens for which more elaborate data are available were taken in daytime, crawling on the ground. The two specimens I collected were both taken early in the morning at 8.30 a.m., one on an asphalt-road in cultivated land, the other along a dirt-road through rain-forest along which a number of trees had been recently uprooted. Thus, from these scanty data it might be concluded that this species is active in the early morning. This agrees well with GANS' (1969) observations on captive specimens. For amphisbaenians in general he noticed a dawn/dusk peak in movement. Also 'the mean rate of movement during the night was generally higher than that during the day'. In the field they seem to 'approach the surface just after dusk and shift to much deeper zones shortly after dawn'. Nothing is known about the breeding habits of this species. According to STAHEL & GEIJSKES (1939) its food consists mainly of the larvae of the beetle *Coelosis biloba* Linnaeus. One specimen (RMNH 3551) has a juvenile snake, *Anilius scytale* (Linnaeus), protruding from its mouth. Three specimens, collected in *Atta cephalotes* Linnaeus nests, had empty stomachs. BEEBE (1945) found in one stomach a mole cricket, a grasshopper and three termites. GANS (1969) reports the amphisbaenian's diet as: annelids, arthropods, reptiles and rodents and suggests that some of them may even be scavengers.

When disturbed this species assumes a defence posture in which both head and tail are lifted from the ground, the tail very much resembling the head (BEEBE, 1945; GANS, 1969, fig. 6).

RANGE.—According to GANS (1962) and GANS & DIEFENBACH (1970): 'Forested lowlands of South-America, from Panama (?) through Venezuela and Guianas; Colombia, Peru and Bolivia east of the Andes; Brazil and northern Paraguay; Trinidad'. In Surinam the species is known only from the area around the lower Surinam River and from some outlying localities on the Lawa, Tapanahoni, Marowijne and Coppename Rivers. This distribution obviously is due to lack of material. Judging by what is known of the localities where it has been found the species probably occurs throughout the country. Nothing is known about the altitudinal distribution, apart from the fact that Benzdorp on the Lawa River has an elevation of 220 m and that the other localities where it was found are only slightly elevated above sea level (fig. 77).

Amphisbaena fuliginosa fuliginosa Linnaeus (figs. 79, 80; pl. 42)

Amphisbaena annulis abdominalibus CC, *caudalibus* XXX Sundius, 1749: 500.
Amphisbaena annulis abdominalibus CCIX *et annulis caudalibus* XXV Gronovius, 1756: 52.
Amphisbaena fuliginosa Linnaeus, 1758: 229; Daudin, 1803: 406; Fitzinger, 1826: 53, Duméril & Duméril, 1851: 148; Van der Hoeven, 1855: 504; Burgersdijk, 1873: 41; Kappler, 1881: 166; Kappler, 1885: 817; Kappler, 1887: 127; Boettger, 1893: 77; Matschie, 1897: 144 (partly); Van Lidth de Jeude, 1898: 24; Van Lidth de Jeude, 1917b: 43; Burt & Burt, 1930: 40; Burt & Burt, 1931: 239; Brongersma, 1932b: 134; Vanzolini, 1951: 10; Gans, 1963: 2.

TYPE-LOCALITY.—'America', restricted to the Guianas by VANZOLINI (1951), here further restricted to the confluence of the Cottica River and the Perica Creek, Surinam.

MATERIAL. – Surinam.

Surinam: 1 ex., RMNH 3556; 4 ex., RMNH 3559, leg. H. H. Dieperink; 1 ex., RMNH 5904, 1867; 2 ex., RMNH 7556, 1904, leg. G. M. Versteeg (Tapanahoni Expedition); 1 ex., RMNH 9508; 3 ex., RMNH 10660, February 1888, leg. J. H. Spitzly; 1♀, RMNH 15267; 2 ex., ZMA 10821; 1 ex., ZMA 10823, 1873, leg. Du Cellier Muller; 2 ex., ZMA 12364–65; 3 ex., ZMA 12366–68; 1 ex., ZMA 12369; 4 ex., ZMA 12370–73, 1887–90, leg. Van Brussel; 1 ex., ZMA 12375; 2 ex., ZMA 12376–77, leg. Schulz; 1 ex., ZMA 12378, 1875; 1 ex., ZMA 12380; 1 ex., ZMA no number, 1922, leg. F. Gonggrijp & G. Stahel; 2 ex., SMNS no number; 2 ex., NMW 12318.2–3; 4 ex., NMW 12319; 3 ex., ANS 9704–06.

Paramaribo: 1 ex., RMNH 5663, November 1911, leg. J. A. Samuels; 1 ex.,

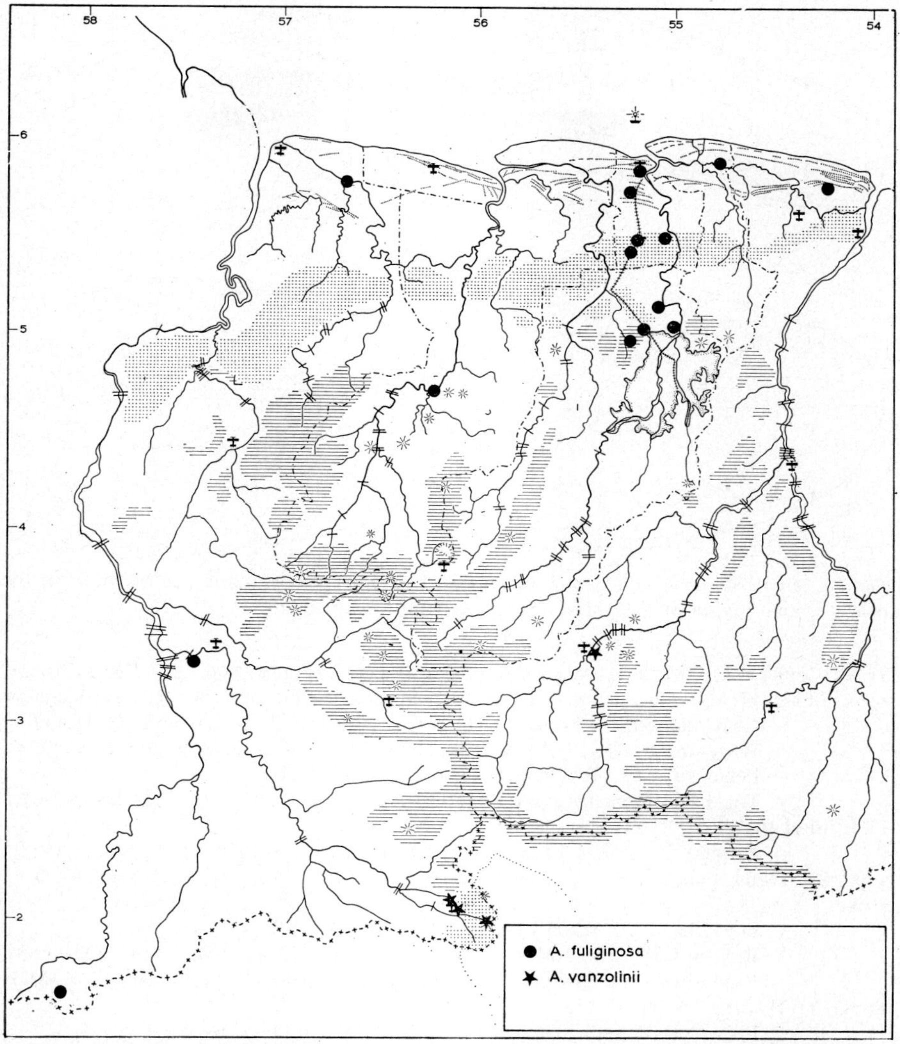

Fig. 79. Distribution of *Amphisbaena f. fuliginosa* Linnaeus and *A. vanzolinii* Gans in Surinam.

RMNH 6714, January 1937, leg. A. L. R. Smit; 1 ex., RMNH 7554, 1911, leg. Jhr. W. C. van Heurn; 5 ex., RMNH 7555, leg. H. van Capelle; 1 ex., RMNH 7583; 1 ex., RMNH 10663, January 1939, 1 ex., RMNH 10664, 22–IV–1939, 1 ex., RMNH 10665, 13–VI–1938, 1♂, RMNH 15261, 10–IX–1963, 1 ex., RMNH 15263, September 1952, 1♂, RMNH 15265, 6–IV–1953, all leg. D. C. Geijskes; 1 ex., RMNH 15260, 19–V–1966, leg. Mrs. Meisenbacher; 1♀, RMNH 15262, January 1962, 1 ex., RMNH 15270, 4–IV–1960, both leg. G. Rijnenberg; 2♀♀, RMNH 15269; 1♀, RMNH 15271, leg. W. N. Polder; 1♂, RMNH 15281, 24–IX–1968, 1♂, RMNH 15282, October 1967, both leg. K. W. R. Zwart; 1 ex., RMNH 16437, 9–IX–1970, leg. Th. R. Deelder; 1♂, RMNH 16442, 6–X–1965, leg. Jhr. A. A. van Heurn; 1♀, RMNH 16443, September 1970, leg. Crew of H. Nl. M. S. Luymes; 1 ex., BM 1946.4.3.86, 3–II–1938, 2 ex., BM 1946.4.3.87, 89, 29–I–1938, 1 ex., BM 1946.4.3.88, 26–III–1938, all leg. I. T. Sanderson; 1 ex., BM 1964.3, leg. C. Rubinstein; 2 ex., ZMB 26282, leg. A. O. C. Heller; 1 ex., CM 44417,

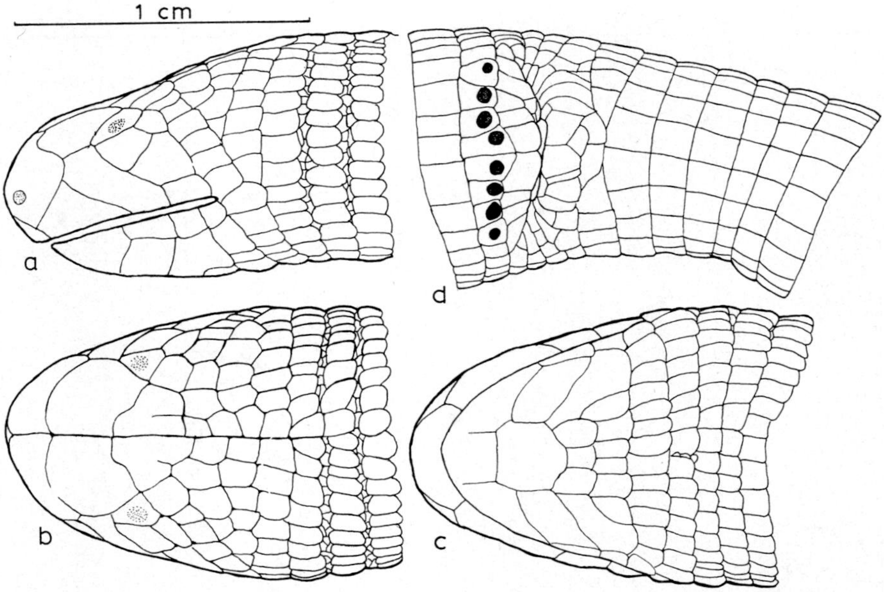

Fig. 80. Amphisbaena f. fuliginosa Linnaeus; a, b, c: lateral, dorsal and ventral views of the head, d: anal region of ♂, RMNH 15246, Gonini Encampment.

25–VII–1966, 1 ex., CM 44418, May 1966, 1 ex., CM 44419, August 1966, 2 ex., CM 44420–21, 1965, all leg. M. H. de la Fuente; 6 ex., AMNH 8134–35, 8649–51, 8678, leg. J. A. Samuels.

Charlottenburg: 2 ex., RMNH 3561, 1861, leg. C. F. Kraepelin & H. Holm.

Wageningen: 2♂♂, RMNH 16457, 11–III–1971, leg. D. C. Geijskes & P. Staffeleu.

Lelydorp: 1 ex., RMNH 10666, 1933, leg. G. Stahel.

Third Encampment: 1 ex., RMNH 10662, km 14.9, 20/21–X–1948, leg. Surinam Expedition 1948–49.

Zanderij: 1 ex., ZMA 12363, Forest Bivouac, 29–IV–1956, leg. J. van der Kamp; 1 ex., BM 1946.4.3.90, 2–VIII–1938, leg. I. T. Sanderson; 1 ex., AMNH 78997, 1958, leg. J. L. Hoke.

Powakka: 1♀, RMNH 15266, 4–IV–1964, leg. M. H. de la Fuente.

Berg en Dal: 1 ex., NMW 12318.1, 1908, leg. A. O. C. Heller; 1 ex., ZMB 7663.

Brownsweg: 1♀, RMNH 15264, February 1964, leg. P. Leentvaar; 1♀, RMNH 17282, 11–II–1972, leg. G. F. Mees.

Brown's Mountain: 1♀, SM 22, 13–X–1970, leg. D. G. Reeder.

Brokopondo: 1♂, RMNH 15268, 22–IX–1964, leg. M. Boeseman; 1 ex., ZMA 12663, 21–VIII–1966, leg. H. Nijssen.

Brokobaka: 1 ex., ZMA 12700, 24–IX–1966, leg. H. Nijssen.

Raleigh Cataracts, Coppename River: 1 ex., RMNH 17280, early 1971, leg. Pomiran.

Gonini Encampment, Coeroeni River: 1♂, RMNH 15246, May 1968, leg. H. J. Kotzebue.

Head of Navigation, Itabu Creek: 2 ex., FMNH 30943–44, 20/21–X–1938, leg. E. R. Blake (Sewell Avery Expedition).

DIAGNOSIS.—A medium-sized amphisbaenid with a relatively long tail. Number of body annuli 198–217, tail annuli 22+1 – 30+1; 42–55 scales around midbody; three supralabials; ocular small; preanal pores seven to ten; a caudal

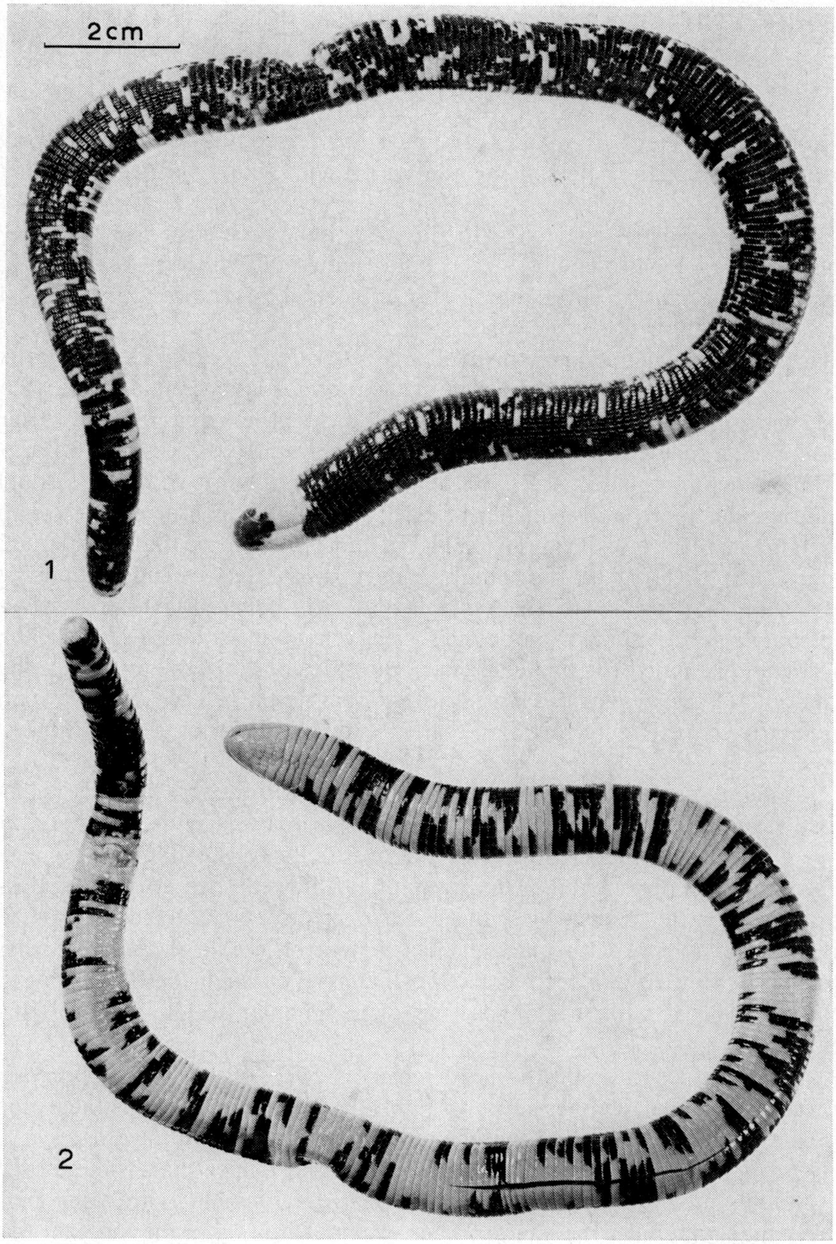

Plate 42. Amphisbaena f. fuliginosa Linnaeus; dorsal and ventral aspect of ♂, RMNH 15281, Paramaribo.

autotomy constriction is present at the level of the fourth to sixth annulus. Black with white blotches in an irregular pattern; head white, with or without black spots.

DESCRIPTION.—Head as wide as the body, depressed; posterior part swollen, with a deep median, dorsal sulcus. Snout rounded, blunt; projecting over the mouth.

Tongue lanceolate, covered with rounded, imbricate, scale-like papillae, in oblique and in longitudinal rows; tip bifid, ending in two short, sharp, smooth points; only slightly protrusible. Teeth conical, the 'canines' larger than the rest; decreasing in size posteriorly.

Rostral large, pentagonal; nearly twice as wide as deep; just visible from above. Three supralabials, the anterior one more or less triangular; the second one vertically elongate; the third one small, more or less quadrangular. Ocular small, wedge-shaped; accompanied by one sub- and two postoculars. A pair of large, quadrangular nasals, bordering on the rostral. A pair of roughly quadrangular prefrontals, smaller than the nasals. A pair of irregularly shaped frontals, smaller than the prefrontals. In most specimens the nasal suture is longer than that of the prefrontals and equal to that of the frontals, also they may all three be of the same length. At the point where nasals and prefrontals meet, a small rhomboidal scale is present in six out of the 21 RMNH specimens examined for this character; BRONGERSMA (1932b) found six specimens out of 31 possessing this scale. Posterior part of the head covered with small, irregularly-shaped to rectangular scales, arranged more or less in transverse rows.

One small mental, anteriorly fused with the first infralabials. Three or four, mostly four, infralabials; the first invariably large, irregularly shaped; the second large or small, rectangular; the third and fourth always small, quadrangular. A large octagonal postmental, wider than long; followed posteriorly by two rows of postgenials, the first row with three to five, mostly three, the second row with five to seven, mostly six, rectangular scales. One or two large malars, bordered posteriorly by a row of rectangular postmalars, of which the lateral ones are enlarged. Throat with transverse rows of rectangular scales, identical to the ventrals.

The scalation pattern on the head region is often very confused because of fusion of some scales, either entirely or only partly, and by the indistinct sutures.

Scales on the nape and on the side of the neck identical to the dorsals.

Dorsals small, rectangular with rounded edges; longer than wide; 22–27 in an annulus at midbody level. Ventrals small, rectangular with right angles; longer than wide near the lateral sulcus, wider than long in the median; 20–28 in an annulus at midbody level. Total number of scales around midbody 42–55. A distinct lateral sulcus with modified small scales, separating the dorsals and the ventrals; a distinct dorsal sulcus without differentiated scales; ventral sulcus absent. Intercalated dorsal annuli occur, mostly between the first and second body annulus and just anterior of the cloacal region; 198–217 body annuli; three to five, mostly four, lateral annuli in the cloacal region.

Preanal pores seven to ten, mostly eight; preanal segments 6–12, mostly 8–10; postanal segments 10–16, mostly 12–14.

Caudal annuli 22+1 – 30+1, a distinct autotomy constriction at the level of the fourth to sixth annulus, the constriction moreover is evident by the annulus having smaller scales than its neighbours.

I did not collect this species myself in Surinam, but fortunately I have colour-slides of living specimens made available to me by Drs. M. BOESEMAN and P. C. H. PRITCHARD. From these slides it is clear that in life this species is black with white blotches in an irregular pattern, the head is white, with or without black spots. Most preserved specimens are also black and white, but quite a few are brown and white. Some specimens are pink instead of white. Most specimens have the head immaculately white, others do have one or more large black spots. BEEBE (1945) gives extensive notes on the colour of living Guyanese specimens, and GANS (1969) gives a coloured plate of this species.

A moderately large amphisbaenian with a thick, cylindrical body and a relatively long, cylindrical, blunt tail. Maximum snout-vent length 424 mm. The tail is about 1/6 of the snout-vent length.

HABITAT.—This is also a burrowing species, occurring both in cultivated land, in the coastal savannah-belt and in undisturbed rain-forest.

NATURAL HISTORY.—Among my material there is one specimen (RMNH 10666) that was taken from an *Atta* nest (without indication of the species) by STAHEL. Dr. D. C. GEIJSKES, an entomologist who stayed in Surinam for 37 years and collected quite a few of the RMNH *A. fuliginosa* specimens, told me that most of the specimens collected by him also came from *Atta* nests, although this is not indicated on the labels. This species also occurs in the refuse pits and might also feed on the *Coelosis biloba* Linnaeus larvae. The contents of eight stomachs were examined. They contained one centipede, two beetles and a beetle larva, one orthopteron, two mole-crickets, a number of small ants, some termites (i.e. *Nasutitermes*), two empty shells of lizard-eggs and some remains of other egg--shells. From these data it is clear that the food is taken from below the surface or directly from the ground. One specimen (RMNH 9508) was taken from the stomach of a snake (*Anilius scytale* (Linnaeus), RMNH 7693).

RANGE.—Surinam, Guyana, French Guiana, Venezuela (Guárico, Monagas, Miranda, Bolívar). In Surinam the species is known from the region south of Paramaribo to the northern part of the Brokopondo artificial lake, on both sides of the coastal savannah-belt, moreover from some localities near the New River, one on the Coppename River and one north of Moengo Tapoe. From this it seems safe to conclude that this species occurs throughout the country on both sides of the coastal savannah-belt, from sea level to about 250 m. The lack of material from more localities in the interior is due to the subterranean way of life of this species (fig. 79).

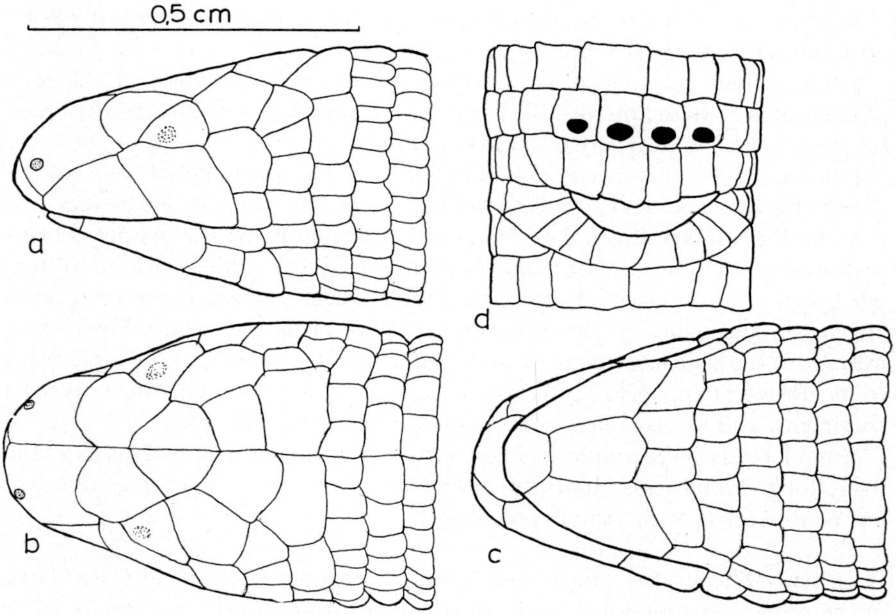

Fig. 81. Amphisbaena vanzolinii Gans; a, b, c: lateral, dorsal and ventral views of the head, d: anal region of ♀, RMNH 15186, Sipaliwini.

Amphisbaena vanzolinii Gans (figs. 79, 81; pl. 41; table 30)

Amphisbaena vanzolinii Gans, 1963: 13.

TYPE-LOCALITY.—Marudi, on a tributary of the Kuyuwini River (2°5′N, 59°E), Guyana.

MATERIAL. – Surinam.

Airstrip Paloemeu: 1♀, RMNH 15144, S., 28–V–1968, leg. M. S. Hoogmoed.

Sipaliwini: 1♀, RMNH 15185, Vier Gebroeders Bivouac, 16–IX–1968, 1♀, RMNH 15186, 2 km E. of airstrip, 5–IX–1968, 1♀, RMNH 15191, Vier Gebroeders Bivouac, 1–X–1968, 1♀, RMNH 15205, 2.5 km N. of Base Bivouac, 2–IX–1968, 1♂, RMNH 15206, 2 km N. of Base Bivouac, 1♀, 1♂, 1 juv., RMNH 15207, forest island on NE. slope of Vier Gebroeders Mountain, 27–IX–1968, all leg. M. S. Hoogmoed; 1♀, RMNH 16438, 3 km N. of Base Bivouac, 16–I–1970, 1♀, 1♂, 1 juv., RMNH 16439, 5 km NW of airstrip, 9–II–1970, all leg. M. S. Hoogmoed & J. J. P. Paats.

Guyana.

Marudi, on a tributary of the Kuyuwini River, at 2°5′N, 59°E: 2♀♀, AMNH 60975 (holotype), 60778 (paratype), March 1938, leg. R. Snedigar.

Rupununi District, N. of Acarai Mountains, W. of New River: 2 ex., KU 69817–18.

DIAGNOSIS.—A small amphisbaenid with a relatively long tail. Number of body annuli 200–210; tail annuli 28 + 1 – 31 + 1; 26–30 scales around midbody; two supralabials; supraocular small; preanal pores four; no distinct autotomy con-

Table 30. Counts and measurements of *Amphisbaena vanzolinii* Gans. For explanation see table 3.

Reg. no.	A	B	C	D	E	F	P'	I	J	K'	L	M	M'	O	P	O'	T	N'	Y
RMNH																			
15144	♀	120	21	5.0	3.2	2.4	10	4	27	204	30+1	14	2-2	2-2	1.5-1.5	2	6	8	13
15185	♀	137	26	5.3	3.5	2.8	11	4	29	201	31+1	15	2-3	2-2	1.5-1.5	2	6	10	14
15186	♀	131	23	5.9	3.8	3.3	11	4	26	202	29+1	14	3-3	2-2	1.5-1.5	2	6	10	12
15191	♀	112	6(+)	?	?	?	–	4	29	203	5+?	15	3-3	2-2	1.5-1.5	2	6	10	14
15205	♀	?	20	?	?	?	10	4	30	?	28+1	16	3-3	?	?	?	6	10	14
15206	♂	130	11(+)	5.5	3.8	2.8	–	4	30	210	12+?	16	2-2	2-2	1.5-1.5	2	6	11	14
15207a	♀	90	4(+)	4.3	3.2	2.4	–	4	30	202	7+?	16	3-2	2-2	1.5-1.5	2	6	10	14
b	juv.	81	14	4.1	2.9	2.2	12	4	30	205	30+1	16	2-2	2-2	1.5-1.5	2	6	10	14
c	♀	126	25	5.4	3.8	3.0	11	4	30	207	31+1	16	2-2	2-2	1.5-1.5	2	6	10	14
16438	♀	123	25	5.2	3.6	2.6	11	4	30	207	31+1	16	2-2	2-2	1.5-1.5	2	6	11	14
16439a	juv.	82	14	4.0	2.8	2.1	11	4	28	200	31+1	14	2-2	2-2	1.5-1.5	2	6	9	14
b	♀	127	24	4.8	3.5	2.6	10	4	28	202	29+1	14	3-3	2-2	1.5-1.5	2	6	8	14
c	♂	133	11(+)	5.5	3.9	3.0	–	4	29	202	11+?	15	2-2	2-2	1.5-1.5	2	6	10	14
AMNH																			
60778	♀	153	14(+)				–	4	30	225	7+?	17	4-4	2-2	1.5-1.5	3	6	12	13
60975	♀	129	7(+)				–	4	28	228	15+?	16	4-5	2-2	1.5-1.5	3	6	12	12
KU																			
69817	?	122	8.5(+)	?	3.4	2.4	–	4	24	207	9+?	12	3-3	2-2	1.5-1.5	2	6	8	12
69818	?	122	23	?	3.0	2.3	10	4	24	214	30+1	12	2-3	2-2	1.5-1.5	2	6	8	12

striction is present, but autotomy is possible. Purplish, light- or dark-brown above; pinkish below; ventral surface of tail white.

DESCRIPTION.—Head as wide as the body, depressed; posterior part with a distinct median dorsal sulcus. Snout rounded, blunt; projecting over the mouth.

Tongue lanceolate, covered with rhomboidal, imbricate, scale-like papillae; tip bifid, ending in two sharp, smooth points. Teeth conical, decreasing in size posteriorly.

Rostral small, pentagonal; wider than deep; just visible from above. Two large supralabials; the anterior one trapezoid; the posterior one roughly quadrangular. A small triangular ocular; subocular absent. One large, irregularly shaped postocular and a smaller temporal. A pair of large, rectangular nasals bordering on the rostral. A pair of smaller, pentagonal prefrontals and a pair of trapezoid frontals, which are smaller than the prefrontals. A pair of irregularly shaped parietals, approximately as large as the prefrontals. Nasal suture as long as or slightly longer than the frontal suture; both are much longer than the prefrontal suture.

Mental small, wider than deep, trapezoid. One and a half infralabials; the first enormous, more or less trapezoid; the second small, rectangular, the posterior half projecting beyond the corner of the mouth. Postmental large, elongate, pentagonal; narrowly separated from the small malars; followed by two rectangular postgenials. Postmalars absent. Throat with transverse rows of rectangular scales identical to the ventrals.

Scales on the nape and on the side of the neck identical to the dorsals.

Dorsals small, rectangular with right angles; longer than wide; 12–14 in an annulus at midbody level. Median ventrals larger than the dorsals, rectangular with right angles; wider than long; 14–16 in an annulus at midbody; as many as, or more in an annulus than the number of dorsals in the same annulus. Total number of scales around midbody 26–30. A lateral sulcus separating the dorsals and the ventrals. Scales bordering on the lateral sulci have rounded lateral margins, the space between the dorsal and the ventral part of an annulus is filled by two small scales. Dorsal sulcus present or absent, without differentiated scales. No intercalated dorsal annuli in the anterior part of the body; some irregularities in the precloacal region are sometimes present; 200–210 body annuli from the back of the second infralabial to and including the preanal pore-bearing row; two or three lateral annuli in the cloacal region.

Preanal pores four; preanal segments six; postanal segments 8–11. Hemipenis cylindrical, smooth, bifid; sulcus spermaticus shallow, on the posterior surface to between the horns, there bifurcating, one branch on the inner surface of each horn.

Caudal annuli 28 + 1 – 31 + 1. No definite autotomy constriction is present, as specimens with broken or healed tails had their tails broken somewhere between the 5th and the 12th caudal annulus. However, in specimens with complete tails the 10th, 11th or 12th caudal annulus is slightly narrower than the other caudal annuli, a character which in *A. fuliginosa* Linnaeus and other species is restricted to the autotomy constriction.

Colour in life variable. RMNH 15144 from Paloemeu was purplish all over, slightly darker dorsally than ventrally. The specimens from Sipaliwini were light- to dark-brown above, with a transparent pink belly, the underside of the tail white.

In preservative RMNH 15144 is dirty white all over, with a brown tinge on the dorsal parts. All Sipaliwini specimens are light- to dark-brown above, white below.

A small amphisbaenian, with a slender, cylindrical body and a relatively long, cylindrical tail with blunt tip. Maximum snout-vent length in females 137 mm, in males 133 mm. The tail is about 1/5–1/6 of the snout-vent length.

HABITAT.—All Surinam specimens (except RMNH 15191) were collected in rotten logs lying on the rain-forest floor. RMNH 15191 (from Sipaliwini) was found while digging in the upper soil-layers. Only a special type of rotten log seems to be favoured by this species, namely logs that are very soft and when pulled apart rendered a pulp of small wood-fragments.

NATURAL HISTORY.—Next to nothing is known about this subject. Never more than one specimen was found in a log. Other inhabitants of the logs in which this species was found include *Bachia cophias* (Schneider), *Leptotyphlops septemstriatus* (Schneider), *Lithodytes lineatus* (Schneider) and often large numbers of annelids. The annelids at first glance are very similar to this amphisbaenid, both in size, colour and behaviour. All specimens were collected in daytime, but this merely reflects the collector's activity pattern. The contents of two stomachs and of one intestine were examined. They contained one pseudoscorpion, one ant and the remains of several termites. As could be expected from the preceding, these are all animals found in the interior of rotten logs.

RANGE.—Thus far known only from the southern parts of Surinam (Sipaliwini, Paloemeu) and Guyana (Marudi, Kuyuwini River 2°5'N, 59°E and N. of Acarai Mountains, W. of New River). In the Sipaliwini area *A. vanzolinii* Gans was present in the forest surrounding the savannah, as well as in at least two of the forest-islands on the savannah (fig. 79).

REMARKS.—Comparison of the types with two other specimens from Guyana (KU 069817–18) from 'Rupununi district, north of Acarai Mountains, west of New River' and with my Surinam material shows several differences. The most important is the difference in total number of body annuli, the types having 228 and 225, the KU specimens 207 and 214 and the Surinam specimens 200–210. In this respect the KU specimens clearly fall in the same group as the Surinam specimens. Another character is the number of lateral annuli in the cloacal region, which, in the KU and Surinam specimens is two or three, in the types four or five. The number of postgenials in both types is three, in the KU and

Surinam specimens it is invariably two. The total number of scales in an annulus around midbody is 28 and 30 in the types, 26–31 in the Surinam specimens, 24 in the KU specimens. The number of postanals in the types is 12, in the KU specimens eight and in the Surinam specimens 8–11, but mostly ten. From these data it is clear that there is a certain amount of geographical variation. The KU specimens agree with the Surinam specimens in most characters, in the number of scales around midbody and in the number of postanals they have somewhat lower values. When more material from southern Guyana becomes available, it will probably be possible to separate the Surinam specimens as a distinct subspecies, the border between the nominate and the Surinam form being somewhere near the New River, but until then it seems highly undesirable to name the Surinam form.

4. Incertae sedis.

DAUDIN (1802a) described a lizard, *Lacerta dumetorum*, which he thought was identical with FERMIN's (1769) 'Huitième espèce' or 'Lézard des broussailles'. The description given by FERMIN is very short and on the basis of it I tentatively placed this species in the synonymy of *Anolis p. punctatus* Daudin. The specimen on which DAUDIN based his description and which, according to him, was sent from Surinam by MARIN DE BÈZE, does not at all resemble *Anolis p. punctatus* Daudin. Neither does it resemble one of the other lizards known from Surinam or the adjacent countries. For that reason I did not include *Lacerta dumetorum* Daudin in one of the synonymies given above. On the basis of the description I incline towards regarding *Lacerta dumetorum* Daudin as a synonym of a lacertid lizard, most probably *Lacerta viridis* (Laurenti). I know only of one other author (CLOQUET, 1823) who used *Lacerta dumetorum*. The name does not appear to have been used since 1823.

V. ADDENDA

Since completion of the manuscript and tables a steady flow of material from Surinam kept arriving in museums. This material has been included in the lists of material and in the distributional maps, but because of technical difficulties material that arrived after December 1, 1971 was not included in the tables. Moreover, this new material did not essentially change the picture.

Ecological data on a number of species discussed in this paper were provided by CRUMP (1971), VANZOLINI (1972) and by VANZOLINI & REBOUÇAS-SPIEKER (1973). These publications came too late in my possession to be dealt with under the pertaining species.

During the proof-reading of this paper I received an article by J. D. LAZELL (The Lizard Genus Iguana in the Lesser Antilles, Bull. Mus. Comp. Zool. 145 (1): 1–28, figs. 1–12), in which the author states that no type ever has been designated for *Iguana iguana* (Linnaeus). As has been pointed out above (pp. 41–44), there are still two syntypes of this species in existence. LAZELL restricts the type-locality of *I. iguana* (Linnaeus) to the island of Terre de Haut, Les Iles des Saintes, Departement de La Guadeloupe, French West Indies. His only argument for this restriction seems to be that the species is extremely abundant on that island, but according to recommendation 72E of the International Code of Zoological Nomenclature (1964) this is no valid argument for such an action. Considering the arguments I provided, we may neglect LAZELL's restriction of the type-locality.

VI. SUMMARY

Forty lizards and three amphisbaenians are presently known from Surinam. In the present paper eight species are reported from Surinam for the first time, whereas two new subspecies are described. The occurrence of three completely (*Gymnophthalmus underwoodi* Grant, *Kentropyx borckianus* W. Peters and *Leposoma percarinatum* (L. Müller)) and one partially (*Cnemidophorus l. lemniscatus* (Linnaeus)) unisexual species is signalized. The confusion around *Kentropyx intermedius* (Gray) is settled, because the types of this species could be located. They proved to belong to *Kentropyx calcaratus* Spix, which makes *intermedius* Gray a junior synonym of that species. The next name available for the species hitherto known as *K. intermedius* (Gray) is *borckianus* W. Peters, and consequently the species has to be renamed *K. borckianus* W. Peters. The name of *Alopoglossus carinicaudatus* (Cope) is changed into *A. angulatus* (Linnaeus). As the holotype of *A. angulatus* (Linnaeus) is apparently lost, a neotype is designated. It is shown that the reported occurrence of *Tupinambis teguixin* (Linnaeus) for northern South America is based on a mistake made by DUMÉRIL & BIBRON (1839).

The zoogeography of the lizards and amphisbaenians occurring in the Guianas is extensively discussed and from the distribution of some savannah-inhabiting lizards conclusions are drawn in connection with the presence of a rain-forest refuge on the Tumuc Humac and Boundary Mountains during the last interpluvial period. Special chapters are dedicated to collectors and expeditions.

Of all species extensive descriptions and ecological data, if available, are presented.

VII. REFERENCES

ANDERSSON, L. G., 1899. Catalogue of Linnean type-specimens of snakes in the Royal Museum in Stockholm.—*Bih. K. Svensk. Vet. Akad. Handl.* 24 (IV, 6): 1–35.

ANDERSSON, L. G., 1900. Catalogue of Linnean type-specimens of Linnaeus's Reptilia in the Royal Museum in Stockholm.—*Bih. K. Svensk. Vet. Akad. Handl.* 26 (IV, 1): 1–29.

ANDERSSON, L. G., 1918. New Lizards from South America, collected by Nils Holmgren and A. Roman.—*Ark. Zool. Stockholm* 11 (16): 1–9, figs. 1–4.

ANONYMUS, 1676. Pertinente Beschrijvinge van Guiana. Gelegen aen de vaste Kust van America. Daer in kortelijck verhaelt wordt/ het aenmerckelijckste dat in en omtrent het Landt van Guiana valt/ als de Limiten, het Klimaet en de stoffen der Landen/ de Mineralen, Edele Gesteenten, Vruchten, Dieren, ende overvloedigheyt der Vissen, nevens der selver Inwoonderen aldaer. Hier is bygevoeght Der Participanten uytschot ende profijten, die daer uyt te volgen staen. Als oock de Conditien van mijn Heeren de Staten van Hollandt en West-Vrieslandt, voor die gene die nae Guiana begeeren te varen: 1–12, 1–55, 1 map.—Amsterdam.

ANONYMUS, 1775. Neuer Schauplatz der Natur nach den richtigsten Beobachtungen und Versuchen in alphabetischer Ordnung durch eine Gesellschaft von Gelehrten.—Vol. 1: I–XIV, 1–1044, 1 pl.—Leipzig.

ANONYMUS, 1841. Het boek der Amphibiën of tweeslachtige en kruipende dieren: I–VIII, 1–199, 45 figs.—Amsterdam.

ANONYMUS, 1886. Reizen en onderzoekingen van Dr. H. TEN Kate.—*Tijdschr. Ned. Aardr. Gen.* (2) 3: 706–710.

ANONYMUS, 1962. A Surinam portfolio.—*Nat. Hist.* 71 (10): 28–41, 17 pls.

ANONYMUS, 1970. Grote expeditie op zuidelijke helft Sipaliwini afgesloten.—*De West* 11124 (18–II–1970).

ANSLIJN N. Z., N., 1827. Systematische Beschrijving der voor ons meest belangrijke Amphibiën: 1–212.—Leyden.

AUDRETSCH, F. C. D', 1965a. Aantekeningen over amphibieën en reptielen in Suriname (IV).—*Lacerta* 23 (7): 53.

AUDRETSCH, F. C. D', 1965b. Aantekeningen over amphibieën en reptielen in Suriname (V).—*Lacerta* 24 (2): 13–16.

AUDRETSCH, F. C. D', 1965c. Aantekeningen over amphibieën en reptielen in Suriname (VI).—*Lacerta* 24 (3): 21–24.

AUDRETSCH, F. C. D', 1967a. Aantekeningen over amfibieën en reptielen in Suriname (VII).—*Lacerta* 25 (10–11): 82–84.

AUDRETSCH, F. C. D', 1967b. Aantekeningen over amfibieën en reptielen in Suriname (VIII).—*Lacerta* 26 (1): 6–7.

AUDRETSCH, F. C. D', 1967c. Aantekeningen over amfibieën en reptielen in Suriname (vervolg en slot van VIII).—*Lacerta* 26 (3): 27.

AUDRETSCH, F. C. D', 1968. Aantekeningen over amfibieën en reptielen in Suriname (IX).—*Lacerta* 26 (6): 51–52.

BAKHUIS, L. A., 1902. Verslag der Coppename-expeditie: 4 pp, 1–158, 31 plates, 1 map, 2 figs., reprinted from: *Tijdschr. Kon. Ned. Aardr. Gen.* (2) 19: 695–842.

BALK, L., 1746. Museum Adolpho-Fridericianum (Museum Principis). In: C. LINNAEUS, 1749, Amoenitates Academicae ......: 277–326, pls. XIII–XIV.—Stockholm and Leipzig.

BANCROFT, E., 1769a. An Essay on the Natural History of Guiana, In South America. Containing A Description of many Curious Productions in the Animal and Vegetable Systems of that Country. Together with an Account of The Religion, Manners, and Customs of several Tribes of its Indian Inhabitants. Interspersed with A Variety of Literary and Medical Observations. In Several Letters from A Gentleman of the Medical Faculty, During his Residence in that Country: I–IV, 2 pp., 1–402.—London.

BANCROFT, E., 1769b. Naturgeschichte von Guiana in Süd-Amerika. Worinn von der natürlichen Beschaffenheit und den vornehmsten Naturproducten des Landes, ingleichen der Religion, Sitten und Gebräuchen verschiedener Stämme der wilden Landes-Einwohner, Nachricht ertheilet wird: I–X, 2 pp., 1–248.

BARBOUR, T., 1934. The Anoles, II. The mainland species from Mexico southward.—*Bull. Mus. Comp. Zoöl.* 77 (4): 121–155.

BARBOUR, T, & G. K. NOBLE, 1915. A revision of the Lizards of the genus Ameiva.—*Bull. Mus. Comp. Zoöl.* 59 (6): 417–479.

BASKIN, J. N. & E. E. WILLIAMS, 1966. The lesser Antillean Ameiva (Sauria, Teiidae) re-evaluation, zoogeography and the effects of predation.—*Stud. Faun. Carib. Isl.* 23 (89): 144–176, figs. 42–43, tabs. 12–19, pls. 1–2.

BATES, M., 1965. De dierenwereld van Zuid-Amerika: 1–200, figs., plates.—Amsterdam.

BAYLIS, H. A., 1947a. Some roundworms and flatworms from the West Indies and Surinam.—I. Nematodes and Acanthocephala.—*Journ. Linn. Soc. London, Zool.* 41 (280): 394–405, figs. 1–12.

BAYLIS, H. A., 1947b. Some roundworms and flatworms from the West Indies and Surinam.—II. Cestodes.—*Journ. Linn. Soc. London, Zool.* 41 (280): 406–414, figs. 1–7.

BECHSTEIN, J. M., 1800a. Herrn DE LA CEPEDE's Naturgeschichte der Amphibien oder der eyerlegenden vierfüssigen Thiere und der Schlangen. Vol. 1: I–XXXII, 1–524, pls. 1–28.—Weimar.

BECHSTEIN, J. M., 1800b. Herrn DE LA CEPEDE's Naturgeschichte der Amphibien oder der eyerlegenden vierfüssigen Thiere und der Schlangen. Vol. 2: I–XX, 1–552, pls. 1–40.—Weimar.

BEEBE, W., 1925. Studies of a tropical jungle, one quarter of a square mile of jungle at Kartabo, British Guiana.—*Zoologica, N.Y.* 6 (1): 1–193, figs. 1–17, pl. 1.

BEEBE, W., 1944a. Field notes on the lizards of Kartabo, British Guiana, and Caripito, Venezuela. Part 1. Gekkonidae.—*Zoologica, N.Y.* 29 (14): 145–160, figs. 1–5, pls. 1–6.

BEEBE, W., 1944b. Field notes on the lizards of Kartabo, British Guiana and Caripito, Venezuela. Part 2. Iguanidae.—*Zoologica, N.Y.* 29 (18): 195–216, figs. 1–17, pls. 1–6.

BEEBE, W., 1945. Field notes on the lizards of Kartabo, British Guiana and Caripito, Venezuela. Part 3. Teiidae, Amphisbaenidae and Scincidae.—*Zoologica, N.Y.* 30 (2): 7–31 figs. 1–16, pls. I–V.

BENJAMINS, H. D., 1898. De Grenzen van Nederlandsch Guiana.—*Tijdschr. Ned. Aardr. Gen.* (2) 15: 797–846, maps IX–XI.

BERGMANS, W., 1965. Enige Cnemidophorus soorten en hun verspreiding.—*Lacerta* 23 (8): 59.

BERKEL, A. VAN, 1695. Amerikaansche Voyagien, Behelzende een Reis na Rio de Berbice, Gelegen op het vaste Land van Guiana, aande Wilde-kust van America, Mitsgaders een andere na de Colonie van Suriname, Gelegen in het Noorder Deel van het gemelde Landschap Guiana. Ondermengd met alle de byzonderheden noopende de Zeden, Gewoonten en Levenswijs der Inboorlingen, Boom- en Aardgewassen, Waaren en Koopmanschappen, en andere aanmerkelijke zaaken: 4 pp., Eerste Deel: Reis na Rio de Berbice: 1–104, 2 pls. Tweede Deel: Reis na Suriname: 105–139.—Amsterdam.

BERTHOLD, A. A., 1840. Ueber verschiedene neue oder seltene Amphibienarten. 3) Ueber das Genus Draconura.—*Gött. gel. Anzeigen.* 1840: 889–902.

BERTHOLD, A. A., 1842. Ueber verschiedene neue oder seltene Amphibienarten. III. Ueber das Genus Draconura Wagl.: 17–21, figs. 7–8.—Göttingen.

BERTHOLD, A. A., 1843. Ueber verschiedene neue, oder seltene Amphibienarten. III. Ueber das Genus Draconura Wagl.—*Abh. Kon. Ges. Wiss. Göttingen* I: 61–65, figs. 7–8.

BLAKE, E. R., 1950. Birds of the Acary Mountains southern British Guiana.—*Fieldiana Zool.* 32 (7): 419–474, map 1.

BOCOURT, M. F., 1873. In: DUMÉRIL, A., M. F. BOCOURT & MOCQUARD, 1870–1909. Mission Scientifique au Mexique et dans l'Amérique Centrale. Recherches Zoologiques. Troisième Partie, 1re Section. Etudes sur les Reptiles: 33–112.—Paris.

BOCOURT, M. F., 1874. In: DUMÉRIL, A., M. F. BOCOURT & MOCQUARD, 1870–1909. Mission Scientifique au Mexique et dans l'Amérique Centrale. Recherches Zoologiques. Troisième Partie, 1re section. Etudes sur les Reptiles: pl. XVI–XVIII.—Paris.

BOESEMAN, M., 1964a. Ichthyological and other zoological observations: Dec. 1963–March 1964.—*Progress Rep. Biol. Brokopondo Res. Project, Surinam* I: 11–15.

BOESEMAN, M., 1964b. Ichthyological and other zoological observations: 1 April–1 July 1964. —*Progress Rep. Biol. Brokopondo Res. Project, Surinam* I: 32–35.

BOESEMAN, M., 1964c. Ichthyological and other zoological observations: July–September, 1964.—*Progress Rep. Biol. Brokopondo Res. Project, Surinam* I: 53–55.

BOETTGER, O., 1893. Katalog der Reptilien-Sammlung im Museum der Senckenbergischen Naturforschenden Gesellschaft in Frankfurt am Main. I Teil. (Rhynchocephalen, Schildkröten, Krokodile, Eidechsen, Chamäleons): I–X, 1–140.—Frankfurt am Main.

BONAPARTE, P., 1886. Note sur les récents voyages du Dr. H. TEN KATE dans l'Amerique du Sud.—*Tijdschr. Ned. Aardr. Gen.* (2) 3: 92–97.

BONNATERRE, J. P., 1789. Tableau encyclopédique et méthodique des Trois Règnes de la Nature. Erpétologie: I–XXVIII, 1–70, pls. 1–43.—Paris and Liège.

BOOS, J. O., & V. C. QUESNEL, 1971. Notes on *Kentropyx striatus* in Trinidad, West Indies.—*Herpetologica* 27 (4): 477–481, fig. 1.

BORY DE ST. VINCENT, 1828. Résumé d'Erpétologie, ou d'histoire naturelle des reptiles, contenant des Notions générales et particulières sur l'antiquité, l'Organisation, les Moeurs de ces animaux, l'Art de les conserver, avec la Description et l'Histoire de leurs familles et de leurs genres, d'après une méthode analytique; précédé d'une Introduction Historique, et suivi d'une Biographie, d'une Bibliographie et d'un Vocabulaire: complétée par une Iconographie de 52 planches: I–VI, 2 pp., 1–292.—Paris.

BOSCHMA, H., 1947. Fauna. In: Ons Koninkrijk in Amerika, West-Indië: 38–44.—'s-Gravenhage.

BOULENGER, G. A., 1885a. Catalogue of the lizards in the British Museum (Natural History).—Vol. I. Geckonidae, Eublepharidae, Uroplatidae, Pygopodidae, Agamidae: I–XII, 1–436, pls. I–XXXII.—London.

BOULENGER, G. A., 1885b. Catalogue of the lizards in the British Museum (Natural History).—Vol. II. Iguanidae, Xenosauridae, Zonuridae, Anguidae, Anniellidae, Helodermatidae, Varanidae, Xantusiidae, Teiidae, Amphisbaenidae: I–XIII, 1–497, pls. I–XXIV.—London.

BOULENGER, G. A., 1887a. Catalogue of the lizards in the British Museum (Natural History).—Vol. III. Lacertidae, Gerrhosauridae, Scincidae, Anelytropidae, Dibamidae, Chamaeleontidae: I–XII, 1–575, pls. I–XL.—London.

BOULENGER, G. A., 1887b. On a new geckoid lizard from British Guiana.—*Proc. Zool. Soc. London* 1887: 153–154, 1 fig.

BOULENGER, G. A., 1900. Report on a collection made by Messrs. F. V. MCCONNEL and J. J. QUELCH at Mt. Roraima in British Guiana.—*Trans. Linn. Soc. London* (2) 8: 51–76.

BOULENGER, G. A., 1912. Descriptions of new reptiles from the Andes of South America, preserved in the British Museum.—*Ann. Mag. Nat. Hist.* (8) 10: 420–424.

BREDER, C. M., 1946. Amphibians and reptiles of the Rio Chucunaque drainage, Darien, Panama, with notes on their life histories and habits.—*Bull. Am. Mus. Nat. Hist.* 86 (8): 379–435, pls. 42–60, figs. 1–24, tabs. 1–11.

BRONGERSMA, L. D., 1927. Neusticurus dejongi, sp. n., a new lizard from Surinam.—*Ann. Mag. Nat. Hist.* (9) 20: 543–545.

BRONGERSMA, L. D., 1928, Arthrosaura dorsistriata Müller, ein Synonym von Arthrosaura kocki (Van Lidth de Jeude).—*Zool. Anz.* 78 (11/12): 333–336.

BRONGERSMA, L. D., 1932a. Notes on the species of Arthrosaura Blgr. (Teiidae).—*Zool. Med. Leiden* 15: 76–88, figs. 1–12.

BRONGERSMA, L. D., 1932b. Abnorme Beschilderung des Kopfes bei Amphisbaenen.—*Zool. Anz.* 99 (5/6): 132–134, figs. 1–2.

BRONGERSMA, L. D., 1934. Contributions to Indo-Australian herpetology.—*Zool. Med. Leiden* 17; I–XVI, 161–251, figs. 1–49, pls. I–II.
BRONGERSMA, L. D., 1935. Herpetological notes X–XII, XII. Note on Arthrosaura reticulata (O'Shaughn.) and Arthrosaura versteegii Lidth.—*Zool. Med. Leiden* 18: 261–264.
BRONGERSMA, L. D., 1946a. On an Alopoglossus from Surinam.—*Zool. Med. Leiden* 26: 231–236, fig. 1.
BRONGERSMA, L. D., 1946b. Some notes on species of the genera Bachia and Scolecosaurus.—*Zool. Med. Leiden*, 26: 237–246, figs. 1–4.
BURGERSDIJK, L. A. J., 1873. De dieren, afgebeeld, beschreven en in hunne levenswijze geschetst. Vol. 3. Kruipende dieren, visschen, ongewervelde dieren: I–XXII, 1–828, 1–12, I–IV, col. pls. 1–110.—Leiden.
BURT, C. E., 1931. A study of the Teiid Lizards of the genus Cnemidophorus with special reference to their phylogenetic relationships.—*U.S. Nat. Mus. Bull.* 154: 1–280, figs. 1–38.
BURT, C. E. & M. D. BURT, 1931, South American lizards in the collection of the American Museum of Natural History.—*Bull. Am. Mus. Nat. Hist.* 61: 227–395, figs. 1–15.
BURT, C. E. & M. D. BURT, 1930. The South American Lizards in the collection of the United States National Museum.—*Proc. U.S. Nat. Mus.* 78 (6): 1–52.
BURT, C. E., & M. D. BURT, 1933. A preliminary check list of the Lizards of South America.—*Trans. Acad. Sci. St. Louis* 28 (1/2): 1–104.
CAPELLE, H. VAN, 1903. De Binnenlanden van het District Nickerie. Lotgevallen en algemeene uitkomsten van eene Expeditie door het westelijk deel der Kolonie Suriname in September en October van het jaar 1900: 2 pp., 1–240, figs., 1 map.—Baarn.
CARROLL, R. L., 1969. Origin of reptiles. In: C. GANS, Biology of the Reptilia, vol. 1: 1–44, figs. 1–14.—London & New York.
C[LOQUET], H., 1820. Gecko. In: Dictionnaire des Sciences Naturelles, dans lequel on traite méthodiquement des differens êtres de la nature, considérés soit en eux-mêmes, d'après l'état actuel de nos connoissances, soit relativement a l'utilité qu'en peuvent retirer la médecine, l'agriculture, le commerce et les arts. Suivi d'une biographie des plus célèbres naturalistes. Vol. 18: 270–279.—Strasbourg and Paris.
C[LOQUET], H., 1823, Lézard. In: Dictionnaire des Sciences Naturelles ......, Vol. 26: 189–200.—Strasbourg and Paris.
C[LOQUET], H., 1827. Sauve-gardes. In: Dictionnaire des Sciences Naturelles ......, Vol. 47: 525–527.—Strasbourg and Paris.
COHEN, A. & J. J. VAN DER EIJK, 1953. Klassificatie en ontstaan van savannen in Suriname.—*Geol. en Mijnb.* N.S. 15 (6): 202–214, pls. 1–10, 1 map, tabs. I–II.
COPE, E. D., 1862. On the genera Panolopus, Centropyx, Aristelliger and Sphaerodactylus.—*Proc. Acad. Nat. Sci. Phil.* 1861: 494–500.
COPE, E. D., 1863. Synopsis of the species of Holcosus and Ameiva, with diagnosis of new West Indian and South-American Colubridae.—*Proc. Acad. Nat. Sci. Phil.* 1862: 60–82.
COPE, E. D., 1869. An examination of the Reptilia and Batrachia obtained by the Orton Expedition to Equador and the Upper Amazon, with notes on other species.—*Proc. Acad. Nat. Sci. Phil.* 1868: 96–140, 5 figs.
COPE, E. D., 1876. Report on the Reptiles brought by Professor James Orton from the Middle and Upper Amazon, and Western Peru.—*Journ. Acad. Nat. Sci. Phil.* (2) 8: 159–183.
COTT, H. B., 1926. Observations on the life-habitats of some batrachians and reptiles from the lower Amazon: and a note on some mammals from Marajó Island.—*Proc. Zool. Soc. Lond.* 4: 1159–1178, pls. I–VI.
CRUMP, M., 1971. Quantitative analysis of the ecological distribution of a tropical herpetofauna.—*Occ. Pap. Mus. Nat. Hist. Univ. Kansas* 3: 1–62, figs 1–29, tables 1–11.
CUNHA, O. R. DA, 1958. I. Lacertílios da Amazônia. Sôbre a ocorrência do gênero Bachia Gray, 1845 na Amazônia Brasileira (Lacertilia-Teiidae).—*Bol. Mus. Goeldi* (*Zool.*) 11: 1–12, figs. 1–2.
CUNHA, O. R. DA, 1961. II. Lacertílios da Amazônia. Os Lagartos da Amazônia Brasileira, com especial referência aos representados na coleção do Museu Goeldi.—*Bol. Mus. Goeldi* (*Zool.*) 39: 1–189.

Cunha, O. R. da, 1967. Lacertílios da Amazônia. III: O gênero 'Arthrosaura' Boulenger, 1885 (Lacertilia, Teiidae).—*Atas Simp. Biota Amazonica*, 5 (*Zoologia*): 141–170, figs. 1–8, tabs. 1–6.

Cunha, O. R. da, 1970. Lacertílios da Amazônia IV.—Um nôvo gênero e espécie de Lagarto de Território Federal do Amapá (Lacertilia-Teiidae).—*Bol. Mus. Goeldi* (*Zool.*) 74: 1–8, figs. 1–2.

Cunha, O. R. da, 1971. Lacertílios da Amazônia. V: Sôbre Gonatodes annularis Boulenger e Neusticurus racenisi Roze no Brazil (Lacertilia, Geckonidae, Teiidae).—*Rev. Brasil. Biol.* 31 (1): 113–117, 1 tab.

Cuvier, L. C. F. D. G., 1816. Le règne animal distribué d'après son organisation, pour servir de base a l'histoire naturelle des animaux et d'introduction a l'anatomie comparée. Vol. 2: I–XVIII, 1–532 (contenant les reptiles, les poissons, les mollusques et les annélides.).—Paris, first edition.

Cuvier, L. C. F. D. G., 1829. Le règne animal distribué d'après son organisation, pour servir de base a l'histoire naturelle des animaux et d'introduction a l'anatomie comparée. Vol. 2: I–XV, 1–406.—Paris, second edition.

Cuvier, L. C. F. D. G., 1834. The animal kingdom, arranged according to its organization, serving as a foundation for the natural history of animals, and an introduction to comparative anatomy. Vol. 2: I–XXXV, 1–412, 17 pls.—London.

Cuvier, L. C. F. D. G., 1836. Le règne animal distribué d'après son organisation pour servir de base a l'histoire naturelle des animaux et d'introduction a l'anatomie comparée. Vol. 1: I–XXII, 1–626, pls. 1–16.—Bruxelles, third edition.

Darevsky, I. S., 1966. Natural parthenogenesis in a polymorphic group of caucasian rock lizards related to Lacerta saxicola Eversmann.—*Journ. Ohio Herp. Soc.* 5 (4): 115–152, figs. 1–24, tabs. 1–7.

Darlington, P. J., 1957. Zoogeography: The geographical distribution of animals. I–XI, 2 pp., 1–675, figs. 1–80.—New York and London.

Daubenton, [L. J. M.], 1784. Encyclopédie Méthodique. Tome II de l'Histoire Naturelle. Les animaux quadrupèdes ovipares et les serpens: 545–712, 26 pls.—Paris and Liège.

Daudin, F. M., 1802a. Histoire naturelle, génerale et particulière des reptiles. Tome III: 1–452, pls. XXIX–XLV.—Paris.

Daudin, F. M., 1802b. Histoire naturelle, générale et particulière des reptiles. Tome IV: 1–397, pls. XLVI–LVIII.—Paris.

Daudin, F. M., 1803. Histoire naturelle, génerale et particulière des reptiles. Tome VII: 1–436, pls. LXXXI–XCII.—Paris.

Desmarest, A. G., 1833. Oeuvres du comte de Lacépède, comprenant l'histoire naturelle des quadrupèdes ovipares, des serpents, des poissons et des cétacés, avec la synonymie des auteurs modernes les plus célèbres. Vol. I: 1–437.—Bruxelles.

Desmarest, E., 1856. Reptiles et poissons. In: Chenu: Encyclopédie d'histoire naturelle ou traité complet de cette science d'après les travaux des naturalistes les plus eminents de tous les pays et de toutes les époques: 8 pp., 1–360, figs. 1–135, pls. 1–48.—Paris.

Dixon, J. R., 1973. A Systematic Review of the Teiid Lizards, Genus Bachia, with Remarks on Heterodactylus and Anotosaura.—*Misc. Publ. Mus. Nat. Hist. Univ. Kansas* 57: 1–47, figs. 1–15.

Donoso-Barros, R., 1968. The Lizards of Venezuela (Check list and key).—*Carib. J. Sci.* 8 (3–4), 105–122.

Donselaar, J. van, 1968. Phytogeographic notes on the savannah flora of Southern Surinam (South America).—*Acta Bot. Neerl.* 17: 393–404.

Dowling, H. G., T. C. Majupuria & F. W. Gibson, 1971a. Hemipenial morphology of the tree lizard, Plica plica (Linnaeus).—*Herpetol. Rev.* 3 (5): 91–92, fig. 1.

Dowling, H. G., T. C. Majupuria & F. W. Gibson, 1971b. The hemipenis of the onion-tail gecko Thecadactylus rapicaudus (Houttuyn).—*Herpetol. Rev.* 3 (6): 110, fig. 1.

Duellman, W. E., 1966. The Central American herpetofauna: An ecological perspective.—*Copeia* 1966 (4): 700–719, figs. 1–8, tabs. 1–5.

DUMÉRIL, A. M. C. & G. BIBRON, 1836. Erpetologie générale ou histoire naturelle complète des reptiles. Tome III: I–IV, 1–517.—Paris.
DUMÉRIL, A. M. C. & G. BIBRON, 1837. Erpétologie générale ou histoire complète des reptiles. Tome IV: I–II, 1–571.—Paris.
DUMÉRIL, A. M. C., & G. BIBRON, 1839. Erpétologie générale ou histoire naturelle complète des reptiles. Tome V: I–VIII, 1–854.—Paris.
DUMÉRIL, C., & A. DUMÉRIL, 1851. Catalogue méthodique de la collection des reptiles: I–IV, 1–224.—Paris.
DUNN, E. R., 1936. Notes on American mabuyas.—*Proc. Acad. Nat. Sci. Phil.* 87: 533–557.
DUNN, E. R., 1944. Los generos de anfibos y reptiles de Colombia, II. Segunda Parte: Reptiles, orden de los saurios.—*Caldasia*, 3: 73–110, figs. 1–9.
EICHELBERG, J. F. A., 1847. Die Hauptformen des Thierreichs in naturgetreuen Abbildungen und ausführlichen Schilderungen. Ein naturgeschichtliches Lesebuch für Schule und Haus. I–IV, 1–709, pls. 1–44.—Stuttgart.
EILERTS DE HAAN, J. G. W. J., 1910. Verslag van de expeditie naar de Suriname-Rivier. (30 Juni–20 November 1908): 1–127, figs. 1–25, 1 map, maps IX–X, reprinted from: *Tijdschr. Kon. Ned. Aardr. Gen.* (2) 27.
ETHERIDGE, R., 1968. A review of the iguanid lizard genera Uracentron and Strobilurus.— *Bull. Br. Mus. nat. Hist.* (*Zool.*) 17 (2): 47–64, figs. 1–2, tab. I.
ETHERIDGE, R,. 1970. A review of the South American Iguanid lizard genus Plica. —*Bull.Br. Mus. nat. Hist.* (*Zool.*) 19 (7): 235–256, figs. 1–4, tabs. 1–2.
FERMIN, P., 1765. Histoire naturelle de la Hollande équinoxiale: ou déscription des animaux, plantes, fruits et autres curiosités naturelles, qui se trouvent dans la colonie de Surinam; avec Leurs Noms différents, tant Francois, que Latins, Hollandois, Indiens & Négre-Anglois: I–XII, 1–240, 1 pl.—Amsterdam.
FERMIN, P., 1769. Description générale, historique, geographique et physique de la colonie de Surinam, Contenant Ce qu'il y a de plus Curieux & de plus Remarquable, touchant sa Situation, ses Rivieres, ses Forteresses; son Gouvernement & sa Police; avec les moeurs et les usages des Habitants Naturels du Païs, & des Européens qui y sont établis; ainsi que des Eclairissements sur l'oeconomie générale des Esclaves Negres, sur les Plantations & leurs Produits, les Arbres Fruitiers, les Plantes Médécinales, & toutes les diverses Especes d'animaux qu'on y trouve, & c. Vol. 2: 1–352, pls. 1–3.—Amsterdam.
FERMIN, P., 1770. Nieuwe algemeene Beschryving van de Colonie van Suriname. Behelzende al het merkwaardige van dezelve, met Betrekkinge tot de Historie, Aardryks- en Natuurkunde. Vol. 2: 1–323, 3 pp., 3 pls.—Harlingen.
FERNANDES, D. L., 1927. Verslag van den zoöloog. In: G. STAHEL, De expeditie naar het Wilhelmina-Gebergte (Suriname) in 1926. V. *Tijdschr. Kon. Ned. Aardr. Gen.* (2) 44: 243–254, 1 fig. Also in: *Bull. no. 86 Maatsch. Bev. Natuurk. Onderz. Ned. Kolon.*: 181–191.
FISCHER, J. G., 1832. Beschreibung aller naturhistorischen Gegenstände, welche auf den in der Grüson'schen Kunst- und Buchhandlung zu Breslau herausgekommener 'neuen Wandtafeln zur Naturgeschichte' abgebildet sind.—*Die Amphibien und Fischen.* Vol. 3: I–X, 1–118.—Breslau.
FITCH, H. S., 1968. Temperature and behaviour of some equatorial lizards.—*Herpetologica* 24 (1): 35–38, table 1.
FITCH, H. S., 1970. Reproduction cycles in lizards and snakes.—*Misc. Publ. Univ. Kansas Mus. Nat. Hist.* 52: 1–247, figs. 1–16.
FITZINGER, L. I., 1826. Neue Classification der Reptilien nach ihren natürlichen Verwandschaften. Nebst einer Verwandschafts-Tafel und einem Verzeichnisse der Reptilien-Sammlung des K. K. Zoologischen Museum's zu Wien: 8 pp., 1–66, 1 tab.—Wien.
FITZINGER, L., 1843. Systema Reptilium: Fasciculus primus: Amblyglossae: 1–106, I–VI.—Wien.
FITZINGER, L. J., 1864. Bilder-Atlas zur wissenschaftlich-populären Naturgeschichte der Amphibien in ihren sämmtlichen Hauptformen: 2 pp., figs. 1–200.—Wien.
FRANSSEN-HERDERSCHEE, A., 1905a. Verslag van de Gonini-expeditie.—*Tijdschr. Kon. Ned. Aardr. Gen.* (2) 22: 1–174, pls. I–III, 19 photo's, map 1.

Franssen-Herderschee, A., 1905b. Verlsag van de Tapanahoni-expeditie.—*Tijdschr. Kon. Ned. Aardr. Gen.* (2) 22: 847–1032, pls. I–III, 17 photo's, maps XIII–XV.

Gans, C., 1960. Studies on amphisbaenids (Amphisbaenia, Reptilia). I. A taxonomic revision of the Trogonophinae, and a functional interpretation of the amphisbaenid adaptive pattern.—*Bull. Am. Mus. Nat. Hist.* 119 (3): 129–204, figs. 1–32, tabs. 1–3, pl. 45.

Gans, C., 1962. Notes on Amphisbaenids (Amphisbaenia, Reptilia). 5. A redefinition and a bibliography of Amphisbaena alba Linné.—*Am. Mus. Nov.* 2105: 1–31, figs. 1–9.

Gans, C., 1963. Notes on Amphisbaenids (Amphisbaenia, Reptilia). 8. A redescription of Amphisbaena stejnegeri and the description of a new species of Amphisbaena from British Guiana.—*Am. Mus. Nov.* 2128: 1–18, figs. 1–14, tab. 1.

Gans, C., 1969. Amphisbaenians-reptiles specialized for a burrowing existence.—*Endeavour* 28 (105): 146–151, figs. 1–8.

Gans, C. & A. Alexander, 1962. Studies on Amphisbaenids (Amphisbaenia, Reptilia). 2. On the Amphisbaenids of the Antilles.—*Bull. Mus. Comp. Zool.* 128 (3): 65–158, figs. 1–39, tab. 1, pls. 1–12.

Gans, C. & C. O. Diefenbach, 1970. Reptilia: Amphisbaenia: Amphisbaenidae. In: J. A. Peters & R. Donoso-Barros: Catalogue of the Neotropical Squamata: Part II. Lizards and amphisbaenians – *Bull. U. S. Nat. Mus.* 297: 26–38.

Geijskes, D. C., 1942. Met de Oajana's op stap: 1–139, 12 figs., 1 map.

Geijskes, D. C., 1946. Resultaten van een onderzoek naar wilde Hevea-boomen in het stroomgebied van de Boven-Coppename.—*Bull. Dept. Landbouwproefst. Suriname* 60: 1–52, 11 pp, 7 maps.

Geijskes, D. C., 1957. Met de Oajana's op stap. Verslag van een reis naar de Litani (Boven Marowijne) in Suriname.—*Vox Guyanae* II (5/6): 193–300, 19 photo's, 1 map.

Geijskes, D. C., 1959. De expeditie naar de Tafelberg in 1958. *Vox Guyanae* III (4): 1–52, 11 photo's, 1 map.

Geijskes, D. C., 1961. De expeditie naar de Tafelberg in 1958.—*Schakels* S45: 2–13, 4 photo's.

Geijskes, D. C., 1961. Luchtverbinding met verre zuiden van Suriname tot stand gebracht.—*Voor elkaar* (*Paramaribo*): 15–V–1961: pp. 3 and 6, 3 photo's.

Geijskes, D. C., 1970. Documentary information about the Surinam Wama or Akurio Indians. —*Nieuwe West-Ind. Gids* 47 (3): 260–285, photo's 1–4.

Gmelin, J. F., 1789. Systema Naturae per Regna tria Naturae, secundum Classes, Ordines, Genera, Species; cum Characteribus, Differentiis, Synonymis, Locis. Tome I, Pars III, 1033–1516.—Lyon.

Goeje, C. H. de, 1908. Verslag der Toemoekhoemak-Expeditie (Tumuc-Humac-Expeditie).—*Tijdschr. Kon. Ned. Aardr. Gen.* (2) 25: 945–1169, pls. XVIII–XXXVI, map XVIII.

Goeldi, E., 1898. Die Eier von 13 brasilischen Reptilien, nebst Bemerkungen über Lebens- und Fortpflanzungsweise letzterer. Beobachtungen aus den Jahren 1884–1897.—*Zool. Jahrb. Syst.* 10: 640–676, pls. 26–27, 1 fig.

Goeldi, E., 1902. Lagartos do Brasil.—*Bol. Mus. Paraense Em. Goeldi* 3: 499–560.

Goeldi, E. A. & G. Hagmann, 1901. Die Eier von Tropidurus torquatus und Ameiva surinamensis.—*Zool. Jahrb. Syst.* 14: 581–589, 3 figs.

Goin, C. J., 1967 (1968). Behavior of Iguana during rain storms.—*Quart. J. Florida Ac. Sci.* 30 (2): 119–120.

Gonggrijp, J. W. & D. Burger, 1948. Bosbouwkundige studiën over Suriname.—*L.E.B. fund publ.* 32: 1–262, pls. 1–64.

Grant, C., 1958. A new Gymnophthalmus (Reptilia, Teidae) from Barbados, B.W.I.—*Herpetologica* 14: 227–228.

Gravenhorst, J. L. C., 1851. Über die im zoologischen Museum der Universität Breslau befindlichen Wirtelschleidchen (Pseudosaura), Krüppelfüssler (Brachypoda), und einige andere, denselben verwandte Reptilien aus den Zünften der Schleichen und Dickzüngler.—*Nov. Act. Ac. Caes. Leop.-Carol. Nat. Cur.* 23 (1): 291–394, pls. 27–45.

Gray, J. E., 1831. A synopsis of the species of the class Reptilia: 1–110. In: E. Griffith & E. Pidgeon: The class Reptilia arranged by the Baron Cuvier, with specific descriptions.—London.

Gray, J. E., 1845. Catalogue of the specimens of lizards in the collection of the British Museum: I–XXVIII, 1–289.—London.

Gray, J. E., 1851. Description of a new genus and family of Cyclosaurian lizards, from Para.—*Proc. Zool. Soc. London*, 1851: 38–39.

Greenberg, B., 1943. Social behavior of the western banded gecko, Coleonyx variegatus Baird.—*Physiol. Zool.* 16 (1): 110–122.

Griffith, E. & E. Pidgeon, 1830–1831. The class Reptilia arranged by the Baron Cuvier, with specific descriptions: 1–481, 55 pl.—London.

Gronovius, L. T., 1756. Amphibiorum Animalium Historia Zoologica exhibens Amphibiorum, qui in Museo Laurentii Theodori Gronovii, J.U.D. adservantur, descriptiones ordine systematico. In: Musei Ichthyologici. Tomus secundus sistens piscium indigenorum & nonnullorum exoticorum, quorum maxima pars in Museo Laurentii Theodori Gronovii, J.U.D. adservatur, nec non quorandam in aliis Museis observatorum descriptiones. Accedunt nonnullorum exoticorum Piscium icones aeri incisae, et Amphibiorum animalium. Historia Zoologica: 47–88.—Leiden.

Guibé, J., 1970. La systematique des reptiles actuels. In: P. Grassé, Traité de Zoologie, Tome XIV, Fasc. III, Reptiles: glandes endocrines-embryologie-systematique-paléontologie: 1054–1160, figs. 707–808.—Paris.

Guichenot, A., 1855. Reptiles in: F. L. de L. de Castelnau, Expédition dans les parties centrales de l'Amérique du Sud, de Rio de Janeiro à Lima, et de Lima au Para; exécutée .... pendant 1843 à 1847, sous la direction de F. de Castelnau. 7 (2): 1–95, pls. 1–18.—Paris.

Haffer, J., 1969. Speciation in Amazonian Forest Birds.—*Science* 165: 131–137.

Haffer, J., 1970. Art-Entstehung bei einigen Waldvögeln Amazoniens.—*J. Orn.* 111 (3/4): 285–331.

Hagmann, G., 1906. Die Eier von Gonatodes humeralis, Tupinambis nigropunctatus und Caiman sclerops. 3. Beitrag zur Kenntnis der Lebens- und Fortpflanzungsweise der brasilianischen Reptilien.—*Zool. Jahrb. Syst.* 24: 307–316, pls. 21–23.

Hagmann, G., 1909. Die Reptilien der Insel Mexiana, Amazonenstrom.—*Zool. Jahrb. Syst.* 28: 473–504, 1 pl.

Hallowell, E., 1856. On several new species of reptiles in the collection of the Academy of Natural Sciences.—*Proc. Acad. Nat. Sci. Phil.* 8: 153–156.

Hammen, T. van der, 1966. De Palynologie van het Guiana-bekken.—*Tijdschr. Kon. Ned. Aardr. Gen.* 83 (2): 128–138, figs. 1–9.

Hartsinck, J. J., 1770. Beschryving van Guiana, of de Wilde Kust, in Zuid-America, Betreffende de Aardrykskunde en Historie des Lands, de Zeeden en Gewoontes der Inwooners, de Dieren, Vogels, Visschen, Boomen en Gewassen, als mede de eerste Ontdekking dier Kust, de Bezittingen der Spanjaarden, Franschen en Portugeezen en voornaamelyk de Volkplantingen der Nederlanderen, als Essequebo, Demerary, Berbice, Suriname, En derzelver Rivieren, met de noodige Kaarten en Afbeeldingen der Forten. Waarby komt eene Verhandeling over den Aart en de Gewoontes der Neger-Slaaven. Alles uit echte stukken opgesteld. Vol. 1: I–XII, 4 pp., 1–520, 9 pls.; Vol. 2: 521–962, 15 pp.—Amsterdam.

Hast, B. R., 1745. Amphibia Gyllenborgiana. In: C. Linnaeus 1749: Amoenitates Academicae ....: 107–140.—Stockholm and Leipzig.

Haverschmidt, F., 1968. Birds of Surinam: I–XXIX, 1–445, pls. I–XXX, 1–40, text-figs.—Edinburgh and London.

Haverschmidt, F., 1970. Barn owls hunting by daylight in Surinam.—*Wilson Bull.* 82 (1): 101.

H[er]l[ein], J. D., 1718. Beschryvinge van de Volk-Plantinge Zuriname: Vertonende De Opkomst dier zelver Colonie, de Aanbouw en Bewerkinge der Zuiker-Plantagien. Neffens den aard der eigene natuurlijke Inwoonders of Indianen; als ook de Slaafsche Afrikaansche Mooren; deze beide Natien haar levens-manieren, Afgoden-dienst, Regering, Zeden, Gewoonten en dagelijksche Bezigheden. Mitsgaders Een vertoog van de Bosch-grond, Water- en Pluim-Gediertens; de veel vuldige heerlijke Vrugten, Melk-agtige Zappen, Gommen, Olyen, en de gehele gesteltheid van de Karaïbaansche Kust: 18 pp., 1–262, pls. 1–4, 1 map.—Leeuwarden.

HIRTH, H. F., 1963. Some aspects of the natural history of Iguana iguana on a tropical strand. —*Ecology*, 44 (3): 613–615.

HOEDEMAN, J. J., 1969. Elseviers aquariumvissen encyclopedie, 6: 1–189, photo's, maps.— Amsterdam and Brussel.

HOEVEN, J. VAN DER, 1855. Handboek der dierkunde. Vol. 2: I–XXVIII, 1–1068, pls. 13–24.— Amsterdam.

HOEVERS, L., 1967. Herpetological collections from the Atkinson-Maduni-Laluni area, east Demerara, Guyana.—*Timehri* (*J. Guyana Mus. Zoo*) 43: 34–50, figs. 1–10.

HOLM, A., 1957. Specimina Linnaeana i Uppsala bevarade zoologiska samlingar från Linnés tid.—*Acta Univ. Upps.*: *Uppsala Universitets Årsskrift* 6: 1–68, figs. 1–14, tabs. 1–8.

HOLTHUIS, L. B., 1958. G. D. COLLIN's Fauna Surinamensis.—*De West-Indische Gids*, 38: 71–85.

HOLTHUIS, L. B., 1959. The Crustacea Decapoda of Suriname (Dutch Guiana).—*Zool. Verh., Leiden*, 44: 1–296, figs. 1–68, pls. I–XVI, maps 1–2.

HOLTHUIS, L. B. & A. M. HUSSON, 1973. Jonkheer Drs. WILLEM CORNELIS VAN HEURN (1887–1972).—*Zool. Bijdr.* (in press).

HOOGMOED, M. S., 1967. Enkele opmerkingen naar aanleiding van eerder verschenen afleveringen van artikelen van Ir. F. C. D'AUDRETSCH.—*Lacerta* 25 (10–11): 82.

HOOGMOED, M. S., 1968. Hagedissen. In: Oosthoeks Encyclopedie 6: 611–612, 613–614.

HOOGMOED, M. S., 1969a. Notes on the herpetofauna of Surinam I.—Itinerary of a herpetological collecting trip in Surinam in 1968.—*Zool. Med., Leiden*, 44 (4): 47–73, figs. 1–11, pls. 1–6, tables 1–3.

HOOGMOED, M. S., 1969b. Notes on the herpetofauna of Surinam III.—A new species of Dendrobates (Amphibia, Salientia, Dendrobatidae) from Surinam.—*Zool. Med., Leiden*, 44 (9): 133–141, figs. 1–3, tab. 1.

HOUTTUYN, M., 1764. Natuurlyke historie of uitvoerige beschryving der dieren, planten en mineralen, volgens het samenstel van den heer LINNAEUS. Vol. I (6): 8 pp., 1–562, pls. 50–56. —Amsterdam.

HOUTTUYN, M., 1782. Het onderscheidt der Salamanderen van de Haagdissen in 't algemeen, en van de Gekkoos in 't byzonder.—*Verh. Zeeuws Gen. Wet. Vlissingen*, 9: 305–336, figs. 1–4.

HUECK, K., 1966. Die Wälder Südamerikas. Ökologie, Zusammensetzung und wirtschaftliche Bedeutung: I–XVIII, 1–422, figs. 1–253.—Stuttgart.

HUEY, R. B. & J. R. DIXON, 1970. A new Pseudogonatodes from Peru with remarks on other species of the genus.—*Copeia* 1970 (3): 538–542.

HUMPHREY, P. S. & R. FREUND, 1962. Notes on a collection of birds from Surinam.—*Postilla* 60: 1–11.

HUSSON, A. M., 1962. The bats of Suriname.—*Zool. Verh., Leiden*, 58: 1–282, figs. 1–39, pls. I–XXX, tabs. I–XXXI.

IM THURN, E. F., 1883. Among the Indians of Guiana: I–XVI, 1–445, figs. 1–43, plates I–X, 1 map.—London.

JONKER, F. P., 1961a. I. De reis naar de Emmaketen.—*Schakels* S 45: 14–18, 2 photo's.

JONKER, F. P., 1961b. II. De noordelijke Emmaketen en de Hendriktop.—*Schakels* S 45: 19–22, 1 photo.

JONKER, F. P., 1961c. III. De zuidelijke Emmaketen en de terugreis.—*Schakels* S 45: 23–29, 6 photo's.

JONKER, F. P., 1961d. IV. Bereikte resultaten en toekomstplannen.—*Schakels* S 45: 30–35, 5 photo's.

JONKER, F. P., 1966. Flora en vegetatie van Suriname.—*Tijdschr. Kon. Ned. Aardr. Gen.* (2) 83: 119–127.

JONKER, F. P. & J. J. WENSINK, 1960. De natuurwetenschappelijke expeditie naar de Emmaketen in Suriname, Juli–Oktober 1959.—*Tijdschr. Kon. Ned. Aardr. Gen.* (2) 77: 145–161, figs. 1–5, photo's 1–7.

KAPPLER, A., 1854a. Zes jaren in Suriname, Schetsen en tafereelen uit het maatschappelijke en militaire leven in deze kolonie.—Vol. I: I–IV, 1–183, 1 pl.—Utrecht.

Kappler, A., 1854b. Zes jaren in Suriname, Schetsen en tafereelen uit het maatschappelijke en militaire leven in deze kolonie.—Vol. II: 1–137.—Utrecht.
Kappler, A., 1862. Holländisch-Französische Expedition ins Innere von Guiana, Septbr. bis Novbr. 1861.—*Petermanns Geogr. Mitth.* 8: 173–179; 246–254.
Kappler, A., 1863. Nederlandsch-Fransche expeditie door de binnenlanden van Guyana, in September tot November 1861.—Album der Natur 1863: 65–80, 97–115.
Kappler, A., 1881. Holländisch-Guiana. Erlebnisse und Erfahrungen während eines 43 jährigen Aufenthalts in der Kolonie Surinam. I–IX, 2 pp. 1–495, 1 map, 1 pl.—Stuttgart.
Kappler, A., 1885. Die Tierwelt im Holländischen Guiana.—*Das Ausland. Wochenschrift für Länder- und Volkerkunde* 58 (41): 815–818.
Kappler, A., 1887. Surinam, sein Land, seine Natur, Bevölkerung und seine Kultur-Verhältnisse mit Bezug auf Kolonisation: 1–383, 1 p., figs., 1 map.—Stuttgart.
Käyser, C. C., 1912. Verslag der Corantijn-Expeditie (19 Juli 1910–1 April 1911).—*Tijdschr. Kon. Ned. Aardr. Gen.* (2) 29: 442–514, figs. 1–27, map III.
Keye, O., 1659. Het waere Onderscheyt tusschen Koude en Warme Landen, Aangewesen In de Nootsakelijckheden die daar vereyscht worden, Ten I. Om beyde die Landen te konnen bewoonen. Ten II. Tot de Culture van die selve Landen. Ende Ten III. In de ongemeene groote voordeelen, dewelcke de Warme Landen, door middel van hare Vruchten, boven de Koude Landen, op-brengen. Voorgestelt en vergeleken met Nieu-Nederlant, als sijnde een Koudt Landt en Guajana sijnde een warm Landt, en beyde gelegen in America, doch onder een besonder Climaet, en dat by gelegentheyt van de bevolckinge dewelcke tegenwoordig, in beyde die gewesten wordt gedaen. Op dat, daer door alle Patroonen mogen onderregt worden in wat Landen sy hare Colonijen met de minste onkosten, ende de spoedichste en grootste profijten konnen leggen: alsmede Op dat alle Luyden dewelcke haer, ofte uyt noot, ofte op hope van groote winsten te doen, met'er woon na alsulcke Nieuwe Landen willen begeven daer uyt mogen bekennen, in welcke van beyde Landen, het voor haer het ghemackelijckste sal sijn te leven, en waer inne zy de spoedigste, grootste en verseeckerste profijten sullen konnen doen: 18 pp., 1–178, 4 pp.—'s-Gravenhage.
Keye, O., 1667. Beschryvinge Van het Heerlijcke ende Gezegende Landt Guajana, Waer inne gelegen is de seer voorname Lant-streke genaemt Serrenamme, Die jegenwoordigh beseten wort by den Staet van de Vereenighde Nederlantsche Provintien. Mitsgaders Aenwysinge Van de groote Vermaeckelijckheden, Voordeelen, ende Rijckdommen die aldaer, ten reguarde van andere Landen, te verkrijgen en te genieten zijn: 10 pp., 1–178, 4 pp.—'s-Gravenhage.
Kleef, H. van, 1963. Cnemidophorus murinus.—*Lacerta* 21 (11): 87.
Kluge, A. G., 1967. Higher taxonomic categories of gekkonid lizards and their evolution.—*Bull. Amer. Mus. Nat. Hist.* 135 (1): 1–59, figs. 1–8, maps 1–2, tabs. 1–3, pls. 1–5.
Kluge, A. G., 1969. The evolution and geographical origin of the New World Hemidactylus mabouia-brookii complex (Gekkonidae, Sauria).—*Misc. Publ. Mus. Zool. Univ. Mich.* 138: 1–78, figs. 1–15, tabs. 1–23.
Kluge, A. G., & M. J. Eckardt, 1969. Hemidactylus garnoti: Duméril and Bibron, a triploid all-female species of gekkonid lizard.—*Copeia* 1969 (4): 651–664.
Kuhl, H., 1820. Beiträge zur Zoologie und vergleichenden Anatomie, Erste Abtheilung. *Beiträge zur Zoologie*: 10 pp., 1–152.
Lacépède, [B. G. E.] de [la V.], 1788a. Histoire naturelle des quadrupèdes ovipares et des serpens. Tome I: 1–651, pls. I–XLI.—Paris, 8°.
Lacépède, [B. G. E.] de [la V.], 1788b. Histoire naturelle des quadrupèdes ovipares et des serpens. Tome I: 1–359, pls. I–XII.—Paris, 12°.
Lacépède, [B. G. E.] de [la V.], 1788c. Histoire naturelle des quadrupèdes ovipares et des serpens, Tome II: I–IV, 1–462, 2 pp., pls. I–XV.—Paris, 12°.
Lacépède, [B. G. E.] de [la V.], 1799. Histoire naturelle des quadrupèdes ovipares et des serpens. Nouvelle édition. Tome I: 1–276, 2 pp., pls. I–XLI.—Dordrecht.
Lacépède, [B. G. E.] de [la V.], 1847. Histoire Naturelle de Lacépède comprenant les Cetacés, les Quadrupèdes ovipares, les Serpents et les Poissons. Nouvelle Édition, précédee de l'Éloge de Lacépède par Cuvier. Avec des Notes et la nouvelle Classification de M.A.-G. Desmarest, Tome 1: 4 pp., I–XII, 1–668, col. pls.—Paris.

LAMMERÉE, L., 1970. Lizards of the genus Cnemidophorus from the Leeward Group and the adjacent mainland of South America.—*Stud. Faun. Cur. Carib. Isl.* 34 (124): 46–72, figs. 22–34, pls. I–XII.

LANJOUW, J., 1936a. Studies on the vegetation of Suriname savannahs and swamps.—*Ned. Kruidk. Arch.* 46: 823–851.

LANJOUW, J., 1936b. Over de vegetatie van de Surinaamsche savannen en zwampen.—*Natuurwetenschappelijk tijdschrift, Gent.* Vol. 18: 181–182.

LANJOUW, J., 1954. The vegetation and the origin of the Suriname savannas.—Rapp. et Comm. 8me Congrès Internat. Bot., Paris, Sect. 7 et 8: 45–48.

LATREILLE, P. A., 1802a. In: C. S. Sonnini & P. A. Latreille: Histoire naturelle des reptiles, avec figures dessinées d'après nature. Vol. I: I–XII, 1–280, 13 pls.—Paris.

LATREILLE, P. A., 1802b. In: C. S. Sonnini & P. A. Latreille: Histoire naturelle des reptiles, avec figures dessinées d'après nature. Vol. II: 1–332, 21 pls.—Paris.

LATREILLE, P. A., 1802c. In: C. S. Sonnini & P. A. Latreille: Histoire naturelle des reptiles, avec figures dessinées d'après nature. Vol. IV: 1–410, 13 pls.—Paris.

LEON, J. R., R. DONOSO-BARROS & A. S. PRIETO, 1970. Alimentacion de tres especies de lagartos de los alrededores de Cumana, Estado Sucre, Venezuela.—*Bol. Soc. Biol. Conception* 42: 349–354, table 1.

LIDTH DE JEUDE, T. W. VAN, 1898. Catalogue ostéologique des poissons, reptiles et amphibiens.—*Mus. Hist. Nat. Pays-Bas* X (2). Reptiles: 2 pp., 1–52.—Leide.

LIDTH DE JEUDE, T. W. VAN, 1904. Reptiles and batrachians from Surinam.—*Notes Leyden Mus.* 25: 83–94, pl. 7.

LIDTH DE JEUDE, T. W. VAN, 1917a. Ameiva surinamensis. In: H. D. BENJAMINS & J. F. SNELLEMAN: Encyclopaedie van Nederlandsch West-Indië: 43.—'s-Gravenhage and Leiden.

LIDTH DE JEUDE, T. W. VAN, 1917b. Amphisbaena. In: H. D. BENJAMINS & J. F. SNELLEMAN: Encyclopaedie van Nederlandsch West-Indië: 43.—'s-Gravenhage and Leiden.

LIDTH DE JEUDE, T. W. VAN, 1917c. Anolis lineatus. In: H. D. BENJAMINS & J. F. SNELLEMAN: Encyclopaedie van Nederlandsch West-Indië: 46.—'s-Gravenhage and Leiden.

LIDTH DE JEUDE, T. W. VAN, 1917d. Cnemidophorus. In: H. D. BENJAMINS & J. F. SNELLEMAN: Encyclopaedie van Nederlandsch West-Indië: 206.—'s-Gravenhage and Leiden.

LIDTH DE JEUDE, T. W. VAN, 1917e. Doryphorus azureus. In: H. D. BENJAMINS & J. F. SNELLEMAN: Encyclopaedie van Nederlandsch West-Indië: 274.—'s-Gravenhage and Leiden.

LIDTH DE JEUDE, T. W. VAN, 1917f. Iguana. In: H. D. BENJAMINS & J. F. SNELLEMAN: Encyclopaedie van Nederlandsch West-Indië: 372–373.—'s-Gravenhage and Leiden.

LIDTH DE JEUDE, T. W. VAN, 1917g. Lacertilia. In: H. D. BENJAMINS & J. F. SNELLEMAN: Encyclopaedie van Nederlandsch West-Indië: 436–438.—'s-Gravenhage and Leiden.

LIDTH DE JEUDE, T. W. VAN, 1917h. Tupinambis nigropunctatus. In: H. D. BENJAMINS & J. F. SNELLEMAN: Encyclopaedie van Nederlandsch West-Indië: 688.—'s-Gravenhage and Leiden.

LINDEMAN, J. C., 1953. The vegetation of Suriname (1). The vegetation of the coastal region of Suriname: 1 p., I–IV, 1–135, 1 pl., figs. 1–32, maps I–II, figs. I–II, tables I–III.—Amsterdam.

LINNAEUS, C., 1749. Amoenitates Academicae seu dissertationes variae physicae, medicae, botanicae antehac seorsim editae, nunc collectae et auctae cum tabulis aeneis. Vol. I: 4 pp., 1–563, tabs. I–V, VII–XVII.—Stockholm and Leipzig.

LINNAEUS, C., 1754. Museum S.R.M. Adolphi Friderici Regis Svecorum, Gothorum, Vandalorumque. Haer. Norv. Duc. Slev. Hols. Storm. Ditm. Com. Oldenb. Delmenhorstiae &c. &c. in quo Animalia rariora imprimis, et exotica: Quadrupedia, Aves, Amphibia, Pisces, Insecta, Vermes describuntur et determinantur, Latine et Svetice cum Iconibus: I–XXX, 1–96, 7 pp., pls. I–XXXIII.—Stockholm.

LINNAEUS, C., 1758. Systema Naturae per Regna tria Naturae, secundum Classes, Ordines, Genera, Species, cum characteribus, differentiis, synonymus, locis. Ed. 10. Tomus I, 1–823. —Stockholm.

LINNAEUS, C., 1760. Systema Naturae per Regna tria Naturae, secundum Classes, Ordines, Genera, Species, cum characteribus, differentiis, synonymus, locis. Ed. 10. Tomus I, 4 pp., 1–823.—Magdeburg.

Linnaeus, C., 1766. Systema Naturae per Regna tria Naturae, secundum Classes, Ordines, Genera, Species, cum characteribus, differentiis, synonymus, locis, Ed. 12. Tomus I: 1–532. —Stockholm.

Lönnberg, E., 1896. Linnean type-specimens of birds, reptiles, batrachians and fishes in the zoological museum of the R. University in Upsala.—*Bih. K. Svensk. Vet. Akad. Handl.* 22 (IV, 1): 1–45.

Lovén, S., 1887. On the species of Echinoidea described by Linnaeus in his work Museum Ludovicae Ulricae. I. The museum Ludovicae Ulricae.—*Bih. K. Svensk. Vet. Akad. Handl.* 13 (IV, 5): 1–185, pls. 1–9.

Lydekker, R., 1896. The Royal Natural History. Vol. 5: I–XV, 1 p., 1–584, figs., pls.—London.

Lynden, A. J. H. van, 1939. Op zoek naar Suriname's Zuidgrens. De grensbepaling tusschen Suriname en Brazilië. 1935–1938.—*Tijdschr. Kon. Ned. Aardr. Gen.* 56: 794–882, photo's 1–91, 1 map, maps I–III.

Maguire, B., 1945. Notes on the geology and geography of Tafelberg, Suriname.—*The Geogr. Rev.* 35 (4): 563–579, figs. 1–14.

Marcuzzi, G., 1950. Breves apuntes sobra algunos lagartos de Venezuela septentrional.—*Mem. Soc. Cienc. nat. La Salle* 10 (26): 73–110, 8 figs.

Marcuzzi, G., 1951. Osservazioni sulla sistematica e vareabilita di alcune specie di Lucertole (Rept. Lacertilia) del Venezuela. *Arch. zool. Torino* 36: 17–56, 3 figs.

Marcuzzi, G., 1954. Notas sobre zoogeografia y ecologia del medio Xerofilo Venezolano.—*Mem. Soc. Cienc. nat. La Salle*, 14 (39): 225–260, figs. 1–26.

Martin, K., 1886. Bericht über eine Reise ins Gebiet des oberen Surinam. Reprint from: *Bijdr. Taal-, Land- en Volkenkunde Ned.-Indië* 5 (1): 1–75, pls. I–IV.

Martin, P. L., 1882. Illustrierte Naturgeschichte der Thiere. Vol. II (1): I–XXVIII, 1–566, textfigs.—Leipzig.

Maslin, T. P., 1962. All-female species of the lizard genus Cnemidophorus, Teiidae.—*Science* 135 (3499): 212–213.

Matschie, P., 1897. Kriechtiere in: Das Tierreich. Vol. II: 41–206, textfigs.

McGinnis, S. M. & C. W. Brown, 1966. Thermal behavior of the green iguana, Iguana iguana.—*Herpetologica*, 22 (3): 189–199.

Mees, G. F., 1967a. Ichthyological and other zoological observations: March–Sept., 1965.—*Progress Rep. Biol. Brokopondo Res. Project, Surinam* III: 161–166.

Mees, G. F., 1967b. Ichthyological and other zoological observations: October 1965–March 1966.—*Progress Rep. Biol. Brokopondo Res. Project, Surinam* III: 167–171.

Mees, G. F., 1968. Enige voor de avifauna van Suriname nieuwe vogelsoorten.—*Giervalk* 58: 101–107.

Menne, H. A. L., 1946. Op zoek naar terrariumdieren in Suriname.—*Lacerta* 5(3):20–22, 1 fig., 2 photo's.

Menne, H. A. L., 1952. Op zoek naar pipa's in Suriname.—*Lacerta* 11 (2/3): 9–11.

Menne, H. A. L., 1954a. Mijn belevenissen in Suriname en de Antillen. I.—*Lacerta* 12 (2/4): 15–19, 4 photo's.

Menne, H. A. L., 1954b. Mijn belevenissen in Suriname en de Antillen. II.—*Lacerta* 12 (8): 62–64, 1 fig.

Merian, M. S., 1705a. Metamorphosis insectorum Surinamensium. In qua Erucae ac Vermes Surinamensis, cum omnibus suis Transformationibus, ad vivum delineantur & describuntur, singulis eorumin Plantas, Flores & Fructus collocatis, in quibus reperta sunt; Tum etiam Generatio Ranarum, Bufonum rariorum, Lacertarum, Serpentum, Araneorum & Formicarum exhibetur; Omnia in America ad vivum naturali magnitudine picta atque descripta: 4 pp., 1–60, pls. 1–60.—Amsterdam.

Merian, M. S., 1705b. Metamorphosis Insectorum Surinamensium. Ofte Verandering der Surinaamsche Insecten. Waar in de Surinaamsche Rupsen en Wormen met alle des zelfs Veranderingen na het leven afgebeeld en beschreeven worden, zynde elk geplaast op die Gewassen, Bloemen en Vruchten, daar sy op gevonden zyn; waar in ook de generatie der Kikvorschen, wonderbaare Padden, Hagedissen, Slangen, Spinnen en Mieren werden vertoond en beschreeven, alles in America na het leven en levensgroote geschildert en beschreeven: 4 pp., 1–60, pls. 1–60.—Amsterdam.

MERIAN, M. S., 1719. Over de Voortteeling en Wonderbaerlyke Veranderingen der Surinaemsche Insecten, Waer in de Surinaemsche Rupsen en Wormen, met alle derzelver Veranderingen, naer het leeven afgebeeldt, en beschreeven worden; zijnde elk geplaest op dezelfde Gewassen, Bloemen, en Vruchten, daer ze op gevonden zijn; Beneffens de Beschrijving dier Gewassen. Waer in ook de wonderbare Padden, Hagedissen, Slangen, Spinnen, en andere zeltzame Gediertens worden vertoont, en beschreeven. Alles in Amerika door den zelve M. S. Meiraen naer het leven, en levensgrootte Geschildert, en nu in 't Koper overgebracht. Benevens een Aenhangsel van de Veranderingen van Visschen in Kikvorschen, en van Kikvorschen in Visschen: 8 pp., 1–72, pls. 1–72.—Amsterdam.

MERIAN, M. S., 1730. Over de Voortteeling en Wonderbaerlyke Veranderingen der Surinaamsche Insecten,...... 6 pp., 1–51, 1 p., pls. 1–72.—Amsterdam.

MERIAN, M. S., 1771. Histoire Générale des Insectes de Surinam et de toute l'Europe, contenant Leurs Descriptions, leurs Figures, leurs differentes Metamorphoses, De même que les Descriptions des Plantes, Fleurs & Fruits, dont ils se nourissent, & sur lesquels on les trouve le plus communément; avec quelques détails sur les Crapauds, Lézards, Serpens, Araignées, & autres petits Animaux de Surinam, peints sur les lieux d'après nature & gravés avec soin. Vol. 1: 6 pp., 1–72, 1 p., pls. 1–72.—Paris.

MERREM, B., 1809. Amphibiologische Beiträge.—*Ann. Wetterauischen Gesell. Ges. Naturk.* 1 (1): 1–20, 1 pl.

MERREM, B., 1820. Versuch eines Systems der Amphibien. (Tentamen systematis Amphibiorum): I–XV, 1–191, 1 pl.—Marburg.

MERTENS, R., 1969. Herpetologische Beobachtungen auf der Insel Tobago.—*Salamandra* 5 (2): 63–70, figs. 1–6.

MERTENS, R., 1972. Herpetofauna tobagana.—*Stuttg. Beitr. Naturk.* 252: 1–22.

MOREAU DE JONNÈS, 1818. Monographie du Mabouia des murailles, ou Gecko Mabouia des Antilles.—*Bull. Sci. Soc. Philom. Paris* 1818: 138–139.

MÜLLER, L., 1923. Neue oder seltene Reptilien und Batrachier der Zoologischen Sammlung des bayerischen Staates.—*Zool. Anz.* 57 (7/8): 145–156.

MÜLLER, L., 1924. Neue oder seltene Reptilien und Batrachier der Zoologischen Sammlung des bayerischen Staates.—*Zool. Anz.* 58 (9/10): 291–297.

MÜLLER, P., 1968. Die Herpetofauna der Insel von São Sebastião (Brasiliën): 1–68, figs. 1–48.—Saarbrücken.

MÜLLER, P., 1969. Vertebratenfaunen brasilianischer Inseln als Indikatoren für glaziale und postglaziale Vegetationsfluktuationen.—*Zool. Anz. suppl. bd.* 33, *Verh. dt. Zool. Ges.*: 97–107, figs. 1–12.

MÜLLER, P., 1970. Durch den Menschen bedingte Arealveränderung brasilianischer Wirbeltiere.—*Natur und Museum* 100 (1): 22–37, figs. 1–12.

MÜLLER, P., 1971a. Herpetologische Reiseeindrücke aus Brasilien.—*Salamandra* 7 (1): 9–30, figs. 1–15.

MÜLLER, P., 1971b. Ausbreitungszentren und Evolution in der Neotropis.—*Mitt. Biog. Abt. Geogr. Inst. Univ. Saarl.* 1: 1–20, figs. 1–6.

MÜLLER, P., & J. SCHMITHÜSEN, 1970. Probleme der Genese südamerikanischer Biota.—*Deutsche geographische Forschung in der Welt von heute*: 109–122, figs. 1–4.

NEBEL, G., 1964. MARIA SIBYLLA MERIAN, die schönsten Tafeln aus dem grossen Buch der Schmetterlinge und Pflanzen Metamorphosis Insectorum Surinamensium: 1–61, 2 pp., 17 pls.—Hamburg.

NEDERMEIJER, J., 1963. Surinaamse daggecko.—*Lacerta* 21 (5): 34.

NIJSSEN, H., 1967a. Ichthyological observations: May and June 1966.—*Progress Rep. Biol. Brokopondo Res. Project, Surinam* III: 172–175.

NIJSSEN, H., 1967b. Ichthyological observations: August–September, 1966.—*Progress Rep. Biol. Brokopondo Res. Project, Surinam* III: 176–178.

NIJSSEN, H., 1967c. Ichthyological observations: October–December 1966.—*Progress Rep. Biol. Brokopondo Res. Project, Surinam* III: 179–180.

NIJSSEN, H., 1969a. Ichthyological observations: January–March 1967.—*Progress Rep. Biol. Brokopondo Res. Project, Surinam* IV: 234–237.

NIJSSEN, H., 1969b. Ichthyological observations: April–June 1967.—*Progress Rep. Biol. Brokopondo Res. Project, Surinam* IV: 238–239.

NIJSSEN, H., 1969c. Final remarks and tentative list of fish species.—*Progress Rep. Biol. Brokopondo Res. Project, Surinam* IV: 240–246.

NOBLE, G. K., 1923. New Lizards from the tropical research station British Guiana. *Zoologica, N.Y.*, III (15): 301–305.

OLDENBURGER, F. H. F., H. T. RIEZEBOS & R. NORDE, 1969. Unpublished, mimeographed. Beknopt verslag van de werkzaamheden en voorlopig overzicht van de resultaten van de Sipaliwini-expeditie '68–'69: 1–8, 1 map.

ORBIGNY, A. D', 1847. Reptiles in: Voyage dans l'Amérique méridionale ...... 5 (1): 1–12, 9 pls.—Paris.

O'SHAUGHNESSY, W. E., 1875. List and revision of the species of Anolidae in the British-Museum collection, with descriptions of new species.—*Ann. Mag. Nat. Hist.* (4) 15, 270–281.

PARKER, H. W., 1926. The Neotropical Lizards of the Genera Lepidoblepharis, Pseudogonatodes, Lathrogecko, and Sphaerodactylus, with the Description of a new Genus.—*Ann. Mag. nat. Hist. London* (9) 17: 291–301, figs. 1–9.

PARKER, H. W., 1935a. The Lizards of Trinidad.—*Trop. Agricult. Trinidad,* 12 (3): 65–70, figs. 1–5.

PARKER, H. W., 1935b. The frogs, lizards and snakes of British Guiana.—*Proc. Zool. Soc. London,* 1935 (3): 505–530.

PENARD, F. P. & A. PH. PENARD, 1906. Grepen uit de Natuurkunde van Suriname. Bijeengebracht uit talrijke wetenschappelijke werken en lokaalbeschrijvingen van Jagers, Visschers enz.—*De Surinamer* 13 (58).

PETERS, J. A., 1964. Dictionary of herpetology. A brief and meaningful definition of words and terms used in herpetology: I–VIII, 4 pp., 1–392, 34 pp., figs. 1–30.—New York and London.

PETERS, J. A., 1967. The Lizards of Ecuador, a check list and key.—*Proc. U.S. Nat. Mus.* 119 (3545): 1–49, figs. 1–6.

PETERS, J. A. & R. DONOSO-BARROS, 1970. Catalogue of the Neotropical Squamata: Part II. Lizards and Amphisbaenians. *U.S. Nat. Mus. Bull.* 297; I–VIII, 1–293, 104 figs.

PETERS, W., 1862. Über Cercosaura und die mit dieser Gattung verwandten Eidechsen aus Südamerika.—*Abh. K. Akad. Wiss. Berlin,* 1862: 165–225, pls. 1–3.

PETERS, W., 1869. Mittheilung über neue Gattungen und Arten von Eidechsen.—*Mon. ber. K. Ak. Wiss. Berlin* 1870: 57–66, 1 pl.

PETERS, W., 1877. Über die von Spix in Brasilien gesammelten Eidechsen des Königlichen Naturalien-Kabinets zu München.—*Mon. ber. K. Ak. Wiss. Berlin* 1877: 407–414.

PISTORIUS, T., 1763. Korte en zakelyke Beschryvinge van de Colonie van Zuriname. Waar in te vinden is Een omstandig Berigt van de gelegenheid deezer Volkplantinge; derzelver Rivieren, Kreeken, Forten, Bergen, Plantagien, Water-Werken en Houtzageryen; opgemaakt volgens de nieuwste Land-Kaarten. Mitsgaders Een kort Verhaal van deszelfs Inwoonderen, inzonderheid de Leevensmanier der Slaven, hunne Huishoudinge, Godsdienst, Huwelyken; zeldzaame Gewoonte by het opvoeden hunner Kinderen, en de wonderbaare begravinge van haare Dooden. Als mede Eene Beschryvinge van de Vrugt- en andere Boomen; Aardgewassen, Wilde en Tamme Land- en Water-Dieren; Boom-; Land- en Water-Vogelen, Visschen, Slangen, Venynige en andere Ongedierten. Benevens Een omstandig Berigt van het Zuiker-Riet, Zuiker- en Koffy-plantagien, Moolens, Kookhuizen en meer andere weetenswaardige zaaken; ook eenige Instructien, zeer dienstig voor alle die eenig Bewind over de Plantagien is aanvertrouwt. Waar agter gevoegt is, Echt, omstandig en naauwkeurig verhaal van de gepleegde Moord aan zyne Excell. den Heer van Sommelsdyk, Gouverneur en het doodelyk kwetzen van de Heer L. Verboom, Commandeur van de Colonie van Zuriname: 8 pp., 1–160, pls. 1–4.—Amsterdam.

RAND, A. S., 1968. A nesting aggregation of iguanas.—*Copeia* 1968 (3): 552–561, figs. 1–5.

RAND, A. S. & S. S. HUMPHREY, 1968. Interspecific competition in the tropical rain forest:

ecological distribution among lizards at Belém, Pará.—*Proc. U.S. Nat. Mus.* 125 (3658): 1–17, figs. 1–2.

RAND, A. S. & P. J. RAND, 1966. Aspects of the ecology of the Iguanid lizard Tropidurus torquatus at Belém, Pará.—*Smiths. Misc. Coll.* 151 (2): 1–16, figs. 1–4, tabs. 1–2.

REED, C. A. & R. BOROWSKY, 1970. The 'World's largest toad' and other herpetological specimens from southern Suriname.—*Stud. fauna Sur. Guyanas* 12 (50): 159–171, figs. 200–207, pl. III.

REESE, A. M., 1922. A note on the breeding habits of Tegu (Tupinambis nigropunctatus).—*Copeia* 110: 69–72.

REKUM, M. VAN, 1962. Cnemidophorus spec.—*Lacerta* 20 (7): 50.

RIEZEBOS, H. T., 1970. Sipaliwini-expeditie.—*Schakels* S 76: 29–35, 6 photo's.

RIVERO-BLANCO, C., 1967. Un genero y dos especies de tuqueques (Sauria: Sphaerodactylinae) citados por primera vez para Venezuela, con notas sobre la distribucion de otras especies poco conocidas.—*Mem. Soc. Cienc. nat. La Salle* 27 (77): 103–119, figs. 1–4.

ROUX, J., 1907. Diagnosen neuerer Reptilien aus Asien und Amerika.—*Zool. Anz.* 31 (24): 762–765.

ROUX, J., 1929. Sur deux reptiles nouveaux du Vénézuéla.—*Verh. Naturf. Ges. Basel* 40 (2): 29–34.

RUIBAL, R., 1952. Revisionary studies of some South American Teiidae.—*Bull. Mus. Comp. Zool.* 106: 477–529, 10 figs.

RUTHVEN, A. G., 1916. Description of a new genus and species of lizard from British Guiana. —*Occ. Pap. Mus. Zool. Mich.* 22: 1–4.

SACK, A. VON, 1821a. Beschreibung einer Reise nach Surinam und des Aufenthaltes daselbst in den Jahren 1805, 1806, 1807 so wie von des Verfassers Rückkehr nach Europa über Nord-Amerika. Erste Reise nach Surinam. Erste Abtheilung: I–XVII, 1–240, pls. I–IV, VI. —Berlin.

SACK, A. VON, 1821b. Reize naar Surinamen, verblijf aldaar, en terugtocht over Noord-Amerika naar Europa. Vol. I: I–XII, 1–250.—Haarlem.

SACK, A. VON, 1821c. Reize naar Surinamen, verblijf aldaar, en terugtocht over Noord-Amerika naar Europa. Vol. II: I–II, 1–263.—Haarlem.

SANDERSON, I. T., 1939. Caribbean Treasure: 1–292, textfigs.—New York.

SCHIERBEEK, A., 1952. MARIA SIBYLLA MERIAN. Reproducties naar haar tekeningen van Surinaamse en Europese insecten met een beschrijving van haar leven en werken en een verklaring van de afgebeeldingen: 1–18, 6 pp., pls. 1–24.—Den Haag.

SCHIFTER, H., 1972. Familie Chamäleons. In: Grzimeks Tierleben 6, Kriechtiere: 229–245, textfigures.—Zürich.

SCHINZ, H. R., 1822. Das Thierreich eingetheilt nach dem Bau der Thiere als Grundlage ihrer Naturgeschichte und der vergleichenden Anatomie von dem Herrn RITTER VON CUVIER. Vol. 2: I–XVI, 1–835.—Stuttgart and Tübingen.

SCHINZ, H. R., 1833. Naturgeschichte und Abbildungen der Reptilien. Nach den neuesten Systemen zum gemeinnützligen Gebrauche entworfen und mit Berücksichtigung für den Unterricht der Jugend bearbeitet: 4 pp., 1–240, I–IV.—Leipzig.

SCHLEGEL, H., 1858. Handleiding tot de beoefening der dierkunde. Vol. II: I–XX, 1–628, 2 pp. —Breda.

SCHNEIDER, J. G., 1801. Historiae Amphibiorum naturalis et literariae. Fasciculus Secundus continens Crocodilos, Scincos, Chamaesauras, Boas, Pseudoboas, Elapes, Angues, Amphisbaenas et Caecilias: I–VI, 1–364, pl. I.—Jena.

SCHNEIDER, J. G., 1812. Kritische Uebersicht der einzelnen Arten aus der Gattung von Eidechsen, welche ich Wandkletterer nenne, Linné aber und andere, Geckonen.—*Denkschr. K. Ak. Wiss. München* 1811–1812: 31–70.

SCHWABE, G. H., 1968. Towards an ecological characterization of the South American continent.—In: E. J. FITTKAU et al.: Biogeography and ecology in South America, 1: 113–136.—The Hague.

SEBA, A., 1734. Locupletissimi rerum naturalium thesauri accurata descriptio, et iconobus artificiossissimis expressio, per universam physices historiam. Vol. I: 34 pp., 1–178, 2 pls., pls. 1–111.

SEBA, A., 1735. Locupletissimi rerum naturalium thesauri accurata descriptio, et iconobus artificiosissimis expressio, per universam physices historiam. Vol. 2: 34 pp., 1–154, 1 pl., pls. 1–114.
SHAW, G., 1802. General Zoology or systematic Natural History. Vol. III. Amphibia: I–VI 1–615, pls. 1–140.—London.
SMITH, H. M., 1946. Handbook of lizards, lizards of the United States and of Canada: I–XXI, 1–557, figs. 1–136, pls. 1–135, maps 1–41.—Ithaca, New York.
SMITH, H. M. & E. H. TAYLOR, 1950. An annotated checklist and key to the reptiles of Mexico exclusive of the snakes.—*Bull. U.S. Nat. Mus.* 199: I–V, 1–253.
SPIX, J. B. DE, 1825. Animalia Nova sive species novae Lacertarum quas in itinere per Brasiliam annis MDCCXVII–MDCCCXX: 1–26, pl. 1–18.—München.
STAHEL, G. & D. C. GEIJSKES, 1939. Ueber den Bau der Nester von Atta caphalotes L. und Atta sexdens L. (Heym. Formicidae).—*Rev. de Entomologia* 10 (1): 27–78, figs. 1–21, pls. 1–26.
STAHEL, G. & D. C. GEIJSKES, 1940. De parasolmieren en hunne bestrijding.—*Dep. Landbouwproefstation in Suriname. Bull.* 56: 1–á8, 4 figs.
STATIUS MÜLLER, P. L., 1774. Von den Amphibien. Des Ritters Carl von Linné Königlich Schwedischen Leibarztes etc. etc. vollständiges Natursystem nach der zwölften lateinischen Ausgabe und nach Anleitung des Holländischen Houttuynischen Werks mit einer ausführlichen Erklärung ausgefertigt. Vol. 3: 1–64, 1–350, 16 pp., tabs. 1–12.—Nürnberg.
STEDMAN, J. G., 1796a. Narrative, of a five years' expedition, against the Revolted Negroes of Surinam, in Guiana, on the Wild Coast of South America; from the year 1772, to 1777: elucidating the History of that Country, and describing its Productions, Viz. Quadrupedes, Birds, Fishes, Reptiles, Trees, Shrubs, Fruits, & Roots; with an account of the Indians of Guiana, & Negroes of Guinea. Vol. 1: I–XVIII, 1–407, 7 pp., pls. I–XL.—London.
STEDMAN, J. G., 1796b. Narrative, of a five years' expedition, against the Revolted Negroes of Surinam, ... Vol. 2: I–IV, 1–404, 7 pp., pls. I–LXXX.—London.
STEDMAN, J. G., 1799a. Reize naar Surinamen, en door de binnenste gedeelten van Guiana. Vol. 1: I–XVI, 1–399, pls. 1–16.—Amsterdam.
STEDMAN, J. G., 1799b. Reize naar Surinamen, en door de binnenste gedeelten van Guiana. Vol. 2: I–VIII, 1–326, pls. 17–29.—Amsterdam.
STEDMAN, J. G., 1800a. Reize naar Surinamen, en door de binnenste gedeelten van Guiana. Vol. 3: I–VIII, 1–350, pls. 30–39.—Amsterdam.
STEJNEGER, L., 1904. The herpetology of Porto Rico.—*Ann. Rep. U.S. Nat. Mus.* 1902: 549–724, figs. 1–197, pl. 1.
STOCKUM, A. J. VAN, 1904. Verslag van de Saramacca-expeditie.—*Tijdschr. Kon. Ned. Aardr. Gen.* (2) 21: 88–122, map III; 226–310, 1 fig., 2 photo's, map IV; 651–721, 2 photo's, map X; 822–878, map XI, 1 photo; 1022–1058, 2 photo's.
STRAUCH, A., 1887. Bemerkungen über die Geckoniden-Sammlung im Zooloogischen Museum der Kayserlichen Akademie der Wissenschaften zu St. Petersburg. *Mem. Ac. Imp. Sc. St. Petersburg* (7) 35 (2): 1–72, I–II, 1 pl.
STRESEMANN, E., 1951. Die Entwicklung der Ornithologie von Aristoteles bis zur Gegenwart: I–XV, 1–431, figs. 1–3, pls. I–XIV,—Berlin.
STUART, L. C., 1939. A description of a new Gymnophthalmus from Guatemala, with notes on other members of the genus.—*Occ. Pap. Mus. Zool. Univ. Mich.* 409: 1–11.
STUHLDREHER-NIENHUIS, J., 1944. Verborgen paradijzen, het leven en werken van MARIA SIBYLLA MERIAN 1647–1717: 1–182, 31 pls.—Arnhem.
SUNDIUS, P., 1749. Surinamensia Grilliana. In: C. LINNAEUS, 1749, Amoenitates Academicae ......: 483–508, pl. 17.—Stockholm and Leipzig.
SWANSON, P. L., 1950. The iguana. Iguana iguana iguana (L.).—*Herpetologica* 6: 187–193.
TAMSITT, J. R. & D. VALDIVIESO, 1963. Teiid lizards of the genus Kentropyx in Colombia.—*Copeia* 1963 (2): 443.
TEENSTRA, M. D., 1835. De landbouw in de Kolonie Suriname, voorafgaande door eene geschied- en natuurkundige beschouwing dier kolonie. Vol. 2: 12 pp., 1–484, 2 pp.—Groningen
TEST, F. H., O. J. SEXTON & H. HEATWOLE, 1966. Reptiles of Rancho Grande and vicinity,

Estado Aragua, Venezuela.—*Misc. Publ. Mus. Zool. Univ. Mich.* 128: 1–63, figs. 1–2, pls. I–II, tabs. 1–7.

THOMAS, R., 1965. The smaller teiid lizards (Gymnophthalmus and Bachia) of the southeastern Caribbean.—*Proc. biol. Soc. Wash.* 78: 141–154, 6 figs.

THUNBERG, C. P., 1787. Museum Naturalium Academiae Upsaliensis I.—Uppsala.

TROSCHEL, F. H., 1848. Amphibien. In: R. SCHOMBURGK: Reisen in British Guiana, in den Jahren 1840–1844. Im Auftrag Sr. Majestät des Königs von Preussen ausgeführt von Richard Schomburgk. Vol. 3. Versuch einer fauna und flora von Britisch-Guiana: 645–661.—Leipzig.

UNDERWOOD, G., 1962. Reptiles of the Eastern Caribbean.—*Caribbean Affairs (N.S.)* 1: 4 pp., 1–192, groups of figs. 14, 1 tab.

UZZELL, T. M., 1966. Teiid Lizards of the genus Neusticurus (Reptilia, Sauria).—*Bull. Am. Mus. Nat. Hist.* 132: 5 pp., 277–327, figs. 1–6, tabs. 1–12.

UZZELL, T. M., 1970. Reptilia: Sauria: Teiidae, Prionodactylus. In: J. A. PETERS & R. DONOSO-BARROS: Catalogue of the Neotropical Squamata: Part II, Lizards and amphisbaenians.—*Bull. U.S. Nat. Museum* 297: 235–236.

UZZELL, T. M., & J. C. BARRY, 1971. Leposoma percarinatum, a unisexual species related to L. guianense; and Leposoma ioanna, a new species from Pacific coastal Colombia (Sauria, Teiidae).—*Postilla* 154: 1–39, figs. 1–17, maps 1–2, tables 1–2.

VANZOLINI, P. E., 1951. Amphisbaena fuliginosa. Contributions to the knowledge of the Brasilian lizards of the family Amphisbaenidae Gray, 1825. 6. On the geographical distribution and differentiation of Amphisbaena fuliginosa Linné.—*Bull. Mus. Comp. Zool.* 106 (1): 1–67, pls. 1–2, maps 1–7, graphs 1–7, tabs. 1–26.

VANZOLINI, P. E., 1955. Sobre Gonatodes varius (Auguste Duméril), com notas sobre outras especies do genero (Sauria, Gekkonidae).—*Pap. Av. Zool.* 12 (3): 119–132, 3 pls.

VANZOLINI, P. E., 1957. O gênero Coleodactylus (Sauria, Gekkonidae).—*Pap. Av. Zool.* 13 (1): 1–17, figs. 1–4, 1 map, graphs 1–3, tabs. 1–4.

VANZOLINI, P. E., 1961. Bachia: Espécies brasileiras e conceito genérico (Sauria, Teiidae).—*Pap. Av. Zool.* 14 (22): 193–209, figs. 1–13, 1 table.

VANZOLINI, P. E., 1963. Problemas faunísticos do Cerrado. In: Simpósio sôbre a Cerrado: 307–321, 1 map.—Sao Paulo.

VANZOLINI, P. E., 1967. Sôbre o gênero Pseudogonatodes, com a descrição de uma espécie nova da Amazônia (Sauria, Gekkonidae).—*Pap. Av. Zool.* 21 (1): 1–12, pls. 1–2, tab. 1, graph 1.

VANZOLINI, P. E., 1968a. Lagartos Brasileiros da família Gekkonidae (Sauria).—*Arq. Zool.* 17 (1): 1–84, figs. 1–9, pls. 1–8.

VANZOLINI, P. E., 1968b. Geography of the South American Gekkonidae (Sauria).—*Arq. Zool.* 17 (2): 85–112, maps 1–9.

VANZOLINI, P. E., 1970a. Zoologia sistemática geografia e a origem das espécies.—*Inst. Geogr. S. Paulo, Série Teses e Monografias* no. 3: 1–56, maps 1–7, graphs 1–2, tabs. 1–3.

VANZOLINI, P. E., 1970b. Unisexual Cnemidophorus lemniscatus in the Amazonas valley: a preliminary note (Sauria, Teiidae).—*Pap. Av. Zool.* 23 (7): 63–68, 1 fig.

VANZOLINI, P. E., 1972. Miscellaneous notes on the ecology of some Brasilian lizards (Sauria).—*Pap. Av. Zool.* 26 (8): 83–115.

VANZOLINI, P. E., & R. REBOUÇAS-SPIEKER, 1969. On a large and surprising sample of Calliscincopus agilis from Brasil, with the invalidation of the genus (Sauria, Teiidae).—*Pap. Av. Zool.* 22 (13): 123–144, pls. 1–3, map 1, graphs 1–4, tabs. 1–16.

VANZOLINI, P. E., & R. REBOUÇAS-SPIEKER, 1973. Notes on the ecology and limb proportions of Amazonian Mabuya mabouya (Sauria, Scincidae).—*Pap. Av. Zool.* 26 (17): 215–226, 1 map, graphs 1–5.

VANZOLINI, P. E., & J. VALENCIA, 1965. The genus Dracaena, with a brief consideration of macroteiid relationships (Sauria, Teiidae).—*Arq. Zool.* 13: 7–35, pls. I–V, 1 map, graphs 1–2, tabs. 1–2.

VANZOLINI, P. E., & E. E. WILLIAMS, 1970. South American Anoles: The geographic differentiation and evolution of the Anolis chrysolepis species group (Sauria, Iguanidae).—*Arq. Zool.* 19 (1–4): 1–298, 1 pl., pls. 1–4, maps A–J, 1–32, graphs 1–81, tabs. 1–146.

WAGLER, J., 1830. Natürliches System der Amphibien mit vorangehender Classification der Säugethiere und Vögel: I–VI, 1–354.—München.

WALSH, J. & R. GANNON, 1967. Time is short and the water rises: 1–224, 2 maps, photographs.

WARREN, G., 1667. An Impartial Description of Surinam upon The Continent of Guiana in America. With a History of several strange Beasts, Birds, Fishes, Serpents, Insects, and Customs of that Colony, &c.: 4 pp., 1–28.—London.

WARREN, G., 1669. Een onpartydige Beschrijvinge van Surinam, Gelegen op het vaste Landt van Guiana in Africa. Mitsgaders een Verhael van alle vreemde Beesten, Vogels, Visschen, Slangen ende Wormen: Gelijck mede van de Gewoonheden ende Manieren van deze Colonie: 1 p., 1–20.—Amsterdam.

WERMUTH, H., 1965. Liste der rezenten Amphibien und Reptilien. Gekkonidae, Pogypodidae, Xantusiidae.—*Das Tierreich* 80: I–XXII, 1–246.—Berlin.

WERNER, F., 1899. Beschreibung neuer Reptilien und Batrachier.—*Zool. Anz.* 22 (602): 479–484.

WESTERMANN, J. H., 1953. Nature preservation in the Caribbean. A Review of Literature on the Destruction and Preservation of Flora and Fauna in the Caribbean Area.—Publ. no. 9, Foundation Scientific Research in Surinam and the Netherlands Antilles: 1–106, figs. 1–25, 1 map.

WESTEROUEN VAN MEETEREN, F. W., 1883. La Guyane Néerlandaise, de la nature de la Guyane Neerlandaise, c'est a dire de son sol, de sa flore et de son faune.—de la population indigène de cette contrée.—des Européens dans cette colonie et de leurs rapports avec les indigènes: 1–79.—Leide.

WIED, M. VON, 1825. Beiträge zur Naturgeschichte von Brasilien. Vol. I: I–XXII, 1–611, pls. I–III.—Weimar.

WIEGMANN, A. F. A., 1834. Herpetologia Mexicana, seu descriptio Amphibiorum Novae Hispaniae, quae itineribus comitis de Sack, Ferdinandi Deppe et Chr. Guil. Schiede in museum zoologicum Berolinense pervenerunt. Pars prima, Saurorum species amplectens. Adiecto systematis saurorum prodomo, additisque multis in hunc Amphibiorum ordinem observationibus: I–VI, 1–54, pls. I–X.—Berlin.

WIJFFELS, L., 1961. Bij de voorplaat.—*Lacerta* 20 (2): 10.

WRETSCHKO, M., 1867. Bilder-Atlas zur wissenschaftlich-populären Naturgeschichte der Wirbelthiere von Dr. LEOPOLD JOSEPH FITZINGER. Kurzer erläuternder Text dazu: I–VI, 2 pp. 1–102.—Wien.

WYMAN, J., 1859. On some unusual modes of Gestation.—*The Amer. J. of Science and Arts* 27: 5–13, figs. 1–5.

ZONNEVELD, J. I. S., 1967. Enkele luchtfoto's van de Sipaliwini-Savanne in Zuid-Suriname.—*Kon. Ned. Aardr. Gen., Geogr. Tijdschr. N.S.* 1: 286–298, figs. 1–2, photo's 1–6.

VIII. INDEX

The valid names of the species and subspecies of lizards occurring in Surinam are in bold type. Page numbers in italics refer to maps; page numbers in bold type to figures or plates; page numbers followed by an a refer to an unnumbered outfolding leaf, following the indicated page number; page numbers followed by an asterisk refer to tables.